Applied Statistical Modelling for Ecologists

A Practical Guide to Bayesian and Likelihood Inference Using R, JAGS, NIMBLE, Stan and TMB

Applied Statistical Modelling for Ecologists

A Practical Guide to Bayesian and Likelihood Inference Using R, JAGS, NIMBLE, Stan and TMB

Marc Kéry
Swiss Ornithological Institute,
Sempach, Switzerland

Kenneth F. Kellner
Department of Fisheries and Wildlife,
Michigan State University, East Lansing,
MI, United States

ELSEVIER

Elsevier
Radarweg 29, PO Box 211, 1000 AE Amsterdam, Netherlands
125 London Wall, London EC2Y 5AS, United Kingdom
50 Hampshire Street, 5th Floor, Cambridge, MA 02139, United States

ISBN: 978-0-443-13715-0

For Information on all Elsevier publications
visit our website at https://www.elsevier.com/books-and-journals

Publisher: Candice Janco
Acquisitions Editor: Maria Elekidou
Editorial Project Manager: Mason Malloy
Production Project Manager: Sruthi Satheesh
Cover Designer: Greg Harris

Typeset by MPS Limited, Chennai, India

Dedicated to Mike Meredith.

Contents

Student resources and helpful ancillaries have been prepared to aid learning and teaching, including all code included in the book available for download. Please visit the companion site for more details: https://www.elsevier.com/books-and-journals/book-companion/9780443137150

Foreword

It was not all that long ago when you could read every important paper written about any branch of ecology. But today, keeping up with the ecological literature is like drinking from a firehose that is fraying at the ends as new journals and preprint servers continually appear. An ecologist could easily dedicate all their research hours to poring through the expanding literature.

And it is not just the ecological literature that is rapidly growing—scientists now struggle to keep up with the tremendous growth that has occurred in the fields of statistics and machine learning. A dizzying array of new quantitative tools have become available thanks to computational advances and the open-source software revolution. But using these methods and understanding them are two different beasts. Proper understanding of quantitative methods is essential to avoid inferential mistakes, but it takes time and careful study, and the primary literature is dense, highly technical, and directed towards subject matter experts, not practitioners. Worse yet, none of the quantitative specialists seem to want to teach their methods at a level accessible to scientists. Or almost none.

Fortunately, there are a few people out there who have explored the depths of the ecological and statistical literatures, and they have emerged with the tools necessary to learn about complex ecological processes from noisy data. Of this group, there is another small fraction that genuinely enjoys the process of shedding light on quantitative tools and helping others apply them. Marc and Ken are among this select group, and their dedication to teaching has been a boon to thousands of ecologists.

Their new book continues with this tradition and fills an important niche covered by few others. Graduate students frequently tell me that there is a lot of good information about basic statistics, and plenty about advanced statistics, but very little in between. In this "gentle introduction and quick-start guide," Marc and Ken masterfully connect the dots between both fundamental and advanced concepts. Pulling off this feat requires hands-on experience, which is what this book is all about. Full of code, simulation examples, and real data, every exercise can be reproduced and studied in detail.

Another highlight is the tremendous amount of practical information on cutting-edge software. Readers will learn how to use high-performance software to fit the most widely used statistical models in ecology, including linear models, generalized linear models, generalized linear mixed models, and occupancy models. If you like Bayesian inference, great, you will find code and examples for drawing posterior samples with JAGS, NIMBLE, and Stan. If you dislike modeling with priors, or you just want the fastest methods available, this book is here for you too. Maximum likelihood methods are demonstrated with Template Model Builder, which uses the same computational tool of automatic differentiation as is employed by Stan and NIMBLE. All of these programs enjoy the speed benefits offered by C and C^{++}, but they can be run from the comfort of R, meaning that you do not have to switch between programs to move from data manipulation to model construction to graphics creation.

Even more important than explaining how to use high-performance software, this book explains what these programs are doing under the hood, and how to interpret the output. Whether you prefer Bayesian inference or classical inference, it is extremely helpful to understand the process of working out the likelihood for a given model, translating the likelihood to computer code, and then maximizing it using an optimizer. This book is unique in the number of do-it-yourself likelihood examples, which should propel understanding to the next level for many quantitative ecologists. Beyond fitting models, there are excellent sections on model selection, fit assessment, prediction, and much more. Perhaps, the best part is that these critical topics are explained in a relaxed style that minimizes technical jargon.

The end result is a treasure of a book that shows why it is an exciting time to be an ecologist in an era when we are not limited by computational or software constraints. Mastering the techniques covered in these pages will open up endless possibilities for studying complex processes and making important discoveries. I am grateful for their contribution and excited to see its impact on the field.

Richard Chandler
Warnell School of Forestry and Natural Resources,
University of Georgia

Acknowledgments

We dedicate this book to our late friend and colleague Mike Meredith, who is one of the most remarkable and generous persons we have ever met. See Kéry et al. (2022)[1] for a very brief summary of Mike's life.

We thank our colleagues who checked an earlier version of the main model-fitting code: Mike Meredith (JAGS), José "Pepe" Jimenez (NIMBLE), Hiroki Itô (Stan), and Mollie Brooks (TMB); any remaining errors are of course ours.

We got valuable comments on one or more chapters from the same Andy Royle as always, Perry de Valpine, Jeff Doser, Florian Hartig, Marc Mazerolle, Michael Schaub, Anne Schlesselmann, and Qing Zhao; thank you very much! In addition, we thank the following persons for help with proofreading: Gabriel Andrade-Ponce, Yan Ru Choo, John Clare, Justin Clarke, Dilsad Dagtekin, Cameron Duquette, Alexandra Evans, Neil Gilbert, Matthew Hyde, Hanna Jackson, Jose Jimenez, Christopher Kilner, Paul Kinas, Quresh Latif, Edward Lavender, Tiago Marques, Frederico Martins, Helen Moor, Jean Nabias, Bryan Nuse, Stephen Parker, Felix Petersma, Rob Robinson, Mahmood Soofi, Evalynn Trumbo, and Gesa von Hirschheydt.

We are grateful to Jean-Marc Delaunay, Alain Georgy, Felix Labhardt, Andi Meier, Andreas Meyer, Thomas Ott, Jordi Rojals, and Niklaus Zbinden, who gave permission to use their great photographs free of charge.

We thank Richard Chandler for writing the Foreword and Jerry Belant for his support of Ken during the writing process. We are indebted to Richard Chandler, Gonçalo Ferraz, Eleni Matechou, Christopher Rota, Andy Royle, and Chris Sutherland, who wrote endorsements for this book.

Special thanks from Marc go to Susana and Gabriel, as always. Ken would like to thank his wonderful family for their support during this project.

Marc & Ken

---

[1] Kéry, M., Juat, N., Ngoprasert, D., Gale, G.A. 2022. In memoriam: Michael (Mike) Meredith (July 19, 1943–January 12, 2023). Conservation Science and Practice, 2023, e12979.

Introduction⊛

1

Chapter outline

The book *Applied Statistical Modeling (ASM) for Ecologists* is a gentle introduction and quick-start guide to the essentials of applied statistical modeling for researchers and professionals in ecology, wildlife management, and related fields. It includes regression and analysis of variance (ANOVA) models and linear, generalized linear, mixed, and hierarchical models, that is, most of the model classes that are widely used in applied statistical modeling these days. They are all parametric statistical models, which are a conceptually simple and powerful tool for learning from data in science and management. In the book we put about equal emphasis on Bayesian and frequentist statistical inference. We introduce you to four of the most powerful, cutting-edge general statistical modeling software currently available in ecology: JAGS (Plummer, 2003), NIMBLE (de Valpine et al., 2017), Stan (Carpenter et al., 2017), and TMB (Kristensen et al., 2016). Finally, we intend this book to serve as a thorough hands-on introduction for nonstatisticians to the foundational method of maximum likelihood estimation, which is demonstrated in detail for every model in the book. Many concepts and indeed some of the contents of the ASM book are identical to those in Kéry (2010). However, there is such a wealth of new and distinct material in the ASM book that it would have been misleading to pitch it as a second edition only.

We wrote this book with multiple audiences in mind. On the one hand are novices to statistical modeling who need to learn about t-test, regression, ANOVA, and analysis of covariance (ANCOVA) models,

---

⊛This book has a companion website hosting complementary materials, including all code for download. Visit this URL to access it: https://www.elsevier.com/books-and-journals/book-companion/9780443137150.

Applied Statistical Modelling for Ecologists. DOI: https://doi.org/10.1016/B978-0-443-13715-0.00002-9

that is, the bread and butter in much of applied statistical modeling. Our goal is to clarify the unity of all these techniques as specific instances of linear models. Then, we present the even greater unifying role of the generalized linear model (GLM), and finally the concept of random effects, which lead to mixed and hierarchical models. These three basic concepts, linear modeling of covariate effects, GLMs, and random effects/mixed models, are often very confusing to novices. The approach in the ASM book to explain them will be extremely useful to budding ecological modelers.

A second audience of the book are more experienced researchers and managers. They may have been using canned functions in R such as `lm()`, `glm()`, `lmer()`, `glmmTMB()` or `brms()` for their statistical analyses for many years. But now they want to get a better understanding of what actually goes on under the hood when using maximum likelihood or Bayesian inference. Working through this book should help them understand much better the workings of these and many similar functions and consequently lead to their improved and more informed use.

A third audience also consists of more seasoned applied statistical modelers who want to learn the JAGS, NIMBLE, Stan, or TMB model-fitting engines. These are often used for far more complex models than those in this book, but it would be very hard to learn these engines at the same time as trying to understand such complex models. In contrast, our approach of introducing JAGS, NIMBLE, Stan, and TMB for mostly dead-simple statistical models should make it very easy for you to learn to "walk" in any of these engines. In addition, our ASM book adopts a "Rosetta stone approach," where we show the code for all these engines side by side. If you know one engine already, then learning another one is typically much easier in this comparative setting.

Finally, a fourth audience is also more advanced and wants to solidify and progress in their understanding of hierarchical models (Royle & Dorazio, 2008; Cressie et al., 2009; Kéry & Schaub, 2012; Royle et al., 2014; Hobbs & Hooten, 2015; Kéry & Royle, 2016; Hooten & Hefley, 2019; Kéry & Royle, 2021). If you belong to this group, you know already quite well how to fit and interpret GLMs and traditional mixed models. Most hierarchical models that you see in current research papers, and which typically are fit with one of the above model-fitting engines or with custom Markov chain Monte Carlo (MCMC) code (Hooten & Hefley, 2019), can be understood conceptually as not much more than a variant of GLM. Only, there is not a single GLM, but there are two or more linked GLMs. Thus, for this type of audience, we aim to explain clearly how the building blocks (i.e., GLMs with linear models for covariate effects) of these more complex hierarchical models are constructed.

In the rest of this chapter, we make additional comments about each of a number of key topics in modeling. In Section 1.10, we give an outline of the book content and describe some additional resources that accompany the book, while in Section 1.11 we offer some advice on how you can use the book for self-study or in courses.

1.1 Statistical models and why you need them

Uncertainty is a hallmark of life and of our attempts to understand life. In the sciences, the inherent uncertainty in natural systems makes it a challenge to generalize from our observations and experiments, and we need a model to separate signal from noise and to draw the "right" conclusions from our data. In management, such as conservation biology, wildlife management, or fisheries, the same uncertainty makes it a challenge for us to take the "right" decisions.

Uncertainty has multiple causes, including the natural variation in living systems: no two years are alike, nor are any two populations or any two individuals, and typically we want to infer features from a collection of years, sites, or individuals from a sample. To describe, understand, and predict living systems, we need a learning framework that formally incorporates this ubiquitous variability and the uncertainty it creates. Statistics is the science that deals with this uncertainty and provides a rigorous method of learning from data. Thus, statistics is a keystone science for all the other sciences and for management. We need statistics to draw conclusions that are sensible in the light of the evidence in our limited data and to take rational decisions in the face of uncertainty.

More specifically, statistical models are the vehicles with which we learn from data. Broadly, they represent a greatly simplified, mathematical description of reality and of life and help us understand, control, and predict life (Pawitan, 2013). Statistical models contain a sketch of the processes that have generated our data set, which itself is usually of much lesser direct interest, regardless of how much it costs us to collect it. The data set simply paves our way to learning about the underlying processes using a model. Statistical models can explicitly accommodate variability and all the other relevant sources of uncertainty. Two such sources of uncertainty, which are completely separate from the inherent variability of natural systems mentioned above, are uncertainty about the model and uncertainty about the parameters. Statistical modeling lets us incorporate and quantify all components of uncertainty when learning from data. It does so by providing estimates of parameters and other estimated quantities that are accompanied by standard errors, confidence intervals, and similar measures of estimation uncertainty.

There are a myriad of possible statistical models for the task of learning from your data. They range from very simple to very complex. For example, they vary in terms of the number of parameters, which may be 10 or in the millions. Similarly, models may differ in the ease with which they may be understood by a user, the ease with which we can fit them to a data set, or with which we may be able to interpret the parameters estimated in them. Finding the right level of complexity of a model for your specific learning task is a key challenge in any attempt to learn from data.

As we will discuss in Chapter 18, a model may have multiple goals. There, we emphasize the distinction between models that are geared towards providing a causal understanding of a process on the one hand and models that provide good predictions of a process for a different place in time or space. This distinction is not absolute, and most models try to achieve at least a little bit of both. Nevertheless, some models are predominantly for causal explanation, while others are predominantly for prediction. Models for pattern searching are an entirely different category still. The goal of a model is a key consideration in our decision about how to choose and build statistical models for learning from data.

In this book, we focus on parametric statistical models, specifically on linear, generalized linear, mixed, and simple hierarchical models. They are all located at the lower end of the complexity gradient and will usually have somewhere between a handful and some hundreds of estimated parameters. Typically, their focus is more on causal explanation and owing to that, some parameters typically have a clear subject-matter interpretation. Examples include population size, the presence or absence of a species at a site, a demographic rate such as survival or fecundity, the growth rate of a population, or the body mass of an individual.

In parametric statistical models, there is a focus on the processes that underlie our data sets. These processes may be biological or they may be procedural or methods-related, that is, describe the measurement processes that "transform" the hidden reality into our error-prone and biased data sets. The parameters that describe these processes form the "hard parts" of a model (Chapter 18), and their presence in the model and their specific form is often stipulated by our science. There may also be "soft parts" in the model, the form of which we don't care so much about, but we want them to adapt flexibly to the data, often to improve the predictions of a model (see Section 18.4).

In summary, the main strength of parametric statistical models is their combination of conceptual simplicity and interpretational clarity and their focus on mechanisms, which is (or should be) a hallmark of any science or science-based activity. Thus, in this book, we give an account of a modeling process that is powerful for ecological science and natural resource management. In it, there is a fairly strong focus on explanation and inference of actual processes, although prediction of observed (i.e., the data) and potentially observable quantities (e.g., the values of some latent variables) is frequently important, too. That is, we emphasize mechanisms and building what Berliner (1996) calls *science-based statistical models*. Accurate prediction is important (Clark et al., 2001; Gerber et al., 2015; Hooten & Hobbs, 2015), but arguably discovery and elucidation of mechanisms is even more important. Finally, we note that all of this leads directly towards *hierarchical models* (Berliner, 1996; Berliner et al., 2003; Wikle, 2003; Royle & Dorazio, 2008; Cressie & Wikle, 2011; see also Chapters 3.5, 19, and 19B). As mentioned above, one goal of ASM is to prepare you for these more complex, and still more powerful and fascinating models; see also Section 21.2.4.

1.2 Why linear statistical models?

A large part in the practice of applied statistical modeling consists in specifying covariate relationships, that is, descriptions of how continuous or categorical (factor) covariates affect a response. Using covariates we can specify mechanisms, test causal hypotheses, improve predictions, and adjust for confounders (see Section 18.6). For the most part in applied statistics, we only adopt linear models for covariates, where the effects of covariates appear as simple sums. In reality, linearity need not hold. However, linear models are often a good approximation, plus, we can also model wiggly (i.e., non-straight line) relationships in what mathematically are linear models (e.g., when we use polynomials), and most of all, linear models are typically easy to interpret and to fit. Thus, they are very often used in statistical modeling. However, in spite of their conceptual simplicity, for novices, they can be challenging, especially when we have factors (categorical explanatory variables) or interactions, topics which we cover in detail in the book.

A good understanding of linear models is essential for applying simple statistical models, such as most of those in this book, but is just as important when you are fitting far more complex models, such as models for species distribution, density, abundance, and demographics (Williams et al., 2002; Royle & Dorazio, 2008; Buckland et al., 2015; Kéry & Royle, 2016, 2021; Schaub & Kéry, 2022). In the context of such models, we will typically have two or more places in the model where we can specify linear models for covariates. For instance, we may have one linear model for expected abundance and another for detection probability in an N-mixture model or a linear model for apparent survival and another for recapture probability in a Cormack-Jolly-Seber survival model. We have frequently observed that a major challenge for ecologists in the application of these complex models to their data and then adding their covariates is exactly that: how do you specify the linear models for these covariates. Hence, what you learn in the context of the simple models covered in this book will be an essential introduction to how you do exactly the same type of modeling for different parameters within more complex models.

1.3 Why go beyond the linear model?

The generalized linear model, or GLM, is a major theme in this book. Here, you model as a linear function of covariates a transformation of the mean response, or in general, of a parameter. GLMs extend the power and conceptual simplicity of linear models to a much larger number of possible analyses. In addition, the GLM concept carries over seamlessly to more complex hierarchical models, except that there the parameters that we model with covariates are typically those that govern the distributions of latent rather than observed random variables, that is, of random effects. In contrast, in a simple GLM we model a parameter that affects an observed random variable, that is, the actual data at hand. However, the concept and the implementation in the model is exactly the same when we model a parameter that is nested more deeply inside the hierarchy of a model: we will often apply a link function (exactly as in a Poisson GLM or a logistic regression) to enforce the range constraints for a parameter, such as a variance or a survival, persistence, colonization, or other probability, and then we specify a linear function of covariates to explain some of the variability in that parameter (e.g., over space or time) by one or more explanatory variables.

We cannot emphasize enough that there is a sense in which most of these big fancy hierarchical models so widely used nowadays in ecology and other sciences can simply be viewed as a combination of two or more GLMs. For instance, a static occupancy model (Chapter 19) can be conceptualized as two linked logistic regressions and a dynamic occupancy model (MacKenzie et al., 2003) as a combination of four logistic regressions. Hence, a big advantage of learning to fit the models in this book is that you then also gain a much deeper understanding of what are the essential building blocks of these more complex models.

1.4 Random effects and why you need them

In our experience, random effects belong to the most confusing topics in applied statistics. We call random effects latent variables or the hidden outcomes of a stochastic process. Alternatively, we can also say they are sets of parameters which are drawn from some distribution, which itself has parameters that we estimate. Typically, we need random effects to specify correlations in the data (e.g., temporal or spatial autocorrelation) and more generally to model patterns in sets of parameters that "belong together" in some way, for example, parameters for the effects of multiple years or sites on some parameter such as a survival rate. As applied statistical modelers we must have a clear understanding of what random effects are, when we need them, and how to specify them in our software. Along with linear models and the GLM, random effects are the third main conceptual theme in applied statistical modeling that we emphasize in this book.

1.5 Why do you need both Bayesian and non-Bayesian statistical inference?

During the last 20 years, use of Bayesian inference has greatly increased in ecology. It is now nearly as common as is non-Bayesian, or frequentist, inference. The latter is typically based on the method of maximum likelihood, or different methods that yield the same types of estimates for certain classes of models (such as least-squares for normal linear models; see Chapter 5). Both have advantages and disadvantages, and what is better in any given project depends on the questions, the data set, the type of model, the modeler and your project deadline. We believe that modern applied statistical modelers must be conversant in both methods of learning from data, and this book is predicated on this belief (see also Inchausti, 2023). In particular, in most chapters we use side by side Bayesian model fitting methods and methods based on maximum likelihood or related techniques. With so-called vague priors and moderate to large sample sizes, Bayesian estimates are typically numerically very similar to the estimates from maximum likelihood (Chapter 2). We think that it is useful to experience this throughout the book, when in the final section of each chapter we always compare the point estimates from all fitting methods.

1.6 More reasons for why you should really understand maximum likelihood

Maximum likelihood is the most widely used estimation method in statistics, and in addition, the likelihood function is also an integral part of Bayesian inference (Chapter 2). As you will see, the likelihood is simply the joint density of the data given the parameters, when interpreted as a function of the parameters. Probability densities, and log-densities, are very challenging concepts for most ecologists without any training in probability, but are central concepts for parametric statistical models, since a model is nothing but a joint probability density for your data set (Chapter 2). Thus, at least some conceptual understanding of probability densities *really* pays for your statistical modeling, because probability densities appear in likelihood functions (both when doing maximum likelihood and in Bayesian inference), in common methods for model checking or goodness-of-fit testing (in the form of the predictive densities; Chapter 18), and in model selection when we use the log-predictive density as a measure of the predictive accuracy of a model (Chapter 18).

In addition, more advanced use of general model-fitting engines such as NIMBLE, Stan, and TMB may require you to specify a statistical model directly via the log-density of the data. As traditional ecologists, this is something we are almost never faced with. This is a major reason for why we put so much emphasis on joint densities, likelihood functions, and minimizing negative log-likelihood functions, in Chapter 2 and then especially in all "do-it-yourself maximum likelihod estimate (DIY-MLE)" sections throughout the book. This is not something you normally need for the types of models covered in this book. But it is something that many of you will need further along in your career as statistical modelers. Since complicated things are best studied in isolation, in as simple settings as possible, we present "likelihood and all that" (Bolker, 2008) mostly in the context of simple models. Trying to learn and understand likelihood and probability densities directly with the complex models for which you may eventually need this knowledge might be a far too steep learning curve.

1.7 The data simulation/model fitting dualism

You will see throughout the book that data simulation is a key method in applied statistical modeling. By data simulation we mean the simulation of stochastic realizations, or "potential data sets," under some model. Interestingly, data simulation and model fitting are almost the same thing, but they apply a model in different directions. Parametric statistical models describe the processes that underlie our data set. Combining this process description with a data set lets us use statistical inference methods (see Chapter 2) to understand this process, for example, by estimating its parameters, and predicting it, for example, in terms of observed and latent variables. Data simulation does exactly the same thing but goes the other way round (Fig. 1.1): we start with a process description and pick, rather than estimate, values of the parameters, and then generate one or more realizations of the stochastic system represented by the model. Thus, data simulation and data analysis are intimately related via a model. We can use the model in a generative way to produce new data sets, and there is a host of things for which data simulation is extremely valuable (DiRenzo et al., 2023).

One of the most important goals for data simulation is to describe a model in a succinct and understandable way, and to enforce on you a complete understanding of what the model is and what its parameters mean. We believe that if you really understand a parametric statistical model, you should be able to simulate data under that model. If you are unable to do so, you probably don't really understand your model. In addition, there are many practical advantages of simulating data. We see throughout the book how we use data simulation to check the goodness-of-fit of a model, for example, using the R package DHARMa (Hartig, 2022; see also Section 18.5). In addition, the analysis of simulated data sets, with known parameter values, is a key method for checking the identifiability of a model as well as the robustness of a model to violations of its parametric assumptions (Section 18.5.6).

Additional advantages of data simulation include the following (Kéry & Royle, 2016: chapter 4):

- Truth is known: Thus, we can check our model fitting code for errors, since normally we expect the estimates to bear at least some similarity to the known input values, i.e., the parameters with which we simulated the data.
- We can calibrate derived model parameters. For instance, we can see what different choices of demographic parameters in a population model imply in terms of the population growth rate or its temporal variance.

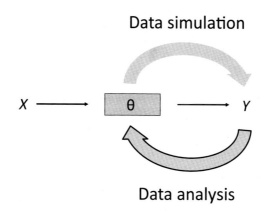

FIGURE 1.1

The duality of data simulation and data analysis by the fitting of a statistical models (see also Fig. 2.1, which shows an analogous duality between probability and statistics). The gray box denotes a stochastic process governed by parameter θ, i.e., a model, while X and Y are some input (e.g., covariates) and output (i.e., observed data), respectively.

- Sampling error: This concept is hard to understand for many non-statisticians, since all we usually have is a single data set; yet, theory lets us compute the variability among this and "all other" data sets that our data-generating mechanism could produce. When you have a data simulation function, you can actually observe the sampling error.
- Frequentist operating characteristics of an estimator: you can study whether an estimation method is biased, and how variable its output is, that is, estimate its precision. You can also gauge its coverage by investigating the proportion of confidence intervals that contain the true value of a parameter.
- Power analysis: What's the minimal sample size to detect an effect of a given magnitude?

Thus, data simulation is a key component of applied statistical modeling. To emphasize this, we exclusively use simulated data sets in this book, with only one exception in Chapter 2. You should get used to seeing data simulated and simulate plenty of data sets yourself.

In addition, in the ASMbook R package (see Section 1.10) we have bundled into R functions the code to simulate all the data sets in this book. For instance, simDat5() contains the bundled code to generate data sets like the one we use to describe simple linear regression in Chapter 5. Likewise, simDat62() and simDat63() let you generate data sets for the two-group comparison of normal responses, with the former assuming equal and the latter unequal variances between the two groups. These functions enable you to quickly generate other data sets than the specific ones generated in each chapter. In addition, they illustrate how useful and easy it is to package R code into functions. Learning to write your own functions is an important part of becoming a proficient modeler.

1.8 The "experimental approach" to statistics

Our passion for data simulation is reflected by what we like to think of as our "experimental approach" to statistics, which is pervasive throughout this book. Basically, it means that you should just "try things out," most often by simulation of data. For instance, in Chapter 2 and elsewhere, we will meet the bootstrap as a highly respected method of this experimental approach: if we don't know how to compute the standard errors around some estimate, we can simply resample our data set, fit the model anew each time, and thus obtain the empirical sampling distribution of the estimator. Thus, we can replace a lot of theory with brute computational power (Millar, 2011). In addition, if you find the standard error a difficult concept, then why not simply simulate replicate data sets, estimate something and then from the frequency distribution of these estimates actually *see* what the standard error is?

The experimental approach to statistics has tremendous power both in practice but also more fundamentally, that is, for your intuitive understanding of things mathematical and statistical. A more formal understanding of the latter may perhaps be difficult for you, because of the limited training of ecologists in probability and mathematical statistics. But you can still get a very strong feel for it by "trying things out." (Here's an important disclaimer: We are *not* saying that we as ecologists should not get a better formal training in statistics. We absolutely should, but as things are now, we simply say that with the experimental approach we can compensate for a lot of deficiencies in our training and obtain at least an intuitive understanding of very many things that might otherwise be completely beyond our reach.)

1.9 The first principle of modeling: start simple!

In scientific problems as well as in modeling, it is always best to first understand simple cases and then increase complexity until you are at the desired level of complexity of your problem or model. Thus, you learn by looking at more and more complex situations, starting with the simplest possible example and then incrementally adding more and more complexity. This step-up approach is what we adopt in this book, where we start with one of the simplest possible models, with one or two parameters, and then add more and more complexity in the form of additional parameters. Also, we start with the conceptually simplest GLM, that with normal errors and identity link function (i.e., the normal linear model), and subsequently progress to the conceptually slightly more complex Poisson and Binomial GLMs. Then, we cover fixed before random effects, and models with a single source of uncertainty (i.e., GLMs) before random-effects or hierarchical models. Finally, to explain the fundamental structure of linear models in Chapter 3, we use a ridiculously simple toy data set. In learning as in modeling the following quote from that famous physicist applies: "*Everything should be made as simple as possible, but not simpler.*" Thus, when choosing toy data sets to study something or when using simulations to answer a statistical question, we choose the simplest scenario that is still able to answer our question and avoid all unnecessary complexity. Likewise, when you start fitting a model to new data, take some code for a very simple version of the model you have in mind (or write it from scratch), for example, with all parameters just constants or intercepts, and make sure it works. After that, iteratively add more covariates, checking that the model works and that you understand it after each step. This is not meant to be a method of model exploration or selection (Chapter 18), but simply a method to avoid making coding mistakes or if you do, to be better able to debug them.

1.10 Overview of the ASM book and additional resources

There are six parts in the ASM book: (1) basics, (2) normal linear models, (3) linear mixed models, (4) generalized linear models, (5) generalized linear mixed models, and (6) more advanced concepts and models.

Part 1 is an introduction to the basic concepts and tools that you need to become an applied statistical modeler. We cover these in Chapters 2–4 and then, somewhat surprisingly perhaps, in Chapter 18. In Chapter 2, we give a very gentle introduction to probability, which is the basis of all statistical modeling. You will learn about random variables and their distributions, probability and density functions, joint densities, and the crucial likelihood function. We give detailed illustrations of how to use the negative log-likelihood function to obtain MLEs, and how we obtain uncertainty assessments around them, such as standard errors or confidence intervals. Then, we cover the principles of Bayesian inference and illustrate their practical implementation using MCMC algorithms. This chapter lays the mathematical and computational groundwork for all the statistical modeling in the book.

In Chapter 3, we present the "ingredients" of parametric statistical models: distributions that we assume for our random variables and the principles of linear modeling to describe the effects of covariates. Also, we cover link functions, which we need when we transition from a normal-response linear model to a GLM with a distribution other than normal. Finally, we give a brief overview of all the classes of statistical models for which we cover examples in this book, and describe the ways in which they are related to each other. Throughout, we emphasize the central role played by the GLM as a building block for mixed, hierarchical, and integrated models.

In Chapter 4, we describe the six model-fitting engines, or methods, that we use throughout the book and directly illustrate each of them by analyzing a first simulated data set. These methods are what we call "canned functions" in R (which use frequentist inference in nearly all our cases), JAGS, NIMBLE, and Stan for Bayesian inference, and then "DIY-MLE" and TMB for frequentist inference by maximum likelihood. To make it easier for you to locate the code for each engine in the book, we use a consistent color coding for the model definitions with each engine. As in all later chapters, we compare point estimates among all engines in the end. We almost always find a strong numerical agreement, supporting our claim that we should be happy to use either frequentist or Bayesian inference, whichever is better for us and our modeling task (see also Section 2.9).

Chapter 18 is the last in the "basics" part of the book. It owes its peculiar placement to our belief that its challenging topics are easier to understand for you if you have worked through most other chapters. This chapter covers the big topics of model checking (also called goodness-of-fit testing) and model selection, and thus deals with the questions of whether our model is any good at all and which one is best if we have more than one. For both, the goal of a model is a key consideration. In addition, all of that is very relevant for how we build a statistical model in the first place. We discuss all these topics. We also deal with model-averaging and regularization as solutions to the problem of "too many covariates" (or the "black hole of statistics"; Tredennick et al., 2021). The former is widely known to ecologists, while the latter is not (but probably should be).

Part 2 covers normal-response linear models without any random effects, that is, the models that underlie ubiquitous analysis techniques like the t-test, linear regression, ANOVA, and ANCOVA, which you find in most introductory books about statistical methods (e.g., Zar, 1998). In Chapter 4, we start with the simplest such model which has just a single parameter (an intercept) in the mean and another parameter for the variance. Then, we progress to models with more parameters in the mean, covering both continuous and categorical (factor) explanatory variables (Chapters 5–7) and finally models with two covariates, for which we may specify main effects or interactions (Chapters 8 and 9). Chapter 9 covers what is known as the *general linear model*. This is the generic prototype of a normal-response linear model with both continuous and categorical explanatory variables and with an interaction.

Part 3 of ASM covers linear mixed models with a normal response, that is, a specific form of hierarchical model, or a normal-response regression model with at least one additional source of random variation in the form of a set of random effects. That is, a group of parameters that are not estimated just as unrelated constants, but subject to the constraint that they are drawn from a common distribution which itself has parameters that we estimate. In traditional mixed models, this random-effects distribution is always a normal distribution. In Chapter 10, we cover the classical example of a mixed model with one factor and one continuous explanatory variable, including the random-intercepts and the random-slopes model. In addition, we somewhat surreptitiously introduce the concept of random effects already in Chapter 7.

Part 4 covers the essentials of the GLM, and comprises Chapters 11–13, 15, and 16. GLMs are nonhierarchical (or "flat") models with only a single source of random variability, but where the response distribution may differ from the normal and where we model a transformation (via the so-called link function) of the mean response by an equation that is linear in the parameters. We cover Poisson GLMs in Chapters 11–13 and binomial GLMs in Chapters 15 and 16; note also that the normal linear models in Chapters 4–9 are simply a special case of GLM. Chapter 12 contains a discussion of three modeling topics that are most relevant to non-normal GLMs: overdispersion, zero inflation, and offsets.

Part 5 covers the extension of the linear mixed models from Chapters 7 and 10 to the general GLM case, that is, where the response distribution may be something other than a normal, but where the random effects are still normally distributed by definition. This is the class of generalized linear mixed models, or GLMMs. We cover a Poisson GLMM in Chapter 14 and a binomial GLMM in Chapter 17.

Part 6 contains more advanced concepts, in Chapters 19, 19B ("B" is for bonus, and this chapter can be downloaded from the ASM book webpage) and 20. The first two of these chapters cover two classical examples of hierarchical models in ecology (Royle & Dorazio, 2008). They can be seen as a generalization of mixed models to the case where the random effects may have a distribution other than the normal, such as a binomial (in the occupancy model in Chapter 19) or the Poisson (in the binomial N-mixture model in Chapter 19B). Hierarchical models are tremendously flexible and powerful in science and management. These two chapters are your gateway in this book to a very large body of published material on hierarchical models in ecology, for example, in Royle & Dorazio (2008), Kéry & Schaub (2012), Royle et al. (2014), Hobbs & Hooten (2015), Kéry & Royle (2016), MacKenzie et al. (2017), Hooten & Hefley (2019), or Kéry & Royle (2021). Chapter 20 introduces integrated models, where we combine two or more data sets that are qualitatively different, but which contain information about one or more shared parameter(s). This chapter is another gateway to the big field of integrated models and integrated population models (Pacifici et al., 2017; Miller et al., 2019; Kéry & Royle, 2021: Chapter 10; Schaub & Kéry, 2022).

Additional resources associated with this book include the ASMbook R package, an ASM book webpage, and a Google group email list. The ASMbook package can be downloaded from CRAN (https://cran.r-project.org/). It contains many useful functions, for example, to simulate data under a wide range of settings (coded as function arguments) for every chapter. The ASM book webpage (https://www.elsevier.com/books-and-journals/book-companion/9780443137150) contains general information about the book, all code including some code not printed in the book, a list of errors found, the bonus Chapter 19B on the binomial N-mixture model for free download, as well as teaching tips for using our book in courses. Finally, please sign up to the *HMecology* Google group mailing list if you want to ask questions about topics in the ASM book. This is a general list covering also the other non-spatial-recapture and non-unmarked parts of the books by Royle, Dorazio, Kéry, Schaub, and other colleagues; see also http://www.hierarchicalmodels.com.

1.11 How to use the ASM book for self-study and in courses

Due to a highly modular structure of most chapters with multiple layers (or subsections), plenty of different ways can be envisioned in which you can use the ASM book either for self-study or to teach a course. While we believe its "Rosetta stone approach" is a great benefit for learning, you can also just pick one particular model-fitting engine and work through all its examples, ignoring the others. Thus, the book may either be used as a whole or else you can choose different layers, corresponding to topics or engines, to give you exactly what you need. For instance:

- If you want to learn or improve your understanding of linear models in terms of how we specify the effects of continuous explanatory variables and of factors, you can study Chapter 3 and then one of the layers in most chapters in the rest of the book, especially in Chapters 4–9.
- If you need an introduction to Bayesian inference and Bayesian statistical modeling, you can read most of Chapter 2 and then select one of the Bayesian layers (typically JAGS or NIMBLE because they're the easiest; see Section 4.9) in most other chapters. And when you do this, don't forget the important Chapter 18.
- If you want to learn program NIMBLE, just take the NIMBLE layer in all chapters (and download the NIMBLE code from the website) and add to it Chapters 2 and 18, depending on whether you do or don't already have experience in Bayesian inference.
- Similarly, if you want to learn program Stan, just take the Stan layer in every chapter.
- If you want to learn how to get likelihood inference for very general models, read Chapter 2 and then work through all DIY-MLE and TMB layers in the other chapters (again, not forgetting Chapter 18).
- If you want to learn simple traditional statistics with linear models, GLMs, mixed models with standard function in R, do the non-Bayesian parts of Chapter 2, all of Chapter 3, and then the "canned functions" layers in all other chapters.
- If you were never sure what the parameter estimates meant that the R functions `lm()` and `glm()` (and many others, too) produce, especially for factors: work through Chapters 3 and 6–8.
- If you were never quite sure what a link function was: read Chapters 2, 3, and then 11–17.
- If you never really understood what "random effects" meant: read Section 3.5 and Chapters 7, 10, 14, 17, 19, and 19B.
- If you were always puzzled by what "hierarchical model" meant: again read Sections 3.5, and Chapters 7, 10, 14, 19, and 19B, especially the first and the last two.
- If you were always confused by what exactly the response variable in a logistic regression was (i.e., is it the probability or is it a count? Or two counts?), read Chapters 15–17 and 19.
- If you always wanted to know what a GLMM was, but never dared to ask: read Sections 3.5 and Chapters 14 and 17, and probably also Chapter 10.
- If you always felt intimidated by likelihood functions: read Chapter 2 and then work through the DIY-MLE layers in most other chapters, especially for the simpler, nonhierarchical models (Chapters 4–9, 11–13, 15, and 16).
- Likewise, if you were wondering what these "densities" were that statisticians always talk about: read Chapters 2 and 18.
- If you wanted to understand Bayesian information criteria (the deviance information criterion, DIC, and the Watanabe-Akaike information criterion, WIC) and were wondering how they are related to the frequentist Akaike information criterion, AIC, read Section 18.6.
- And if you want to read all of what Ken and Marc would have liked to have in an introductory stats book when they were in grad school: read the whole book and work through all the examples.

We emphasize that *by far* the most effective manner to work through this book is to read the text, too, yes, but especially to sit at your computer and re-run all the analyses... and then start to modify them. You can download all the R and other computer code from the book website; see above.

We don't have any exercises in the book, but we encourage you to invent your own by doing what we like to call "addition and deletion exercises." For instance, you can fit a new model which lacks one of the covariates that are included in a particular model in the book. On the other hand, you can invent a covariate with the right format (e.g., a vector of the right length) and fill it with random numbers in R (e.g., using `rnorm()`) and then add this as another covariate into one of the models in the book. In another type of exercise you may change the distribution assumed for a response, for example, to turn a Poisson GLM into a normal-response GLM or vice versa. By doing your own exercises in this way, you can greatly boost the learning effect produced by working through the ASM book.

1.12 Summary and outlook

In this chapter, we have set the scene for the *Applied Statistical Modeling*, or ASM, book. ASM is a gentle introduction for ecologists to parametric statistical modeling by linear, generalized linear, linear mixed, generalized linear mixed, hierarchical, and integrated models. We also cover the essentials of probability, statistics, and computation that ecologists usually miss in their training, such as random variables, probability densities, the all-important likelihood function, Bayes theorem, posterior distributions, likelihood optimization, and MCMC. We motivate the coverage of multiple model-fitting engines or methods, both Bayesian (JAGS, NIMBLE, and Stan) and maximum likelihood ("DIY MLE" and TMB) in what we call a Rosetta stone approach. Here, fluency in one method may greatly help with learning another, new one. Due to the highly modular concept of the book, it can either be worked through all in one or else by chosing particular layers to meet your exact needs in self-study or when designing a course.

In this introductory book, we don't cover any of the following four topics: the traditional least-squares narrative for linear models, more complex linear models for more complex designs, nonlinear models, and study design. First, linear models are traditionally presented and fit in the least-squares framework, as also in the R function `lm()` (Zar, 1998; Quinn & Keough, 2002). This has some practical advantages and is likely the way how you learned normal linear models. However, it does not fit so well into the much broader framework of parametric statistical models that we adopt throughout this book. For this reason and also because you will literally find dozens of books about least squares, we have not included this perspective here.

Second, in terms of the covariates, throughout the book we will cover simple prototypes with up to two covariates. We don't show linear models with more than two covariates, nor (with the exception of Chapter 18) with polynomial terms such as quadratic or cubic terms of a continuous covariate, for example, to model "wiggly" covariate relationships. We show only the simplest form of a nested design (Quinn & Keough, 2002; Schielzeth & Nakagawa, 2013) —it appears in every case where we have a random effect, for example, in Chapters 7, 10, 14, 17, 19, and 19B. But there are many more complex situations of nesting, such as split-plot designs with two or more levels of nesting (Steel et al., 1996; Mead et al., 2012). We refer to the useful book by Quinn & Keough (2002) who discuss many more complex study designs and give linear models for these. Armed with what you learn in the ASM book, you should be able to fit also models with such more complex designs.

Third, we don't cover models that are nonlinear in the parameters. An example would be the von Bertalanffy growth model, which relates the length l_t of an individual (that's the response) to its age t (that's the covariate) via $l_t = L_\infty (1 - e^{-K(t-t_0)})$. The three unknown constants are the asymptotic length L_∞, the growth coefficient K, and t_0, which is the hypothetical age at which the individual would have had size 0. There is no possible transformation by which one could re-arrange this nonlinear equation into the form of a linear model. Nonlinear models are often more mechanistic because they can arise by considering the chemical, physical, or biological processes that underlie the variation in a response variable. In this sense, they may be preferable over the more phenomenological and descriptive linear models, which can rarely ever be developed from first principles. But linear models are statistically and computationally far less challenging than nonlinear models. Most parametric statistical models used in applied statistics belong to the class of linear models. Thus, in this book we restrict ourselves to linear models. But it would be perfectly possible to also specify nonlinear models and fit them to data using the methods covered in this book; see Bolker (2008) for examples with maximum likelihood and Oldham et al. (2016) to experience some bafflement about the wide range of possible nonlinear functions that one might also entertain in a parametric statistical model.

Finally, we don't say anything about study design and experimental design (Quinn & Keough, 2002; Thompson, 2002; Williams et al., 2002). We did this mostly for lack of space, because also in the era of citizen-science and big data, we think that these remain fundamental topics which, by the way, should be very much more widely known by ecological analysts. We make some more comments about this topic in Chapter 21.

Introduction to statistical inference

Chapter outline

In this chapter, we give a gentle introduction to the key statistical concepts and computational techniques used in statistical modeling. That is, we explain what goes on under the hood when we fit a statistical model to data to estimate parameters and assess their associated uncertainty, for example, by

---

⊛This book has a companion website hosting complementary materials, including all code for download. Visit this URL to access it: https://www.elsevier.com/books-and-journals/book-companion/9780443137150.

Applied Statistical Modelling for Ecologists. DOI: https://doi.org/10.1016/B978-0-443-13715-0.00023-6

computing standard errors (SEs) and confidence intervals (CIs). We do this for the two dominant schools of model-based inference, classical, or maximum likelihood (ML), and Bayesian posterior inference. They have far more in common than what we are sometimes made to believe. For instance, they are both model- as opposed to design-based (Little, 2004) and they have a strong focus on the likelihood function (see below). We think that to be a productive researcher you should be versed in both. However, these are truly vast topics. Hence, we can necessarily only scratch the surface.

We describe the main underlying concepts and techniques involved in the fitting of a parametric statistical model. Often, this model is hidden under the hood when we use software such as R, SPSS, or SAS, but it is nevertheless the basis of our analysis. Thus, it is essential that we understand what our software does. More importantly, an understanding of these basics allows us to apply more general model-fitting engines in order to fit statistical models in a completely flexible manner. This greatly extends our modeling options compared to what we can do with canned functions.

We begin by briefly describing probability as the foundation on which statistical modeling rests. Then, we introduce parametric models for statistical inference, that is, for parameter estimation, uncertainty assessment, hypothesis testing, goodness-of-fit, model selection, prediction, or missing value estimation. The likelihood function is a central concept in both the classical, or frequentist, and in the Bayesian schools of statistical inference. Hence, we spend most of the chapter covering parameter estimation and uncertainty assessment using either the likelihood function alone (that is, by ML) or by combining the likelihood function with prior distributions and obtaining Bayesian posterior inference.

2.1 **Probability as the basis for statistical inference**

In Chapter 1, we argued that there are only a few things in nature or life that are perfectly known, completely invariable, and thus fully understandable and predictable. That is, hardly anything is entirely deterministic; rather, there is an element of chance in almost everything. But on the other hand, very few things are completely irregular and unpredictable; rather, most things can be predicted at least in some broad, average sense. Thus, we need a way of drawing conclusions from noisy and incomplete observations and of making decisions in the face of the resulting uncertainty. Our methods for learning from data in an uncertain world must be able to describe phenomena both in an average sense as well as in terms of their variability, which is a major source of our uncertainty about any random phenomenon. A powerful basis for such a description is *probability*: the branch of mathematics that deals with chance processes and their outcomes (Pishro-Nik, 2014; Blitzstein & Hwang, 2019). Probability is a fascinating subject in its own right, because it enables you to better understand and navigate an uncertain world. Indeed, probability is the extension of logic from certain (i.e., true or false) events, which are extremely rare in life, to all events (Lindley, 2006). Probability is also the basis for statistical modeling and inference, the topics of this book.

A statistical model is a description of a probability mechanism that could have produced as one possible outcome the data set we have in hand. Thus, the same chance mechanism could also have produced a different data set or indeed an infinity of potential data sets. This view of our particular data set as being just one out of many possible data sets is the conceptual basis of the frequentist school of inference in statistics. It defines probability as the frequency of an event in hypothetical replicates (Casella & Berger, 2002). For now, we won't discuss where our models come from and how we build them; we say more about this in Chapters 3 and 18. Instead, we will cover inference, that is, the use of formal methods to infer unknown things once we have a model and some relevant data. "Statistical modeling" comprises both the development of a statistical model and its analysis using some method of inference, such as maximum likelihood or Bayesian posterior inference.

Statistical inference is based on probability, but in a sense it goes "the other way round" (Link & Barker, 2010). While probability assumes a stochastic process and then looks at its possible outcomes, that is, the data it may produce, in statistical inference we start with a data set and make assumptions about the stochastic process(es) that could have produced it. Combining data and model enables us to learn about features of the stochastic process (Fig. 2.1). This inductive process is called *statistical inference*. Its key task is the use of the data to infer the likely values of parameters that govern a stochastic process and to assess our uncertainty in these estimates. Statistical modeling is a powerful formal way of learning from data. In this sense, statistics could be called a meta-science that overarches all empirical sciences. Statisticians have also been called the *"custodians of the scientific method"* (Hooke, 1980).

A key concept in probability is that of the *random experiment*: the process of observing or measuring something unpredictable. A particular observation made in such an experiment is called an *outcome* (Pishro-Nik, 2014). Associated with each random experiment is the set of all possible outcomes, called a *sample space* or *S*. Favorite examples of random experiments for statisticians include the toss of a coin and the throw of a die, which have sample spaces with two or six possible outcomes, respectively. In ecology, more directly relevant random experiments include the following:

- We choose an animal or plant population and select, or capture, an individual and take some measurement on it, such as the body mass of peregrines, snout-vent length or body mass of snakes or box turtles, wing-length of butterflies, color morph of snakes, or the number of ectoparasites on dragonflies. In all these cases, we often assume that sampling is at random, that is, that every individual in the biological population has the same chance to be selected and to appear in our sample.
- We select a sample of survey sites or "local populations" and make an assessment of whether a study species is present or absent, of its population size, or the proportion of pairs that

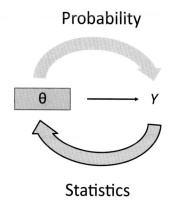

Probability

θ ⟶ Y

Statistics

FIGURE 2.1

The duality between probability and statistics (compare with Fig. 1.1, which shows an analogous duality between data simulation and data analysis). The gray box denotes a stochastic process, or model, governed by parameter θ, and *Y* denotes some output, that is, observed data. In statistical modeling, we treat the observed data as the output from a stochastic process and use probability to describe this process in the form of a statistical model. Combining data and model lets us infer features of the data-generating process, such as parameter θ.

reproduce successfully, or take any other measurement at the scale of that spatial sample site. The sampled sites will typically again be assumed to be selected at random from a larger statistical population of sites that could have been selected and about which we want to make an inference. That could be the mean proportion or the total number of sites in the statistical population of sites that are inhabited or the average number of individuals per site.

In the former case, where we select individuals in a population, the sample space of these random experiments may be all possible measurements of: body mass of peregrines (e.g., some value between 400 and 1600 g), snout-vent length of snakes (e.g., some value in the range of 20–100 cm), butterfly wing length (e.g., 6–10 cm), adder color morph ("black" vs "gray"), or the number of dragonfly ecto-parasites (a non-negative integer). In the latter case, where we select a number of study sites or local populations, our sample space may consist of the values "presence" or "absence" of our target species or a count such as 0, 1, ... for an abundance at each site.

In all these examples, there will be chance involved in the particular outcomes observed. Thus, the actual outcome of a random experiment will vary when we repeat it (because we happen to select a different individual or because of measurement error) or when different people conduct it. This *sampling variability* creates an important part of the uncertainty that we have to deal with in statistical modeling. We note in passing that in probability the term *experiment* has a different meaning than in experimental design, where it always involves some manipulation in the form of a treatment (Snedecor & Cochran, 1980; Cochran & Cox, 1992; Quinn & Keough, 2002; Mead et al. 2012). In probability, the term *experiment* simply denotes the formal manner in which we take some observation or measurement in a chance process.

Often we are not interested in all possible outcomes of a random experiment, but only in some summary, or in an *event*, which is a subspace of the entire sample space. For instance, we may want to distinguish heavy from lean peregrines and take 800 g as a threshold for such a classification, or we simply add up the number of black and gray adders observed in a population. Similarly, we may summarize an assessment of presence and absence of a species at a set of sites by the total number of sites where a species is detected. Working with suitably defined events will be more directly relevant to our questions, and the randomness in the outcome of the original random experiment carries over to the events such defined.

To describe the effects of chance on the particular outcomes obtained in a random experiment, and therefore on the events observed, we use probability. A *probability function* for an event A is defined by the following three axioms (Pishro-Nik, 2014; Blitzstein & Hwang, 2019):

- Probability is a number between 0 and 1, where $P(A) = 0$ denotes an impossible event and $P(A) = 1$ is a certain event.
- The probability of the entire sample space $P(S)$ is 1, since by definition the outcome of a random experiment must be part of the set S.
- The probability of the union of mutually exclusive events (e.g., either or both of A and B) is the sum of the probability of the individual events, that is, $P(A \cup B) = P(A) + P(B)$. This is also called the addition rule.

Often the multiplication rule is also added, which describes the probability of the intersection of two events, i.e., that both of them occur or are true: $P(A \cap B) = P(A)P(B|A)$. Statistics is built on these axioms of probability, regardless of the interpretation that we attach to probability (Lindley, 2006; Spiegelhalter, 2019). Hence, these axioms and a host of further probability rules that can be deduced from them hold regardless of whether we interpret probability as a relative frequency in a classical, or frequentist, analysis, or as a measure of uncertainty or an expression of the amount of knowledge about something uncertain in a Bayesian analysis (see Sections 2.5–2.8).

2.2 **Random variables and their description by probability distributions**

Typically, in a study we are interested in a particular aspect of the sample obtained, that is, of the particular event recorded in a random experiment. Examples may include the proportion of black versus gray adders in our study populations, the number of occupied versus unoccupied grid cells, the number of plants in each of a number of populations, or the body mass recorded on each of the captured peregrines. All of these features of the studied samples are examples of *random variables*, abbreviated RVs. A RV is a real-valued function defined on the sample space of a random experiment (Pishro-Nik, 2014), and it is a key concept in probability and statistical modeling. While RVs may hardly even feature in the statistical training of most ecologists, they are the fundamental manner in which statisticians conceptualize a data set at the start of a statistical analysis. That is, the observed data (and, for hierarchical models, also unobserved data or random effects; see Section 3.5) will be treated as the outcomes of RVs. Our statistical model then describes the stochastic mechanisms that gave rise to the particular observed outcomes of these random processes.

The RV concept is critical for our understanding of statistical modeling, but meaning and use of the term may be confusing at first. Statisticians often distinguish RV Y, when meant as a name for a stochastic process, from the realized value or outcome of that RV, y. That is, they denote the abstract process in upper case and the realized value produced by the process in lower case. The realized value y denotes a single measurement, or observation, produced by the random process denoted Y. For instance, the body mass of a male peregrine falcon may be our RV Y, but when we measure the mass of our first peregrine, then we treat that measurement as our RV y. Worse yet, when we take multiple measurements of Y, say one from each of n examined peregrines, we will distinguish RVs Y_1, Y_2, ..., Y_n, and each will have an associated realized value y_1, y_2, ..., y_n. Moreover, we may distinguish RV Y from RV X, where Y may be the body mass of a captured peregrine and X its sex. The outcome of a random experiment consisting of capturing and examining n peregrines can be summarized by one random vector $\{x_1, x_2, ..., x_n\}$ for the sex of each bird and another, $\{y_1, y_2, ..., y_n\}$, for its mass.

As we will see below, for inference about non-hierarchical models we construct a likelihood function from the joint probability of all observed RVs. By this we mean for instance the set of all peregrine body mass measurements in our study, that is, the vector $\mathbf{y} = \{y_1, y_2, ..., y_n\}$. But in a hierarchical model (see Chapter 3), say, in a study of the detection or nondetection of a species in an occupancy species distribution model (Chapter 19), we may distinguish two RVs Z and Y: the former is the latent, i.e., unobserved, presence or absence of a species at a site, while the latter is the observed detection or nondetection at a site during one particular survey (see Chapter 19 for why presence/absence and detection/nondetection are not in general the same thing). Together, they produce the observed data $\{y_1, y_2, ..., y_n\}$. A probability expression defining our likelihood function will at first have to include all RVs, both unobserved/latent (i.e., Z) and observed (i.e., Y), see Kéry & Royle (2016: Section 2.4.6). See Section 3.5 for more about what is considered a RV in a hierarchical model and how this affects the likelihood construction. In summary, the specific meaning of "random variable" may vary somewhat depending on the context.

RVs are the essential building blocks for parametric statistical models. It is therefore paramount that you obtain a sound working knowledge of them. We strongly advocate thinking about your data and observations from the "random variable perspective." This is very natural in the context of learning about and understanding most things in nature, which can almost always be conceptualized as the outcome of a chance process, and hence as some RV Y, producing one or multiple observations y or **y**, respectively.

2.2.1 Discrete and continuous random variables

We can distinguish two classes of RVs: discrete-valued and continuous. To some degree, this distinction coincides with the nominal, ordinal, and metric or interval measurement scales (Zar, 1998): the first two produce discrete-valued RVs, and the latter two continuous RVs.

Examples of discrete RVs include descriptions, labels, or names of classes, such as sex; color of hair, fur or plumage; geographic strata such as "Population A", "Population B" and "Population C"; or the identity of the field technician doing a bird point count. There is no sense of order in these class labels. In contrast, the other typical case of a discrete RV is a count, which does have a natural ordering. Examples of counts include the number of female nestlings in a nest with a known number of young, the number of asp vipers observed per visit to a site, or the number of whales migrating past a point on the coast during some time interval.

Continuous RVs can take on any value within some range, at least up to measurement accuracy. Measurements of length, distance, area, duration or of mass provide the typical examples of a continuous RV. Some measurements may be either discrete or continuous, depending on the specific type. For instance, the location of an animal in a movement study may be either continuous, when you record its exact coordinates at a point in time, or discrete, when you just note the identity of a grid cell wherein the animal is located. In general, you may translate between continuous and discrete RVs by "binning" or grouping outcomes of the former (see Fig. 2.3). Note that measurement accuracy is always finite. Hence, strictly, every continuous RV is measured in a finite number of discrete classes, but in practice this can usually be ignored when modeling continuous RVs.

Many types of RVs have natural bounds. For instance, a count cannot be negative, nor can most measurements, but a RV formed as the difference between two counts or measurements can. The range of a RV is also called its *support* and can be an important consideration when choosing a suitable probability model for a data set (see Section 3.2).

2.2.2 Description of random variables by probability distributions

A parametric statistical model (see Section 2.3) for a data set must describe all of its RVs, and for this we use probability distributions. They describe how the total probability of 1 in a random experiment is distributed across the sample space, that is, among all possible realizations of a RV. We use a *probability mass function* (PMF) for discrete RVs and a *probability density function* (PDF) for continuous RVs. Note that in practice we often use the terms "PDF" or "density" interchangeably for either a PMF or a PDF and the context determines which is meant. We begin by describing a PMF because this is arguably simpler to grasp.

A PMF $f(Y)$ gives the probability of every possible observable value y of a discrete RV Y:

$$f(Y) = P(Y = y) \tag{2.1}$$

The function value of a PMF lies always between 0 and 1, since it's a probability, and the sum of the PMF over all possible values is equal to 1. Indeed, these are the defining features of a PMF. Some

RVs are binary and have only two possible values, for example, a survival process and its associated RV Y can only produce an individual that is either dead or alive. Therefore, if we define $y = 1$ as an animal that is alive and $y = 0$ as one that is dead, then it is sufficient to define the probability $P(y = 1) = \phi$, i.e., survival probability, to fully describe the binary RV (Royle & Dorazio, 2008). Since the sample space of the underlying random experiment contains only two values, and since the probability of the entire sample space is equal to 1, we know that $P(y = 0) = 1 - \phi$. To describe a discrete RV with more than two possible outcomes such as hair color, where there are n mutually exclusive events in a finite set, we need only n-1 free probabilities, since the probability of one event can always be expressed as 1 minus the sum of the probabilities of all others. The binomial, multinomial, or categorical distributions are examples of such PMFs (Royle & Dorazio, 2008: chapter 2; Schaub & Kéry, 2022: chapter 2).

A PDF gives the *probability density* $f(Y)$ of every possible observable value of a continuous RV Y. This density is a non-negative value that is the limit of the area of a rectangle with height $f(Y)$ and base $(y - \delta, y + \delta)$, as the rectangle half-width δ goes to 0. A PDF has the peculiar feature that the probability (but not the density) of any particular value of Y is 0, and that the density may be > 1 for some values of Y. To obtain the probability of a range of values of Y, say for (y_1, y_2) for density $f(Y)$, we must integrate f between y_1 and y_2, that is, $P(y_1 < Y < y_2) = \int_{y_1}^{y_2} f(Y) dY$. The integral over the entire support of a PDF is again equal to 1 by definition.

In statistics, there is a large number of stochastic processes that occur so frequently that their associated probability distributions have been given names. These named *statistical distributions* include the Bernoulli, binomial, Poisson, uniform, and the normal (or Gaussian), which we describe in more detail in Chapter 3. They are the typical building blocks of our statistical models; hence, it is important that you obtain a good understanding of them. All distributions are governed by a small number of constants, called *parameters*, which determine the specific form of a distribution (see also Fig. 2.5). For instance, a Poisson distribution is a typical stochastic model for unbounded counts. The PMF of a Poisson RV is as follows (remember that the exclamation mark is the factorial function):

$$f(y|\lambda) = P(Y = y|\lambda) = \frac{\lambda^y e^{-\lambda}}{y!} \tag{2.2}$$

A PMF gives the probability of observing any possible value of the RV Y as a function of some parameter(s). For a Poisson PMF, λ (lambda) is the only parameter, and it is both the mean and the variance of Y. In the expressions for f and P we see λ on the right side of a conditioning bar (|). This means that the values of these functions depend on the specific value chosen for λ. We suspect that the eyes of many ecologists will glaze over when confronted with this explicit way of writing a statistical distribution. However, it is extremely valuable to learn to understand what distribution functions are and how we use them. Play around with them by filling in some values and see what comes out, and also produce plenty of graphs, i.e., adopt the experimental approach to statistics (see Section 1.8). You should note that λ in the Poisson PMF is just a customary name, and you might just as well denote the Poisson mean with any other Greek or Roman letter or write it out as a word, as we do in our model-fitting engines (see Chapter 4) (In contrast, e is a mathematical "system letter"; it is is Euler's number 2.7183.).

Thus, the Poisson PMF enables you to compute the probability of observing any of the possible outcomes of a random experiment for which we define a RV that can be adequately approximated by the model of a Poisson distribution. This assumes that λ is known. So what if we assume for now $\lambda = 2$? Then, we can for instance obtain the probability of a count of 1, which is $\lambda^y e^{-\lambda}/y! = 2^1 e^{-2}/1! \approx 0.27$. What about the probability of a count of zero? We can evaluate the PMF for $Y = 0$ and obtain $2^0 e^{-2}/0! \approx 0.14$. Also, we can use this probability model to determine that the probability of a count of either 2 or of 3 is $2^2 e^{-2}/2! + 2^3 e^{-2}/3! \approx 0.27 + 0.18 = 0.45$ (that we can add these probabilities follows directly from the axioms of probability).

The definition of the Poisson PMF is Eq. 2.2, but we will frequently use shorthand. Thus, you will see a Poisson RV written in many different forms, including these (and remember the comment about the somewhat arbitrary naming of the Poisson parameter from above):

- $y \sim f(Y|\lambda)$
- $y \sim f_Y(Y|\lambda)$
- $y \sim Poisson(\lambda)$ or $y \sim Pois(\lambda)$
- $y \sim Poisson(Y|\lambda)$ or $y \sim Pois(Y|\lambda)$

The BUGS language is extremely simple and powerful for describing statistical models (Gilks et al., 1994). We use it to specify statistical models in software JAGS and NIMBLE and in a dialect also in Stan (see Chapter 4). In BUGS we define a Poisson RV in a similar way:

```
y ~ dpois(lambda)
```

In addition, the simplest statistical model for a set of Poisson RVs is a Poisson generalized linear model (see Chapter 3) which in R can be fit by issuing the following command:

```
glm(y ~ 1, family = poisson, data = list(y = y))
```

Although not the definition of a Poisson PMF, this comes close to it: it defines the values in vector y to be Poisson RVs and estimates a mean parameter for them.

In R, we have four functions for each of a large number of probability distributions. They are all named with a short form of the distribution name, and have a letter d, p, q, or r added in front. These letters denote, in this order, the probability density (or probability mass) function, the cumulative distribution function (or CDF), the quantile function, and the random number generation (RNG) function. What do these mean? Here's an illustration for all four with our example of a Poisson RV with lambda equal to 2:

```
# Look up help text for the Poisson
?dpois                           # Also '?Poisson'

# Poisson probability mass function (PMF) evaluated for y = 0
dpois(0, lambda = 2)             # Probability of getting a value 0
dpois(0, lambda = 2, log = TRUE) # Same on log scale

# Get density 'by hand' to emphasize dpois() is just shorthand!
lam <- 2; y <- 0
(lam^y)*exp(-lam) / factorial(y)     # Probability of getting a value 0

# Cumulative distribution function (CDF), or just 'distribution function'
ppois(3, lambda = 2)             # Probability of getting a value of 0, 1, 2 or 3
sum(dpois(0:3, 2))               # Same, summing up probs 'by hand'
plot(ppois(0:9, 2), type = 'h', lend = 'butt', lwd = 20, ylim = c(0, 1),
   xlab = 'Value of Y', ylab = 'P(Y <= y)', main = "CDF of Y")     # not shown

# Quantile function: value of y for which CDF(y) has a certain value
qpois(0.85, lambda = 2)
```

```
# Random number generator (RNG) function
set.seed(2016)                          # Set seed if want same numbers as we have
rpois(n = 10, lambda = 2)               # 10 Poisson(lambda = 2) random numbers

[1] 0.1353353                           # P(y = 0)
[1] -2                                  # log(P(y = 0))
[1] 0.1353353                           # same 'by hand'
[1] 0.8571235                           # P(y <= 3)
[1] 0.8571235                           # same 'by hand'
[1] 3                                   # Value of y for which CDF evaluates to 0.85
[1] 1 1 3 0 2 0 2 4 0 0                 # 10 random Poisson numbers with lambda = 2
```

We have emphasized how dpois(..., log = TRUE) yields the log-probability, since this will become very important in model fitting and for many methods of doing model selection. Note also that the CDF is an alternative to the PMF or PDF for defining a probability distribution and works equally for discrete and continuous RVs. The CDF is often denoted $F(Y)$ and it gives the probability that Y takes on a value less or equal than y, that is, $P(Y \leq y)$. We don't cover the CDF in this book but will briefly encounter it when discussing quantile residuals (Dunn & Smyth, 1996) in Chapters 5 and 18.

We can easily produce a plot of a PMF, and we highly encourage you to do this to train your intuition about probability distributions. Fig. 2.2 shows the PMF of our Poisson(lambda = 2) RV.

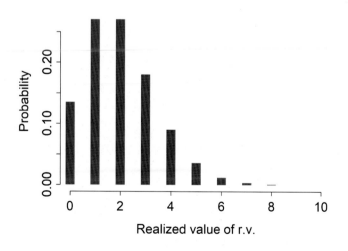

FIGURE 2.2

PMF of a Poisson RV with parameter $\lambda = 2$. The y-axis shows the probability of each possible value for this discrete RV. The probability of values greater than about 10 is not 0, but is simply too small to be seen at the scale of the y-axis.

```
# Plot PMF of Poisson with lambda = 2 (Fig. 2.2)
par(mar = c(6,8,5,4), cex.lab = 1.5, cex.axis = 1.5, cex.main = 1.5)
plot(0:10, dpois(0:10, lambda = 2), xlab = 'Realized value of RV',
   ylab = 'Probability', main = 'Poisson(2) PMF', type = 'h', lend = 'butt',
   lwd = 20, col = 'gray30', frame = FALSE, las = 1)
```

As an example for the probability distribution of a continuous RV, let's take the peregrine body mass example for which a normal distribution is our usual statistical model. The PDF of a normal, or Gaussian, RV is as follows:

$$f(y|\mu,\sigma) = \frac{1}{\sigma\sqrt{2\pi}} e^{-\frac{(y-\mu)^2}{2\sigma^2}} = \frac{1}{\sigma\sqrt{2\pi}} \exp\left(-\frac{(y-\mu)^2}{2\sigma^2}\right) \tag{2.3}$$

This distribution has two parameters, μ and σ, which are the mean and the standard deviation (SD) of a normal RV. If we pick a value for each, then the PDF enables us to compute the probability density (or alternatively, the log-density) for any possible observation and, by integration over a range of values, to obtain the probability that our RV will take on a value within this range. We next give an illustration for all four variants of the normal distribution functions in R with a Normal(600, 30) RV, as in Chapter 4 where we assume this describes body mass of male peregrine falcons.

```
?dnorm                                    # Also '?Normal': Look up the help
                                          # text for the Normal


# Gaussian, or normal, probability density function (PDF) for y = 650
dnorm(650, mean = 600, sd = 30)
dnorm(650, mean = 600, sd = 30, log = TRUE)      # Same on log scale


# Get the density 'by hand' to emphasize dnorm() is just a shortcut
mu <- 600; sig <- 30; y <- 650
1/(sig*sqrt(2*pi)) * exp(-(y - mu)^2 / (2 * sig^2))


# Cumulative distribution function (CDF), or just 'distribution function'
pnorm(600, mean = 600, sd = 30)               # Prob. of value between -Inf and 600
# Next is same, but integrating the PDF 'by hand'
integrate(dnorm, lower = -Inf, upper = 600, mean = 600, sd = 30)
plot(pnorm(seq(400, 800, by = 1), 600, 30),
   type = 'l', lwd = 5, ylim = c(0, 1),
   xlab = 'Value of Y', ylab = 'Prob. density',
   main = "CDF of Y", frame = FALSE)          # not shown


# Quantile function: value of y for which CDF(y) has a certain value
qnorm(0.95, mean = 600, sd = 30)


# Random number generator (RNG) function
set.seed(2016)
rnorm(n = 10, mean = 600, sd = 30)            # 10 Normal(600, 30) random numbers
```

```
[1] 0.003315905          # f(y = 650)
[1] -5.709025            # log-density: log(f(y = 650))
[1] 0.003315905          # same 'by hand'
[1] 0.5                  # P(y <= 600)
0.4995709 with absolute error < 1.4e-14    # same 'by hand'
[1] 649.3456             # value for which 95% of values
                         # are smaller
[1] 572.5577 630.0374 598.3073 608.8994 516.2559    # random numbers
[6] 591.5178 577.0947 579.4497 611.0122 605.4899
```

We can't directly use the PDF to obtain the probability of a particular measurement, but we can do so for a range of values. For illustration, let's compute the probabilities for a large number of 5g-bins of body mass. We will produce two plots, one for the density and the other for the probabilities of the binned body mass RV (Fig. 2.3). You can imagine the density plot on the left as what we would get if, in the plot on the right, we make the bins smaller and smaller, for example, by choosing a bin width of 0.1 g instead of 5 g. As always, we highly encourage you to adopt the experimental approach to statistics and try this out.

```
# Compute probabilities for classes of binned Normal RV
limits <- seq(500, 700, by = 5)
midpts <- seq(500 + 5/2, 700 - 5/2, by = 5)
cumProb <- pnorm(limits, mean = 600, sd = 30)
probs <- diff(cumProb)

# Plot probability density function (PDF) of Normal(600, 30)
par(mfrow = c(1, 2), mar = c(6,6,5,2), cex.lab = 1.5, cex.axis = 1.5,
  cex.main = 1.5)                                              # Fig. 2.3
curve(dnorm(x, mean = 600, sd = 30), 500, 700, xlab = 'Body mass (g)',
  ylab = 'Probability density', main = 'PDF of Normal(600, 30) RV', type = 'l',
  lwd = 3, col = 'gray30', frame = FALSE, las = 1)
```

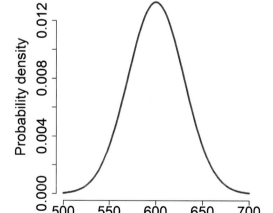
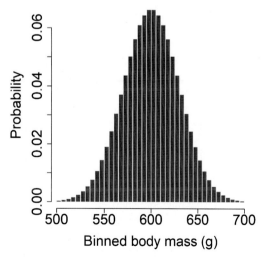

FIGURE 2.3

Left: PDF of a normal random variable with parameters $\mu = 600$ and $\sigma = 30$. Right: Corresponding PMF of binned body mass, with a bin width of 5 g. Note the different scales of the y-axis. When you recompute and draw the plot on the right with much narrower bins, you will approximate the density plot on the left.

```
# Plot probabilities for 5g-binned mass random variable
plot(midpts, probs, xlab = 'Binned body mass (g)', ylab = 'Probability',
  main = 'PMF of binned Normal RV', type = 'h', lend = 'butt', lwd = 5,
  col = 'gray30', frame = FALSE, las = 1)
```

2.2.3 Location and spread of a random variable, and what "modeling a parameter" means

The probability density (PMF or PDF) yields a complete description of a RV, but we may want to characterize it in a simpler way. Two important properties of a RV are its central tendency, or location, and its dispersion, or spread. The former is also called the expectation E (or mean) and can be envisioned as the average over a large number of realizations of the RV. Thus, for a discrete RV the expectation is a weighted mean of all possible outcomes, where the weights correspond to the probability of each: $E(Y) = \sum_{i=1}^{\infty} p(y_i)y_i$. For a continuous RV, we replace the sum by an integral. The variance of the RV is a function that quantifies its dispersion or variability. It is defined as another expectation $Var(Y) = E[(Y-E(Y))^2]$. Expectation and variance can be obtained for most RVs, and for some distributions, they correspond to a parameter of the distribution.

For instance, for both Poisson and normal distribution, the expectation corresponds to a parameter of the density: it is λ in the Poisson and μ in the normal distribution. In most statistical models we will want to "model" a response variable, that is, an observed or possibly an unobserved RV, by expressing its mean as some function of a covariate (Section 3.4). This simply means that we replace the parameter for the mean in the PDF/PMF by a function of these covariates, with new parameters that replace the old parameter, that is, the mean. Such a function relating the mean to covariates may be nonlinear or linear. In this book, we focus on mathematically linear relationships between the mean of a response or some transformation of it in the case of non-normal generalized linear models (GLMs; Section 3.3) and a set of covariates, that is, on linear models. However, conceptually, the step towards the modeling of nonlinear functions in the mean is straightforward; see Pinheiro & Bates (2000), Bolker (2008), Hobbs & Hooten (2015) or Hooten & Hefley (2019). Note that the formal definition of a linear model is a model where a parameter can be expressed as the sum of a series of explanatory variables (i.e., covariates) multiplied by their effects, and "linear" refers to the effects of these explanatory variables. We can easily describe very wiggly curves by what mathematically are linear models; thus, a linear model does not restrict us to mere straightline relationships. Models with higher-order polynomial terms (i.e., squares, cubes, etc.) of a covariate offer a good example of wiggly relationships represented by a model that is mathematically linear.

In this book, we focus on the normal, Poisson, and Bernoulli/binomial distributions; see Chapter 3. They all have one parameter which represents the mean response. Obviously, this makes it straightforward to model the response by modeling the mean of the distribution as a function of covariates. However, in many other distributions there is no correspondence between the mean and any of the parameters. Then, to express the mean response as a function of covariates, you must first reparameterize such that the mean becomes a parameter by what is known as *moment matching* (Hobbs & Hooten, 2015; Kéry & Royle, 2016).

A normal distribution has a parameter for the dispersion (i.e., the variance, inverse precision, or SD), but in many other distributions there is no free parameter for the dispersion. Rather, the dispersion is a function of the mean, as in the binomial and the Poisson distributions. The dispersion of a RV is often treated as a sort of a nuisance in many statistical models: that is, a model must account for the unexplained variability in the RVs to get the uncertainty assessments right, but the dispersion is rarely a parameter of interest in its own right. However, there may be exceptions to this, and it is also possible to specify a model for the dispersion, that is, to explain variability in the variance with covariates (see also Section 6.4). In addition, in a hierarchical model (Chapter 3) we also model structure in the variance of a response, but do so in a more implicit manner.

2.2.4 Short summary on random variables and probability distributions

This concludes our introduction to that foundational topic of the RV and its probability distribution. RVs are the defining features of statistical models, because they represent the stochastic building blocks of our models. Since explanation is often easier with concrete examples, we have singled out the Poisson and the normal distributions. In Chapter 3, we will say more about when these statistical models are suitable for a data set and also discuss a small number of other commonly used probability distributions. You can find long lists of discrete and continuous distribution functions in almost any statistics book, including Bolker (2008) or Royle & Dorazio (2008), and of course on the internet. In addition, many of these distributions are related to each other such that one arises as a special or limiting case of another; see the famous flowchart at the end of the *Table of Common Distributions* section in Casella & Berger (2002). Please read up on these fundamental pieces of statistical modeling because the small list of distributions in this book is far from sufficient to cover all stochastic systems that you are likely to encounter in your research.

We make one final remark on terminology: typically, "probability distribution" can mean either a CDF, PMF, or PDF. "Probability density" is frequently used synonymously for a PMF or a PDF. In contrast, "distribution function" is normally reserved for the CDF.

2.3 Statistical models and their usages

What is a parametric statistical model? According to Millar (2011):

> "*A parametric statistical model is a collection of joint density functions, $f(\mathbf{y}; \boldsymbol{\theta})$ ....*"

The bold face font for data $\mathbf{y}$ and parameter(s) $\boldsymbol{\theta}$ defines both as vectors or matrices. In addition, Lee et al. (2017) say that "*the model ... should specify how the data could have been generated probabilistically.*" Hence, a statistical model is a probabilistic representation of a data-generating mechanism which provides a density for every datum in an analysis. A shorter and more approximate definition is "*a parametric statistical model is a collection of density functions for its random variables.*"

To clarify, let's define a model for our Poisson example above, where we drew 10 RVs that we interpret as counts, by constructing the joint density for this data set. In the simplest case, we assume that these 10 numbers were produced independently from stochastic processes with identical parameter lambda. In statistics, the resulting RVs are called *independent and identically distributed*, or i.i.d. Statistical independence means that the value of one count contains no information about the value of any other, once we account for the Poisson parameter lambda (for more on independence, you can look up *statistical independence* in any book on probability ... and while you're at it, you may also read about *conditional independence*). The "identical" in i.i.d. means that all 10 numbers were produced by a Poisson RV with the same value of the intensity parameter. Under i.i.d. assumptions, the joint density of the 10 RVs is a simple product of the densities of each constituent RV in the vector $\mathbf{y}$.

$$f(\mathbf{y}|\boldsymbol{\theta}) = f(\{y_1, ..., y_{10}\}|\boldsymbol{\theta}) = \prod_{i=1}^{10} f(y_i|\theta) \tag{2.4}$$

A heuristic depiction of a statistical model is shown in Fig. 2.4. It emphasizes that the data at hand are assumed to be the outcome from a stochastic process, whose main characteristics we try to emulate in our model, which is the gray box. Of course, we will never be able (nor even want) to describe the data-generating processes exactly; a useful model must be an abstraction and hence a simplification. But we aim for a representation that is useful given the goals of our modeling, which may be prediction, explanation, or summarization (Tredennick et al., 2021; see also Section 1.1).

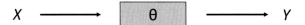

FIGURE 2.4

Schematic of a model (gray box) with parameter θ; X and Y denote some input (e.g., covariates) and output (i.e., observed data), respectively. In parametric statistical modeling, the model typically represents the main processes that generate the observed data (modified from Breimann (2001)).

Parametric modeling, that is, the act of building a model and then fitting it to a data set, is at once a rigid and a flexible process. It has a rigid component, because we make some very specific choices. For example, we must choose the distribution assumed for the observed RVs, and in more complex (i.e., hierarchical or mixed) models, we also must make choices about latent RVs and the distributions that govern them. Furthermore, we typically have covariates and will assume specific functional forms for the linkage with the parameters in the model (although this may be relaxed in "semi-parametric" models such as generalized additive models, or GAMs; Wood, 2017).

But in another sense, parametric modeling is very flexible, since an important part of the model will be adjusted (i.e., fitted) to the specific data as closely as possible. This is where the parameters come in. Parameters give the distributions assumed for the RVs their specific form. It is astonishing how different the same distribution can look depending on the values of its parameters. Fig. 2.5 illustrates this for a beta distribution, which is used to model continuous variables between 0 and 1. These arise for instance as the ratio of two continuous variables, such as the proportion of a leaf that is eaten by an insect.

```
# Look how different a beta can look depending on the values chosen for its parameters
par(mfrow = c(2,2))                           # Fig. 2.5
curve(dbeta(x,10,10), 0, 1, lwd = 3, col = 1, main = "beta(10, 10)")
curve(dbeta(x,0.1,0.1), 0, 1, lwd = 3, col = 1, main = "beta(0.1, 0.1)")
curve(dbeta(x,1,100), 0, 1, lwd = 3, col = 1, main = "beta(1, 100)")
curve(dbeta(x,100,1), 0, 1, lwd = 3, col = 1, main = "beta(100, 1)")
```

A parametric statistical model has many uses. Most importantly, we can apply the model to our observed data y to infer the key features of the process that we suppose has produced these data (at least as an approximation, which is why Royle & Dorazio (2008) write about "approximating models"). The most immediate such inference will be estimation of the parameter(s) $\boldsymbol{\theta}$ and assessment of the associated uncertainty of these estimates, typically in the form of some quantification of error

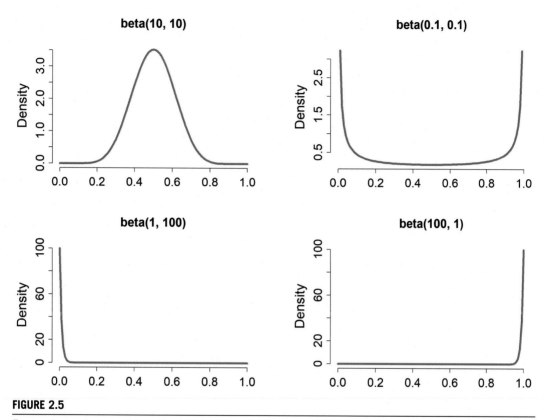

FIGURE 2.5

Four variants of a beta distribution which differ solely in terms of the values of its two parameters.

(e.g., SEs, or posterior SDs) or uncertainty intervals, such as CIs or credible intervals (CRIs). An example might be a capture-recapture study where a key parameter of interest is population size, which we want to estimate along with a SE and 95% CI. In addition, we may also want to test hypotheses about likely values of the parameters, or assess the goodness-of-fit to gauge the degree to which the observed data agree with the key assumptions made by our model. Yet another use is to compare one model variant with another, that is, to do model selection. Further, sometimes we may have observed some output and we may have estimated some model parameters, but we lack any information about the inputs, that is, about the covariates X in Fig. 2.4. We may then use the model to estimate (or "impute") the values of the missing covariates. And finally, we may have observed the input and then along with parameter estimates use the model to estimate (or "predict") missing responses from a model. A typical example of such "missing response estimation" is in species distribution modeling (SDM) (Elith & Leathwick, 2009), where we fit a model to a smaller data set that includes both environmental and species occurrence or abundance data. If we know covariate values also for some larger area, we may use them along with the estimated parameters to extrapolate the expected species occurrence or abundance to this wider area to yield a species distribution map.

All this is known as *statistical inference* and, to summarize, includes: parameter estimation, uncertainty assessment of the estimates, goodness-of-fit assessment, model selection, missing value estimation (imputation), and missing response estimation (prediction). Fundamentally, these are all inductive goals: we take a sample of observations and from them and a model infer features of the wider statistical population from which our data was drawn. Models are our vehicles for inference.

There are two main schools in statistics for making inferences: classical or frequentist, and Bayesian inference. We cover both, because we are convinced that to become an effective statistical modeler you need to understand and be able to use both. Furthermore, their apparent differences are often exaggerated. In an important way they are closely related: both are clearly based on a parametric statistical model and for both likelihood is a central concept. Therefore, we next introduce the likelihood function, and then in Sections 2.5 and 2.6, describe statistical inference that is directly based on the likelihood function, that is, maximum likelihood estimation (MLE). In Sections 2.7 and 2.8, we describe Bayesian posterior inference, which adds to the likelihood function a probability statement about how likely parameter values are by specification of so-called prior distributions and computes the posterior distribution of a parameter.

2.4 The likelihood function

The likelihood function is a key concept for statistical inference. It is a main ingredient of both the classical and the Bayesian schools of learning from data. The likelihood function represents the formal connection between what you observe (the data) and what you don't observe (the unknowns) and especially the parameters you want to estimate.

To quote Millar (2011) again, *"the likelihood function is the (joint) density function evaluated at the observed data, and regarded as a function of* θ *..." That is,*

$$L(\theta) \equiv L(\theta; \mathbf{y}) = f(\mathbf{y}; \theta) \qquad (2.5)$$

The likelihood function provides a measure of the relative support, in terms of the probability of the observed data under a model, for different possible values of θ (Edwards, 1992; Pawitan, 2013).

Likelihood is conceptually simple, yet it can be surprisingly hard to grasp. As ecologists, we are not normally used to thinking about models as the joint densities of the data (Eq. 2.4). Furthermore, algebraically the likelihood function is the exact same thing as the joint density function of the data, but in likelihood the density is "read in the opposite direction," that is, the role of data and parameters is reversed. While for the joint density we take the parameters as given, in the likelihood we take the data as given and evaluate the density as a function of varying values of the parameters.

One consequence of this "turning the joint density around" is that the likelihood function is *not* a PDF over the parameter space. Remember the two defining features of a probability density: it evaluates to a non-negative value everywhere and its sum (for a PMF) or integral (for a PDF) is equal to 1. This is usually not the case when a joint density is used as a likelihood function.

The likelihood function of an i.i.d. sample of n data points is constructed as a product of n densities, which usually (but not always) each evaluate to a number between 0 and 1. Thus, the value of a likelihood function when evaluated for a data set will be relatively small, and the likelihood will be smaller for a larger data set (e.g., with $n = 100$) than for a smaller one (e.g., with $n = 10$) ... which however does not mean that a larger data set is less likely than a smaller. To make calculation of the likelihood easier, we usually work with the log (more specifically, the natural log) of the likelihood function, where the product in the joint density function of the data (Eq. 2.4) is replaced by a sum:

$$\log L(\theta) = l(\theta) = \sum_{i=1}^{n} \log(f(y_i|\theta)) \tag{2.6}$$

The likelihood function is easiest to understand in the simple case of i.i.d. data. However, in most of our analyses, you won't have data that are strictly i.i.d., but instead will have covariates or random effects. How do we then construct the likelihood functions for our data? The simpler of the two cases is when we have covariates … which is virtually always. In this case, we simply replace the parameter θ with a function of these covariates plus a new set of parameters. We will see this in most other chapters in the book where we have covariates in a model, and also in Section 2.6.

Often we will have unobserved RVs, or random effects, in a model. In that case, the construction of a likelihood function is a little more involved. In a frequentist analysis the likelihood is now constructed by starting with the *joint density function of both the observed and the unobserved RVs*. From this, the unobserved RVs (i.e., the random effects) are removed by summation (for discrete random effects) or by integration for the more typical case where random effects are continuous; see Berger et al. (1999), Hobert (2000), Pinheiro & Bates (2000), Royle & Dorazio (2008: chapter 2); Royle et al. (2014: Chapter 2) and Kéry & Royle (2016: chapter 2). The result is the integrated, or marginal, likelihood, wherein the latent variables or random effects are replaced by one or two parameters of the distribution that we assume for them. Typical examples would be a mean and a variance for random effects for which we assume a normal distribution, as in all linear or generalized linear mixed models; see Chapters 10, 14, and 17. However, the random-effects distribution might also be a Bernoulli in an occupancy model (see Chapter 19) or a Poisson in an *N*-mixture model (Chapter 19B).

The likelihood (or alternatively, the log-likelihood [LL]) function contains most or all the information about the parameters that is contained in our data set. Look up the *likelihood principle* for more details. Both statistical inference methods that we will apply use likelihood, but in a different manner. We next cover them both, beginning with classical inference and followed by Bayesian posterior inference.

2.5 Classical inference by maximum likelihood and its application to a single-parameter model

The crucial difference between classical and Bayesian statistics is their different use of probability. In classical inference, probability is used to describe variability in the data among hypothetical replicate data sets. As a consequence, probability is also used to describe the variability of things that we compute from the data, such as estimates of parameters or CIs. In classical statistics, probability is *never* directly used to make an inferential statement involving uncertainty about likely values of an estimated parameter.

That is, classical statistics defines probability as a long-run relative frequency of data and functions of data in hypothetical replicates of a study. This relative-frequency interpretation of probability has also given rise to the label "frequentist." An important concept in classical statistics is that of the *estimator*: a method for computing an estimate ($\hat{\theta}$, the hat indicates an estimate) of a parameter (θ) from data. Conceptually, the RV characteristic of the data carries over to that of the estimator. Hence, in classical inference we imagine that a single estimate of a parameter obtained from application of an estimator to our data set is just one out of an infinitely large population of such estimates that we could have obtained in hypothetical replicates of our study. This hypothetical distribution of estimates across replicates is called the *sampling distribution* of the estimator. Often, there is theory that tells us about the form of this distribution. How wide or narrow the sampling distribution is determines the precision of our estimates as assessed by the SE or CI. The SE of an estimate is simply the SD of the hypothetical replicate estimates, while the CI is given by appropriate percentiles of the sampling distribution of the estimator. Statistical theory enables us to make a statement about that distribution of the estimates even though the only thing we may ever have in hand is a single estimate. As a result, the concept of a SE or CI can be extremely elusive to nonstatisticians. Resampling approaches such as the bootstrap (Section 2.5.4) can be very helpful in this respect to build intuition.

In much of classical statistics, we work directly with the likelihood function when we do ML inference. The likelihood based on the density of the data, or the LL based on the log-density, provides a numerical measure for how likely the observed data are under different assumed values of the parameters in a statistical model. By picking those values of the parameters that maximize the value of the likelihood function when evaluated for our data set, we hope to obtain a good guess for these unknown parameters. This is the principle of maximum likelihood estimation (Edwards, 1992; Millar, 2011; Pawitan, 2013; Lee et al., 2017). The resulting maximum likelihood estimates (MLEs) make the observed data set the most probable. ML is the most widely used estimation method in all of statistics. In addition, many other estimation methods such as least-squares for normal linear models (see Chapters 4–9) or iterative reweighted least-squares (IRLS) for generalized linear models (see Chapter 11–15) yield MLEs for these particular model classes.

Next, we show how to obtain the MLEs for a given model and data set and illustrate three methods for assessing the uncertainty associated with these estimates. We illustrate ML with two examples. First, in the rest of Section 2.5, we use the counts simulated in Section 2.2.2 and assume a one-parameter Poisson model for them. In this simple setting the principle of ML will be easiest to understand. Our second example, in Section 2.6, will use an actual data set for which we will assume a two-parameter Poisson model. Both models can be viewed as a Poisson regression or Poisson GLM; see Chapters 3 and 11–14. The first example is the Poisson equivalent of the simplest normal model in Chapter 4, while the second is the Poisson equivalent to ordinary linear regression in Chapter 5.

2.5.1 Introduction to maximum likelihood

We start with the simple case where our model has only a single parameter that we want to estimate from our data using a likelihood function. For illustration, we work with two data sets with different sample sizes, to see how the different amount of information affects our inferences. We create these data sets by respectively drawing 10 and 100 Poisson RVs in R. Owing to sampling variability, the likelihood function evaluated for these two data sets typically will not have its maximum at the same place. More importantly, the shape of the likelihood function around the maximum will be different, and this is crucial for the assessment of the uncertainty of the estimates.

```
# Two data sets representing counts with expected count lam
set.seed(2)
lam <- 2
y1 <- rpois(10, lambda = lam)      # Small data set (n = 10)
y2 <- rpois(100, lambda = lam)     # Large data set (n = 100)
table(y1) ; table(y2)

y1                                 # Small data set
0 1 2 3 4
1 2 3 2 2

y2                                 # Large data set
 0  1  2  3 4 5 6
12 36 21 16 7 7 1
```

To use the method of ML, we need to choose a suitable probability model for our data. We will assume that we have Poisson RVs, which is a standard assumption for counts without any natural upper bound (Chapter 3). That is, we make the strong claim that the data-generating process for our data is adequately described by the Poisson PMF, that is, $p(y|\lambda) = \exp(-\lambda)\lambda^y/y!$, and that the only thing unknown is the value of λ (this claim is exactly true for our simulated data, but for real data it is never exactly true). The likelihood function is algebraically identical to the joint density of the data under that model. Under i.i.d. assumptions, the joint density is a product of the densities of each single datum. To obtain the MLE, we combine the density function and the data and see which parameter value leads to the highest likelihood function value when evaluated for our data set.

In practice, we will always work with the natural log of the likelihood function, because that takes us from a product of likelihoods to a sum of log-likelihoods. This tends to avoid numerical problems with very small or very large values when the function is evaluated for a data set. In addition, we will use *function minimization* to obtain the MLEs; hence, instead of the log-likelihood function, we will work with the negative of the log-likelihood function. All three functions lead to the same MLEs and actually, we may often say "we maximize the likelihood" when in fact we are minimizing the negative log-likelihood (NLL).

```
# Define likelihood function in R
L <- function(lambda, y) {
  Li <- dpois(y, lambda)           # likelihood contribution of each data point i
  L <- prod(Li)                    # Likelihood for entire data set is a product
  return(L)
}

# Define log-likelihood function in R
LL <- function(lambda, y) {
  LLi <- dpois(y, lambda, log = TRUE) # log-likelihood contribution of i
  LL <- sum(LLi)                   # Log-likelihood for entire data set is a sum
  return(LL)
}

# Define negative log-likelihood function in R
NLL <- function(lambda, y) {
  LL <- dpois(y, lambda, log = TRUE) # log-likelihood contribution of i
  NLL <- -sum(LL)                  # *neg* log-likelihood for entire data set is a sum
  return(NLL)
}
```

Next, we evaluate all three R functions for the smaller data set `y1` and for a large range of possible values for the Poisson parameter `lambda`, draw a plot of each function, and determine the MLEs by hand, or by eye. Fig. 2.6 shows that all three variants of the likelihood function lead to the same value of `lambda` that either maximizes the function (for the likelihood and the log-likelihood) or minimizes it (for the negative log-likelihood). Thus, we can use any of the three functions to find the MLEs. However, it becomes obvious from looking at the very small values along the y-axes that working with the likelihood directly can be numerically challenging. Hence, we usually work with the log-likelihood or the negative log-likelihood function when doing ML inference.

```
# Evaluate all three functions for large range of possible values for lambda
possible.lam <- seq(0.001, 4, by = 0.001)
like1 <- loglike1 <- negloglike1 <- numeric(length(possible.lam))
for(i in 1:length(like1)){
  like1[i] <- L(possible.lam[i], y1)            # Likelihood
  loglike1[i] <- LL(possible.lam[i], y1)        # log-Likelihood
  negloglike1[i] <- NLL(possible.lam[i], y1)    # negative log-likelihood
}

# Plot profiles (this is Fig. 2.6)
par(mfrow = c(1, 3), mar = c(6,6,5,3), cex.lab = 1.5, cex.axis = 1.5, cex.main = 2)
plot(possible.lam, like1, xlab = 'lambda', ylab = 'L', main = 'Likelihood', frame = FALSE)
abline(v = possible.lam[which(like1 == max(like1))], col = 'gray', lwd = 3, lty = 3)
plot(possible.lam, loglike1, xlab = 'lambda', ylab = 'LL', main = 'log-Likelihood',
    frame = FALSE)
abline(v = possible.lam[which(loglike1 == max(loglike1))], col = 'gray', lwd = 3, lty = 3)
plot(possible.lam, negloglike1, xlab = 'lambda', ylab = '-LL', main = 'Negative log-Likelihood',
    frame = FALSE)
abline(v = possible.lam[which(negloglike1 == min(negloglike1))], col = 'gray', lwd = 3, lty = 3)
```

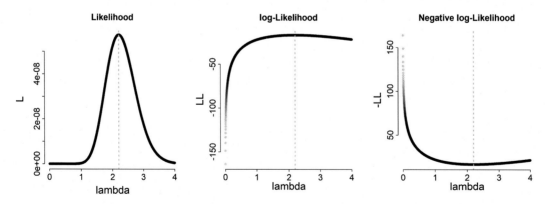

FIGURE 2.6

Curves of the likelihood (left), the log-likelihood (middle), and the negative log-likelihood (right) functions when assuming our 10 counts are i.i.d. Poisson RVs. Vertical gray line shows the value of lambda of 2.2 which is the extremum in all three profiles. We call this the *maximum likelihood estimate*, or MLE. This figure shows a very large range of values along both axes, see Fig. 2.7 for a zoomed-in version of the log-likelihood. Be prepared to be astonished at how the shape can look different for a different scaling of the *y*-axis.

```
# Determine MLE 'by hand'
possible.lam[which(like1 == max(like1))]              # Maximum likelihood
possible.lam[which(loglike1 == max(loglike1))]        # Maximum log-likelihood
possible.lam[which(negloglike1 == min(negloglike1))]  # Minimum negative log-likelihood
```

To get acquainted with the principle of ML, it is good to show things in a super-simple case where we can evaluate the functions essentially "by hand," that is, by trying out a large number of possible values for a parameter and then picking the one that leads to the highest function value of the LL or the lowest for the NLL. But it's hard or impossible to do this with more complex models, for which we use function minimization instead, for example, through `optim()`. This is the R workhorse for function minimization and thus for obtaining our MLEs (see also Section 4.7).

Next, we repeat our search for the MLE for our data set `y1` using `optim()`. We can directly use our R function for the NLL and provide it as an argument to `optim()`, along with an initial value for the start of the iterative search. Function miminization is a huge topic, and indeed function `optim()` enables you to choose among several optimization methods. Here, we simply work with the function's default method. Function minimisation is an iterative process, so we have to initialize it somewhere.

```
# Determine MLE using function minimization in R with optim()
inits <- c('lambda' = 1)
out <- optim(inits, NLL, y = y1)   # Optimize function for y1 over lambda
out
(MLE <- out$par)                   # Grab the MLE

> out
$par
lambda
   2.2

$value
[1] 16.67301

$counts
function gradient
      30       NA

$convergence
[1] 0

$message
NULL

> (MLE <- out$par)               # Grab the MLE
lambda
   2.2
```

We get a warning message (not shown), which we can ignore here. And we see that minimizing the NLL yields the same value for the MLE as did our earlier search "by hand" in the computations for Fig. 2.6.

ML is the gold standard in statistical estimation, and has become extremely well known during the first 100 years since its discovery (Fisher, 1922; Aldrich, 1997). In a sense, it is an automatic inference method: one simply defines a likelihood function for a model and data set and then uses function minimization to obtain the MLEs (Efron, 1986). Importantly, MLEs have several desirable properties, such as invariance to transformation, efficiency, consistency, and asymptotic normality; see Royle & Dorazio (2008): chapter 2, and Kéry & Royle (2016): also chapter 2. Transformation invariance means that we can maximize a Poisson likelihood in terms of parameter `lambda` or in terms of the log of `lambda`, and when we exponentiate the MLE of the latter, we will get the same MLE as in the former case. This is useful, since it is good to avoid constraints on parameters by transforming them in a suitable way when minimizing a likelihood function. Examples of parameters with natural constraints are the Poisson mean or a variance, both of which we may log-transform, or a probability, which we may logit-transform (see Chapter 3). Efficiency and consistency can simply be subsumed under the claim that MLEs are "good" when your sample size is big enough (which is often, but not always). Asymptotic normality is a big deal, since it is the basis for the most common method of computation of SEs and CIs around MLEs, as we will see below.

Any estimate without an associated uncertainty assessment, such as a SE or a CI, is of dubious value. Maximization of the (log-)likelihood yields point estimates that are easy to get, but how can we obtain uncertainty assessments for our MLEs? We next illustrate three such methods:

- CIs based on a likelihood ratio test (LRT): These are based on LRT theory and use the actual shape of the likelihood function around its maximum.
- Asymptotic normality of the MLE ("inverting the Hessian"): SEs and Wald-test-based CIs can be obtained under the assumption that the MLEs are distributed as normal around the true parameter value.
- The bootstrap (parametric or nonparametric).

The information about estimation uncertainty (or put in another way, about the precision of an estimate) comes from the shape of the likelihood function in the vicinity of the MLEs in the case of the first two, and from the spread of the experimental (re-)sampling distribution of the MLEs when using the bootstrap.

Turning to the first two methods, we note that the maximum values of the LL function are about −17 for the smaller data set and −170, that is, 10 times more, for the 10 times larger data set. This does *not* mean that the small data set is more likely than the larger one, but is simply a consequence of the different sample sizes: the sum of a larger number of negative values will be more negative than the sum of a smaller number of such values. The important thing is how much the LL *changes* as we vary the values assumed for the parameters. That is, proportional likelihoods are equivalent and carry the same information about the parameters (Pawitan, 2013; Lee et al., 2017).

Let us first zoom in to the middle panel in Fig. 2.6 and plot the LL function when evaluated for both the small and large data set (Fig. 2.7). When comparing LLs it is customary to set the maximum to 0 and to check a range of the parameter such that the range of LL is roughly between −4 and 0 (Lee et al., 2017). We already computed these values for the smaller data set, but we yet have to do this for the large data set.

```
# Evaluate L, LL and NLL for large sample
like2 <- loglike2 <- negloglike2 <- numeric(length(possible.lam))
for(i in 1:length(like2)){
  like2[i] <- L(possible.lam[i], y2)                  # Likelihood
  loglike2[i] <- LL(possible.lam[i], y2)              # log-Likelihood
  negloglike2[i] <- NLL(possible.lam[i], y2)          # negative log-likelihood
}
```

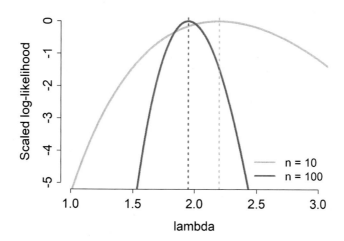

Curves of the scaled log-likelihood in the vicinity of the MLE for the small ($n = 10$) and the large ($n = 100$) variant of our data set. Both curves were scaled to a maximum of 0 to make them comparable. Vertical gray lines show the two MLEs. Note that the gray curve is just a zoom into the middle panel of Fig. 2.6.

```
# Compute value of the maximized log-likelihood for both data sets
(lmax1 <- max(loglike1, na.rm = TRUE))
(lmax2 <- max(loglike2, na.rm = TRUE))

# Plot LL in the vicinity of the MLE for both samples (Fig. 2.7)
ylim <- c(-5, 0)                                      # Choose scaling of y axis
plot(possible.lam, loglike1-max(loglike1), xlab = 'lambda',
   ylab = 'Scaled Log-likelihood', type = 'l', frame = FALSE, col = 'gray',
   lwd = 5, ylim = ylim, xlim = c(1, 3))
lines(possible.lam, loglike2-max(loglike2), col = 'gray40', lwd = 5)
abline(v = possible.lam[which(loglike1 == max(loglike1))],
   lwd = 3, lty = 3, col = 'gray')                    # MLE for small data set
abline(v = possible.lam[which(loglike2 == max(loglike2))],
   lwd = 3, lty = 3, col = 'grey40')                  # MLE for large data set
legend('bottomright', col = c('gray', 'gray40'), lwd = 4,
   bty = 'n', legend = c("n = 10", "n = 100"), cex = 1.5)
```

We note that the MLE obtained from the larger sample happens to be closer to the known truth in our case than the MLE from the smaller sample. But let us instead focus on the shape of the two likelihood functions. Remember that a likelihood function tells us how much support by the observed data different assumed values of a parameter have under a given model. The more concentrated the likelihood around its maximum, the better will be our "guess" about the parameter value, that is, the more precise will be our estimate. Thus, the curvature and, in general, the shape, of the LL near its maximum is a key for uncertainty assessments of MLEs. Thus, we clearly see there is much less uncertainty about the estimate of λ when $n = 100$ than when $n = 10$.

The first method for characterizing uncertainty of MLEs that we present is based directly on the shape of the LL function near its maximum and uses theory about the distribution of differences of LLs (called a likelihood ratio test) when sample sizes are large. This method is more accurate, but (much) more cumbersome than the subsequent two. The second method uses the expected curvature of the LL function near its maximum and is based on the assumption of asymptotic normality of the MLEs. In practice, it is by far the most common of these approaches. For moderate to large sample sizes and for likelihood functions that are symmetric and approximately quadratic, the resulting estimates of SEs and CIs are typically completely adequate. Moreover, the CIs will then be very similar under the first two methods. Finally, the bootstrap is an under-used method which is extremely flexible and can often be very easy to use. We illustrate it for uncertainty assessment of MLEs, using the two variants of a parametric and a nonparametric bootstrap.

2.5.2 Confidence intervals by inversion of the likelihood ratio test

The LRT is the basic significance test in the likelihood framework (Millar, 2011). When we are interested in comparing an estimate $\hat{\theta}$ with some hypothetical parameter value, for example, $\theta_0 = 0$, theory says that for large samples $2(LL(\hat{\theta}) - LL(\theta_0)) \sim \chi_r^2$, where r is 1 for a scalar parameter. This expression can be inverted to obtain a CI around the estimate at some test level α, that is, for a $100(1 - \alpha)\%$ CI. For instance, an approximate 95% CI is defined by $2(LL(\hat{\theta}) - LL(\theta_0)) < 3.84$ and therefore by $LL(\hat{\theta}) - LL(\theta_0) < 3.84/2 = 1.92$ (Bolker, 2008; Millar, 2011; Pawitan, 2013). Note that 3.84 is the 95th quantile of a Chi-squared distribution with 1 d.f., as you can check out by typing in R qchisq(0.95, 1).

To apply this method for obtaining a 95% CI, we choose the two values of lambda which lie vertically below the intersections of the LL curve with a horizontal line drawn at 1.92 units below the maximum (Fig. 2.8). For a 99% CI we would place that horizontal line at $6.635/2 = 3.32$ units below the maximum. For our one-parameter model, the resulting CI is equivalent to a so-called profile interval (Bolker, 2008).

```
# Method 1 for CIs: profile-likelihood- or LRT-based 95% CI
# -----------------------------------------------------------
(lmax <- max(loglike1, na.rm = TRUE))          # Maximized ll
(llim <- lmax - 1.92)                           # value of that is 3.84/2 units below

# Plot profile focused on some range around MLE (Fig. 2.8)
par(mar = c(6, 6, 6, 3), cex.lab = 1.5, cex.axis = 1.5, cex.main = 1.5)
ooo <- which(abs(loglike1-lmax) < 3.5)
xlim <- c(0.9 * possible.lam[min(ooo)], 1.1*possible.lam[max(ooo)])
ylim <- c(-20, -16)
plot(possible.lam[ooo], loglike1[ooo], type = 'l', lwd = 3,
    xlab = 'lambda', ylab = 'Log-likelihood', frame = FALSE, xlim = xlim,
    ylim = ylim, main = 'Log-likelihood curve in vicinity of the MLE\n with LRT-based 95% CI')

# Add to graph vertical line that is chisq(1) / 2 + max(ll)
idx <- which(loglike1 == max(loglike1))         # order which contains max
arrows(possible.lam[idx], llim, possible.lam[idx], lmax,
    length - 0.25, angle = 30, code = 3, lwd = 2)
abline(h = llim, lty = 2, lwd = 3)
text(2.4, mean(c(llim, lmax)), "1.92", cex = 2, srt = 90)

# Compute two CI limits separate for left and right branch of profile
l.left <- loglike1[1:idx]                       # Left branch
l.right <- loglike1[(idx+1):length(loglike1)]   # Right branch
```

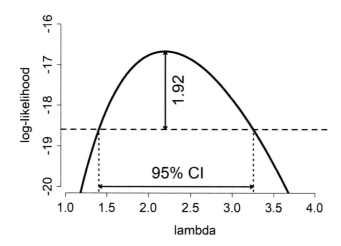

FIGURE 2.8

A likelihood curve showing a LRT-based 95% CI for lambda computed for the smaller data set with $n = 10$. The value of 1.92 is one half of the 0.95 quantile of a Chi-squared distribution with 1 d.f., that is, 3.84/2. Adapted from Millar (2011). The resulting interval goes from 1.404 to 3.251. Note also the slight asymmetry of the curve.

```
# Compare lower limit (llim) with values in l.left and then with those in l.right
idx.left <- which(abs(l.left - llim) == min(abs(l.left - llim), na.rm = TRUE))
idx.right <- which(abs(l.right - llim) == min(abs(l.right - llim), na.rm = TRUE))

# Compute profile confidence limits...
(LCL <- possible.lam[1:idx][idx.left])
(UCL <- possible.lam[(idx+1):length(loglike1)][idx.right])

# add the confidence limits into the plot
segments(possible.lam[idx.left], -21, possible.lam[idx.left], llim, lwd = 2, lty = 3)
segments(possible.lam[idx+idx.right], -21, possible.lam[idx+idx.right], llim, lwd = 2, lty = 3)
arrows(possible.lam[idx.left], -20, possible.lam[idx+idx.right], -20, length = 0.25,
   angle = 30, code = 3, lwd = 2)
text(2.4, -19.5, "95% CI", cex = 2, srt = 0)

# Compare with canned profile CI method in R for our Poisson GLM
exp(confint(glm(y1 ~ 1, family = poisson)))

Waiting for profiling to be done...
    2.5%     97.5%
1.403978 3.251454
```

This code is long-winded and the method fairly complicated, but we show it here to give you an intuition for how such a profile interval is computed. For the specific example with our simple Poisson counts, we can also use R's `confint()` function which gives the same result in a more concise but far less transparent way.

2.5.3 Standard errors and confidence intervals based on approximate normality of the maximum likelihood estimates

Statistical theory tells us that asymptotically, that is, as sample size goes to infinity, the sampling distribution of the MLE becomes normal (Royle & Dorazio, 2008: chapter 2, Pawitan, 2013: chapter 2, Lee et al., 2017: chapter 1), that is:

$$\hat{\theta} \sim Normal(\theta, I(\hat{\theta})^{-1}) \tag{2.7}$$

Thus, with increasing sample sizes, our MLEs will get closer and closer to the true value of the parameter they're estimating, and their variance will be the inverse of $I(\hat{\theta})$, which is called the observed Fisher information (Chapter 2 in Millar, 2011; Royle & Dorazio, 2008). $I(\hat{\theta})$ is the negative of the second partial derivative of the LL function with respect to the parameters in the model, when evaluated at the MLE and for our data set. For a single-parameter model $I(\hat{\theta})$ is a scalar, but more generally, for a model with s parameters it will be an $s \times s$ matrix and becomes the observed Fisher information matrix. To see this, note that the first derivative of the LL function with respect to the parameters gives the slope in each direction, while the second partial derivative quantifies the curvature of the LL around its maximum, that is, the rate of change of the slope at each point of the curve. At the MLEs, the latter function must be negative, since the slope changes from positive to negative. The Fisher information is expressed as the negative of the function value, so information increases when the curvature is greater.

The matrix of second partial derivatives of the LL is called the Hessian matrix. Hence, when obtaining the MLEs from the LL in a model with two or more parameters, we get the variance-covariance (VC) matrix of the MLEs by taking the inverse of the negative of the Hessian matrix $\boldsymbol{H}$ when evaluated for our data set at the MLEs.

$$\mathbf{VC}(\hat{\theta}) = [\mathbf{I}(\hat{\theta})]^{-1} = -\mathbf{H}(\hat{\theta})^{-1} \tag{2.8}$$

The variances are on the diagonal of that matrix. When using `optim()` to get the MLEs, the argument `hessian = TRUE` asks for the Hessian to be computed. Interestingly, since we optimize the negative of the LL function with `optim()` to obtain the MLEs, what `optim()` reports as "the Hessian" is in fact the negative of the Hessian with respect to the LL. Hence, to get the VC matrix of the estimated parameters, we then simply take the inverse of what is reported as the Hessian. To obtain SEs, we take the square root of the inverse of the diagonal of the Hessian matrix, see Section 2.6.

In this way it is very easy to compute asymptotic SEs, where "asymptotic" should always remind us that this is a large-sample approximation. Moreover, we can use the resulting SEs for a so-called Wald-based $100 * (1 - \alpha)\%$ CI as the MLE plus/minus $z_{1-\alpha/2}$ times the asymptotic SE of the estimate, that is, $\hat{\theta} \pm z_{1-\alpha/2} SE(\hat{\theta})$. Here, $z_{1-\alpha/2}$ is the $1 - \alpha/2$ quantile of a standard normal distribution, so for a 95% CI, we will use $z = 1.96$. We next demonstrate all of this in R.

```
# Method 2 for SEs and CIs: 'inverting the Hessian'
# ---------------------------------------------------
inits <- c(lambda = 1)

# Small data set (n = 10)
(out1 <- optim(inits, NLL, y = y1, hessian = TRUE))
(MLE1 <- out1$par)              # Grab MLE
(VC1 <- solve(out1$hessian))    # Get variance-covariance matrix
(VC1 <- 1/out1$hessian)         # Same for a one-parameter model !
(ASE1 <- sqrt(diag(VC1)))       # Extract asymptotic SEs
```

```
# Large data set (n = 100)
(out2 <- optim(inits, NLL, y = y2, hessian=TRUE))
(MLE2 <- out2$par)              # Grab MLE
(VC2 <- solve(out2$hessian))    # Get variance-covariance matrix
(VC2 <- 1/out2$hessian)         # Same for a one-parameter model !
(ASE2 <- sqrt(diag(VC2)))       # Extract asymptotic SEs

# Compare with the truth MLEs and ASE's for small and large data set
print(cbind('truth' = lam, MLE1, ASE1, MLE2, ASE2), 5)

        truth MLE1    ASE1   MLE2    ASE2
lambda      2  2.2 0.46904   1.95 0.13964
```

We note a decent agreement between the estimates and the truth, that is, the value used to simulate our data, and we see that the estimated SEs are much smaller for the larger data set. We can use them to get Wald-type CIs, which we compare with the LRT-based intervals computed by R.

```
# Do-it-yourself Wald-type CI for smaller data set
print(CI1 <- c(MLE1 - 1.96 * ASE1, MLE1 + 1.96 * ASE1), 4)

# LRT-based CI for smaller data set (from R)
print(exp(confint(glm(y1 ~ 1, family = poisson))), 4)

# DIY Wald-type CI for larger data set
print(CI2 <- c(MLE2 - 1.96 * ASE2, MLE2 + 1.96 * ASE2), 4)

# LRT-based CI for larger data set (from R)
print(exp(confint(glm(y2 ~ 1, family = 'poisson'))), 4)

> # Do-it-yourself Wald-type CI for smaller data set
lambda lambda
 1.281  3.119

> # LRT-based CI for smaller data set (from R)
 2.5% 97.5%
1.404 3.251

> # DIY Wald-type CI for larger data set
lambda lambda
 1.676  2.224

> # LRT-based CI for larger data set (from R)
 2.5% 97.5%
1.689 2.237
```

We note a better agreement between the more precise LRT-based intervals and the Wald-type intervals for the larger sample size. This is not surprising, because the distribution of the MLEs is more approximately normal for larger than for small samples.

2.5.4 Obtaining standard errors and confidence intervals using the bootstrap

The bootstrap (Efron, 1979) is a very useful statistical method that is not nearly as well known among ecologists as it should be. Millar (2011) succinctly describes it as follows:

"Frequentist inferential procedures are based on the notion of repeat sampling and so, in the likelihood context, it is necessary to determine the properties and behavior of ML-based inference under repetition of the experiment. The general tools and techniques ... have been obtained from a well established

body of theory that required large doses of calculus, probability theory and mathematical statistics... The bootstrap effectively replaces this calculus and theory with pure computational effort. The essential concept of bootstrapping is to emulate repetition of the experiment by simulating new data on the computer, followed by recalculation of the MLE using the simulated data."

The bootstrap is perhaps most useful for obtaining uncertainty assessments of estimates and derived quantities such as predictions and can also be used for goodness-of-fit assessments (Chapter 18, Kéry & Royle, 2016: chapter 2). Here, we first bootstrap the SEs and CIs parametrically, and then nonparametrically. We will first conduct all the computations for the two bootstraps and then compare the resulting estimates.

For a *parametric bootstrap* with the smaller data set we repeat the following a large number of times (remember that our MLE in the Poisson model is 2.2 for the small, and 1.95 for the large data set):

(1) We simulate another data set of size n (i.e., the same size as the original dataset) by drawing from a Poisson(2.2) or from a Poisson(1.95),

(2) we fit a Poisson model with an intercept only using R function `glm()` and

(3) we save the resulting new MLE.

For a *nonparametric bootstrap*, we repeat the following a large number of times:

(1) We draw another data set of size n *from our actual data* (i.e., y1 or y2). This sampling is with replacement, hence some values will occur more than once, while others will appear not at all in the resampled data sets.

(2) We fit a Poisson model with an intercept only using R function `glm()`.

(3) We save the resulting new MLE.

In either case, once we're done, we take the SD of the bootstrapped sampling distribution of the MLEs as the bootstrapped SE of the estimate, and we take the 2.5 and 97.5th percentile of the distribution as our bootstrapped 95% CI.

The following code produces a histogram of the bootstrapped sampling distribution of the MLE. We don't show this for both, but Fig. 2.9 shows this for the parametric bootstrap. To get the bootstrapped SEs and CIs, we simply summarize these two distributions by their sample SD and by a symmetric 95% percentile interval.

```
# Method 3.1 for SEs and CIs: Parametric bootstrap
# ---------------------------------------------------
simrep <- 10000                                 # Number of bootstrap samples:
                                                # a large number
estiPB <- array(NA, dim = c(simrep, 2))         # Array to hold estimates
colnames(estiPB) <- c('Small data set', 'Large data set')

for(i in 1:simrep){
    if(i %% 500 == 0) cat(paste("iter", i, "\n"))   # Counter
    yb1 <- rpois(10, lambda = MLE1)             # Draw another small data set
    tm1 <- summary(glm(yb1~1, family = 'poisson'))  # re-fit the same model
    estiPB[i,1] <- exp(fm1$coef[1])             # Save estimate on natural scale
    yb2 <- rpois(100, lambda = MLE2)            # Draw another large data set
    fm2 <- summary(glm(yb2~1, family = 'poisson'))  # re-fit the same model
    estiPB[i,2] <- exp(fm2$coef[1])             # Save estimates
}
```

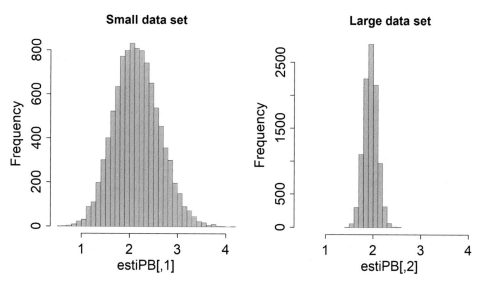

FIGURE 2.9

Two examples of a bootstrapped sampling distribution of the MLE of a model (this is for the parametric bootstrap); left ($n = 10$), right ($n = 100$).

```
# Parametrically bootstrapped sampling distribution of the MLE (Fig. 2.9)
par(mfrow = c(1, 2), mar = c(5,5,4,3), cex.lab = 1.5, cex.axis = 1.5, cex.main = 1.6)
xlim <- range(estiPB[,1])
hist(estiPB[,1], col = 'gray', main = 'Bootstrapped sampling distribution of MLE
  of lambda (small data set; parametric bootstrap)', xlim = xlim)
hist(estiPB[,2], col = 'gray', main = 'Bootstrapped sampling distribution of MLE
  of lambda (large data set; parametric bootstrap)', xlim = xlim)

# Parametric-bootstrapped standard errors
se1.pb <- sd(estiPB[,1])                            # Small data set
se2.pb <- sd(estiPB[,2])                            # Large data set

# Parametric-bootstrapped CIs
CI1.pb <- quantile(estiPB[,1], c(0.025, 0.975))    # Small data set
CI2.pb <- quantile(estiPB[,2], c(0.025, 0.975))    # Large data set

# Method 3.2 for SEs and CIs: Nonparametric bootstrap
# ----------------------------------------------------
simrep <- 10000                                     # Number of bootstrap samples:
                                                    # a large number
estiNPB <- array(NA, dim = c(simrep, 2))            # Array to hold estimates
colnames(estiNPB) <- c('Small data set', 'Large data set')

for(i in 1:simrep){
  if(i %% 500 == 0) cat(paste("iter", i, "\n"))     # Counter
  yb1 <- sample(y1, 10, replace = TRUE)             # Re-sample a small data set
  fm1 <- summary(glm(yb1~1, family = 'poisson'))    # Re-fit model
  estiNPB[i,1] <- exp(fm1$coef[1])                  # Save estimate
  yb2 <- sample(y2, 100, replace = TRUE)            # Re-sample a large data set
  fm2 <- summary(glm(yb2~1, family = 'poisson'))    # Re-fit model
  estiNPB[i,2] <- exp(fm2$coef[1])                  # Save estimate
}
```

```
# Non-parametrically bootstrapped sampling distributions of the MLE
par(mfrow = c(1, 2), mar = c(5,5,4,3), cex.lab = 1.5,
  cex.axis = 1.5, cex.main = 1.6)                    # not shown
xlim <- range(estiNPB[,1])
hist(estiNPB[,1], col = 'gray', main = 'Bootstrapped sampling distribution of MLE
  of lambda (small data set; non-parametric bootstrap)', xlim = xlim)
hist(estiNPB[,2], col = 'gray', main = 'Bootstrapped sampling distribution of MLE
  of lambda (large data set; non-parametric bootstrap)', xlim = xlim)

# Nonparametric-bootstrapped standard errors
se1.npb <- sd(estiNPB[,1])                           # Small data set
se2.npb <- sd(estiNPB[,2])                           # Large data set

# Nonparametric-bootstrapped CIs
CI1.npb <- quantile(estiNPB[,1], c(0.025, 0.975))    # Small data set
CI2.npb <- quantile(estiNPB[,2], c(0.025, 0.975))    # Large data set
```

Once we're done, we compare the SEs and the 95% CIs using all methods in Sections 2.5.2–2.5.4.

```
# Standard errors
# ----------------
# Small data set: asymptotic normality and the two bootstraps
print(cbind(ASE1, 'PB-SE' = se1.pb, 'NPB-SE' = se1.npb), 4)

# Large data set: asymptotic normality and the two bootstraps
print(cbind(ASE2, 'PB-SE' = se2.pb, 'NPB-SE' = se2.npb), 4)

# 95% Confidence intervals
# ------------------------
# Quickly re-calculate LRT-based CI
CIp1 <- exp(confint(glm(y1 ~ 1, family = 'poisson')))
CIp2 <- exp(confint(glm(y2 ~ 1, family = 'poisson')))

# CIs for small data set: LRT-based, normal approximation, bootstraps
print(cbind(CIp1, CI1, CI1.pb, CI1.npb), 4)

# CIs for large data set: LRT-based, normal approximation, bootstraps
print(cbind(CIp2, CI2, CI2.pb, CI2.npb), 4)

> # Small data set: asymptotic normality and the two bootstraps
> print(cbind(ASE1, 'PB-SE' = se1.pb, 'NPB-SE' = se1.npb), 4)
        ASE1  PB-SE NPB-SE
lambda 0.469 0.4725 0.3926

> # Large data set: asymptotic normality and the two bootstraps
> print(cbind(ASE2, 'PB-SE' = se2.pb, 'NPB-SE' = se2.npb), 4)
         ASE2  PB-SE NPB-SE
lambda 0.1396 0.1398 0.1455

> # 95% Confidence intervals
> # ------------------------
> # CIs for small data set: LRT-based, normal approximation, bootstraps
> print(cbind(CIp1, CI1, CI1.pb, CI1.npb), 4)
       CIp1   CI1 CI1.pb CI1.npb
2.5%  1.404 1.281    1.3     1.4
97.5% 3.251 3.119    3.2     3.0
```

```
> # CIs for large data set: LRT-based, normal approximation, bootstraps
> print(cbind(CIp2, CI2, CI2.pb, CI2.npb), 4)
        CIp2    CI2 CI2.pb CI2.npb
2.5%   1.689  1.676   1.69    1.67
97.5% 2.237  2.224   2.23    2.24
```

Despite the very small size of the smaller sample, we see remarkable agreement between the uncertainty assessments, with the exception of those from the nonparametric bootstrap, for which the SE is smaller and the CI narrower for the smaller data set. This may illustrate a limitation of the nonparametric bootstrap when applied to very small sample sizes. When comparing the LRT-based CIs with the others, we find the former a little wider for the small data set, because it properly accounts for the asymmetry of the likelihood profile in that case. In general, we can't say that one type of bootstrap is always better than another. A parametric bootstrap may be preferable if the parametric model is adequate, but a nonparametric bootstrap may be more robust, exactly because it does not make any parametric assumptions to create replicate data sets. If you're interested to learn more about this powerful method, read up some in the vast literature on it (e.g., Efron, 1979; Efron & Tibshirani, 1993; Dixon, 2006; Manly, 2006).

2.5.5 A short summary on maximum likelihood estimation

This concludes our first exposition of that most widely used method of inference for parametric statistical models in classical or frequentist statistics: maximum likelihood estimation. The likelihood function is simply the joint density of the data when viewed as a function of the parameters. Both point estimation and uncertainty assessments are based directly on the likelihood function: the parameter values associated with the maximum of the likelihood function evaluated for the observed data set are taken as a point estimate (i.e., are the MLEs), and the shape and the curvature of the function around its maximum provides the information about the uncertainty around the estimates. The bootstrap offers a powerful alternative method of uncertainty assessment (Manly, 2006).

We have omitted the delta method (Dorfman, 1938) for variance estimation of functions of MLEs, including transformations of an MLE, or when forming predictions based on the linear predictor of a GLM (see Chapter 3). You may need this when you obtain the MLE for a transformed parameter, such as the log of the Poisson mean or the logit of a binomial probability, and then want to obtain the SE not on the log-scale, but on the natural scale of the parameter. You can read up on the delta method, also called "Taylor series expansion," at many places, including Powell (2007), Millar (2011): Chapter 4, ver Hoef & Boveng (2015), Cooch & White (2021): Appendix B. For an example application, see Chapter 19B. For the two specific transformations mentioned, see Kéry & Royle (2016: p. 35).

2.6 Maximum likelihood estimation in a two-parameter model

In practice you will rarely ever have a statistical model with just a single parameter. To illustrate a slightly more complex case, this time with 2 (!) parameters, we repeat some of the preceding analyses. Specifically, we will *model a parameter* with a covariate (Section 2.2.3), in the context of a Poisson model as in Section 2.5. In so doing, we get a bit ahead of ourselves, because we will develop a likelihood implementation of a Poisson log-linear model, or Poisson GLM. We will cover this type of model in depth in Chapters 3 and 11–14. For once in the book, we will use a real data set. We describe every step that underlies our approach, both conceptually and procedurally. This will be a re-cap of most of what we have covered so far in this chapter. You will find more extensive code for the analyses in this section on the book website https://www.elsevier.com/books-and-journals/book-companion/9780443137150.

We use a time series of counts of the Swiss population of Bee-eaters (Fig. 2.10) between 1990 and 2020 from Müller (2021); see Fig. 2.11. Bee-eaters live predominantly in Southern Europe and in Switzerland have benefitted greatly from climate warming over the last several decades, when their known population increased from 0 to 200 pairs. In our analysis, we will gloss over several potentially important issues in this data set. First, we ignore any possible sampling issues having to do with detection probability (see Section 11.2 and Chapters 19 and 19B). Second, there is strong evidence for over-dispersion in these counts (see additional analyses on the website). And third, in Fig. 2.11 we see some intriguing patterns of temporal autocorrelation in the form of periodicities. Imperfect detection, when not taken account of, will lead to an underestimation of population size, while neglecting to accommodate overdispersion or serial correlation in a model typically leads to an underestimate of the uncertainty, that is, too small SEs and too narrow CIs. Here, we use these data for a simple procedural illustration of ML and ignore these issues. For more on overdispersion, see Chapter 12.

FIGURE 2.10

Two sensational Bee-eaters (*Merops apiaster;* photo by Dominique Delfino). This is one of the most colorful bird species in Europe.

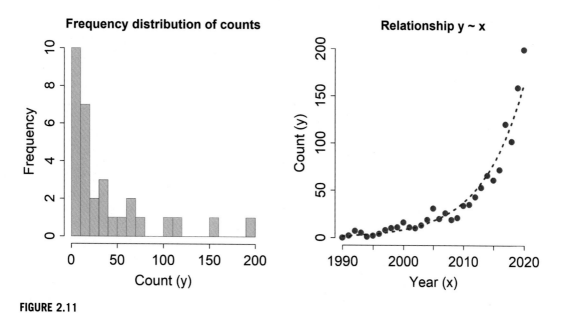

FIGURE 2.11

Two summaries of counts of Swiss Bee-eaters 1990–2020. Red curve on the right is the line of best fit from a Poisson GLM fit with `glm()` in R (Data from Müller, 2021).

```
# Swiss Bee-eater data set (national counts of pairs 1990-2020)
# ---------------------------------------------------------------
# Counts of known pairs in the country 1990-2020
y <- c(0, 2, 7, 5, 1, 2, 4, 8, 10, 11, 16, 11, 10, 13, 19, 31, 20, 26, 19, 21, 34,
  35, 43, 53, 66, 61, 72, 120, 102, 159, 199)
year <- 1990:2020                 # Define year range
x <- (year-1989)                  # Scaled, but not centered, year covariate
x <- x-16                         # Now year covariate is also centered

# Plot bee-eater data (Fig. 2.11)
par(mfrow = c(1, 2), mar = c(5,5,5,2), cex.lab = 1.5, cex.axis = 1.5, cex.main = 1.5)
hist(y, xlab = 'Count (y)', ylab = 'Frequency', breaks = 20,
  col = 'gray', main = 'Frequency distribution of counts')
plot(year, y, xlab = 'Year (x)', ylab = 'Count (y)', frame = FALSE, cex = 1.5,
  pch = 16, col = 'gray20', main = 'Relationship y ~ x')
fm <- glm(y ~ x, family = poisson)     # Add Poisson GLM line of best fit
lines(year, predict(fm, type = 'response'), lwd = 3, col = 'red', lty = 3)
```

   To use parametric statistical inference to learn about the Swiss bee-eater population trend based on our sample of noisy data from 31 years, we view these 31 numbers as outcomes of a stochastic process that we must describe. The counts are non-negative integers without any obvious ceiling. Among the probability distributions covered in this book (see Chapter 3), a Poisson distribution looks like the right choice. Thus, we work with the same basic model as in Section 2.5, but will now relax the assumption of a constant mean. Fig. 2.11 makes it clear that the counts cannot have been produced by a Poisson random process with the same mean λ. Instead, we want to fit a model where the counts, through their expectation, depend on the year x_i. That is, we will estimate a population trend, or growth rate, for Swiss bee-eaters during 1990–2020.

We will *model the Poisson parameter* λ, that is, we allow the Poisson mean to differ as a function of year x_i for every observation i, thus becoming λ_i. But we can't just assume $\lambda_i = \alpha + \beta x_i$ as in a simple linear regression, since this might get us into trouble with negative expected counts, which are of course impossible. Hence, we do what we often do when modeling a parameter that cannot be negative: we model it on the log-transformed scale (and we mean the natural log). Our model for Swiss bee-eaters can be described in algebra as follows, where $i = 1, 2..., 31$ is an index for year in count y_i and in the year covariate x_i:

$$y_i \sim Poisson(\lambda_i)$$

$$\log(\lambda_i) = \alpha + \beta x_i$$

Note that we could also write the second expression as $\lambda_i = e^{\alpha + \beta x_i}$ or as $\lambda_i = \exp(\alpha + \beta x_i)$. When defining the joint density of our data set, we replace λ_i by $e^{\alpha + \beta x_i}$. Thus, we can say that we replace the old parameter, the constant λ from Section 2.5, with two new ones, α and β. These are the intercept and the slope of a linear regression for the Poisson mean on the log scale.

Thus, we treat each count y_i as a realization from a Poisson RV with PMF:

$$P(Y = y_i | \lambda_i) = \lambda_i^{y_i} e^{-\lambda_i} / y_i!$$

Replacing λ_i by $e^{\alpha + \beta x_i}$, we get

$$P(Y = y_i | \alpha, \beta, x_i) = (e^{\alpha + \beta x_i})^{y_i} e^{-(e^{\alpha + \beta x_i})} / y_i!,$$

where on the left side we make explicit the dependency of the PMF on the two new parameters, α and β, and on the values of the year covariate x. Under our independence assumptions, the joint density of the data set is a product of 31 such terms, one for each pair of count y_i and year covariate x_i. Remember that the R function `dpois()` greatly simplifies our life because it saves us from having to type the explicit expression for the Poisson PMF.

In practice, we always work with the log-density when doing likelihood inference. For a single pair of y_i and x_i this yields

$$\log P(Y = y_i | \alpha, \beta, x_i) = y_i(\alpha + \beta x_i) - e^{\alpha + \beta x_i} - \log(y_i!).$$

The joint density and therefore the LL of the entire data set is a sum of 31 such terms. Taking the negative of the sum yields the NLL, which is what we minimize with `optim()`. In R, we get the LL by using function `dpois()`, setting the argument `log = TRUE`.

Next, we define the NLL function for our Poisson GLM as an R function that we minimize for the bee-eater data over the parameter space of α and β. We will also compare our do-it-yourself MLEs with estimates obtained by fitting the same model with the R function `glm()`. This uses a different algorithm, but for GLMs it yields the MLEs. It is always a good idea to double-check your solutions for more complex models or when you start to write your own code for fitting models.

When preparing the material in this section, we discovered that `optim()` is unable, at least with default settings, to find the global minimum of the NLL when working directly with the year covariate in the form of $x = \{1, 2, ..., 31\}$. The reason is that the LL of this model is almost flat over a large range of values of α near the maximum (you might want to try that out!). When working with continuous covariates it is often a good idea to transform them in some way since this reduces numerical challenges in likelihood maximization. Therefore, we will work with the centered version of the year covariate x, for which we did not encounter any problems (although we still select a non-default optimization algorithm; see below). Here's a summary of what we do:

(1) We start by defining an R function for the NLL for the Poisson GLM, where we will use R's density function for the log-density of the Poisson `dpois(..., log = TRUE)`. To emphasize once more that this is only a shorthand for an explicit algebraic expression which should not make us panic, we also write the log-density explicitly in comments on the function definition.

(2) Then, we will use that function in a brute-force, or grid, search for the maximum of the likelihood surface. It is instructive to plot the resulting likelihood surface, or landscape, in part because you can view a 2d likelihood surface as a conceptual model also for higher-dimensional likelihood "clouds" and later for joint posterior densities (see below).

(3) Finally, we call `optim()`, providing as arguments our new NLL function along with a vector of starting values and possibly other arguments, too. The function internally uses one of a number of optimization algorithms. The default is `method = "Nelder-Mead,"` which often works well, but sometimes setting `method = "BFGS"` works better; see the function's help file. We also set `hessian = TRUE`, which lets `optim()` report a numerical estimate of the Hessian matrix. Remember that the Hessian is the second partial derivative matrix of the LL function evaluated at the MLEs, and that the negative of the Hessian matrix is called the observed Fisher information matrix. However, since we work with the NLL function, what `optim()` reports as "the Hessian" is in fact directly the information matrix. Inverting it yields our usual estimate of the VC matrix of the estimated parameter vector. Taking square roots of the diagonal gives us SEs of the estimates, and the MLE plus/minus 1.96 times these SEs provides a Wald-type 95% CI.

For comparison, we will fit the same model with `glm()` and also produce 95% LRT-based CIs with function `confint()`, which for a fitted GLM uses profile likelihood analogous to what we did in Section 2.5.2. On the website you will find further code for bootstrapping (parametric and nonparametric) both the SEs and CIs.

We start by defining the NLL function for the Poisson log-linear model and organize the parameters in a vector which we call `param`. Then, we calculate the expected abundance as a function of a log-linear model and use it for evaluating the LL of every single observation in data vector `y`; this is stored in vector `LLi` on executing the function. Taking the sum and negating the result yields the NLL for the entire data set.

```
# Define an R function for the NLL of a Poisson log-linear model
NLL <- function(param, y, x) {
  alpha <- param[1]                         # Intercept
  beta <- param[2]                          # Slope
  lambda <- exp(alpha + beta * x)
  LLi <- dpois(y, lambda, log = TRUE)       # log-likelihood for an observation
# LLi <- y*log(lambda)-lambda-lfactorial(y) # Same 'by hand' !
  LL <- -sum(LLi)                           # NLL for all observations in vector y
  return(LL)
}
```

```
# Try out, evaluate a few candidates for alpha, beta (lower is better)
NLL(c(0, 0), y, x)
NLL(c(0, 1), y, x)
NLL(c(1, 0), y, x)

> NLL(c(0, 0), y, x)
[1] 3930.635
> NLL(c(0, 1), y, x)
[1] 5164763
> NLL(c(1, 0), y, x)
[1] 2803.901
```

We see that among the three pairs of trial values, assuming `alpha = 1` and `beta = 0` comes closest to the MLEs that we're looking for. We continue this same approach of simply trying out the NLL function for different pairs of values for `alpha` and `beta`, and conduct a systematic search across a regular grid. Our choice in the next code box was informed by an earlier search over a much wider range of possible values for the two parameters. Thus, here we are "zooming in" into that part of the likelihood landscape where we know the maximum is located.

```
# Brute-force search for the MLEs in a grid
nside <- 1000                              # Grid resolution governs
                                           # quality of approximation
try.alpha <- seq(2.8, 2.96, length.out = nside)  # intercept
try.beta <- seq(0.135, 0.16, length.out = nside) # slope
nll.grid <- array(NA, dim = c(nside, nside))

# Evaluate NLL over the grid: try 1 Million of (alpha, beta) pairs
for(i in 1:nside){
  for(j in 1:nside){
    nll.grid[i,j] <- NLL(c(try.alpha[i], try.beta[j]), y, x)
  }
}

# Get pair of values associated with minimum function value
(best <- which(nll.grid == min(nll.grid)))
(best.point <- cbind(rep(try.alpha, nside), rep(try.beta, each = nside))[best,])

[1] 2.8868068 0.1472372                    # Best pair of values for (alpha, beta)

# Plot the likelihood surface, or likelihood landscape (Fig. 2.12)
par(mar = c(6,6,5,3), cex.lab = 1.5, cex.axis = 1.5, cex.main = 1.5)
mapPalette <- colorRampPalette(c("gray", "yellow", "orange", "red"))
image(x = try.alpha, y = try.beta, z = nll.grid, col = mapPalette(100),
    axes = FALSE, xlab = "Intercept (alpha)", ylab = "Slope (beta)")
contour(x = try.alpha, y = try.beta, z = nll.grid, add = TRUE, lwd = 1.5, labcex = 1.5)
axis(1, at = seq(min(try.alpha), max(try.alpha), by = 0.05))
axis(2, at = seq(min(try.beta), max(try.beta), by = 0.005))
box()
points(best.point[1], best.point[2],
  col = 'black', pch = 'X', cex = 1.5)       # Best point from our grid search
```

We identify the minimum over the grid as `alpha = 2.8868` and `beta = 0.1472`, respectively. The map of the likelihood surface in Fig. 2.12 shows that the two estimates are negatively

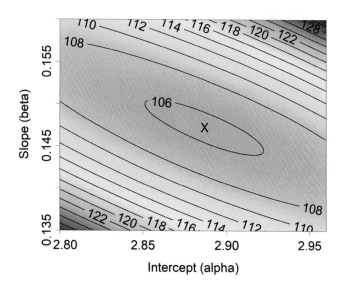

FIGURE 2.12

(Negative log-) Likelihood surface for the two parameters of a Poisson log-linear regression fit to the Swiss bee-eater data obtained by evaluating the NLL function over a grid with a total of 1 million pairs of possible values of the intercept and the slope. The minimum is marked with an X. These are the MLEs of the two parameters as obtained by this fairly crude method (compare with Fig. 2.17 for Bayesian inference for the same model).

correlated, since there is a ridge going from north-west to south-east. Moreover, the curvature around the maximum is greater in direction of the y-axis than in that of the x-axis. This indicates greater absolute precision of the estimate of the slope than of the intercept.

Finally, we use `optim()` to minimize the NLL function for our data set over the parameter space. This is the customary method with which we produce what we call the "DIY-MLEs" (see Section 4.7). We choose `method="BFGS,"` since the default choice in `optim()` does not quite find the true MLEs.

```
# Minimize NLL and get VC matrix by inverting the Hessian
inits <- c(alpha = 0, beta = 0)          # Inits
(out <- optim(inits, NLL, y = y, x = x, hessian = TRUE, method = 'BFGS'))
MLE <- out$par                           # Grab MLEs
(VC <- solve(out$hessian))               # Get variance-covariance matrix
ASE <- sqrt(diag(VC))                    # Extract asymptotic SEs
LCL <- MLE - 1.96 * ASE                  # Lower limit of Wald-type CI
UCL <- MLE + 1.96 * ASE                  # Upper limit of Wald-type CI
print(cbind(MLE, ASE, LCL, UCL), 4)      # Print MLEs, SEs and CI
```

```
# Compare with function glm() and get 95% profile CIs
fm <- glm(y~x, family = 'poisson')
summary(fm)
confint(fm)
```

There is a lot of output here. First, we have the output produced by function optim(). We have the MLEs first, that is, those values of alpha and beta that lead to the highest function value of NLL when evaluated for our data set. Then, we have the minimum value of the NLL, which, when you put a minus sign in front, is the maximum value of the LL for the data set. What is labeled the Hessian matrix gives the negative curvature of the LL function with respect to the parameters when evaluated at the MLEs (it's the negative curvature since we work with the NLL rather than the LL). That is, this is directly the observed information matrix.

```
$par
    alpha       beta
2.8868266  0.1472298

$value
[1] 105.774

$counts
function gradient
      62       13

$convergence
[1] 0

$message
NULL

$hessian
            alpha       beta
alpha    1180.003   10646.29
beta    10646.288  138329.53
```

This last table just above gives the observed Fisher information matrix. Notice how the information is much larger in absolute terms for beta than for alpha, something that we saw already in Fig. 2.12. Inverting the information matrix yields the VC matrix of the estimates. The off-diagonal is negative, illustrating what we saw in Fig. 2.12: the two estimates are negatively correlated.

```
> (VC <- solve(out$hessian))   # Get variance-covariance matrix
            alpha           beta
alpha 0.002772921    -2.134130e-04
beta -0.000213413     2.365407e-05
```

And here is a summary of our DIY ML inference, showing the MLEs, the asymptotic SEs, and the lower and upper limits of a Wald-type 95% CI for both parameters.

```
        MLE       ASE      LCL     UCL
alpha 2.8868  0.052659  2.7836  2.9900
beta  0.1472  0.004864  0.1377  0.1568
```

Comparing these estimates with those from function `glm()` we find an excellent agreement. The normality assumption of the MLEs, on which the Wald-type CIs rely upon, appears to be fairly adequate here, when judged by a comparison with the profile intervals produced by `confint()`.

```
Coefficients:
            Estimate  Std. Error  z value  Pr(>|z|)
(Intercept) 2.886821  0.052659    54.82    <2e-16 ***
x           0.147230  0.004864    30.27    <2e-16 ***

> confint(fm)
                2.5%      97.5%
(Intercept) 2.7814779 2.9879605
x           0.1378238 0.1568931
```

This concludes our demonstration of what is probably the most common way of conducting statistical inferences for a parametric statistical model, using the ML method. We find the MLEs by numerical minimization of the NLL and invert the Hessian matrix to obtain an estimate of the VC matrix. This yields asymptotic SEs, which enable us to construct Wald-type CIs. Throughout the book, we use this method to produce our "DIY-MLEs."

Ah, and before we forget, we have also learned something biological in this statistical exercise. We estimate that over the last 31 years, the Swiss bee-eater population has grown every year by a factor of $\exp(0.1472) = 1.169$, that is, by about 17% annually. Compounded over the 30 interannual intervals, this corresponds to a total population growth of $\exp(0.1472)^{30} = 82.76$. That is, the population has multiplied by a factor of 83 during 1990–2020. A smooth representation of the trajectory of the population is shown by the red line in Fig. 2.11.

2.7 Bayesian inference using posterior distributions

Next, we discover the Bayesian way of learning from a data set via a statistical model. Conceptually, the start of any Bayesian analysis is exactly the same as in a frequentist analysis with ML: a parametric statistical model. That means the data we want to analyze are viewed as the outcome of a stochastic process, and we describe this data-generating process in a manner that is adequate for the aims of our analysis, for example, description, prediction, or mechanistic understanding (Tredennick et al., 2021). We use probability to describe the data-generating process. Hence, we treat all observed data, and for hierarchical models (see Chapter 3) also all unobserved variables, as RVs endowed with a probability distribution governed by a small number of unknown constants called parameters. This is our parametric statistical model wherein every RV is fully described by a probability distribution that we claim to know except for its parameters. We can do a number of things with the model, such as estimating unknown inputs (missing covariates) or unknown outputs (making predictions). Most importantly, when we combine a data set with a model, we can estimate parameters and assess the uncertainty associated with these estimates. It is only from this last point on that Bayesian inference differs fundamentally from frequentist statistical inference.

2.7.1 **Probability as a general measure of uncertainty, or of imperfect knowledge**

Bayesians divide the world into things that are known and things that are at least partially unknown and therefore uncertain (Hooten & Hefley, 2019). The defining feature of Bayesian inference is the use of probability as a measure of knowledge about these uncertain things. In Bayesian inference, as with inference by ML, we do use probability as a measure of variability in a stochastic process, that is, we use probability distributions to characterize RVs and to build our statistical models. But in addition, probability is used in a more general way to quantify the amount of knowledge about anything that is not perfectly known to us. Spiegelhalter (2019) calls these two sources of uncertainty 'stochastic' and 'epistemic', both of which are quantified by probability when using Bayesian inference. Often, the Bayesian interpretation of probability is described as a *degree of belief probability*. This is perhaps a little unfortunate, since "belief" has a connotation of being subjective and therefore perhaps unscientific. A better description of the Bayesian usage of probability is arguably that it is a *degree of personal knowledge* or *a measure of uncertainty*.

Lindley (2006) stresses the fact that under this view, probability becomes personal, since what is well known to one person may be partly or wholly unknown to another. For instance, as a conservation biologist you may wonder whether the asp viper (*Vipera aspis*) still occurs at some site where it was known to occur 30 years ago. This is uncertain to you, and therefore as a Bayesian you will use a probability distribution to quantify your incomplete knowledge about the current presence or absence of the viper. But imagine another biologist who happens to have recently seen an asp viper at that site, so to him there is no more uncertainty. That the probability of an event, when probability is used as a measure of knowledge, may be different for one person than for another does not mean that such use of probability is unscientific.

The world is full with things that we don't know for certain. In fact, if you think about it, there are very few things that you know for certain (see Section 2.1). Unknown things include the following:

* In conservation, whether the asp viper still occurs at a site it was earlier known to occur;
* in a survival analysis, the fate and body mass of an individual during an occasion when it is not observed or caught;
* in a SDM exercise, presence/absence or abundance of a study species at an unsurveyed site;
* in another SDM application, the number of grid cells occupied by a study species in a region;
* and in a population viability analysis (PVA), the fate (extant vs. extinct) and the size of a population some time into the future.

In the context of modeling, the value of a parameter is an outstanding example of something we are uncertain about. Remember that in our statistical models we claim to be at once omniscient and completely ignorant: we know exactly the type of the data-generating process, for example, we claim that our bee-eater counts were produced exactly by a random process known as a Poisson distribution. But we feign complete ignorance about the value of the unknown constants in this process, that is, about parameter λ or, more specifically, α and β that replace λ in the Poisson GLM above.

When we conduct a Bayesian analysis, we use probability as a measure of our knowledge about anything that is not known for certain, by placing a probability distribution on it. Hence, we deal with our uncertainty about unknown things in the same way as with the variability of observable things such as our data. As a consequence, we treat parameters and other unknown quantities *as if* they were RVs. This does not mean that *a priori* the value of a parameter may not be a constant also to a Bayesian. In fact most parameters in a Bayesian analysis are fundamentally identical to parameters in a frequentist analysis: they are simply unknown constants and we are 'epistemically uncertain' about their values (Spiegelhalter, 2019). It is only due to our use of probability as a measure of uncertainty that they are treated as RVs, but their true values are actually fixed. Link & Barker (2010) cite the example of the millionth digit of the number π, which is an unknown constant to most of us, but which we would treat as a RV in a Bayesian analysis that aimed at guessing it.

In the context of our examples above, this means:

- In the asp viper example, if we denote as A the event that there still are snakes at the site 30 years after the last known sighting, then we can write $p(A)$ as the probability that there is still at least one snake living there.
- In the survival analysis, for an individual i that is not detected during occasion t, we will place a probability distribution on both its fate (which is binary: dead or alive) and on its body mass (which is continuous), for example, $p(z_{i,t} = 1)$ and $p(x_{i,t})$ where z is the alive state and x is mass.
- In the two SDM examples, we place probability distributions on the unknown presence/absence or on the abundance of the modeled species at an unsurveyed site, and on the unknown number of occupied grid cells in the region.
- Similarly, in a PVA we place a probability distribution on the fate or the size of our study population at every time step up to our prediction horizon (Petchey et al., 2015).

In the context of our second SDM example, Fig. 2.13 summarizes the use of probability as a general measure of uncertainty. A probability density enables you to describe both complete lack of knowledge (that's the blue horizontal line) and complete certainty (that's the red vertical line), as well as anything in between (e.g., the gray curves). When we learn, our probability distribution for the value of some parameter will become more and more pointed. Thus, how wide or how narrow

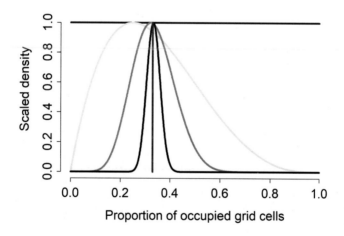

FIGURE 2.13

The use of probability as a measure of knowledge about uncertain things, in the context of a hypothetical study that asks about the proportion of grid cells in a region that are occupied by some species. With probability we can describe any state of knowledge ranging from ignorance (this is the blackish-blue horizontal line) to certainty (red) as well as any state of knowledge in between (gray). The amount of our information is represented by how narrow the probability density is (note densities are scaled for ease of comparison).

our probability distribution is quantifies the amount of our knowledge about a parameter. Again, this use of probability as a measure of uncertainty is *the* defining feature of Bayesian inference. Interestingly, this coincides with how we humans like to express uncertainty, when we say things such as "*I am 99% certain that tonight it's going to rain,*" or "*I am 100% certain that this or that is going to happen.*" The only difference is that Bayesians do calculations with such probability assessments.

2.7.2 The posterior distribution as a fundamental target of inference

In an inference problem with a data set y and a model with unknown parameter(s) θ, a Bayesian would argue that the most sensible thing to do is to estimate θ based on the information in the data using the conditional probability expression $p(\theta|y)$. That is, to go after the probability of what is unknown (i.e., the value of parameter θ) given what is known (i.e., data y). Hence, even though the data are considered as the outcome of a random process, once we have observed them, they are perfectly known and there is no longer any uncertainty about them. So how should this conditional probability of the parameters given the data be computed?

That's where Bayes rule comes in, which for two events A and B can be written as

$$p(A|B) = \frac{p(B|A)p(A)}{p(B)}. \tag{2.7}$$

Bayes rule (which for aesthetic reasons we prefer to write without an apostrophe) states that the probability of A, given that B is true (or has been observed), is the product of the probability that B is true, given that A is true, times the probability that A is true, divided by the probability that B is true. This is the same as $p(A|B) = p(A, B)/p(B)$, that is, the joint probability of A and B (or that both A and B are true), divided by the probability of B.

Bayes rule is a mathematical fact that can easily be derived from the definition of conditional probability (Blitzstein & Hwang, 2019; Pishro-Nik, 2014), which itself is given by a re-arrangement of the multiplication rule (Section 2.1). There is nothing intrinsically Bayesian about it and the rule has many non-Bayesian applications. These include the assessment of how likely you have some disease when you receive a positive test result, and the estimation of the realized values of random effects in a non-Bayesian analysis, such as the site-specific population size in an N-mixture model (Kéry & Royle, 2016: chapter 6) and the presence or absence at a site where a species was not detected in an occupancy model (Kéry & Royle, 2016: chapter 10). Indeed, Eq. 2.7 is the general statistical expression by which we update knowledge in the light of new information (Blitzstein & Hwang, 2019).

The English minister and mathematician Thomas Bayes (1702–1761) and others such as Pierre-Simon Laplace (1749–1827) used what later became to be known as Bayes rule for inference about unobservable things, such as parameters in statistical models. In this setting, Bayes rule can be written as follows:

$$p(\theta|y) = \frac{p(y|\theta)p(\theta)}{p(y)} = \frac{p(y, \theta)}{p(y)}. \tag{2.8}$$

The four components of Bayes rule in the context of Bayesian statistical inference are as follows:

- the posterior distribution $p(\theta|y)$ is the probability (or probability density) of the parameters given the data and the prior,
- the likelihood $p(y|\theta)$ is the probability (or probability density) of the data (i.e., a PMF or a PDF) viewed as a function of the parameters,

- the prior distribution $p(\theta)$,
- the marginal distribution of the data $p(y) = \int p(y|\theta)p(\theta)d\theta$ is the integral of the numerator over the parameter(s) θ, that is, what we get when we average the numerator over all possible values of parameter θ, weighted by its plausibility given in $p(\theta)$.

Note that the posterior distribution is also given by the ratio of the joint distribution of the data and the parameters, and the marginal distribution of the data.

You can see that hallmark of Bayesian inference, the use of probability as a measure for uncertain knowledge, at two places in Bayes rule: in the prior and in the posterior. Because of this we can make formal probability expressions about specific values of parameters in a model. For instance, in a PVA we may rightfully say things like *"we are 95% certain that the rate of decline of the population is steeper than 20% per year."* Such expressions are impossible in non-Bayesian ways of learning from data, or at least cannot be obtained in such a straightforward way.

The formal steps underlying every Bayesian analysis are the following:

- We use probability as a measure of uncertainty about unknown quantities such as parameter θ.
- We treat all statistical inference (e.g., parameter estimation, testing, imputation of missing values, or predictions) as a probability calculation by application of Bayes rule.
- We start by expressing our initial, or prior, knowledge about parameter θ by one probability distribution, which is the prior $p(\theta)$. Importantly, no information from the data must be used to form the prior; this would be using the data twice and amount to cheating.
- Then, we use Bayes rule to update our prior knowledge with the information contained in the data and embodied by the likelihood function $p(y|\theta)$.
- The result is another probability distribution, the posterior distribution $p(\theta|y)$, which represents our new state of knowledge about θ.
- All these steps can be repeated and Bayes rule is a natural way for updating knowledge in a sequential manner.

You see that the likelihood is a central piece also of Bayesian inference, since it is the likelihood (via the data y) which transforms our prior knowledge into our posterior knowledge about parameter θ. However, to do straight ML inference, we use *only* the likelihood function along with the data and optimize it over the parameter space to find the MLEs. In contrast, in Bayesian inference our target is always a distribution for every unknown.

There is a lot of heuristic appeal to Bayes rule as a method for learning about uncertain things. First, it is based on a "human concept" of probability, that is, we formally use probability as a measure of knowledge similar to what we humans like to do in everyday language. Second, Bayes rule is often described as $p(\theta|y) \propto p(y|\theta)p(\theta)$, that is, the posterior is proportional to the product of the likelihood and the prior. And this is exactly how we humans learn: when faced with a new piece of information we hardly ever consider it in isolation. Rather, we almost always consider it in the context of what we know already and then base our conclusion or decision on a combination of both the new and the old information. Thus, the new information updates our old information, and the result is a new and (hopefully) improved state of our knowledge.

Bird identification provides a good illustration of the Bayesian nature of human thinking. When we see a bird and are unsure about its identity, our final conclusion may be "it's species A" or "it's very probably species B," or perhaps "it's more likely to be species C than species A." In coming to this conclusion, in addition to the features noted in the field or visible on a photograph that we took, we will *always* also use external (i.e., *prior*) knowledge about how likely a species is to occur at the given place and time. For instance, when seeing a large soaring raptor in Europe it is exceedingly unlikely that this is a condor, since the two condor species are endemic to the Americas. Likewise, it will be highly unlikely to see a neotropical long-distance migrant in the Canadian arctic during winter, when such a species is expected to be in its winter quarters in Latin America.

Also, when we learn according to Bayes rule, it is clear that every scientific position or every opinion, as embodied by the prior, can be modified by the arrival of new information. Therefore, in science we should avoid 0/1 priors, which place a probability of 0 (representing impossibility) or 1 (representing absolute certainty) on some event, or on the specific value of a quantitative hypothesis (Lindley, 2006). This would be the end of learning and would go counter that defining activity of science, which is to test our theories (loosely represented by the prior distribution in Bayes rule) with empirical data, which we can similarly loosely imagine as equivalent to the likelihood. The confrontation between thoughts about how the world might be, and observations about how it actually is, will up- or downweigh our trust in our theories and thus lend them more or less credibility. Thus, over time, this procedure lets us eliminate less apt explanations of the world and keep the better ones instead.

When estimating parameters we are used to having some number as a point estimate (i.e., as our best guess of what the parameter value is) and then one or two other numbers that quantify the likely margin of error of this estimate, see Sections 2.5–2.6. How can we get these when the result of our analysis is an entire (posterior) probability distribution? The answer is that we can simply use any measure of location and spread, as in the most basic descriptive statistics. As a point estimate we can use the mean, median, or mode of the posterior distribution, and as an uncertainty assessment we may use the SD or a range given by some percentile, such as 95 or 99 ... or even 89 if you *really* want to insist on being different from the rest.

2.7.3 Prior distributions

In any Bayesian analysis we must specify prior distributions for all unknown parameters. Historically, the use of prior information in statistical analyses has been controversial. Arguments in favor of the use of priors focus on their value as a formal way of incorporating information from outside your study. You can think of the priors almost as a sort of plug with a sign attached to it saying "*If you have external information and want to use it in your estimation then put it in here.*" This looks like a good thing, because often you may have quite good contextual or even quantitative knowledge about a parameter. Bayes rule makes it relatively straightforward to combine this information with the information in your data set. For instance, in the examples of uncertain things in Section 2.7.1, we may perhaps think that the proportion of recently visited historic asp viper sites in the same region where the viper was still found to survive represents a sensible expectation also for the site in question. In a survival analysis, available estimates for similarly sized species may provide a reasonable expectation for our study species (McCarthy & Masters, 2005). In general, results from "similar studies nearby in space, time or taxonomy" may provide sensible *a priori* information about the likely value of a parameter. Bayes rule then lets us use that information as long as we can express it as a probability distribution.

Priors that are meant to bring into estimation some amount of external information are called *informative*. Arguably their use makes a lot of sense, since you can imagine them as if they were additional data. In any estimation task it wouldn't be sensible to ignore part of the available data. Hence, it does not make sense to ignore *a priori* information when available. There can be major benefits in the use of informative priors. Exactly as additional data will make your posterior distributions narrower (i.e., increase your knowledge about a parameter), so will informative priors tend to increase your knowledge about a parameter. You will then get smaller posterior SDs and narrower Bayesian credible intervals. In addition, you may be able to estimate additional parameters which may not be estimable without the use of informative priors.

On the other hand, those who resist the formal use of prior information (and by extension, Bayesian inference) are motivated by the belief that Bayesian inference is subjective, since priors must be chosen actively and are not suggested by seemingly objective considerations. This appears to contradict the tenet that science must be objective and ought to minimize personal influences by the experimenter or analyst. While it is probably true that we ought to try and make science as objective as possible, we feel that the issue of the subjectivity of Bayesian inference has often been oversold dramatically: the choice of prior is only one out of literally dozens or even hundreds of arbitrary choices that we make when we study a scientific question. For instance, when we are interested in the effects of density dependence on population dynamics, which species should we choose to work with? A plant or an animal? Should we study it in the field, in a cage or in the greenhouse? In North America or in Australia? During which set of years? And on and on and on. Every single one of our subjective choices to these questions will be guaranteed to influence our results to some degree, but nobody seems to get nervous about that. When we use Bayesian inference, we just add one more such choice to our study. We really don't think that this jeopardizes the scientific value of our study.

However, it is true that in a Bayesian analysis the priors always affect our estimates at least to some small degree; there is no escaping that. It is true that most of the time, the influence of the prior will be small unless we use informative priors, and unless the amount of information in the data is really small. But this is not always so. In addition, having to specify priors does induce an additional ambiguity, since one prior may lead to one result, while another may lead to a slightly different result. This is of course not wrong, but it may make reporting a study more complex, because we might have to report two or even more different sets of results, one for each prior.

For better or worse, what people do in the vast majority of applied Bayesian analyses is to use so-called vague priors, which are meant to minimize their influence on the posteriors. That is, we choose low-information priors which are almost uniform over the range of parameter values where the likelihood has non-negligible support. Vague priors give similar *a priori* weight to a large range of plausible parameter values. For instance, we might specify a uniform(0, 1) prior for the probability that a historic asp viper location is still occupied, as we might do also for any parameter representing a probability, such as survival or occupancy. In practice, we often use a uniform prior within some suitable range or a "flat normal" distribution, that is, a normal distribution with a large variance chosen again so that the *a priori* probability of parameter values are approximately similar over a wide relevant range. For another view, wherein the choice of weakly or moderately informative priors merits much more thought, see Gelman et al. (2020).

Another issue with Bayesian inference that we think is often oversold is the lack of transformation invariance with respect to the choice of prior (Royle & Dorazio, 2008). For instance, in a Poisson GLM we will typically model the mean on the log-transformed scale. In an analysis with vague priors, should we then put a uniform on log(lambda) or on lambda? This choice will have some effect on the resulting posterior distribution, though most times it will be minor and negligible. Another famous example is the choice of a uniform prior on the logit-scale intercept of a logistic regression (see Chapter 3 and 15–17). This implies a very informative prior for the intercept on the probability scale, with far higher mass placed on values near 0 and 1 (Link & Barker, 2010; Northrup & Gerber, 2018). In this book as in most of our research, we follow Royle & Dorazio (2008) and specify vague priors on a "natural scale," which is one we like to think about a parameter. That would be lambda in a Poisson GLM and the probability scale in a binomial GLM, because most of us don't like to think in terms of the log number of bee-eaters or logit probabilities.

With vague priors, we will find that the Bayesian estimates (both point and uncertainty) are usually extremely similar numerically to the the same estimates using ML for a model, unless sample sizes are really small. We find this comforting and see it as a strong argument in favor of an opportunistic use of ML and Bayes. Technically, it is the mode of the posterior distribution (sometimes called the maximum *a posteriori*, or MAP, estimate) in an analysis with vague priors which corresponds to the MLE. However, with approximately symmetric posteriors, the mode, median, and mean all co-incide. In practice, the posterior mean is used most often as a Bayesian point estimate, along with the posterior SD as a Bayesian analog of the SE and the 95% percentiles of the posterior which give a 95% Bayesian credible interval. This is called a credible interval, or CRI, to distinguish it from the frequentist CI. Note though that for more complex models (specifically those with parameters that have posteriors with non-Gaussian shapes, for example, one-tailed distributions), the median is often a preferred point estimate, since it is better than the mean and mode at representing the middle of the distribution (Jeff Doser, pers. comm.).

Interestingly, the use of prior information in learning from data is not unique to Bayesian inference. For instance, penalized likelihood (Lele et al., 2012; Moreno & Lele 2010; see also Chapter 18) is a conceptually equivalent frequentist estimation method in which we assume that some values of a parameter are *a priori* less likely than others. Nevertheless, prior distributions are a salient feature of Bayesian inference because we can't do Bayesian inference without them. And as a final remark we note that in spite of our preference for vague priors, we often use what are also called weakly informative priors. Such priors will be a bit more informative than very vague ones, but still produce estimates that numerically strongly resemble estimates obtained without use of prior information (e.g., MLEs). But weakly informative priors often produce Markov chains with much better convergence and mixing (see the next section).

2.8 Bayesian computation by Markov chain Monte Carlo (MCMC)

When we're doing Bayesian inference we're always after the posterior distribution of the unknowns in our statistical models, especially the posterior distributions of the parameters, $p(\theta|y)$. How can we obtain these posterior distributions? For centuries, this was the Achilles' heel of Bayesian inference. Mathematically, Bayes rule can only be evaluated in exceptional cases because the required integrals over every parameter in the model are usually intractable (Brooks, 2003; Hobbs & Hooten, 2015; Plummer, 2023b). Thus, Bayesian inference needed to either rely on often restrictive approximations or (in special cases) on the use of so-called conjugate priors. These priors are of a similar "form" as the likelihood and this combination leads to posterior distributions of a known form. The famous textbook example is the combination of a binomial likelihood with a beta prior when observing r successes and

n failures for a Bernoulli RV such as 'presence' or 'absence' in a SDM context. Use of a beta(a,b) prior leads to a posterior that is also a beta distribution with parameters $a + r$ and $b + n$. But such "tricks" were never sufficient to make Bayesian inference a general-purpose inference tool for widespread use.

This changed dramatically in the early 1990, when statisticians re-discovered work conducted by physicists in the 1950s (Metropolis & Ulam, 1949; Metropolis et al., 1953). This work showed how one can develop simulation algorithms to explore the shape of unknown distributions (Geman & Geman, 1984; Tanner & Wong, 1987; Gelfand & Smith, 1990), for example, of posterior distributions that could not be derived analytically. The rediscovery and subsequent further development of what is now called Markov chain Monte Carlo (MCMC) algorithms has caused a statistical revolution since then (Brooks, 2003). MCMC algorithms coupled with greatly increased computing power have been primarily responsible for the wide adoption of Bayesian inference both in the statistical sciences and, with the advent of user-friendly software such as WinBUGS and OpenBUGS (Lunn et al., 2013), also in ecology. At least conceptually, these ground-breaking software engines represent the precursors and inspiration for all three Bayesian inference engines covered in this book: JAGS, NIMBLE, and Stan (Chapter 4).

For several reasons, we would not advocate learning how to code up your own MCMC algorithms to everyone in ecology. First, the learning curve for these methods for complex models may be too steep for many. Second, debugging your custom MCMC software would take a lot of time which might be better spent elsewhere. And third, the available general-purpose engines such as JAGS, NIMBLE, and Stan are incredibly powerful and let an expert statistician do virtually everything that he or she might be able to achieve with custom-written MCMC code.

Nevertheless, it is important to develop an intuition about what goes on under the hood when running such engines because it may help you diagnose causes for problems, interpret MCMC output, and simply be intellectually satisfying. Finally, some of you may come into the situation where you have to use some custom MCMC code that somebody else wrote. And a small proportion of you may need and want to learn how to write their own MCMC algorithms in R or some other statistical programming language. For these reasons we want to give you a flavor of "how MCMC works." We will show one of the simplest possible variants of MCMC: a random-walk Metropolis algorithm. We believe that even just understanding this simple prototypical MCMC algorithm will help you achieve some of the aims just formulated. Read Hooten & Hefley (2019) and Zhao (2024) when you want to learn much more about MCMC in the context of models used in ecology. Chapter 17 in Royle et al. (2014) is a neat introduction for ecologists to several classes of MCMC algorithms.

To help understand what you do when you use JAGS, NIMBLE, or Stan, we focus on two topics: Monte Carlo integration and a simple MCMC algorithm for implementation of the Poisson log-linear regression model for the Swiss bee-eater data from Section 2.6. In the former, we use sample statistics computed from random numbers drawn from a distribution to learn about features of that distribution. An MCMC algorithm can be viewed as a highly customized random number generator (RNG) for the posterior distributions of the parameters in our models using Bayesian inference.

2.8.1 Monte Carlo integration

An MCMC algorithm produces random numbers from a posterior distribution. Once we have a large amount of these random numbers, for example, some 100s or 1000s, we can describe them to learn about the features of the posterior distribution. Drawing a larger number of values will provide a better approximation. For instance, we can take the sample mean as an approximation of the mean of the distribution and the sample SD as an approximation of its SD. Technically, in these and other cases we use simulation to solve an integral. For instance, for the mean of the distribution, we use simulation to solve the integral $E(x) = \int p(x)x dx$.

To see how a sample of random numbers from a distribution can be used to learn about the mean and the SD or other features of that distribution, consider the following artificial case. We use the Gaussian RNG `rnorm()` and draw an increasing number of values from a normal distribution with known parameters for the mean and the SD. Then we see how well we can estimate these two parameters from the samples of values produced.

```
# Illustration of the concept of Monte Carlo integration (Fig. 2.14)
set.seed(2)
par(mfrow = c(2,2), mar = c(3,5,4,2))

# 10 draws from distribution
out1 <- rnorm(10, mean = 5, sd = 2)
hist(out1, breaks = 50, col = "gray", main = "Can't believe it's normal !")
mean(out1) ; sd(out1)

# 100 draws from distribution
out2 <- rnorm(100, mean = 5, sd = 2)
hist(out2, breaks = 50, col = "gray", main = 'Well, perhaps ...')
mean(out2) ; sd(out2)

# 1000 draws from distribution
out3 <- rnorm(1000, mean = 5, sd = 2)
hist(out3, breaks = 50, col = "gray", main = 'Oh !')
mean(out3) ; sd(out3)

# 1,000,000 draws from distribution
out4 <- rnorm(10^6, mean = 5, sd = 2)
hist(out4, breaks = 100, col = "gray", main = 'Oh, wow!')
mean(out4) ; sd(out4)
```

With increasingly large samples we can get an essentially arbitrarily good approximation for the distribution from which these random numbers were drawn (Fig. 2.14). But perhaps even more astonishing is how close the numerical summaries may come to the true parameters even in small samples. Look at this:

```
>    # 10 draws from distribution
[1] 5.422303
[1] 1.969975
>
>    # 100 draws from distribution
[1] 4.969041
[1] 2.345771
>
>    # 1000 draws from distribution
[1] 5.134862
[1] 1.982934
>
>    # 1,000,000 draws from distribution
[1] 5.000644
[1] 1.999467
```

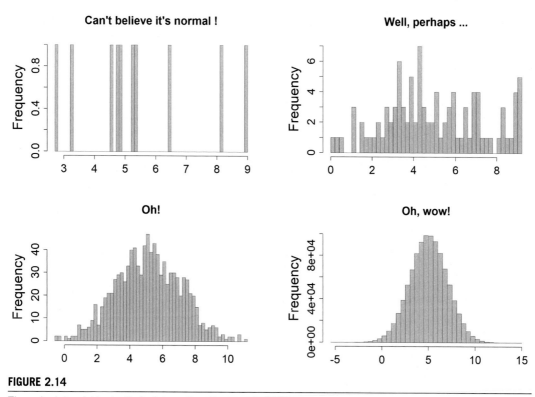

FIGURE 2.14

The principle of Monte Carlo integration illustrated with four samples of different size (10, 100, 1000, and 10^6) from the same normal(mean = 5, sd = 2) distribution. The sample mean and the sample SD provide an increasingly good estimate of the true mean and standard deviation of the distribution.

Thus, with as few as 100 or 1000 draws, we can obtain a pretty good picture of what the parent distribution looks like, either graphically or numerically. This is what we always do with Bayesian inference based on iterative simulation algorithms: we produce a large number of random numbers and then summarize them in suitable ways to infer the distributions from which they were drawn. But this is of course not all, because with real MCMC we must ensure that the distributions we draw random numbers from are in fact the posterior distributions for our parameters of interest. In the next section, we develop our own MCMC algorithm for fitting a Poisson log-linear regression to the Swiss bee-eaters from Section 2.6.

2.8.2 MCMC for the Swiss bee-eaters

We will now develop a random-walk Metropolis algorithm for our two-parameter model. This algorithm serves essentially as a RNG for the joint posterior distribution of the two parameters, that is, the intercept (`alpha`) and the slope (`beta`) of the log-linear regression of the Swiss bee-eater population counts on the years 1990–2020. We will see that in this magic algorithm the denominator of Bayes rule vanishes. Thus, one need no longer evaluate any integrals and Bayesian inference becomes possible even for very complex models such as those in Royle & Dorazio (2008), Kéry & Royle (2016, 2021), Hooten & Hefley (2019), Schaub & Kéry (2022) or Zhao (2024). For many of these, inference would have been impossible before the advent of MCMC.

We will construct a simple MCMC algorithm that in essence is nothing but a glorified RNG for the joint posterior distribution of the two parameters in our log-linear Poisson model. But in contrast to the RNG functions such as `rnorm()` in R, our "posterior distribution RNG" will not produce independent draws from a distribution, but instead the draws produced have a serial dependence by design. In addition, the draws from multiple parameters may be correlated. An MCMC algorithm iteratively produces a series of numbers where the output at each iteration is stochastically dependent on the output at the immediately preceding iteration: that's the "Markov chain" in MCMC. A Markov chain of order 1 is a stochastic process, or a series of RVs, where at each time t the output is independent from the past when the output at time t-1 is known. The "Monte Carlo" in MCMC simply indicates a simulation algorithm.

Here is the recipe for a random-walk Metropolis algorithm for the posterior distribution of parameter θ (note we use superscripts for indexing only).

(1) Starting at some arbitrary initial value for the parameter, θ^0, we use a stochastic rule for obtaining a proposed new value for it. We call the new value θ^*.

(2) We compare the proposed θ^* with the current value of θ in the chain. At the start of the chain, this means that we compare θ^* and θ^0, while at iteration t we will compare θ^* with θ^{t-1}. Specifically, we compare the two values of θ in each iteration t by calculating their posterior density ratio:

$$r = \frac{p(y|\theta^*)p(\theta^*)/p(y)}{p(y|\theta^{t-1})p(\theta^{t-1})/p(y)} \tag{2.9}$$

(3) We accept the proposed new value into our sample of draws with probability $\min(1, r)$. That is, if the posterior density of the proposed new value of the parameter, θ^*, is greater than the posterior density of the current value, θ^{t-1}, that is, if $r > 1$, we accept the proposed new value right away: we add θ^* to our sample and it becomes θ^t, which will be the "current value" in the next iteration. In contrast, if the posterior density of the proposed new value is less than that of the current value, that is, if $r < 1$, we may still accept the proposed new value, but only with probability equal to r, while with probability $1 - r$ we make a copy of the current value of θ and add that to our sample as our new current value θ^t.

And that's all there is for a very basic MCMC algorithm! We note four important things.

First, the stochastic rule in (1) is called the proposal or candidate-generating distribution h. It is often taken as a normal density, such that the proposed new value at iteration t is drawn as $\theta^* \sim Normal(\theta^{t-1}, \sigma_h)$, that is, from a normal density centered on the current value and with some SD σ_h. This normal density is NOT part of the model, rather the normal RNG is simply a convenient device to produce a stochastic proposal for parameter θ. It is the centering of the proposal distribution on the current values which builds into the output some degree of temporal autocorrelation. Thus, the proposal SD σ_h is NOT a parameter in the model, rather, it is best considered a tuning parameter or a "setting" of the algorithm. We will see that with a larger value of σ_h, successive values in the Markov chain will be less similar to each other than with a smaller value of the proposal SD. We may also say that σ_h governs the step length of the algorithm. The

MCMC algorithm is a random walk, since the proposals are produced in a "blind" manner by randomly disturbing the current value. This makes for a simple algorithm, but can lead to reduced efficiency compared to more complex algorithms such as Hamiltonian Monte Carlo (HMC), which use as information for their proposals the local gradient of the posterior in question (Section 2.10).

Second, in Eq. 2.9 the marginal distribution of the data $p(y)$ cancels from the numerator and the denominator, that is, it drops out from the equation! That is, we don't need to evaluate any integrals. These integrals were the main stumbling blocks for implementing practical Bayesian inference for most interesting models for centuries, so this feature of the MCMC algorithm is a huge relief.

Third, if we repeat this simple recipe a large number of times (e.g., 100, 1000, or more), then after some initial transient phase, the Markov chain of values produced will move towards some equilibrium distribution. There, values of θ are drawn in proportion to their posterior density $p(\theta|y)$. In other words, at equilibrium our algorithm produces serially correlated random numbers from the target posterior distribution.

Fourth, you may wonder why we may also accept proposals resulting in a lower posterior density. This is because we also want to characterize the tails of the posterior distribution, and not just the region with the highest density.

For illustration, we now fit the Poisson GLM to the Swiss bee-eater data using this type of algorithm. As for the MLE algorithms earlier, we again work in the log space. We start by defining two R functions. The first function is for the LL of the model (see also Chapter 13 for a similar model).

```
# Define R function for log-likelihood of Poisson GLM
LL <- function(param, y, x) {
  alpha <- param[1]                  # Intercept
  beta <- param[2]                   # Slope
  lambda <- exp(alpha + beta * x)    # Log-linear model
  LLi <- dpois(y, lambda, log=TRUE)  # LL contribution of each datum i
  LL <- sum(LLi)                     # LL for all observations in the data vector y
  return(LL)
}
```

The second function is for the unnormalized log-posterior density, which means it calculates only the numerator in Bayes rule, ignoring the denominator. We can ignore the marginal distribution of the data, as we have just seen, since it drops out from the posterior density ratio. As for the likelihood, working with the log posterior means that we sum LL and log-priors instead of getting products of likelihood and priors. As priors, we specify zero-mean normal densities with SD = 100. These are meant to be vague in this analysis.

```
# Define R function for un-normalized log-posterior of the model
log.posterior <- function(alpha, beta, y, x){
  loglike <- LL(c(alpha, beta), y, x)
  logprior.alpha <- dnorm(alpha, mean = 0, sd = 100, log = TRUE)
  logprior.beta <- dnorm(beta, mean = 0, sd = 100, log = TRUE)
  return(loglike + logprior.alpha + logprior.beta)
}
```

Next, we make some additional preparations: At each iteration we will keep track of acceptance/rejection of the proposed new values of alpha and beta and store them as binary indicators in an R object. The proportion of proposals that are accepted is an important indicator of the efficiency of an MCMC algorithm. The efficiency is related to the SD of the normal proposal distribution, σ_h, which we call sd.tune here.

```r
# Choose number of iterations for which to run the algorithm
niter <- 10000

# Create an R object to hold the posterior draws produced
out.mcmc <- matrix(NA, niter, 2, dimnames = list(NULL, c("alpha", "beta")))

# Initialize chains for both parameters: these are the initial values
alpha <- rnorm(1)              # Init for alpha
beta <- rnorm(1)               # Init for beta

# R object to hold acceptance indicator for both params
acc <- matrix(0, niter, 2, dimnames = list(NULL, c("alpha", "beta")))

# Evaluate current value of the log(posterior density) (for the inits)
logpost.curr <- log.posterior(alpha, beta, y, x)

# Choose values for the tuning parameters of the algorithm
sd.tune <- c(0.1, 0.01)        # first is for alpha, second for beta
```

We are ready now to run the MCMC algorithm for the desired number of iterations. You will observe how we cycle through the parameters in the model. At every iteration we will first update the intercept and then the slope, but in this model the order does not matter.

```r
# Run MCMC algorithm
for(t in 1:niter){
  if(t %% 1000 == 0)                        # counter
    cat("iter", t, "\n")
# First, update log-linear intercept (alpha)
# --------------------------------------------
# Propose candidate value of alpha
alpha.cand <- rnorm(1, alpha, sd.tune[1])    # note tuning 'parameter'

# Evaluate log(posterior) for proposed new alpha and
#   for current beta for the data y and covs x
logpost.cand <- log.posterior(alpha.cand, beta, y, x)
# Compute Metropolis acceptance ratio r
r <- exp(logpost.cand - logpost.curr)
# Keep candidate alpha if it meets criterion (u < r)
if(runif(1) < r){
  alpha <- alpha.cand
  logpost.curr <- logpost.cand
  acc[t,1] <- 1                             # Indicator for whether candidate alpha accepted
}

# Second, update log-linear slope (beta)
# --------------------------------------------
beta.cand <- rnorm(1, beta, sd.tune[2])    # note again tuning parameter
```

```
# Evaluate the log(posterior) for proposed new beta and
#  for the current alpha for the data y with covs x
logpost.cand <- log.posterior(alpha, beta.cand, y, x)
# Compute Metropolis acceptance ratio
r <- exp(logpost.cand - logpost.curr)
# Keep candidate if it meets criterion (u < r)
if(runif(1) < r){
  beta <- beta.cand
  logpost.curr <- logpost.cand
  acc[t,2] <- 1                         # Indicator for whether candidate beta accepted
}
out.mcmc[t,] <- c(alpha, beta)          # Save samples for iteration i
# NOTE: if proposed new values not accepted, then in this step
# we will copy their 'old' values and insert them into object 'out'
}
```

In reading this code, remember that `exp(log(a) - log(b))` is equal to a/b, but less likely to cause calculation problems, i.e. more numerically stable. In addition, the line `if(runif(1) < r)` is a shortcut for the condition in the third step of the algorithm as outlined above.

For our small data set and simple model, the algorithm runs very swiftly. We get two main outputs: one is a matrix containing 10,000 posterior draws for `alpha` and `beta`, and the other is a matrix containing the two acceptance indicators. We compute the acceptance ratio over all iterations of the algorithm and produce two time-series plots of the Markov chains (Fig. 2.15). These are called traceplots.

```
# Compute overall acceptance ratio separately for alpha and beta
(acc.ratio <- apply(acc, 2, mean))          # Should be in range 25%-40%

# Check the traceplots for convergence of chains
# Plot all MC draws right from the start (Fig. 2.15)
par(mfrow = c(2, 1))
plot(1:niter, out.mcmc[,1], main = paste('alpha (acc. ratio = ', round(acc.ratio[1], 2),')'),
  xlab = "MC iteration", ylab = 'Posterior draw', type = 'l')
plot(1:niter, out.mcmc[,2], main = paste('beta (acc. ratio = ', round(acc.ratio[2], 2),')'),
  xlab = "MC iteration", ylab = 'Posterior draw', type = 'l')
```

We see that during the initial couple of 100 iterations, the chains have not yet reached their equilibrium distribution. For inference, we discard this initial, transient phase as a "burnin," and produce another plot which takes account of a burnin of 500 iterations. Fig. 2.16 indicates that both chains have nicely converged after iteration 500. The acceptance ratio is at about one third, which is well within the range of 25%–40% that is a rule of thumb for MCMC algorithms with high efficiency (Gelman et al., 2014).

```
# Based on Fig. 2.15 choose the amount of burnin
nb <- 500                      # Discard this number of draws at the start

# Compute post-burnin acceptance ratio
(acc.ratio.pb <- apply(acc[nb:niter,], 2, mean))
```

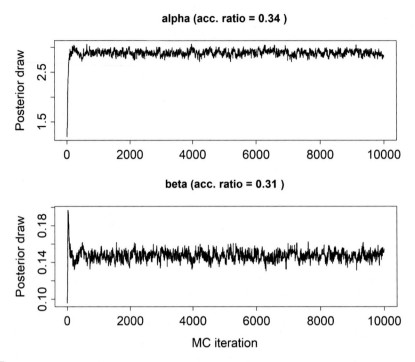

FIGURE 2.15

Traceplots of the posterior draws produced for `alpha` and `beta` right from the start of the MCMC algorithm. Acceptance ratio for each is given in the header.

```
# Repeat traceplots only for post-burnin draws (Fig. 2.16)
par(mfrow = c(2, 1))
plot(nb:niter, out.mcmc[nb:niter,1], main = paste('alpha (post-burnin acc. ratio = ',
    round(acc.ratio.pb[1], 2),')'), xlab = "Iteration", ylab = 'Posterior draw', type = 'l')
plot(nb:niter, out.mcmc[nb:niter,2], main = paste('beta (post-burnin acc. ratio = ',
    round(acc.ratio.pb[2], 2),')'), xlab = "Iteration", ylab = 'Posterior draw', type = 'l')
```

We go on to summarize the post-burnin draws from `alpha` and `beta` to obtain inferences. We compute posterior means as point estimates and posterior SDs and percentiles as the Bayesian analog of a SE and a 95% CI. We also compare our Bayesian inferences on the intercept and the slope of the Poisson GLM fit to the Swiss bee-eater data with the inferences from ML in Section 2.6. (For this you need to have the result from Section 2.6, from the call to `optim()` called `out`, in your R workspace.)

```
# Get posterior mean, sd and 95% CRI (all post-burnin)
post.mn <- apply(out.mcmc[nb:niter,], 2, mean)      # posterior mean
post.sd <- apply(out.mcmc[nb:niter,], 2, sd)        # posterior sd
CRI <- apply(out.mcmc[nb:niter,], 2,
    function(x) quantile(x, c(0.025, 0.975)))       # 95% CRI
```

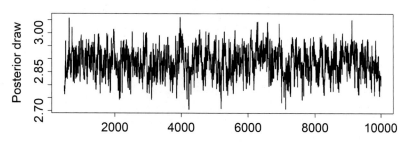

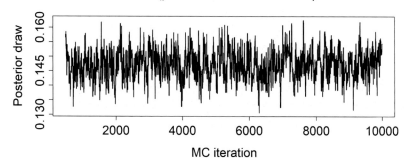

FIGURE 2.16

Post-burnin traceplots of the posterior draws produced for `alpha` and `beta`.

```
# Produce marginal posterior summary for post-burnin part of chains
tmp <- cbind(post.mn, post.sd, t(CRI))
print(tmp, 4)

# Compare Bayesian with DIY-MLE inferences
library(ASMbook)
get_MLE(out, 4)

> # Produce posterior summary for post-burnin part of chains
      post.mn  post.sd   2.5%    97.5%
alpha    2.889 0.053663 2.7795  2.9910
beta     0.147 0.004989 0.1373  0.1568
>
> # Compare with DIY-MLE inferences
          MLE       ASE  LCL.95  UCL.95
alpha  2.8868 0.052659 2.7836  2.9900
beta   0.1472 0.004864 0.1377  0.1568
```

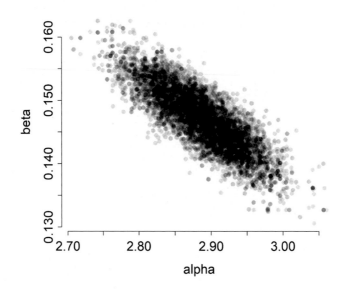

FIGURE 2.17

Joint posterior density for the two parameters of a Poisson log-linear regression fit to the Swiss bee-eater data (compare with likelihood surface in Fig. 2.12).

As is typical for a Bayesian analysis with vague priors, we find that inferences based on ML and posterior distributions are numerically practically identical. This holds both for the point estimates and for the uncertainty assessment in form of SE versus posterior SD and for 95% CI versus 95% CRI.

Finally, we want to emphasize that fundamentally what we get in the Bayesian analysis is the joint posterior density of the two parameters in the model (Fig. 2.17). Owing to our use of vague priors, this joint density will be driven by the likelihood and thus resemble very much the likelihood surface that we plotted in Fig. 2.12. We see once more that the estimates of the two parameters are negatively correlated.

```
# Plot the joint posterior distribution of alpha and beta (Fig. 2.17)
par(mar = c(5,5,4,3), cex.lab = 1.5, cex.axis = 1.5)
plot(out.mcmc[nb:niter,1], out.mcmc[nb:niter,2], xlab = 'alpha', ylab = 'beta',
   frame = FALSE, cex = 1.2, pch = 16, col = rgb(0,0,0,0.1))
```

Summarizing the posterior distribution of one parameter in a multiparameter model involves an integration over the distribution of the other(s) (Royle & Dorazio, 2008: chapter 2). With MCMC-based inferences, we can easily obtain such marginal posterior inferences by simply summarizing the joint posterior for one of its dimensions, while ignoring the others. This is simply another instance of Monte Carlo integration. Fig. 2.18 provides what is the most complete depiction of the inference about each parameter alone. For `alpha` and `beta` this is a probability distribution that conveys our knowledge about the magnitude of these parameters after we have used the information in our bee-eater data set with the Poisson GLM used as an inferential vehicle. Comparing the

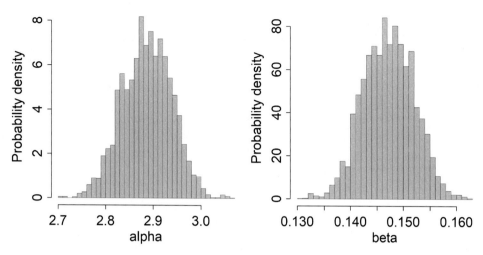

FIGURE 2.18

Marginal posterior distributions for the two parameters of a Poisson log-linear regression fit to the Swiss bee-eater data. These are achieved by integrating the joint distribution in Fig. 2.17 over the other dimensions for each parameter.

figure with the numerical marginal posterior summaries above, you can probably recognize posterior mean, SDs, and the 95% CRIs.

```
# Plot the marginal posterior distributions of alpha, beta (Fig. 2.18)
par(mfrow = c(1, 2), mar = c(5,5,4,3), cex.lab = 1.5, cex.axis = 1.5)
hist(out.mcmc[nb:niter,1], xlab = 'alpha', col = 'gray', main = '')
hist(out.mcmc[nb:niter,2], xlab = 'beta', col = 'gray', main = '')
```

2.8.3 A practical in MCMC for the bee-eaters with function demoMCMC() (*)

We have packaged in the R function demoMCMC() the code for the MCMC algorithm in Section 2.8.2 so that we can play around with it and explore some features of these algorithms. We don't have any real exercises in this book, and yet here is a practical which we ask you to do for yourself. Please do it: it's very simple and what we ask is that you just execute the code and observe what happens ... and there is a really cool animation. Here is the function with all its arguments at their defaults.

```
str(tmp <- demoMCMC(y=y, x=x, true.vals = c(2.5, 0.14), inits = c(0, 0),
   prior.sd.alpha = 100, prior.sd.beta = 100, tuning.params = c(0.1, 0.1),
   niter = 10000, nburn = 1000, quiet = FALSE, show.plots = TRUE))
```

The inits are the initial values, the two arguments `prior.sd.alpha` and `prior.sd.beta` are the SDs of zero-mean normal priors for the two parameters, while the `tuning.params` are the SDs of the normal proposal distributions, called σ_h above. Finally, `niter` and `nburn` specify to the total length of the single chain and the number of iterations discarded as a burnin. When you execute the code in the next box, close the graphics window after each model run. Also,

---

*For this section, RStudio users are advised to switch to base R to experience the full power of R.

don't initialize at more extreme places than $(-40, 40)$, since otherwise you will get a numerical failure. Note that this function just implements an MCMC algorithm for a Poisson log-linear model with a single continuous covariate; thus you could of course use it to fit this model to other data sets than just the bee-eaters.

```
# Shorter run than default
# (can use for true.vals the MLEs from above for graphics)
str(tmp <- demoMCMC(y = y, x = x, true.vals = MLE, inits = c(0, 0),
  prior.sd.alpha = 100, prior.sd.beta = 100, tuning.params = c(0.1, 0.1),
  niter = 2000, nburn = 1000))          # Bad mixing, strong autocorrelation

# Smaller tuning param to increase acceptance
str(tmp <- demoMCMC(y = y, x = x, true.vals = MLE, inits = c(0,0),
  prior.sd.alpha = 100, prior.sd.beta = 100, tuning.params = c(0.1, 0.01),
  niter = 2000, nburn = 1000))

# Larger step length (=tuning.params)
str(tmp <- demoMCMC(y = y, x = x, true.vals = MLE, inits = c(0, 0),
  prior.sd.alpha = 100, prior.sd.beta = 100, tuning.params = c(0.5, 0.5),
  niter = 2000, nburn = 1000))          # Ugly Manhattan traceplots, not efficient,
                                        # no convergence after 1000 steps

# Try to break by picking far-out inits (pretty amazing!)
str(tmp <- demoMCMC(y = y, x = x, true.vals = MLE, inits = c(40, 40),
  prior.sd.alpha = 100, prior.sd.beta = 100, tuning.params = c(0.1, 0.1),
  niter = 5000, nburn = 2000) )

# Try to break some more ... (perhaps even more amazing)
str(tmp <- demoMCMC(y = y, x = x, true.vals = MLE, inits = c(-40, 40),
  prior.sd.alpha = 100, prior.sd.beta = 100, tuning.params = c(0.1, 0.1),
  niter = 5000, nburn = 2000) )

# ... and more
str(tmp <- demoMCMC(y = y, x = x, true.vals = MLE, inits = c(40, 0),
  prior.sd.alpha = 100, prior.sd.beta = 100, tuning.params = c(0.1, 0.1),
  niter = 5000, nburn = 2000) )         # Remarkable search trajectory!

# Same with bigger step length (and shorter chain)
str(tmp <- demoMCMC(y = y, x = x, true.vals = MLE, inits = c(40, 0),
  prior.sd.alpha = 100, prior.sd.beta = 100, tuning.params = c(0.5, 0.2),
  niter = 3000, nburn = 2000) )
```

We believe that by experimenting with this function you can develop some valuable intuition for how basic MCMC algorithms work, and how settings such as the initial values, priors, tuning parameters, and MCMC settings (here, `niter` and `nburn`) affect what they produce. We also hope that you share our own astonishment at how incredibly powerful such a very simple algorithm can be for exploring posterior densities. In this respect, especially the last few calls to the function are really interesting, where we make the job of the function much harder by initializing the search at far-out places in the parameter space. And yet the algorithm finds the solution, that is, that part of the parameter space where most of the posterior density is located.

2.9 So should you now be a Bayesian or a frequentist?

Our short answer is: you should be both! Here is why. Maximum likelihood and Bayesian posterior inference are both very powerful and general methods for fitting models to data, that is, for conducting statistical inferences based on a model and a data set. There are a number of advantages and of disadvantages to both ML and Bayes, which we summarize next.

ML is an extremely well-understood method with a very long history (Fisher, 1922). It is "automatic" since you simply define the joint density of the data, which links the unknowns (the parameters) with the knowns (the data). Then, you treat that density as a likelihood function and minimize the NLL to obtain the MLEs. There exist several methods to obtain uncertainty assessments such as SEs and CIs associated with the MLEs. In addition, MLEs have a number of desirable properties which in general make them "good." ML is arguably the most widely used estimation method.

But ML has disadvantages, too. First, MLEs can be hard or impossible to obtain for complex models, such as those with lots of random effects, especially when we model correlations as in spatial or time-series models. Second, SEs and CIs are only valid asymptotically, that is, for "large" samples, and we rarely have truly large samples in ecology. Third, we often want to compute functions of parameters, such as transformations of one parameter or ratios of two parameters. The delta rule and the bootstrap may be used to compute SEs for those (Kéry & Royle, 2016: chapter 2), but both may become hard or impossible for more complicated functions and models. Fourth, ML is fundamentally based on a probability statement about the data, not about the parameters. That is, a probability such as the "95%" for a CI is *not* about the parameter in question, but about the reliability of the method used to compute the interval. Also, you don't have the right to say things such as *"I am 95% certain that my population is declining"* if 0 is outside of a 95% CI for a population growth rate parameter. And finally, the appeal to hypothetical replicate data may appear ridiculous: for instance, what could we possibly mean with "1 million replicate populations of panda bears"?

There are at least three major advantages of using Bayesian inference using tools such as JAGS, NIMBLE, and Stan. Interestingly, they are fairly independent reasons for choosing Bayesian inference. First is the Bayesian paradigm for inference, which interprets probability exactly like most humans use it: as a measure of knowledge about uncertain things. Also, when we want to introduce into our model fitting external information, then it is very clear how to do this via the prior distributions. The second reason for wanting to go Bayes has to do with Bayesian computation, which typically means use of MCMC and other simulation algorithms. These make it easy to fit hierarchical models to accommodate multiple sources of uncertainty and correlations in the data (see Chapter 3). These days a large part of the models fit in applied statistics fall under this category. Also, simulation-based algorithms make it really easy to compute derived quantities or functions of parameters along with a full assessment of their uncertainty. Thus, you no longer have to worry about the right delta-rule approximation for the SE of functions of parameters. The third good reason for going Bayes has to do with the model-definition language adopted in JAGS and NIMBLE, and earlier in BUGS, WinBUGS, and OpenBUGS (Lunn et al., 2013), and to a lesser degree also in Stan. The BUGS language has proven to be superbly accessible to statisticians and nonstatisticians alike for specifying even very complex statistical models. It is the BUGS language that brings within the reach of ecologists the implementation of really complex, custom models. With BUGS you can develop models that you would never have been able to fit if you had to use ML. Thus, we like to say that BUGS frees the modeler in you.

But Bayesian inference also has some disadvantages. First, the prior will always exert some influence on your results, and in data-poor situations, this influence will be greater. We will see this in several chapters later in the book. This may make reporting of your results more complex since you may need to tell multiple stories. Second, for complex models and big data sets simulation-based methods such as MCMC may take a very long time to run, for example, days or even weeks. Worse yet, sometimes you may never achieve convergence within a reasonable time. Third, the final disadvantage is that BUGS engines may be so flexible and powerful that it may be easy to fit nonsensical models. By this we mean especially models with parameters that are unidentifiable, that is, for which your data contain no information. True, parameter identifiability is a very big issue for any complex statistical model and it may plague inferences by ML just as well. However, lack of identifiability of a parameter may be even harder to diagnose when using Bayesian inference than when using ML. There are formal methods to confront this (Cole, 2021), but they may be hard to implement for complex models and for nonstatisticians. In practice, the best method to judge whether a parameter is identifiable is simulation (Kéry & Royle, 2016: chapter 4): you simulate a large number of data sets with varying values of all parameters, especially of the suspect parameter, fit your model, and compare estimates and true values. For an indentifiable parameter, there ought to be a clear numerical resemblance between estimates and true values.

In summary, we completely agree with Royle & Dorazio (2008) that ecologists need a good working knowledge of both ML and Bayesian methods in statistical modeling. Indeed, this whole book is predicated on this idea. Sometimes, or for some of you, one is the better choice, while another time or for others it's the other. In addition, the two methods of inference have far more in common than what one is sometimes made to believe: ML and Bayes share their basis as model-based approaches to inference, with a central role played by the likelihood function. In addition, by far the most common way in which Bayesian analyses are conducted in practice is such that we use the powerful computing machinery associated with Bayesian MCMC methods to solve difficult estimation problems, while minimizing the amount of information contained in our priors. That is, we work with vague priors. As a result, at least numerically we are then almost always back to something that greatly resembles the same old MLEs. We demonstrate this throughout the book.

2.10 Summary and outlook

In this chapter, we have seen that probability is the foundation of statistical modeling and inference. We then covered the key concept of a RV, which is defined as a real-valued function defined on the sample space of a random experiment. RVs are described by PMFs when they have discrete values only or by PDFs when continuous. A statistical model is a collection of PDFs or PMFs for all your response variables. We have also seen that models have multiple uses in statistical inference, but that a key thing we use them for is to "fit them to data" to estimate parameters (we note in passing that we always fit a model to data, but we never fit data to a model). The two dominant methods of parameter estimation are both based on the likelihood function, which is algebraically identical to the joint density (or probability) of all RVs, but viewed as a function of the parameters. The likelihood function links what we observe (the data) with what we don't observe but would like to estimate (the parameters). It carries the information about the parameters contained in our data set.

With ML we work directly with the likelihood function for statistical inference, by maximizing it over the parameter space to obtain the MLEs and using the curvature of the likelihood function around the MLEs for uncertainty assessments. The concept of probability in ML is based on a relative-frequency interpretation in a large number of hypothetal replicates of our data set, that is, as a measure of the variability of observable quantities or essentially of the variability of data. In contrast, in addition to the use of probability as a measure of variability of observable things, Bayesians adopt a more general definition of probability as a metric for our knowledge about any kind of uncertain things, including parameters in a model. They use the likelihood function to update their prior knowledge about the parameter(s) via Bayes rule, in a simple application of conditional probability to statistical inference. The result is another probability distribution, the posterior distribution, which contains all information contributed by the data (via the likelihood) and our prior assumptions about a parameter (represented by the prior). We can then summarize the posterior by its mean for a point estimate and by its spread for an uncertainty assessment of that estimate. We have also seen that with vague priors, Bayesian point estimates and uncertainty assessments are typically very similar, numerically, to those from ML.

In practice, Bayesian inference is most often obtained with MCMC methods, for which we have described and studied the prototypical case of a random-walk Metropolis algorithm. Even with this simple algorithm, we can obtain a lot of useful intuition about the workings of more complex simulation-based algorithms which are used in JAGS, NIMBLE, Stan, and also other Bayesian model-fitting software. MCMC algorithms are incredibly powerful and yet conceptually simple algorithms. For plenty of inference problems, they work extremely well if you are prepared to wait some, but for others, convergence may be harder to achieve or even never at all.

At the end of the chapter, we have argued that both ML and Bayes have a number of advantages and disadvantages. We are convinced that to become a successful quantitative ecologist, you need to understand both ML and Bayesian inference, and this is indeed a major reason for why we wrote this book. But even if you want to go Bayes only, you still need to develop a sound understanding of concepts that underlie both methods. Most of all, you need to understand the concepts of the joint density, likelihood and log-likelihood, which are essential ingredients of every Bayesian analysis. Arguably, these may be easier to grasp in the context of ML, and this is one of the reasons for which we emphasize DIY-MLEs (see Section 4.7) so much throughout the book.

Much of the material in this chapter is conceptual and meant to demonstrate the workings of ML and Bayesian inference, along with their typical computational implementations. However, you could probably use the simple MCMC code in this chapter as a starting point for writing your own code for more complex or different models, for example, by translating the Poisson into a binomial likelihood, or by adding more covariates. As always, it's usually by making such small incremental changes to existing code that you learn and write new code. For much deeper and comprehensive introductions to MCMC methods for ecologists, see the recent books by Hooten & Hefley (2019) and Zhao (2024). For a deeper introduction to ML geared at biologists, excellent references include Bolker (2008), Royle & Dorazio (2008: chapter 2), and Millar (2011).

Now that we have covered two powerful methods for fitting statistical models to data, we will first (in Chapter 3) cover some of the most typical parametric statistical models applied in ecology. Then, in Chapter 4 we will give an overview of several extremely powerful, general-purpose model-fitting engines that we use throughout the book, namely JAGS, NIMBLE, and Stan for Bayesian inference (mostly) and R and TMB for ML inference (mostly). In all later chapters, we will also continue getting our own MLEs "by hand", because we believe it is so important that ecologists get a better understanding of ML. For Bayesian inference we will no longer write our own algorithms, but use JAGS, NIMBLE, and Stan.

Further reading

- Probability and statistics: Casella & Berger (2002), Pishro-Nik (2014), Blitzstein & Hwang (2019); with focus on biology/ecology: Royle & Dorazio (2008: chapter 2), Royle et al. (2014: chapter 2), Kéry & Royle (2016: chapter 2).
- Bayesian inference: Lindley (2006), Carlin & Louis (2009), Gelman et al. (2014a), Kruschke (2015), Spiegelhalter (2019), McElreath (2020); with focus on biology/ecology: Royle & Dorazio (2008: chapter 2), King et al. (2009), Link & Barker (2010), Royle et al. (2014: chapter 2), Hobbs & Hooten (2015), Kéry & Royle (2016: chapter 2), Inchausti (2023), Gimenez (2025).
- Maximum likelihood (with focus on biology/ecology): Bolker (2008), Royle & Dorazio (2008: chapter 2), Millar (2011), Lee et al. (2017: chapter 1 [without biology]), Inchausti (2023).
- Markov chain Monte Carlo (with focus on biology/ecology): Royle & Dorazio (2008: chapter 2), Royle et al. (2014: chapter 2), Kéry & Royle (2016: chapter 2), Hooten & Hefley (2019), Zhao (2024).

Linear regression models and their extensions to generalized linear, hierarchical, and integrated models[★]

3

Chapter outline

3.1 Introduction

The principles of probability covered in Chapter 2 let us build almost any statistical model, depending on the type of random variables (RVs), the probability distributions assumed for them, and the architecture chosen for combining different sets of RVs. In this chapter, we first describe some key model-building choices, including how to pick an appropriate statistical distribution for our RVs, how to specify effects of covariates on a model parameter, and how to model these parameters on a transformed scale if necessary. Indeed, these three criteria of a *response distribution*, a *link function*, and a linear model (LM), or *linear predictor*, define the famous *generalized linear model*, or GLM (McCullagh & Nelder, 1989; Dobson & Barnett, 2018). The GLM is a key model in applied statistical modeling, and all the other model classes in this book can be viewed either as a special

---

[★]This book has a companion website hosting complementary materials, including all code for download. Visit this URL to access it: https://www.elsevier.com/books-and-journals/book-companion/9780443137150.

Applied Statistical Modelling for Ecologists. DOI: https://doi.org/10.1016/B978-0-443-13715-0.00012-1

case or as a generalization which combines two or more GLMs in sequence, leading to a mixed model or hierarchical model, or HM (Berliner, 1996; Royle & Dorazio, 2008; Cressie & Wikle, 2011), or with a branching structure, leading to an integrated model, or IM (Besbeas et al., 2002; Chapter 10 in Kéry & Royle, 2021; Schaub & Kéry, 2022).

In Section 3.2, we present a set of just four statistical distributions: normal, uniform, binomial, and Poisson. Astonishingly, we will build all models in this book with just these basic ingredients. In Section 3.3, we cover link functions, that is, transformations for a parameter to enforce adequate range constraints, such as nonnegativity for counts or a (0, 1) range for probabilities. We again get very far with only three link functions: the identity, log, and logit.

In practice, much of what we do when we "model" is simply specifying ways in which we think our covariates are related to a response. Thus we define LMs with covariates of interest, such as when we write in R `lm(y ~ A * B)` to fit a normal LM with both main effects and interactions of `A` and `B`. The principles of specifying LMs for covariate effects are identical regardless of whether we work with a normal LM, a generalized linear mixed model (GLMM), or a complex HM. Thus it is essential to know how to specify LMs and how to interpret their parameters. In Section 3.4, we give a catalog of prototypical LMs with explanatory variables that may be continuous or categorical (=factors).

Finally, in Section 3.5, we give a brief overview of the six classes of regression models for which we illustrate examples in this book: LMs, GLMs, linear mixed models (LMMs), GLMMs, HMs, and IMs. These may appear fairly disparate at first, but in fact they differ mainly in the type of statistical distribution assumed for a response, in whether or not they also have latent variables (or random effects) and if so, in the manner in which different types of RVs are connected in the model. Thus, understanding statistical distributions adopted for both observed and latent RVs, link functions and linear modeling of covariate effects are all key for your understanding of these vast classes of regression models that are so widely used in applied statistical modeling.

3.2 Statistical distributions for the random variables in our model

In Chapter 2, we emphasized the central place of RVs in statistical inference, that is, of real-valued functions defined on the sample space of a random experiment. Identifying the RVs and picking a probability distribution for them lies at the core of statistical modeling. In this section, we cover a small, but powerful set of statistical distributions: normal, continuous uniform, binomial (with the Bernoulli as a special case), and Poisson. Astonishingly, we will be able to build all the models in this book either by one of them alone, or by combining two or more of them in a mixed, hierarchical, or integrated model.

Two key considerations when choosing a distribution for a set of RVs are their support and whether they are discrete or continuous. For instance, some distributions are defined for the entire real line (i.e., from $-\infty$ to ∞), some only for nonnegative numbers, while yet others only between 0 and 1. In Section 2.2, we have seen discrete and continuous RVs and likewise there are discrete and continuous distributions. Nevertheless, although the divide between discrete and continuous RVs may appear deep, there is some overlap in practice. First, measurement accuracy is always finite; hence, in practice every continuous RV is *recorded* in a discrete way. Usually, this is of no consequence. On the other hand, several discrete distributions can sometimes be well approximated by a normal distribution. For instance, large counts may often be modeled with a normal distribution. In addition, mathematical or computational convenience may sometimes stipulate choice of one distribution, especially a normal, over another which might a priori appear more adequate. For instance, the normal is often used for count data in the observation submodels of many integrated population models (Schaub & Kéry, 2022).

Next, we give brief vignettes of the four distributions that we will use throughout the book. We describe them mathematically and give typical circumstances in which they may be used in a model. In addition, we provide R code to create random numbers from each distribution and plot them in a histogram, for checking how selected parameter values influence the shape of a distribution. You can play around with different parameter values and so get a feel for how the distributions change as parameters are varied. You may study further features of these and of many other distributions in R; type `?rnorm`, `?runif`, `?rbinom`, or `?rpois` to find out more.

3.2.1 Normal distribution

The normal or Gaussian distribution is the *de facto* default choice in applied statistical regression modeling for RVs that are continuous, not bounded by 0 and 1, and, when plotted in a histogram, cluster in a symmetrical way around some "typical," central value.

$$\text{Mathematical description (PDF):} \quad f(y|\mu, \sigma) = \frac{1}{\sigma\sqrt{2\pi}} \exp\left(-\frac{(y-\mu)^2}{2\sigma^2}\right)$$

The two parameters are the mean (μ) and the standard deviation σ.

Typical sampling situation: We take measurements which are affected by a large number of effects that act in an additive way.

Classical examples: Body size or many other linear measurements. In Bayesian inference the normal is also used to specify ignorance about the quantitative value of a parameter via a vague, zero-mean normal prior with a large variance (i.e., a "flat normal"). A normal with small variance can be used as an informative prior or for regularization (see Section 18.8).

Variants and relatives: When effects are multiplicative instead of additive, we get a log-normal distribution. A *t*-distribution can be seen as an overdispersed normal which has more extreme values, or outliers (Chapter 18). A normal RV can be transformed into a log-normal RV by taking the log. In addition, the normal can be used as an approximation of the binomial when N is large and p is not too close to 0 or 1, as well as to a Poisson when λ is not too close to 0.

Typical picture: The Gaussian bell curve, that is, a symmetrical, single hump, with more or less long tails. In small samples can look remarkably irregular, even skewed; see Fig. 2.14.

R code to draw n random number with specified parameter(s) and plot a histogram (see Fig. 3.1):

```
n <- 100000                                    # Sample size
mu <- 600                                      # Value of population mean
sigma <- 30                                    # Value of population SD

sample <- rnorm(n = n, mean = mu, sd = sigma)  # Draw random numbers
print(sample, dig = 4)                         # ...print them
hist(sample, col = "grey", breaks = 60, freq = F)  # ...plot them
```

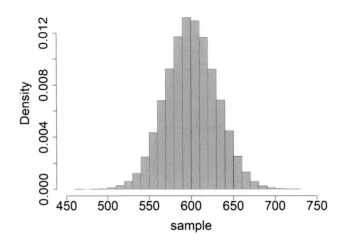

FIGURE 3.1

Histogram of a large sample from a normal distribution with $\mu = 600$ and $\sigma = 30$.

Mean and standard deviation:

$$E(y) = \mu \quad \text{mean}$$

$$sd(y) = \sigma \quad \text{sd}$$

3.2.2 Uniform distribution

Mathematical description(PDF): $f(y) = \dfrac{1}{b-a}$ if $a \le y \le b$ and $f(y) = 0$ otherwise.

The two parameters of the continuous uniform distribution are a and b, but in most applications these limits are known and thus not considered parameters. In that case, we have the peculiarity of a distribution without any unknown parameter.

Typical sampling situation: Measurements are taken within a range of values a and b, and they are all equally likely to occur anywhere in that range.

Classical examples: In a Bayesian analysis often used to specify ignorance in a vague prior. Also, common prior distribution for the activity centers in a Bayesian analysis of a spatial capture-recapture model (Royle et al., 2014).

Variants and relatives: Discrete uniform. For example, when rolling a die, the response is an integer between 1 and 6.

Typical picture: A rectangle with some ruggedness due to sampling variability.

R code to draw n random number with specified parameter(s) and plot histogram (see Fig. 3.2):

```
n <- 100000                              # Sample size
a <- lower.limit <- 0                    # Lower limit of RV
b <- upper.limit <- 10                   # Upper limit of RV

sample <- runif(n = n, min = a, max = b) # Draw random numbers
print(sample, dig = 4)                   # ...print them
hist(sample, col = "grey", breaks = 30, freq = F)  # ...plot them
```

Mean and standard deviation: $\quad E(y) = (a+b)/2 \qquad$ mean

$$sd(y) = \sqrt{(b-a)^2/12} \quad \text{sd}$$

3.2.3 Binomial distribution: the "coin-flip distribution"

The binomial and its special case, the Bernoulli, is often our first choice for a count RV that has a known upper bound and that is governed by a probability. It describes a stochastic process for which a coin flip is often a good physical analogy. If we flip a coin which produces heads with probability p, then for a single flip the probability of heads is p and that of tails $1-p$; this corresponds to the Bernoulli distribution. For N independent flips of that coin, the probability of y heads is given by a binomial distribution with sample size N and probability p. Thus, the Bernoulli is a special case of a binomial with sample size $N=1$. Note that a binomial RV is always bounded by 0 and N, and the presence of the upper bound is a key distinction to a Poisson RV. The binomial distribution is very important in

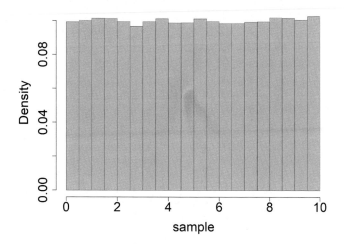

FIGURE 3.2

Histogram of a large sample from a continuous uniform distribution with lower bound of 0 and upper bound of 10.

ecological statistics, where it is the classical measurement error model in capture-recapture, occupancy, *N*-mixture, and related models (Borchers et al., 2002; Williams et al., 2002; Royle & Dorazio, 2008; Buckland et al., 2015; Seber & Schofield, 2019, 2024).

Mathematical description (PMF):

$$f(y|p) = p^y(1-p)^{1-y} \qquad \text{Bernoulli}$$

$$f(y|N,p) = \frac{N!}{y!(N-y)!} p^y(1-p)^{N-y} \quad \text{Binomial}$$

The two parameters of the binomial are the sample or trial size *N* and the success probability *p*. In most applications of a binomial, the sample size is known and thus not an estimated parameter (but see Chapter 19B for a model where *N* is estimated as a parameter).

Typical sampling situation: When *N* independent things all have an identical probability *p* of belonging to a certain class (example: being counted, or having a certain attribute like being male or dead), then the number *y* that is actually counted in that sample, or has that attribute, is binomially distributed with *N* and *p*.

Classical examples: Number of males in a clutch or in a school class or in a herd of size *N*, number of times heads show up among *N* flips of a coin, number of times you get a six among *N* = 10 rolls of a die, number of animals among those *N* present that you actually detect.

Variants and relatives: The binomial can be approximated by a normal when the product *Np* is not too small. The beta-binomial is an overdispersed version of a binomial (Royle & Dorazio, 2008).

Typical picture: Varies a lot, but always discrete. Normally skewed, but skewness depends on the value of the parameter *p*; symmetrical for *p* = 0.5.

Important feature: The binomial comes with a "built-in" variance equal to $N \cdot p \cdot (1-p)$, that is, the variance is a function of the mean, which is $N \cdot p$.

R code to draw n random number with specified parameter(s) and plot histogram (Fig. 3.3):

```
n <- 100000                           # Sample size
N <- 16                               # Number of individuals that flip a coin
p <- 0.8                              # Probability of being counted (seen),
                                      # dead or a male

sample <- rbinom(n = n, size = N, prob = p)    # Draw random numbers
print(sample, dig = 4)                # ...print them
plot(table(sample)/n, ylab = 'Probability', type = 'h', lwd = 20, col = rgb(0, 0, 0, 0.5),
   xlim = c(0, 16), lend = 'butt', frame = F)    # ...plot them
```

Mean and standard deviation: $\qquad E(y) = N \cdot p \qquad\qquad$ mean

$$sd(y) = \sqrt{N \cdot p \cdot (1-p)} \quad \text{sd}$$

Mean and standard deviation for the Bernoulli are p and $\sqrt{p \cdot (1-p)}$, respectively.

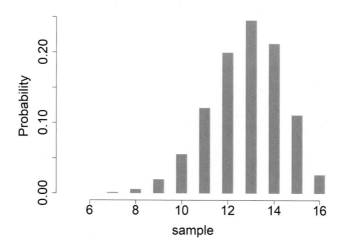

FIGURE 3.3

Histogram of a large sample from a binomial distribution with $N = 16$ and $p = 0.8$. The probability for realizations less than 5 is not zero, but just too small to be seen.

3.2.4 Poisson distribution

While the binomial is our typical first choice for a RV that corresponds to a *bounded* count, the Poisson distribution is typically our first choice of a counting RV that has no theoretical upper bound.

> *Mathematical description (PDF):* $f(y|\lambda) = \dfrac{\lambda^y e^{-\lambda}}{y!}$, where e is Euler's number (2.71282...)

The Poisson has a single parameter, typically denoted λ, which is both the mean and the variance of a Poisson RV and is also known as the intensity. The assumption of a variance/mean ratio of 1 is commonly violated in the real world. Frequently the empirical variance is greater than the mean, in which case we speak of overdispersion (see Section 12.2).

Typical sampling situation: When objects (e.g., birds, cars, erythrocytes) are independently and uniformly distributed along one, two, or more dimensions with intensity λ and we randomly place a "counting window" along these dimensions and record the number of objects, then that number is a Poisson RV. The intensity is the expected number of objects per unit length/area/volume of the counting window.

Classical examples: Number of cars passing at a street corner during 10 min, number of birds that fly by you at a migration site, number of car accidents per day, month, or year, number of some animal or plant species per sample quadrat.

Variants and relatives: The Poisson is an approximation to a binomial when N is large and p small (with $\lambda \approx Np$), and can itself be approximated by a normal when λ is large (e.g., > 10).

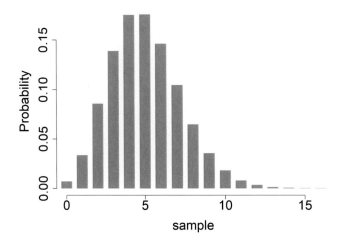

FIGURE 3.4

Histogram of a large sample from a Poisson distribution with intensity $\lambda = 5$. The probability of values greater than about 16 is not zero, but just too small to be seen.

The negative binomial distribution can be described as an overdispersed version of the Poisson and can be derived by assuming that the Poisson mean, λ, is a RV with a gamma distribution (Royle & Dorazio, 2008).

Typical picture: Varies a lot, but strictly speaking always discrete. Typically skewed, but degree of skewness declines with increasing value of λ.

R code to draw n random number with specified parameter(s) and plot histogram (see Fig. 3.4):

```
n <- 100000                          # Sample size
lambda <- 5                          # Average count per window (density or intensity)

sample <- rpois(n = n, lambda = lambda) # Draw random numbers
print(sample, dig = 4)               # ...print them
plot(table(sample)/n, ylab = 'Probability', type = 'h', lwd = 20, col = rgb(0, 0, 0, 0.5),
  lend = 'butt', frame = F)          # ...plot them
```

Mean and standard deviation: $E(y) = \lambda$ mean

$$sd(y) = \sqrt{\lambda} \quad \text{sd}$$

3.2.5 Summary remarks on statistical distributions

This concludes our overview of four named distributions that we will use to describe the RVs of all models in this book. There are many, many more and you can find dozens tabulated in statistics books and on the internet, and in more advanced modeling you will often need other distributions. Several key features are shared by all statistical distributions:

- We use them to quantitatively describe the RVs in our models; thus they form the essential building blocks for our statistical models.

- Statistical distributions let us express the joint density of all RVs (data, latent variables/random effects) in our model and thus permit construction of a likelihood function, which we use for statistical inference in both maximum likelihood or Bayesian posterior inference.
- We often use distributions in a generative way to simulate data to do a large number of useful things; see Chapter 1 and also Chapter 4 in Kéry & Royle (2016).

Typically, however, we will not work directly with the regular parameters of a distribution, but instead want to study potential effects of covariates on them. The concept of the *link function* then becomes important.

3.3 **Link functions to model parameters on a transformed scale**

Typically, we want to study how the regular parameters in a probability distribution, such as the means in a normal (μ), Bernoulli (p), or a Poisson (λ), are affected by covariates. These covariates then determine a RV in some average sense. To achieve this, we simply replace the regular parameter with a function of the values of these covariates and of some new parameters in what then becomes a sort of regression model. Parameters that are real numbers (i.e., that can take on any real value) may be directly modeled as a linear function of covariates. However, often we have parameters that are defined on a restricted range. For instance, expected counts for Poisson RVs or variances for a Gaussian RV are both nonnegative, and the success probability in a binomial RV must lie between 0 and 1. To meet these restrictions, we can specify our covariate models for a transformation of a parameter. A typical transformation for a nonnegative parameter is the natural logarithm, or log. Thus, we usually specify covariate models for the mean of a Poisson RV as $\log(\lambda_i) = \alpha + \beta x_i$. Now, regardless of the values of α, β, and x, the mean λ_i (and also the realized values of the RV) will always be nonnegative. Such a transformation is called a *link function*. Depending on the type of RV, we may choose different link functions to enforce the proper range constraints of the parameter in its probability distribution.

As for distributions, you may be surprised to learn that in most applied settings we work with just two transformations: the log link and the logit link. There is a third common link function called the "identity" (McCullagh & Nelder, 1989), but this is just GLM jargon for "don't transform," as when we model the mean response in a normal LM directly. The log link is the typical link for the Poisson mean parameter. The logit is used for parameters that represent probabilities. It "stretches out" the regular parameter, which lives in the range of 0–1, to the full real line. The logit is the natural logarithm of the odds of the event of interest, where the latter is the ratio of p over $1-p$:

$$\text{logit}(p) = \log\left(\frac{p}{1-p}\right)$$

The logit is the most widely used link function for the probability in a binomial distribution and results in a model known as *logistic regression* (Hosmer et al., 2013). In Chapter 20, you will discover another link function for a probability, the complementary log-log or cloglog. It can be motivated by the assumption of an abundance distribution which underlies binary data and for which we model covariate effects via a log link. In some contexts, this results in an eminently interpretable model (Scharf et al., 2022).

Link functions make your life easier when modeling covariate effects for parameters with range constraints and thus are very commonly used. However, they may make your life miserable when you want to calculate predictions, for example, to plot the expected binomial probability as a function of a wide range of values of an explanatory variable X. Typically, we want predictions not on the link-transformed scale, but on a more "natural" scale. For this, you must first apply the linear model on the link scale to obtain predictions and then apply the *inverse* link transformation to back-transform predictions from the link scale to the "natural" scale. Obtaining predictions for a covariate model with a link function like the log or the logit can be confusing. We give an example of this with a log link in Chapter 13, where the inverse link is the exponential function. For predictions in a logistic regression, you must apply the inverse-logit link to the linear predictor η, that is, do $1/(1 + \exp(-\eta))$, which in R can be done by `plogis($\eta$)`.

3.4 **Linear modeling of covariate effects**

In our statistical models, we almost always want to "explain" a response **y** by one or more covariates $\mathbf{x}_1$, $\mathbf{x}_2$ ... These covariates may be continuous, categorical (also called factors), or a combination of both. There is an infinite number of ways in which a functional relationship between **y** and the covariates could be specified. However, by far the most common way in which we model covariate effects is by a *linear model*. That is, we model the response as a weighted sum of the values of covariates, where the weights, also called coefficients, are the parameters which we want to estimate. A typical example of a linear model might be $E(y_i) = \mu_i = \alpha + \beta_1 x_i + \beta_2 x_i^2$, where the expected value of response y_i is the sum of an intercept α and of coefficients β_1 and β_2, each multiplied with the linear and quadratic terms of covariate x, respectively. Learning to specify appropriate LMs is crucial for testing hypotheses, making accurate predictions or simply obtaining a model that provides a parsimonious description of a data set (Tredennick et al., 2021).

Most ecologists have become accustomed to using the Wilkinson-Rogers (1973) formula-based notation for specifying LMs. For instance, to specify an LM for **y** with two covariates and their interaction in R, we just type the formula `y ~ A * B` and provide this as the first argument to functions such as `lm()` or `glm()`. Many of us don't know exactly what this causes R to do internally, and this may not *always* be a problem. However, to fit an LM in the more general model-fitting engines in this book (see Chapter 4), you must know exactly what this and many other LM short-hands mean, and how they can be adapted to match your hypotheses. Therefore we next describe how several prototypical LMs can be specified and how we interpret their parameters.

In particular, we introduce the *design matrix* (also called *model matrix* or *X matrix*) of an LM, which is what R functions such as `lm()` are building in the background using the formula we provide. We use `model.matrix()` to inspect the design matrix of every LM in this section. The design matrix has one row per data point and one column per parameter of the LM and contains the values of all explanatory variables for each response y_i in a modeled data set. The design matrix comprises a complete description of the type of LM adopted and defines the meaning of the estimated parameters. For factor covariates, the corresponding columns in the design matrix indicate which factor levels are present for each data point; for continuous covariates, the column indicates what "amount" of the covariate is present for each data point, i.e., it gives its value. When the design matrix is matrix-multiplied with the parameter vector, a new vector called the *linear predictor* is the result (you may want to quickly look up matrix multiplication if you need a reminder). It contains the expected value of the response on the link scale, given the values

Table 3.1 Our toy data set for six imaginary snakes.				
mass	pop	reg	hab	length
60	1	1	1	40
80	1	1	2	45
50	2	1	3	39
70	2	1	1	50
90	3	2	2	52
110	3	2	3	56

of all covariates in the model. For models with factors, a key concept is that of a *dummy* or *indicator variable*, with which we code the effects of factor levels. We will see how an LM involving a factor can be described in different ways, corresponding to different *parameterizations* of the model.

To illustrate the design matrices for a wide range of LMs, we introduce a toy data set with just six data points (Table 3.1). We imagine we measured body mass (mass, in grams; a continuous response variable) of six snakes in three populations (pop), two regions (reg), and three habitat types (hab). These three categorical explanatory variables are factors with 3, 2, and 3 levels (i.e., distinct values representing types), respectively. We also have a continuous explanatory variable for each snake, snout-vent length (length, measured in centimeters).

Here is the R code for setting up these variables:

```
mass <- c(60, 80, 50, 70, 90, 110)     # Response variable (continuous)
pop <- factor(c(1, 1, 2, 2, 3, 3))     # A categorical covariate (factor)
reg <- factor(c(1, 1, 1, 1, 2, 2))     # Another factor
hab <- factor(c(1, 2, 3, 1, 2, 3))     # Yet another factor
length <- c(40, 45, 39, 50, 52, 56)    # A continuous covariate
```

We use factor() to tell R that the numbers in an explanatory variable are to be treated simply as names or labels for the factor levels and do not have any quantitative meaning. You can see this when printing out a factor.

```
# A continuous covariate with quantitative values
mass
# A categorical covariate or factor
pop

> mass
[1] 60 80 50 70 90 110

> pop
[1] 1 1 2 2 3 3
Levels: 1 2 3
```

In the remainder of Section 3.4, we use the R function lm() to specify different LMs for these data using the Wilkinson-Rogers short-hand. Each time we will inspect what this exactly means in terms of the design matrix of the LM, which itself determines the interpretation of the parameters in a model, as we will see. We also look at how each model can be defined using vector/matrix notation and in algebra. We work with `lm()` for illustration only and because you may be most likely to have fitted LMs with this function. However, all concepts of linear modeling carry over to the other classes of models we discuss in this book, such as GLMs, mixed models, HMs, or IMs. They also carry over regardless of whether you fit a model in R or any other software, or with maximum likelihood or Bayesian inference. Thus the material in Section 3.4 is fundamental for applied statistical modeling with any type of regression model.

3.4.1 The "model of the mean"

What if we just wanted to fit a common mean to the mass of all six snakes? That is, fit an LM with an intercept only (see Chapter 4). This is the simplest linear statistical model possible with a single estimated parameter in the mean. In R, we can do this by issuing the following command, where the 1 implies a covariate with a single value of 1 for every snake:

```
lm(mass ~ 1)

Coefficients:
(Intercept)
      76.67
```

We see that the average snake weighs about 77 g.

Let's look at the design matrix of this LM. It consists of a vector of ones, which is termed the intercept by R.

```
model.matrix(mass ~ 1)

  (Intercept)
1           1
2           1
3           1
4           1
5           1
6           1
[ ... ]
```

We can write this model algebraically as

$$mass_i = \mu + \varepsilon_i.$$

Thus, we imagine that the mass measured for snake i is composed of a deterministic part, which here is simply the overall mean μ, plus a stochastic part, which is the individual deviation from that mean, that is, residual ε_i. If we want to estimate the mean using maximum likelihood, we also need a distributional assumption, such as $\varepsilon_i \sim Normal(0, \sigma^2)$. Note we could write exactly the same model as $mass_i \sim Normal(\mu, \sigma^2)$, where the residuals appear only implicitly. Finally, to emphasize that the only component in the deterministic part of the response is the intercept, we could write the same model also as $mass_i = \alpha + \varepsilon_i$, with $\varepsilon_i \sim Normal(0, \sigma^2)$. This is more directly comparable with how we write models with covariates, which we examine next.

3.4.2 Comparison of two groups

We now fit an LM that describes the effects of a binary explanatory variable such as `reg`. As a least-squares procedure for a normal response, this is the LM that underlies the famous t-test (Zar, 1998). Here is the specification of this two-group comparison as an LM in R:

```
lm(mass ~ reg)    # not shown
```

To understand the interpretation of the parameters, we inspect the design matrix for this model. When we fit `reg` as a factor, then R internally expands this categorical explanatory variable into a design matrix with two binary vectors: one contains only ones and codes the intercept, while the other is a single *indicator, or dummy, variable* for the effect of `reg`. It indicates which snake was caught in the region indicated in R's name for that column in the design matrix. In the presence of an intercept in an LM, a factor with k levels is always expressed by $k - 1$ dummy variables. In this way, all LMs are fit as multiple linear regressions, regardless of whether they have continuous or categorical covariates or both.

```
model.matrix( ~ reg)

> model.matrix( ~ reg)
  (Intercept)    reg2
1           1       0
2           1       0
3           1       0
4           1       0
5           1       1
6           1       1
attr(,"assign")
[1] 0 1
attr(,"contrasts")
attr(,"contrasts")$reg
[1] "contr.treatment"
```

This is what the R default *treatment contrasts* yields; type `?model.matrix` to find out what other contrasts are possible in R. We see that indicator variable `reg2` contains a 1 for the snakes in region 2 only. Thus on fitting the model in R, the first parameter, the intercept, will denote the expected mass of snakes in region 1, while the second parameter will denote the difference between the expected mass of snakes in region 2 and that in region 1. Region 1 thus serves as a baseline or reference level in the model, and the *effect* of region is expressed as a difference from that baseline for snakes in region 2.

Fitting the two-group model to our toy data set with `reg` defined in this way implies a system of equations which is instructive to write down.

$$60 = \alpha \cdot 1 + \beta \cdot 0 + \varepsilon_1$$
$$80 = \alpha \cdot 1 + \beta \cdot 0 + \varepsilon_2$$
$$50 = \alpha \cdot 1 + \beta \cdot 0 + \varepsilon_3$$
$$70 = \alpha \cdot 1 + \beta \cdot 0 + \varepsilon_4$$
$$90 = \alpha \cdot 1 + \beta \cdot 1 + \varepsilon_5$$
$$110 = \alpha \cdot 1 + \beta \cdot 1 + \varepsilon_6$$

We see why the intercept is represented by a 1: it can be interpreted like a covariate with an identical value of 1 for all snakes. To get a solution for this system of equations, that is, to obtain values for the unknowns α and β that are "good" in some way, we need to define a criterion for dealing with the residuals ε_i. Choosing the unknowns α and β such that the sum of the squared residuals becomes minimal is the least-squares criterion. For GLMs with a normal response and identity link, it yields parameter estimates for α and β (though not for the residual variance) that are identical to those obtained using maximum likelihood; see Section 5.9.

We can write the same system of equations using vector/matrix notation as:

$$\begin{pmatrix} 60 \\ 80 \\ 50 \\ 70 \\ 90 \\ 110 \end{pmatrix} = \begin{pmatrix} 1 & 0 \\ 1 & 0 \\ 1 & 0 \\ 1 & 0 \\ 1 & 1 \\ 1 & 1 \end{pmatrix} \begin{pmatrix} \alpha \\ \beta \end{pmatrix} + \begin{pmatrix} \varepsilon_1 \\ \varepsilon_2 \\ \varepsilon_3 \\ \varepsilon_4 \\ \varepsilon_5 \\ \varepsilon_6 \end{pmatrix}$$

or, in shorter form, as $\mathbf{Y} = \mathbf{X}\gamma + \varepsilon$, where $\mathbf{Y}$ is the response vector, $\mathbf{X}$ the design matrix, $\gamma = (\alpha, \beta)$ the parameter vector, and ε the residual vector. Note the matrix multiplication between the design matrix and the parameter vector here, and in many later examples in this chapter. Please look up how it works if you are unsure about this.

Let us now write this model in algebra.

$$mass_i = \alpha + \beta \cdot reg_i + \varepsilon_i$$

In this notation, reg is the dummy variable for region 2. We see that the mass of snake i is composed of the sum of three components: a constant α, the product of another constant (β) with an indicator for region 2, and a noise term ε_i specific to each snake i, for which we assume a zero-mean normal distribution with a constant variance, that is, $\varepsilon_i \sim Normal(0, \sigma^2)$.

This clarifies what exactly needs to be normally distributed in normal LMs: the residuals and *not* the raw response! If you are concerned about the normality of a response variable, you must *first* fit your model and *then* inspect the distribution of the residuals.

Another, equivalent way to write the same model is this:

$$mass_i \sim Normal(\mu_i, \sigma^2), \text{ with } \mu_i = \alpha + \beta \cdot reg_i.$$

This resembles more the way in which we will usually specify the model in our model-fitting engines (see Chapter 4). It emphasizes that the expected response μ_i is not identical for every snake. We also see that $\alpha + \beta \cdot reg_i$ represents the deterministic part of the response, while $Normal(\ldots, \sigma^2)$ denotes its stochastic or noise part.

Finally, we actually fit the model in R:

```
lm(mass ~ reg)

Coefficients:
 (Intercept)    reg2
          65      35
```

We recognize the value of the intercept as the mean mass of snakes in region 1 and the parameter called `reg2` as the difference between the mean mass in region 2 and that in region 1, that is, the *effect* of region 2, or here with two groups only, the effect of region. We may call this an *effects parameterization* (though the official name in R is *treatment contrast parameterization*).

Importantly, this is not the only way in which the LM for our two-group comparison can be specified. We could also write in R the following (and immediately inspect the design matrix):

```
lm(mass ~ reg - 1)    # not shown
model.matrix( ~ reg - 1)

> model.matrix( ~ reg-1)
        reg1    reg2
1         1       0
2         1       0
3         1       0
4         1       0
5         0       1
6         0       1
[...]
```

This is a different *parameterization* of the same basic model. It corresponds to the following set of equations:

$$\begin{pmatrix} 60 \\ 80 \\ 50 \\ 70 \\ 90 \\ 110 \end{pmatrix} = \begin{pmatrix} 1 & 0 \\ 1 & 0 \\ 1 & 0 \\ 1 & 0 \\ 0 & 1 \\ 0 & 1 \end{pmatrix} \begin{pmatrix} \alpha \\ \beta \end{pmatrix} + \begin{pmatrix} \varepsilon_1 \\ \varepsilon_2 \\ \varepsilon_3 \\ \varepsilon_4 \\ \varepsilon_5 \\ \varepsilon_6 \end{pmatrix}$$

What do the parameters α and β mean now? The interpretation of α has not changed: it is still the expected mass of a snake in region 1. However, the interpretation of parameter β is now different: it gives directly the expected mass of a snake in region 2. Thus, in this parameterization of the same LM, the two parameters directly represent the group means. Therefore we call this a *means parameterization* of the LM.

We fit the model in this form:

```
lm(mass ~ reg - 1)   # Fit the model to get parameter estimates

Call:
lm(formula = mass ~ reg - 1)

Coefficients:
reg1    reg2
  65     100
```

Why would we want to use different parameterizations of the same model? The answer is that they serve different aims: the effects parameterization is more useful for *testing for a difference* between the means in the two regions. This is equivalent to testing whether the effect of region 2, that is, parameter β in the effects parameterization, is equal to zero. In contrast, for a summary of the analysis, the means parameterization might be more useful for reporting the estimated mean snake mass for each region. However, the two models are equivalent and the sum of the two effects in the effects parameterization is equal to the value of the second parameter in the means parameterization, that is, $65 + 35 = 100$.

It is valuable to see the following connection between the two parameterizations of this model: they are both constrained versions of a more general, but overparameterized model, which has an intercept (α) and two parameters for reg, one for region 1 (β_1) and the other for region 2 (β_2).

$$
\begin{pmatrix} 60 \\ 80 \\ 50 \\ 70 \\ 90 \\ 110 \end{pmatrix} = \begin{pmatrix} 1 & 1 & 0 \\ 1 & 1 & 0 \\ 1 & 1 & 0 \\ 1 & 1 & 0 \\ 1 & 0 & 1 \\ 1 & 0 & 1 \end{pmatrix} \begin{pmatrix} \alpha \\ \beta_1 \\ \beta_2 \end{pmatrix} + \begin{pmatrix} \varepsilon_1 \\ \varepsilon_2 \\ \varepsilon_3 \\ \varepsilon_4 \\ \varepsilon_5 \\ \varepsilon_6 \end{pmatrix}
$$

This model is overparameterized, since with it we would try to estimate three parameters from only two groups (in addition to the residual standard deviation σ). Thus we need to reduce the number of estimated parameters by 1 by adding a constraint, which means that we fix one of the parameters at some value rather than estimating it. One possible constraint is to set to 0 one of the effects of reg. A common choice, and the default in R, is to set to 0 the effect of the first level in a factor, which here effectively drops the second column in the design matrix and the second element in the parameter vector. This results in the effects parameterization. Another choice is to set to 0 the intercept and thus remove it from the model, and this produces the means parameterization.

The same applies in LMs for any number of factors and with any number of levels: we cannot fit both an intercept and one parameter for each level in each factor, since this would result in an overparameterized model. Instead, constraints must be imposed by setting to 0 some parameters. This is dealt with automatically in R when using functions such as lm(). In contrast, when fitting these models in software such as JAGS or TMB, we must specify the appropriate constraints ourselves (see Chapter 8).

3.4.3 Simple linear regression: modeling the effect of one continuous covariate

To examine the relationship between a continuous response (mass) and a continuous explanatory variable such as snout-vent length (length), we would specify a simple linear regression in R (see also Chapter 5). It's called simple because it only has one such covariate, while it would be called multiple linear regression when there are two or more covariates.

```
lm(mass ~ length)
```

We inspect the design matrix that R builds on issuing the above call to `lm()`:

```
model.matrix(~ length)
```

```
  (Intercept)   length
1           1       40
2           1       45
3           1       39
4           1       50
5           1       52
6           1       56
[...]
```

The design matrix contains an intercept and another column with the values of the covariate `length`. The interpretation of the parameters α and β, respectively, is thus that of a baseline, representing the expected value of the response (`mass`) at a covariate value of `length` = 0, and an effect of `length`, representing the change in `mass` for each unit change in `length`. Equivalently, this effect of `length` is the slope of the regression of `mass` on `length`, while α is the intercept.

Fitting the simple linear regression model implies solving the following system of equations, subject to minimizing the sum of the squared residuals or maximizing the associated likelihood:

$$\begin{pmatrix} 60 \\ 80 \\ 50 \\ 70 \\ 90 \\ 110 \end{pmatrix} = \begin{pmatrix} 1 & 40 \\ 1 & 45 \\ 1 & 39 \\ 1 & 50 \\ 1 & 52 \\ 1 & 56 \end{pmatrix} \begin{pmatrix} \alpha \\ \beta \end{pmatrix} + \begin{pmatrix} \varepsilon_1 \\ \varepsilon_2 \\ \varepsilon_3 \\ \varepsilon_4 \\ \varepsilon_5 \\ \varepsilon_6 \end{pmatrix}$$

This model can be written algebraically in the same way as that underlying the two-group comparison we just saw:

$$mass_i = \alpha + \beta \cdot length_i + \varepsilon_i, \quad \text{with} \quad \varepsilon_i \sim Normal(0, \sigma^2)$$

Or as this:

$$mass_i \sim Normal(\mu_i, \sigma^2), \quad \text{with} \quad \mu_i = \alpha + \beta \cdot length_i$$

The only difference to the two-group model lies in the contents of the explanatory variable, `length`, which may now contain any real number (in the specific case of length, any positive number) rather than just the two values of 0 and 1. Here, `length` contains the lengths for each of the six snakes, while in the previous section it was a dummy variable for region 2.

Finally, we fit the model in R:

```
lm(mass ~ length)

Call:
lm(formula = mass ~ length)

Coefficients:
(Intercept)       length
    -59.066        2.888
```

Note that the intercept is biological nonsense, since it tells us that a snake of zero length weighs on average minus 59 grams! This illustrates that often an LM is only a useful characterization of a biological relationship over a restricted range of the explanatory variables. To give the intercept a more meaningful interpretation, we could reparameterize the model by transforming `length` to `length-mean(length)`. Fitting such a centered version of `length` will cause the intercept to become the expected mass of a snake at the *average* of the observed size distribution. We can do this by using `I()` inside the model formula to do the centering on the fly.

```
lm(mass ~ I(length - mean(length)))

Call:
lm(formula = mass ~ I(length - mean(length)))

Coefficients:
   (Intercept)   I(length - mean(length))
        76.667                      2.888
```

Now we see that a snake with average length is expected to weigh 76.7 grams, which corresponds to the parameter estimate in the "model of the mean" above.

We have seen above that in a model with a factor we can simply switch between different parameterizations by subtracting the intercept. In contrast, removing the intercept in a simple linear regression changes the model by forcing the regression line to go through the origin, that is, a snake with length 0 then has a mass of 0 by assumption. While this may appear to be a sensible model at a first glance, it assumes that the relationship between response and continuous explanatory variable X is linear all the way from zero and up to the maximum value of X. This is rarely the case and thus this variant of an LM should be used only when you have strong reasons for adopting it.

```
model.matrix( ~ length - 1)

  length
1     40
2     45
3     39
4     50
5     52
6     56
```

3.4.4 Modeling the effects of one categorical covariate, or factor

This is just a variant of the two-group comparison above, but now our factor has more than two levels. In the context of least-squares analyses, this LM is also known as a one-way analysis of variance or one-way ANOVA (Zar, 1998, see also Chapter 7). For illustration, let's examine the relationship between `mass` and `pop`, which is a factor with three levels. Once again our focus will be the differences between the effects and the means parameterizations.

In R, when we ask for a one-way ANOVA with factor `pop`, we get the default effects parameterization:

```
lm(mass ~ pop)   # not shown
```

Inspecting the design matrix clarifies the precise nature of the model fitted and shows that in order to obtain an identifiable model, R has automatically dropped (i.e., set to 0) the effect of `pop1` (see Section 3.4.2).

```
model.matrix(~pop)
```

```
  (Intercept)   pop2   pop3
1           1      0      0
2           1      0      0
3           1      1      0
4           1      1      0
5           1      0      1
6           1      0      1
[...]
```

Here is the effects parameterization of the ANOVA model for our toy snakes in vector/matrix notation:

$$\begin{pmatrix} 60 \\ 80 \\ 50 \\ 70 \\ 90 \\ 110 \end{pmatrix} = \begin{pmatrix} 1 & 0 & 0 \\ 1 & 0 & 0 \\ 1 & 1 & 0 \\ 1 & 1 & 0 \\ 1 & 0 & 1 \\ 1 & 0 & 1 \end{pmatrix} \begin{pmatrix} \alpha \\ \beta_2 \\ \beta_3 \end{pmatrix} + \begin{pmatrix} \varepsilon_1 \\ \varepsilon_2 \\ \varepsilon_3 \\ \varepsilon_4 \\ \varepsilon_5 \\ \varepsilon_6 \end{pmatrix}$$

Both make clear the meaning of the parameters α, β_2, and β_3, as the expected mass in population 1 and then the differences in the expected mass in populations 2 and 3 with respect to population 1. They also clarify that we cannot simultaneously have a parameter β_1 for population 1 and also an intercept, since this would result in an overparameterized model—to describe three groups, we can only have three, not four parameters (see Section 3.4.2).

There are different equivalent ways to write this model algebraically; see also Chapter 7. We could write the effects parameterization as follows, where `pop2` and `pop3` are indices (dummy variables) for membership of snake i in population 2 and 3, respectively:

$$mass_i = \alpha + \beta_2 \cdot pop2_i + \beta_3 \cdot pop3_i + \varepsilon_i, \quad \text{with} \quad \varepsilon_i \sim Normal(0, \sigma^2)$$

Another way to specify the same model is this:

$$mass_i \sim Normal(\alpha + \beta_2 \cdot pop2_i + \beta_3 \cdot pop3_i, \sigma^2).$$

We now fit the one-way ANOVA linear model in the effects parameterization.

```
lm(mass ~ pop)   # Fit model in effects parameterization (R default)
```

```
Coefficients:
  (Intercept)    pop2   pop3
           70     -10     30
```

Here is how we can fit the means parameterization of a one-way ANOVA with factor `pop`:

```
lm(mass ~ pop - 1)   # not shown
```

The design matrix shows that now we drop the intercept to obtain an identifiable model with three parameters for three group means.

```
model.matrix(~ pop - 1)

    pop1   pop2   pop3
1     1      0      0
2     1      0      0
3     0      1      0
4     0      1      0
5     0      0      1
6     0      0      1
[...]
```

Translated into vector/matrix notation we get this:

$$\begin{pmatrix} 60 \\ 80 \\ 50 \\ 70 \\ 90 \\ 110 \end{pmatrix} = \begin{pmatrix} 1 & 0 & 0 \\ 1 & 0 & 0 \\ 0 & 1 & 0 \\ 0 & 1 & 0 \\ 0 & 0 & 1 \\ 0 & 0 & 1 \end{pmatrix} \begin{pmatrix} \beta_1 \\ \beta_2 \\ \beta_3 \end{pmatrix} + \begin{pmatrix} \varepsilon_1 \\ \varepsilon_2 \\ \varepsilon_3 \\ \varepsilon_4 \\ \varepsilon_5 \\ \varepsilon_6 \end{pmatrix}$$

Here now, the three parameters β_1, β_2, and β_3 directly represent the expected mean mass of a snake in populations 1–3.

We can write this model in algebra as follows, where pop1, pop2, and pop3 are the indices for membership of snake i in population 1, 2, or 3, respectively, and where the intercept has implicitly been set to 0 (i.e., $\alpha = 0$) to avoid overparameterization.

$$mass_i = \beta_1 \cdot pop1_i + \beta_2 \cdot pop2_i + \beta_3 \cdot pop3_i + \varepsilon_i, \quad \text{with} \quad \varepsilon_i \sim Normal(0, \sigma^2)$$

Or as

$$mass_i \sim Normal(\beta_1 \cdot pop1_i + \beta_2 \cdot pop2_i + \beta_3 \cdot pop3_i, \sigma^2).$$

Here is the one-way ANOVA linear model fit in the means parameterization.

```
lm(mass ~ pop - 1)   # Fit model in means parameterization

Coefficients:
    pop1   pop2   pop3
      70     60    100
```

3.4.5 Modeling the effects of two factors

This model is also known as a two-way ANOVA model in the context of an LM for a normal response fit with least squares (Zar, 1998). A two-way ANOVA linear model serves to examine the relationships between a continuous response such as mass and two factors, such as region (reg) and habitat (hab) in our example (see also Chapter 8). Importantly, there are two different ways in which to combine the effects of two explanatory variables: additive (resulting in what is also called a *main-effects model*) or multiplicative (resulting in an *interaction-effects model*). In addition, we can specify these models using an effects or a means parameterization. We will consider each one in turn and for once begin with fitting each model directly.

Here is how we fit a main-effects ANOVA with reg and hab in R:

```
lm(mass ~ reg + hab)

Call:
lm(formula = mass ~ reg + hab)

Coefficients:
  (Intercept)    reg2    hab2    hab3
         65.0    35.0     2.5    -2.5
```

We look at the design matrix of the main-effects model.

```
model.matrix( ~ reg + hab)

  (Intercept)    reg2    hab2    hab3
1           1       0       0       0
2           1       0       1       0
3           1       0       0       1
4           1       0       0       0
5           1       1       1       0
6           1       1       0       1
[...]
```

To avoid overparameterization, the effects of one level for each factor need to be set to zero. The effects of the remaining levels then get the interpretation of *differences* relative to the base level. In principle, it does not matter which level is used as a baseline or reference, but often stats programs use the first or the last level of each factor. By default, R always sets the effect of the first level to zero.

Here is the system of equations that R solves for us when fitting the model with lm():

$$\begin{pmatrix} 60 \\ 80 \\ 50 \\ 70 \\ 90 \\ 110 \end{pmatrix} = \begin{pmatrix} 1 & 0 & 0 & 0 \\ 1 & 0 & 1 & 0 \\ 1 & 0 & 0 & 1 \\ 1 & 0 & 0 & 0 \\ 1 & 1 & 1 & 0 \\ 1 & 1 & 0 & 1 \end{pmatrix} \begin{pmatrix} \alpha \\ \beta_2 \\ \delta_2 \\ \delta_3 \end{pmatrix} + \begin{pmatrix} \varepsilon_1 \\ \varepsilon_2 \\ \varepsilon_3 \\ \varepsilon_4 \\ \varepsilon_5 \\ \varepsilon_6 \end{pmatrix}$$

In our snake toy data set, for the mass of individual i we can write this model as follows:

$$mass_i = \alpha + \beta_2 \cdot reg2_i + \delta_2 \cdot hab2_i + \delta_3 \cdot hab3_i + \varepsilon_i, \text{ with } \varepsilon_i \sim Normal(0, \sigma^2)$$

Here, α is the expected mass of a snake in habitat 1 and region 1. Since we have only two regions, there is only one parameter for factor reg, which is β_2. It specifies the difference in the expected mass between snakes in region 2 and those in region 1. We need two parameters (δ_2, δ_3) to specify the differences in the expected mass for snakes in habitats 2 and 3, respectively, relative to those in habitat 1 and region 1, that is, the intercept α.

Next, we inspect the model with interactive effects, which lets the effect of one factor level depend on the level of the other factor. In R, we write the default effects parameterization like this:

```
lm(mass ~ reg * hab)

Call:
lm(formula = mass ~ reg * hab)

Coefficients:
  (Intercept)        reg2   hab2   hab3
           65          60     15    -15
    reg2:hab2   reg2:hab3
          -50          NA
```

The design matrix for this model is shown next:

```
model.matrix(~ reg * hab)

    (Intercept)   reg2   hab2   hab3   reg2:hab2   reg2:hab3
1             1      0      0      0           0           0
2             1      0      1      0           0           0
3             1      0      0      1           0           0
4             1      0      0      0           0           0
5             1      1      1      0           1           0
6             1      1      0      1           0           1
[...]
```

And here is the model written in vector/matrix notation:

$$
\begin{pmatrix} 60 \\ 80 \\ 50 \\ 70 \\ 90 \\ 110 \end{pmatrix} = \begin{pmatrix} 1 & 0 & 0 & 0 & 0 & 0 \\ 1 & 0 & 1 & 0 & 0 & 0 \\ 1 & 0 & 0 & 1 & 0 & 0 \\ 1 & 0 & 0 & 0 & 0 & 0 \\ 1 & 1 & 1 & 0 & 1 & 0 \\ 1 & 1 & 0 & 1 & 0 & 1 \end{pmatrix} \begin{pmatrix} \alpha \\ \beta_2 \\ \delta_2 \\ \delta_3 \\ \gamma_{22} \\ \gamma_{23} \end{pmatrix} + \begin{pmatrix} \varepsilon_1 \\ \varepsilon_2 \\ \varepsilon_3 \\ \varepsilon_4 \\ \varepsilon_5 \\ \varepsilon_6 \end{pmatrix}
$$

We see that one parameter is not estimable. The reason for this is that in the 2-by-3 table of effects of region crossed with habitat, we lack snake observations for habitat 1 in region 2; do `table(reg, hab)` to check this out. Algebraically, we assume that the mass of individual i can be broken down into the following sum:

$$
mass_i = \alpha + \beta_2 \cdot reg2_i + \delta_2 \cdot hab2_i + \delta_3 \cdot hab3_i
$$
$$
+ \gamma_{22} \cdot reg2 \cdot hab2_i + \gamma_{23} \cdot reg2 \cdot hab3_i + \varepsilon_i,
$$
$$
\text{with } \varepsilon_i \sim Normal(0, \sigma^2).
$$

To obtain a model that is identifiable "in principle" for data obtained for the six combinations of the levels of the two factors `reg` and `hab`, R has automatically set to 0 the first levels of main and interaction effects, that is, imposed the constraints $\beta_1 = 0$, $\delta_1 = 0$, and $\gamma_{21} = 0$. The meanings of parameters α, β_2, δ_2, and δ_3 remain as before, that is, they specify the main effects of the levels for the `habitat` and `reg` factors other than the first. The new coefficients, γ_{22} and γ_{23}, specify the *interaction effects* between these two factors. Since one of the cells in the cross-classification of `reg` and `hab` has no observations, we cannot estimate both interaction effects. As a consequence, one of them is given as missing on fitting the model.

You may ask why one of γ_{22} and γ_{23} cannot be estimated when in fact we have no observations for the combination of region 2 and habitat 1 and this would correspond to a parameter γ_{21}. The reason is that since we lack information on γ_{21} in our data set, setting that term to 0 does not prevent overparameterization. Indeed, the term for `reg2:hab3` is completely linearly dependent on `reg2` and `reg2:hab2`, as you find out when doing `alias(lm(mass ~ reg*hab))`; thus it cannot be estimated. The details of this are beyond the scope of our book, and if you want to go deeper, read up on *aliasing* in linear models (McCullagh & Nelder, 1989).

Finally, here is the means parameterization of the interaction-effects model:

```
lm(mass ~ reg * hab - 1 - reg - hab)
```

```
Coefficients:
   reg1:hab1    reg2:hab1    reg1:hab2    reg2:hab2    reg1:hab3    reg2:hab3
          65           NA           80           90           50          110
```

Here is the design matrix of this model:

```
model.matrix(mass ~ reg * hab - 1 - reg - hab)
```

```
   reg1:hab1    reg2:hab1    reg1:hab2    reg2:hab2    reg1:hab3    reg2:hab3
1          1            0            0            0            0            0
2          0            0            1            0            0            0
3          0            0            0            0            1            0
4          1            0            0            0            0            0
5          0            0            0            1            0            0
6          0            0            0            0            0            1
[ ... ]
```

We can translate this in vector/matrix notation as follows:

$$
\begin{pmatrix} 60 \\ 80 \\ 50 \\ 70 \\ 90 \\ 110 \end{pmatrix} = \begin{pmatrix} 1 & 0 & 0 & 0 & 0 & 0 \\ 0 & 0 & 1 & 0 & 0 & 0 \\ 0 & 0 & 0 & 0 & 1 & 0 \\ 1 & 0 & 0 & 0 & 0 & 0 \\ 0 & 0 & 0 & 1 & 0 & 0 \\ 0 & 0 & 0 & 0 & 0 & 1 \end{pmatrix} \begin{pmatrix} \alpha_{11} \\ \alpha_{21} \\ \alpha_{12} \\ \alpha_{22} \\ \alpha_{13} \\ \alpha_{23} \end{pmatrix} + \begin{pmatrix} \varepsilon_1 \\ \varepsilon_2 \\ \varepsilon_3 \\ \varepsilon_4 \\ \varepsilon_5 \\ \varepsilon_6 \end{pmatrix}
$$

We see clearly the lack of any information about effect α_{21} in this parameterization of the model. This parameter would be the expected mass of a snake in region 2 and habitat 1, represented by the second column in the design matrix, which contains zeroes only.

This model can be written algebraically like this:

$$mass_i = \alpha_{11} \cdot reg1_i \cdot hab1_i + \alpha_{21} \cdot reg2_i \cdot hab1_i +$$
$$\alpha_{12} \cdot reg1_i \cdot hab2_i + \alpha_{22} \cdot reg2_i \cdot hab2_i + \quad , \text{ with } \varepsilon_i \sim Normal(0, \sigma^2)$$
$$\alpha_{13} \cdot reg1_i \cdot hab3_i + \alpha_{23} \cdot reg2_i \cdot hab3_i + \varepsilon_i$$

There are six parameters α, corresponding to the six ways in which the levels of the two factors reg and habitat can be combined, and all covariates in the model are dummy covariables.

3.4.6 Modeling the effects of a factor and a continuous covariate

As a final case in our catalog of prototypical LMs, we now examine how we can specify the combined effects on mass of a factor (e.g., pop) and a continuous covariate like length. For a normal LM fit with least squares, this type of LM is called an analysis of covariance, or ANCOVA, model (Zar, 1998, see also Chapter 9). There are again two ways in which we might specify such a model. First, we may assume that the relationship between mass and length is the same in all populations, or alternatively, that the mass differences among populations do not depend on the length of a snake. In statistical terms, this would be represented by a *main-effects model*. Second, if we believed that the mass–length relationship might differ among populations or that the differences in mass among populations might depend on the length of a snake, we would fit an *interaction-effects model*. (Note that here "effects" has a slightly different meaning from that in effects *vs.* means parameterization.)

We start by looking at the main-effects ANCOVA model in the default effects parameterization. We can specify it in R as follows:

```
lm(mass ~ pop + length)    # not shown
```

It has the following design matrix:

```
model.matrix( ~ pop + length)

  (Intercept)  pop2  pop3  length
1           1     0     0      40
2           1     0     0      45
3           1     1     0      39
4           1     1     0      50
5           1     0     1      52
6           1     0     1      56
  [ ... ]
```

The parameter associated with the first column is the intercept and signifies the expected mass in population 1 for a snake of length zero. The parameters associated with the design matrix columns named pop2 and pop3 quantify the differences in the intercept between these populations and population 1. The parameter associated with the last column in the design matrix measures the common slope of mass on length for all snakes, regardless of population.

This translates into the following vector/matrix notation:

$$\begin{pmatrix} 60 \\ 80 \\ 50 \\ 70 \\ 90 \\ 110 \end{pmatrix} = \begin{pmatrix} 1 & 0 & 0 & 40 \\ 1 & 0 & 0 & 45 \\ 1 & 1 & 0 & 39 \\ 1 & 1 & 0 & 50 \\ 1 & 0 & 1 & 52 \\ 1 & 0 & 1 & 56 \end{pmatrix} \begin{pmatrix} \alpha \\ \beta_2 \\ \beta_3 \\ \delta \end{pmatrix} + \begin{pmatrix} \varepsilon_1 \\ \varepsilon_2 \\ \varepsilon_3 \\ \varepsilon_4 \\ \varepsilon_5 \\ \varepsilon_6 \end{pmatrix}$$

In algebra we can write the main-effects model in the effects parameterization as follows:

$$mass_i = \alpha + \beta_2 \cdot pop2_i + \beta_3 \cdot pop3_i + \delta \cdot length_i + \varepsilon_i, \text{ with } \varepsilon_i \sim Normal(0, \sigma^2)$$

This model says that the mass of snake i consists of a constant α plus the effects β_2 and β_3, respectively, when in population 2 and 3, plus a constant δ times length, plus the residual. The two β's are the differences of the intercepts in these populations from the intercept in population 1, while δ is the common slope of the mass–length relationship in all three populations.

We now fit this model in R. In addition, we illustrate how we can directly fit a design matrix with lm(), though we have to subtract the intercept, since lm() automatically adds one in a LM.

```
lm(mass ~ pop + length)                  # Default model fitting
DM <- model.matrix(~pop + length)        # create design matrix DM
lm(mass ~ DM-1)                          # Fit same model via explicit DM

Call:
lm(formula = mass ~ pop + length)

Coefficients:
  (Intercept)         pop2      pop3    length
      -34.938      -14.938     1.605     2.469

Call:
lm(formula = mass ~ DM - 1)

Coefficients:
  DM(Intercept)       DMpop2    DMpop3   DMlength
        -34.938      -14.938     1.605      2.469
```

So, the intercept is the expected mass of a snake with length zero in population 1. Parameters pop2 and pop3 are the differences in the intercept between populations 2 and 3 and that in population 1, and parameter length quantifies the slope of the mass–length relationship common to snakes in all populations. It is instructive to plot the estimates of the relationships for each population under this model (Fig. 3.5 left).

```
par(mfrow = c(1, 2))                              # Fig. 3.5 left
fm <- lm(mass ~ pop + length)                     # Refit model
plot(length, mass, col = c(rep("red", 2), rep("blue", 2), rep("green", 2)),
    pch = 16, cex = 1.5, ylim = c(40, 120), frame = FALSE)
abline(fm$coef[1], fm$coef[4], col = rgb(1, 0, 0, 0.5), lwd = 2)
abline(fm$coef[1] + fm$coef[2], fm$coef[4], col = rgb(0, 0, 1, 0.5), lwd = 2)
abline(fm$coef[1] + fm$coef[3], fm$coef[4], col = rgb(0, 1, 0, 0.5), lwd = 2)
```

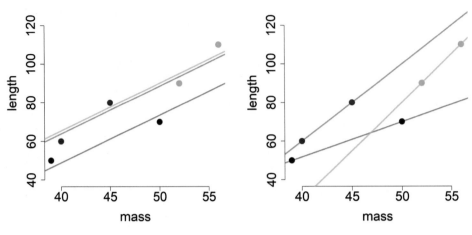

FIGURE 3.5

Lines of best fit under the main-effects (left) and the interaction-effects ANCOVA models (right) fit to our toy snake data set. The former has additive, or main, effects of factor pop and covariate length, while in the latter we have multiplicative, or interaction, effects.

This model assumes that the mass-length relationship differs among populations only in the average or baseline level, that is, the intercept, hence, the three lines are parallel. What we see then is that population 1 (red) hardly differs from population 3 (green), but that snakes in population 2 (blue) weigh less at a given length than do snakes in populations 1 or 3.

Next, we look at the interaction-effects ANCOVA model in R's default effects parameterization. We can specify it in R in a number of ways, including these:

```
lm(mass ~ pop * length)                    # Interaction-effects (not shown)
lm(mass ~ pop + length + pop:length)   # same, with R syntax for int. term
```

This model has the following design matrix:

```
model.matrix( ~ pop * length)

  (Intercept)  pop2  pop3  length  pop2:length  pop3:length
1           1     1     0      40            0            0
2           1     1     0      45            0            0
3           1     1     1      39           39            0
4           1     1     1      50           50            0
5           1     0     1      52            0           52
6           1     0     1      56            0           56
[ ... ]
```

The parameters associated with the first three columns in this design matrix signify the population 1 intercept and the effects of populations 2 and 3, respectively, that is, the difference in intercepts between the three regression lines. The parameter associated with the fourth column, length, is the slope of the mass–length regression in the first population, and the parameters associated with the last two columns are the differences between the slopes in populations 2 and 3 relative to the slope in population 1.

This translates into the following vector/matrix notation:

$$
\begin{pmatrix} 60 \\ 80 \\ 50 \\ 70 \\ 90 \\ 110 \end{pmatrix} = \begin{pmatrix} 1 & 0 & 0 & 40 & 0 & 0 \\ 1 & 0 & 0 & 45 & 0 & 0 \\ 1 & 1 & 0 & 39 & 39 & 0 \\ 1 & 1 & 0 & 50 & 50 & 0 \\ 1 & 0 & 1 & 52 & 0 & 52 \\ 1 & 0 & 1 & 56 & 0 & 56 \end{pmatrix} \begin{pmatrix} \alpha \\ \beta_2 \\ \beta_3 \\ \delta \\ \gamma_2 \\ \gamma_3 \end{pmatrix} + \begin{pmatrix} \varepsilon_1 \\ \varepsilon_2 \\ \varepsilon_3 \\ \varepsilon_4 \\ \varepsilon_5 \\ \varepsilon_6 \end{pmatrix}
$$

In algebra we can write the interaction-effects ANCOVA model in the effects parameterization as follows:

$$mass_i = \alpha + \beta_2 \cdot pop2_i + \beta_3 \cdot pop3_i + \delta \cdot length_i +$$
$$\gamma_2 \cdot length_i \cdot pop2_i + \gamma_3 \cdot length_i \cdot pop3_i + \varepsilon_i$$
, with $\varepsilon_i \sim Normal(0, \sigma^2)$

In this model, α is the intercept for population 1 and vector β_2 and β_3 represent the difference in the intercept between population 1, and 2 and 3, respectively. Parameter δ is the slope of the mass–length relationship in the first population. Parameters γ_2 and γ_3 correspond to the difference in the slope between population 1, and 2 and 3, respectively.

Next, we fit the interaction-effects models in the effects parameterization:

```
lm(mass ~ pop * length)

Call:
lm(formula = mass ~ pop * length)

Coefficients:
  (Intercept)       pop2       pop3   length   pop2:length   pop3:length
     -100.000     79.091    -70.000    4.000        -2.182         1.000
```

The first and the fourth parameters describe intercept and slope of the relationship between mass and length in the first population, while the remainder refers to intercept and slope *differences* between the other two populations and those of the baseline population.

We plot the estimated relationships also under the interaction model (Fig. 3.5 right). We see a cluster of regression lines that are not parallel:

```
fm <- lm(mass ~ pop * length)              # Refit model, Fig. 3.5 right
plot(length, mass, col = c(rep("red", 2), rep("blue", 2), rep("green", 2)),
   pch = 16, cex = 1.5, ylim = c(40, 120), frame = FALSE)
abline(fm$coef[1], fm$coef[4], col = rgb(1, 0, 0, 0.5), lwd = 2)
abline(fm$coef[1]+ fm$coef[2], fm$coef[4] + fm$coef[5], col = rgb(0, 0, 1, 0.5), lwd = 2)
abline(fm$coef[1]+ fm$coef[3], fm$coef[4] + fm$coef[6], col = rgb(0, 1, 0, 0.5), lwd = 2)
```

As an aside, the interaction-effects ANCOVA is not a very useful statistical model for our toy data set, since it has $R^2 = 1$ and thus explains all the variability in the response. But the example does serve to illustrate the two kinds of assumptions about homogeneity (or not) of slopes that one may examine in an ANCOVA model.

Next, we fit the main-effects ANCOVA model in the means parameterization. In R, we code this LM as follows.

```
lm(mass ~ pop + length - 1)    # not shown
```

Its design matrix looks like this:

```
model.matrix( ~ pop + length - 1)

     pop1   pop2   pop3   length
1     1      0      0       40
2     1      0      0       45
3     0      1      0       39
4     0      1      0       50
5     0      0      1       52
6     0      0      1       56
[...]
```

The parameters associated with the first three columns in the design matrix now directly represent the intercepts for each population, while that associated with the fourth column denotes the common slope of the mass-length relationship.

In vector/matrix notation, we can write this model as follows:

$$\begin{pmatrix} 60 \\ 80 \\ 50 \\ 70 \\ 90 \\ 110 \end{pmatrix} = \begin{pmatrix} 1 & 0 & 0 & 40 \\ 1 & 0 & 0 & 45 \\ 0 & 1 & 0 & 39 \\ 0 & 1 & 0 & 50 \\ 0 & 0 & 1 & 52 \\ 0 & 0 & 1 & 56 \end{pmatrix} \begin{pmatrix} \alpha_1 \\ \alpha_2 \\ \alpha_3 \\ \delta \end{pmatrix} + \begin{pmatrix} \varepsilon_1 \\ \varepsilon_2 \\ \varepsilon_3 \\ \varepsilon_4 \\ \varepsilon_5 \\ \varepsilon_6 \end{pmatrix}$$

In algebra, we can write it as follows:

$$mass_i = \alpha_1 \cdot pop1_i + \alpha_2 \cdot pop2_i + \alpha_3 \cdot pop3_i + \delta \cdot length_i + \varepsilon_i, \text{ with } \varepsilon_i \sim Normal(0, \sigma^2)$$

The three α's represent the intercepts in each population, while δ is the common slope. To fit this model in R, we can type this:

```
lm(mass ~ pop + length - 1)

Coefficients:
      pop1       pop2       pop3   length
   -34.938    -49.877    -33.333    2.469
```

This gives us the estimates of each individual intercept, plus the slope common to all three populations.

As a fourth and final case of an ANCOVA linear model, we now examine the interaction-effects model in the means parameterization. In R, we code this LM as follows.

```
lm(mass ~ pop * length - 1 - length)   # not shown
```

Its design matrix looks like this:

```
model.matrix( ~ pop * length - 1 - length)
```

	pop1	pop2	pop3	pop1:length	pop2:length	pop3:length
1	1	0	0	40	0	0
2	1	0	0	45	0	0
3	0	1	0	0	39	0
4	0	1	0	0	50	0
5	0	0	1	0	0	52
6	0	0	1	0	0	56
[...]						

This parameterization of the ANCOVA model with interaction between `pop` and `length` defines parameters that have the direct interpretation as the three sets of intercepts and slopes for the three mass-length regression lines, one in each population.

In vector/matrix notation, we can write this model as follows:

$$
\begin{pmatrix} 60 \\ 80 \\ 50 \\ 70 \\ 90 \\ 110 \end{pmatrix} = \begin{pmatrix} 1 & 0 & 0 & 40 & 0 & 0 \\ 1 & 0 & 0 & 45 & 0 & 0 \\ 0 & 1 & 0 & 0 & 39 & 0 \\ 0 & 1 & 0 & 0 & 50 & 0 \\ 0 & 0 & 1 & 0 & 0 & 52 \\ 0 & 0 & 1 & 0 & 0 & 56 \end{pmatrix} \begin{pmatrix} \alpha_1 \\ \alpha_2 \\ \alpha_3 \\ \delta_1 \\ \delta_2 \\ \delta_3 \end{pmatrix} + \begin{pmatrix} \varepsilon_1 \\ \varepsilon_2 \\ \varepsilon_3 \\ \varepsilon_4 \\ \varepsilon_5 \\ \varepsilon_6 \end{pmatrix}
$$

In algebra, we can write it as follows:

$$mass_i = \alpha_1 \cdot pop1_i + \alpha_2 \cdot pop2_i + \alpha_3 \cdot pop3_i +$$
$$\delta_1 \cdot length_i \cdot pop1_i + \delta_2 \cdot length_i \cdot pop2_i + \delta_3 \cdot length_i \cdot pop3_i + \varepsilon_i$$

, with $\varepsilon_i \sim \text{Normal}(0, \sigma^2)$

The three α's represent the intercepts in each population, while the three δ's represent the three slopes. To fit this model in R, we can type this:

```
lm(mass ~ pop * length - 1 - length)
summary(lm(mass ~ pop * length - 1 - length))   # not shown

Call:
lm(formula = mass ~ pop * length - 1 - length)

Coefficients:
         pop1        pop2        pop3   pop1:length   pop2:length   pop3:length
     -100.000     -20.909    -170.000         4.000         1.818         5.000
```

These estimates have direct interpretations as the intercept and the slope of the three regressions of `mass` on `length`. And we must remember that we cannot estimate any SEs from this saturated model. You can see this when executing the second line in the R code above.

3.5 Brief overview of linear, generalized linear, (generalized) linear mixed, hierarchical, and integrated models

This book covers six large classes of statistical regression models that are widely used in ecology, wildlife management, and related sciences: ordinary, or normal, linear models (LMs), GLMs, LMMs, GLMMs, HMs, and IMs. When you add to this the different statistical distributions that we may adopt for a response and the different kinds of explanatory variables (e.g., continuous or factors) and ways in which we can combine them (e.g., with main effects or with interactions), then a huge and bewildering zoo of models arises. It is easy to become overwhelmed by such a multitude of models if you see them just as unrelated procedures. Key to not getting lost in this zoo is to recognize the relationships among these models. This lets you make smart decisions as to which model is useful for you and allows you to pick, or build from scratch, exactly the model that you need. Here, we provide a brief overview of these models, of how they are similar and of what makes them distinct.

In the previous sections, we have described three fundamental statistical modeling concepts: *statistical distributions* that we use to describe the stochastic nature of our RVs, *link functions* to transform the parameters in these distributions, and *linear models* for the effects of covariates on the parameters. Virtually all the differences among the six classes of models in this book can be explained by differences in terms of these three criteria, plus, by whether or not we put RVs in sequence, or into a hierarchy, leading to a mixed model or HM. Table 3.2 shows how the first four classes of these models are related. They occupy the cells of a cross-classification obtained with just two criteria:

- Whether a response is normal (Gaussian) or whether it may also follow a different statistical distribution from the so-called exponential family (McCullagh & Nelder, 1989) such as a binomial or Poisson. This distinguishes the top from the bottom row in the body of the table.
- Whether there is only a single source of random variability in the data or whether there are two or more random processes. This distinguishes the left from the right column in the body.

Table 3.2 Classification of four core classes of regression models used for applied statistical modeling. Also shown are the book chapters/sections wherein examples of each class of model are covered. Modified from Kéry, M. (2010). *Introduction to WinBUGS for ecologists. - A Bayesian approach to regression, ANOVA, mixed models and related analyses.* **Academic Press, Burlington.**

	Single random process	Two or more random processes
Normal response	Linear model (LM): chap. 4–9 (not 7.3)	Linear mixed model (LMM): chap. 7.3, 10
Exponential family response	Generalized linear model (GLM): chap. 11–13, 15, 16	Generalized linear mixed model (GLMM): chap. 14, 17

Going from top to bottom we go from a simpler or more special case to a more general case, while from left to right we take a non-HM and add one or more sets of random effects to arrive at an HM. The relationships among LMs, GLMs, LMMs, and GLMMs include the following:

- An LM is a special case of a GLM, where the distribution of the response must be normal (Gaussian) and where the link function is the identity (see Section 3.3).
- A GLM is a generalization of an LM, as the name says, where the response may follow a distribution other than a normal, and where we may transform the parameter of the distribution by a link function other than the identity.
- A LMM is an LM with at least one additional set of normal RVs, or random effects, which appear additively in the linear predictor of the response.
- Similarly, a GLMM is a GLM with at least one additional set of normal random effects which appear additively in the linear predictor of the response.
- If we allow the distribution of the response (but not of the random effects) in an LMM to be something other than a normal, for example, a Poisson or a binomial, we arrive at a GLMM.

As we have already argued earlier, the GLM has a very special place in applied statistical modeling (McCullagh & Nelder, 1989): it describes a set of RVs by a statistical distribution, a link function, and an additive combination of the effects of covariates in a model that is linear in the parameters at the link scale. These three defining concepts of a GLM appear over and over again also in far more complex statistical models, such as most HMs, which can be viewed as a combination of two or more GLMs in sequence. And at each level of an HM, we have to make the same decisions as in a regular GLM: we need to pick a suitable distribution for the RVs, a link function, and then we may specify a LM for the effects of covariates on a parameter in a submodel.

The final two classes of models in the book, general HMs and IMs, fall outside of this scheme. An HM is a statistical model for two or more sets of linked RVs, of which typically one is latent, that is, a random effect. As such, all mixed models are special cases of an HM. However, random effects in general HMs may be non-Gaussian (e.g., Poisson or binomial) and they may appear in the model for the observed data, that is, the response, in a nonadditive fashion. We will see such HMs illustrated in occupancy and N-mixture models in Chapters 19 and 19B.

Finally, while typical HMs have just a single response variable, IMs by definition have two or more. Thus, in IMs we combine multiple data sets which contain some information about one or more shared parameters. Many IMs are also HMs, but this need not always be so. In fact, in Chapter 20 we feature an IM that is not an HM. We like the metaphor of an HM as a sequence of two or more GLMs. An analogous metaphor for many IMs would be that of multiple GLMs branching out from a common root.

3.6 Summary and outlook

We have covered three basic regression modeling choices: statistical distributions to describe RVs, link functions to transform parameters, and LMs to specify the effects of covariates. All three are extremely general and occur in all models in this book, as well as in far more complex regression models all over ecology and wildlife science. In addition, they are also the defining features of the celebrated GLM, which forms the essential building block for most HMs.

Depending on how we make our choices for these three criteria, we may end up with any of the six big classes of regression models that we have briefly characterized in the last section. We have described the defining features of each in the hope that you will no longer have to memorize them just as unrelated specimens in a huge and bewildering zoo of statistical models, but instead can see their relationships and distinctions, or their family tree. To fledge as a creative statistical modeler, you will need to use all of them and from them tailor those models that best address your research questions or management problems.

After choosing a class of model for our analysis, such as GLM or mixed model, we also must specify the covariates of interest that may affect the response variable(s) in our study. Covariate relationships are by far most commonly specified in the form of LMs, that is, in linear combinations. LMs for covariate effects can be represented by the design or model matrix. Model matrices, and more generally the structure of LMs, are essential to understand. This is particularly true when you want to fit these models in general model-fitting engines such as those in our book, see Chapter 4. The LMs in this chapter were presented in a progression from simple to complex. Chapters 4–9 in the book follow that structure and show how to fit these models for normal responses, that is, for normal LMs, while in Chapters 11–13 we do the same for Poisson GLMs and in Chapters 15–17 for binomial GLMs.

Introduction to general-purpose model fitting engines and the "model of the mean" ✪

4

Chapter outline

4.1 Introduction

As modern ecologists, we have access to a vast library of ready-made (or as we will call them, "canned") fitting functions for statistical models that cover a wide range of data types, experimental designs, and hypotheses. Applying the software to fit such models typically requires only our data, perhaps setting a few custom options, and, crucially, only a limited understanding of the statistical model that is fit. The growth of the R package ecosystem (R Core Team, 2023) has been the most important contributor to the current library of canned statistical model-fitting functions available to us. Two or more decades ago the average ecologist was largely limited to a small set of models available in expensive programs such as SAS, SPSS, JMP, GenStat or Minitab; e.g., *t*-tests,

---

✪This book has a companion website hosting complementary materials, including all code for download. Visit this URL to access it: https://www.elsevier.com/books-and-journals/book-companion/9780443137150.

Applied Statistical Modelling for Ecologists. DOI: https://doi.org/10.1016/B978-0-443-13715-0.00009-1
109

regression, and analysis of variance. The available models in these programs to a large extent prescribed the types of analyses that we as ecologists would conduct, because if our data or question did not fit into one of the available 'boxes', then you either had to (rarely) discard the data set or (much more commonly) shoehorn it into one of the models that your software let you fit.

Now, there are free R packages available for a very wide range of analyses, both general and discipline-specific, without much need for a deep understanding of the underlying statistical and computational aspects. Without question, this increase in available analysis tools has largely benefited the quality and scope of ecological research. And yet, despite the diversity of canned models now available to ecologists in R and other software, they cannot cover every possible modeling need. If we find our research question requires a more bespoke treatment, one of a suite of what we will call *general-purpose model fitting engines* may be the solution. These engines are varied in their design, but all present us with the advantage, and challenge, of explicitly specifying the statistical model we wish to fit. This specification can take a variety of forms, such as writing out the math of a likelihood function, or writing the model in a sort of algebraic short-form. For example, in the BUGS language, we specify likelihoods, prior distributions and the relationships between all deterministic and stochastic components of a model. In all cases this requires a more thorough understanding of the statistical model to be applied than when fitting the same model in an equivalent canned R function, but it also allows for significantly more flexibility. If you master a general-purpose model fitting engine, this will free the modeler in you!

In this chapter, we begin by showing an example analysis with a canned R function. Next, we will describe and demonstrate five general-purpose model fitting engines that will be used throughout the book: JAGS (Plummer, 2003), NIMBLE (de Valpine et al., 2017), Stan (Carpenter et al., 2017), maximum likelihood in R, and Template Model Builder (TMB; Kristensen et al., 2016). Three of these (JAGS, NIMBLE, and Stan) predominantly fit models using Bayesian inference with simulation methods similar to MCMC shown in Chapter 2, while the other two use maximum likelihood. All five can be accessed from within R. We believe that there is value in presenting and comparing multiple engines rather than focusing on a single engine. Each engine has advantages and disadvantages, as well as different approaches to describing statistical models. Certain types of models will be easier to fit with some engines than with others, and you may find you have a preference for one particular syntax. We will show that with properly specified models, reasonable sample sizes, and vague priors for the Bayesian analyses, all engines should yield approximately the same numerical results.

We will demonstrate each engine with an analysis of one of the simplest possible models for a normal response – for what we call the "model of the mean", a model with just two parameters. More specifically, we estimate the mean and the dispersion (standard deviation or variance) parameter of a normal distribution, which we assume as a statistical model for a sample of measurements taken from a population. As you have seen already in Chapter 3, this is a normal linear regression model with just an intercept in the deterministic part of the response. Our first example will deal with body mass of male peregrines (Fig. 4.1).

4.2 Data generation

We simulate our example dataset in R. The code to generate a data set can be viewed as a very accurate description of a parametric statistical model, written in the R language. We believe that R

FIGURE 4.1

Male peregrine falcon (*Falco peregrinus calidus*) wintering on the balcony of a condominium in the French Mediterranean, Sète, 2008 (Photo by Jean-Marc Delaunay).

code to simulate a statistical model provides an incredibly powerful explanation of the model, especially for nonstatisticians.

The model in algebra looks like this, where y_i is the mass measurement for the i-th male peregrine ($i = 1, 2...n$):

$$y_i \sim Normal(\mu, \sigma^2)$$

Here, the mass measurements are defined to be independent and identically distributed (i.i.d) draws from a normal distribution with mean μ and variance σ^2 (we could alternatively specify the dispersion in terms of the square root of the variance, i.e., in terms of the standard deviation σ).

Male peregrines in Western Europe weigh on average about 600 g and Monneret (2006) gives a range of 500–680 g. The assumption of a normal distribution of body mass therefore implies a standard deviation of about 30 g, since half the range corresponds to about two standard deviations. We will create two variations of the dataset: one with a small sample size ($n = 10$) and the other with a large sample size ($n = 1000$) (Fig. 4.2).

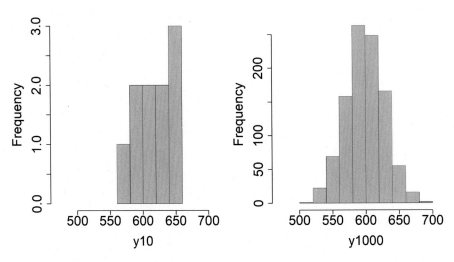

FIGURE 4.2

Histograms of the small (left) and the large (right) sample of weights of male peregrine falcons, which were all simulated as i.i.d. draws from a normal distribution with mean 600 and standard deviation 30.

```
# Generate two samples of body mass measurements of male peregrines
set.seed(4)
y10 <- rnorm(n = 10, mean = 600, sd = 30)          # Sample of 10 birds
y1000 <- rnorm(n = 1000, mean = 600, sd = 30)      # Sample of 1000 birds

# Save the data-generating values of the parameters for later comparisons
truth <- c(mean = 600, sd = 30)

# Plot data (Fig. 4.2)
xlim = c(450, 750)
par(mfrow = c(1, 2), mar = c(6, 6, 6, 3), cex.lab = 1.5, cex.axis = 1.5)
hist(y10, col = 'grey ', xlim = xlim, main = 'Body mass (g) of 10 male peregrines')
hist(y1000, col = 'grey', xlim = xlim, main = 'Body mass (g) of 1000 male peregrines')
```

If you want an identical simulated dataset, as well as identical analysis results, then you must use the same seed as we do. However, here, and indeed in all further examples, it is *extremely instructive* to execute the data-simulation code repeatedly and without a seed to experience *sampling error* – the variation in your data that stems from the fact that only part but not the whole of a heterogeneous population has been measured and different samples will thus differ. Sampling error, or sampling variance, is something absolutely central to statistics; and yet, it is among the most difficult concepts for ecologists to grasp, especially, since in practice we only ever observe a single sample from the distribution that characterizes this variation! It is astonishing to observe how *different* multiple realizations from the exact same random process can be – here, the sampling of 10 or 1000 male peregrines from an assumed infinite population of male peregrines. Also surprising is how far from normal the distribution of values in the smaller sample may look.

4.3 Analysis using canned functions in R

Before diving in to the general-purpose model fitting engines, we first demonstrate fitting the model of the mean using a canned function in R: `lm()`. This function uses a formula-based interface to define the specific linear model to be fit to the data. We use this so-called Wilkinson-Rogers (1973) linear model notation repeatedly with canned functions in later chapters. The left-hand side of the formula (i.e., on the left side of the tilde ~) denotes the response variable, or what we treat as the observed random variable in the analysis, and on the right-hand side we have the covariates and the structure of their effects in a linear model. Here we only estimate a single parameter in the linear model (the mean, or the intercept), so the right-hand side of the formula contains only the number 1, representing the intercept.

Technically, the function `lm()` uses least squares, not maximum likelihood. However for normal linear models such as ours, the solutions for the parameters in the deterministic part of the model correspond exactly to the MLEs, while the solution for the residual variance is almost the same as the MLE of the variance (the former divides the sums of squares by n-1, and the latter by n; Millar, 2011, and see also Section 5.9).

```
summary(out4.3 <- lm(y10 ~ 1))      # small data set: not shown
summary(out4.3 <- lm(y1000 ~ 1))    # large data set

[...]

Coefficients:
            Estimate Std. Error   t value  Pr(>|t|)
(Intercept) 598.8461     0.9176     652.6   <2e-16 ***
---
Signif. codes: 0 '***' 0.001 '**' 0.01 '*' 0.05 '.' 0.1 ' ' 1
Residual standard error: 29.02 on 999 degrees of freedom

# Save estimates
lm_est <- c(coef(out4.3), sigma = sigma(out4.3))
```

Estimates of the population mean (the intercept) and of the population standard deviation (here called the residual standard error) are very similar to the values we used to simulate the dataset for the larger, but a little less so for the smaller data set (not shown). In the next section, we will fit the same model using our first general-purpose model fitting engine: JAGS.

4.4 JAGS

4.4.1 Introduction to JAGS

JAGS stands for **J**ust **A**nother **G**ibbs **S**ampler and is an engine for fitting models in a Bayesian mode of inference. Before talking about JAGS, we will briefly discuss the historical context that led to its development (Lunn et al., 2013; Plummer, 2023b).

JAGS applies a lightly modified version of the original BUGS (**B**ayesian analysis **U**sing **G**ibbs **S**ampling) language. The BUGS language and command-line program was developed by epidemiologists in Cambridge, UK, in the 1990s (Gilks et al., 1994; Lunn et al., 2009). A Windows version of the software, called WinBUGS, was then developed (Spiegelhalter et al., 2003), after which the BUGS language began to see widespread use also among biologists. (Win)BUGS was a groundbreaking program, which for the first time made flexible and powerful Bayesian statistical modeling available to a large, and also non-statistical, audience.

Although the BUGS language is very simple, we can specify almost arbitrarily complex statistical models with it. We first specify in our BUGS code a series of *nodes* or quantities which represent things such as the input data and the parameters we want to estimate. Each node must either be a constant (i.e., data) or be described as a function of other "parent" nodes by defining either a stochastic or a deterministic relationship (Gilks et al., 1994; Lunn et al., 2013). This system of nodes and relations among them can be visualized as a *directed acyclic graph (DAG)*, and indeed generating such a graph is a major feature of WinBUGS. A DAG can be powerful for clarifying the structure of a statistical model (Hobbs & Hooten, 2015) or hypothesized cause-effect relationships in scientific research (Pearl, 2017, 2021). From the DAG, an MCMC algorithm is developed and used to produce random samples from the joint posterior distribution of all parameters (Lunn et al., 2013; Hooten & Hefley, 2019).

WinBUGS development ceased in 2005, and it was succeeded by an open-source version called OpenBUGS (Spiegelhalter et al., 2007) which runs on both Windows and Linux. At the time of writing, development of OpenBUGS has ceased as well, but a new branch of the BUGS project emerged as multiBUGS (Goudie et al., 2020). At the turn of the century, development of new software applying the BUGS language called JAGS began, motivated by the desire to add new features (Plummer, 2003). Currently, JAGS has become the dominant strain of BUGS software used among ecologists (but see Section 4.5). We suggest this is due to its smooth integration with R, because it is supported on Mac OS in addition to Windows and Linux, and because it has proved to be a robust program.

JAGS and a simple manual are both freely available online (https://mcmc-jags.sourceforge.io/), and must be downloaded and installed separately from R. JAGS is most commonly accessed from R via an intermediary R package. Several such packages are available to be installed from CRAN, including the "official" `rjags` package (Plummer, 2023a) as well as `jagsUI` (Kellner & Meredith, 2021) and `runjags` (Denwood, 2016). We prefer `jagsUI` (although we may of course be biased) and will use it throughout the book.

The standard workflow for JAGS analyses consists of the following steps, which will be similar for the other engines and for other models.

1. Organize the input data in R into the format required by JAGS; typically this is a list of numeric scalars, vectors, and arrays
2. Write the BUGS code describing our model and save it to a text file
3. Specify settings for Markov Chain Monte Carlo (MCMC) algorithms, such as number of chains, iterations, burn-in, and thinning rate
4. Supply data, model code, and settings to the `jagsUI` R package
5. A call to function `jags()` sends the inputs to JAGS, which conducts MCMC – this may take a long time depending on your data and model complexity
6. Resulting random draws from the joint posterior distribution of the model parameters are returned to R and summarized in various ways, ready to be used for inference

4.4.2 **Fit the model with JAGS**

We begin by loading the `jagsUI` package.

```
library(jagsUI)
```

To run a model in JAGS from R using the `jagsUI` interface we use a function that is called `jags()` (no surprise here). Here is its usage, with only those arguments shown which we most often or always deal with when running JAGS (run `?jags` in your R console to see information about all arguments).

```
jags(data, inits, parameters.to.save, model.file, n.chains, n.adapt = NULL,
    n.iter, n.burnin = 0, n.thin = 1, parallel = FALSE)
```

Before we call `jags`, we will need to specify the input for each argument:

- `data` is an R list of the data for the analysis
- `inits` is either a list or else a function that generates random starting values for some or all parameters
- `parameters.to.save` is a vector with the names of the model parameters for which we want to save samples from the joint posterior distribution
- The `model.file` is the name of a text file that contains the model description written in the BUGS language
- The next five arguments define what we call the MCMC settings and include: `n.chains`, the number of Markov chains to run; `n.adapt`, the length of the adaptive phase before the production run starts; `n.iter`, the total number of iterations per chain; `n.burnin`, the number of post-adaptation iterations that are discarded as a burn-in; and `n.thin`, the thinning rate.
- `parallel` is a switch to indicate whether the Markov chains should be run in parallel.

First, we will package (or "bundle") into an R list the data that JAGS uses for the analysis. In this case, we have two components to our data: the measurements of body mass (`y1000`), and the total number of such measurements (the sample size, 1000). As we will see, it is very common to have to explicitly tell JAGS information about our sample size(s). We combine these two data components into a named R list. Keep track of the names you assign to each list element: they will need to exactly match the names we use in the model description. A final consideration when building the data list is that each element *must be numeric*: either a scalar (one value), a vector of values, a matrix of values (2 dimensions), or an array of values (>2 dimensions). Our list contains a numeric vector (the mass values) and a numeric scalar (the sample size). We like to wrap our data list into a call to `str()` since this provides a useful overview of our data, along with information about their dimensions. For budding BUGS modelers, keeping track of the dimensions of all data objects in a model is often a challenge.

```
# Bundle and summarize data
str(dataList <- list(mass = y1000, n = length(y1000)))

List of 2
 $ mass: num [1:1000] 617 600 611 599 601 ...
 $ n   : int 1000
```

The next step is to write the model we want to fit in the BUGS language. A BUGS model is composed of *nodes* (our data and parameters we want to estimate, such as the mean in our model) and *relationships* (either stochastic or deterministic) which define nodes as functions of other nodes or as draws from specified distributions. Superficially, the BUGS language has many similarities to R: the basic mathematical operators (addition, multiplication) are available, you can define for loops to iterate over a series of repeated structures, and the assignment operator for deterministic relationships, like R, is the left-pointing arrow <- (and not an equal sign). For example, to specify that the parameter pop.var (the variance) is equal to the square of pop.sd (the standard deviation), one possible line of BUGS code is pop.var <- pop.sd * pop.sd. To describe a distribution for data or a set of parameters in BUGS, we use the tilde (~) instead of <-. For example, to specify that the prior distribution of pop.mean (the population mean) parameter should be normal with mean 0 and precision 0.01, the equivalent BUGS code is pop.mean ~ dnorm(0, 0.01). This highlights an important difference between R and BUGS: whereas R parameterizes the normal distribution with the mean and standard deviation, the variant of BUGS used by JAGS uses mean and precision instead for reasons that are beyond the scope of this book (precision is the inverse of the variance, or $1/\sigma^2$). Many other mathematical functions and distributions are available in BUGS/JAGS; see the JAGS manual, chapter 2, for more details (Plummer, 2017).

A great advantage of BUGS is that we can organize our model almost any which way we want: we don't necessarily have to describe the nodes and relationships in any specific order. However, to keep things organized for this book, we will divide our BUGS models into two sections. The first section (containing mainly stochastic relationships) will describe the prior distributions of top-level model parameters, and the second section (containing mainly deterministic relationships/mathematical equations, but also the observed data) will describe the model likelihood. We comment our code frequently, to help you navigate. Comment lines begin with hash sign, as in R. We'll write models to separate text files from within R using the R cat function, but you could also edit the model files separately.

We're finally ready to write the model of the mean in the BUGS language. We'll first show the model code, and describe it in detail afterwards.

```
# Write JAGS model file
cat(file = "model4.4.txt", "        # This code line is R
model {                            # Starting here, we have BUGS code
# Priors
pop.mean ~ dunif(0, 5000)         # Population mean
precision <- 1 / pop.var          # Precision = 1/variance
pop.var <- pop.sd * pop.sd
pop.sd ~ dunif(0, 100)

# Likelihood
for(i in 1:n){
  mass[i] ~ dnorm(pop.mean, precision)
}
}                                  # This is the last line of BUGS code
")                                 # ... and this is R again
```

The first and last lines of the code block above tell R to save in the R working directory the enclosed code to a text file called `model4.4.txt` – these lines are not part of the JAGS model. The first and last lines of the model code are marked by `model{` and `}` respectively. Every JAGS model we write will need to include these two lines to mark the start and end of the model code. In this book, we always highlight the JAGS model by a light-grey box.

The first section in our model is the definition of the priors. We have two model parameters we want to estimate, the mean (`pop.mean`) and standard deviation (`pop.sd`) and we need a prior for each. In both cases we have specified very wide (and thus vague) uniform priors, using `~ dunif()`. The prior section also contains some extra lines of code to define the precision, which we will need later, as a function of the standard deviation. Notice that these are in the opposite order from what you might expect – illustrating that BUGS doesn't require things to be in a specific order, since the code is not executed line by line, but used to create an internal representation of the model as a graph.

The likelihood section is next. Here we iterate over each of our n mass values using a for loop, which should look similar to loops you have probably seen in R. Note that in R, we are often able to perform *vectorized* calculations, i.e., we don't need to explicitly iterate over each element of a vector, matrix, or array. For example, with function `lm()` in R we can fit the model simply by specifying y ~ 1 for vector y – no need to iterate explicitly over each element of y. However, in BUGS we generally have to specify calculations or distributions for every single element in a vector or other structure (JAGS does support a limited degree of vectorization, see Section 6 in the manual, but we will not make use of it in this book). You will later see that when we have 2- or higher-dimensional response variables, we will see two or more loops that are nested (see Chapters 19 and 19B). In some complex models involving multiple dimensions such as space, time, or species, you may see response variables organized in arrays of 4 dimensions or more (Kéry & Royle, 2016, 2021, Schaub & Kéry 2022).

Both `n` and `mass` came from our list of data, and the names of the elements in the data list match the names we used in the model description. We specify that each value of `mass` should be distributed as normal with mean `pop.mean` and precision `precision` (this is why we needed to calculate precision in the prior section). That's it! Notice the similarities between the likelihood section of the model and our original model equation written in algebra, as well as with the simulation code in R in Section 4.2. Indeed, BUGS code, algebra, and R data simulation code describe the statistical model in nearly identical ways. This is one of the reasons for which it is important that you learn to write your models in algebra (Chapter 3).

With the model code written, we next need to tell JAGS how to initialize the MCMC chains. That is, what should be our starting values for each parameter in each chain? There are various ways to do this, but we will always define an R function called `inits()` which for most or all parameters initializes the chains randomly, ensuring that each chain starts at different places. This is desirable for convergence assessment: if the chains then move to cover the same range of values (when they "mix" well), we have increased confidence that they have converged to their stationary distribution, which is our desired posterior distribution. Our `inits()` function must return an R list with named values for each parameter for which we want to set initial values. Note that you don't necessarily need to set initial values for every parameter – if you leave some out, JAGS will initialize them for you at the center of the prior chosen for them in the model. This will often work, but can cause problems due to a data-prior-inits conflicts for some models; see the occupancy and the N-mixture models in Chapters 19 and 19B. In addition, this default will lead to initial values which are identical for all chains, which is undesirable for convergence assessments.

```
# Function to generate starting values
inits <- function(){
  list(pop.mean = rnorm(1, 600), pop.sd = runif(1, 1, 30))
}
```

We must also tell JAGS for which parameters it should save the draws, i.e., the random numbers, from the joint posterior distribution. Let's say we want to obtain estimates of the variance, in addition to the population standard deviation and population mean.

```
# Parameters monitored
params <- c("pop.mean", "pop.sd", "pop.var")
```

Finally, we need to select the MCMC settings. At the beginning of a model run, JAGS performs an initial "adaptive" sampling phase in which the algorithm changes its behavior in order to optimize performance, similar to what we did 'by hand' by choosing the standard deviation of the proposal distribution in Section 2.8. Typically a relatively small number of adaptive iterations (n.adapt) are required, but you should not skip them. Leaving this argument empty in the call to jags() will cause JAGS to automatically add adaptive iterations until JAGS reports they are sufficient. The total number of iterations for each chain n.iter sets the total length of each chain – larger values increase your chance of reaching convergence and a stationary distribution, but also increase the runtime. The number of burn-in iterations n.burnin specifies how many iterations should be discarded at the beginning of each chain (not including adaptive iterations). Too high and you won't have enough samples from the posterior; too low and you risk including some samples taken before the MCMC reaches its stationary distribution. Thinning rate n.thin controls how many samples from the posterior are retained; e.g. if n.thin = 5, only one of every five samples is retained. This reduces the size of the resulting set of posterior samples and the size of the file that you have to save, as well as the correlation between the remaining samples, though it always throws out some information (that's why the original draft title of the paper by Link & Eaton (2012) was "*Stop the thinning*"). Finally, n.chains controls how many independent MCMC chains are run. Adding more chains facilitates assessment of convergence as well as diagnosis of possible multiple optima in the likelihood. However, it also increases runtime, though this depends on the number of cores in your system. Our usual choice is 4 chains.

```
# MCMC settings
na <- 1000    # Number of iterations in the adaptive phase
ni <- 12000   # Number of draws from the posterior (in each chain)
nb <- 2000    # Number of draws to discard as burn-in
nc <- 4       # Number of chains
nt <- 1       # Thinning rate (nt = 1 means we do not thin)
```

Now, we have completed all the preparations required to run the analysis. We call function jags() to carry out the analysis in JAGS and afterwards to put the results into an R object which we call out4.4, because we're in Chapter 4, Section 4.4. Note how we always give the approximate run time (ART), which may be helpful for to decide whether you should take a coffee break now.

```
# Call JAGS (ART 1 min) and marvel at JAGS' progress bar
out4.4 <- jags(data = dataList, inits = inits, parameters.to.save = params,
    model.file = "model4.4.txt", n.iter = ni, n.burnin = nb, n.chains = nc,
    n.thin = nt, n.adapt = na, parallel = FALSE)
```

Executing JAGS with `parallel = FALSE` lets us see the wonderful progress bar, staring at which gives an exciting sense of achievement. However, for any analysis that takes longer than just a couple of minutes, we will usually run JAGS in parallel. This will distribute each chain to its own core on your computer (you can set the number of cores with an argument in the function). With four chains (and a computer with at least 4 real cores) you will typically reduce the run time by a factor of about 3.5; usually not quite 4x, since there's some overhead. All that we need to run JAGS in parallel is to set the parallel 'switch' to `TRUE`.

```
# Call JAGS in parallel (ART <1 min) and check convergence
out4.4 <- jags(data = dataList, inits = inits, parameters.to.save = params,
  model.file = "model4.4.txt", n.iter = ni, n.burnin = nb, n.chains = nc,
  n.thin = nt, n.adapt = na, parallel = TRUE)
jagsUI::traceplot(out4.4)                # Produce Fig. 4.3
```

The first thing that we should always do after JAGS (or any MCMC algorithm) has finished is to assess whether the chains have converged. We can do this visually by examining trace plots, such as with the `traceplot()` function (Fig. 4.3). Our trace plots oscillate around a horizontal average level, without any long- or even short-term trends being visible, suggesting that the chains have

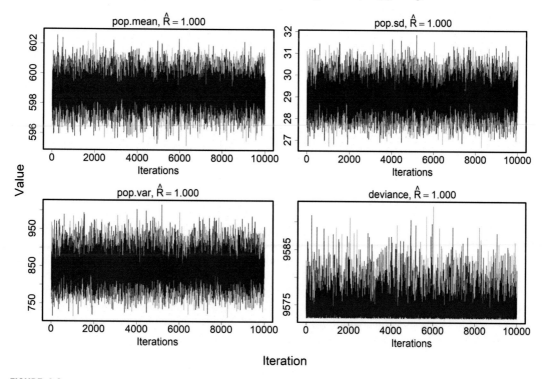

FIGURE 4.3

Trace plots for the parameters in the model of the mean, including the variance as a derived quantity, and the deviance. In the header of each plot we see the value of the Brooks-Gelman-Rubin (BGR) convergence statistic, called $\hat{R}$ (Rhat) here. At convergence Rhat is near 1.

converged and indeed mixed very well. This impression is confirmed also by the value of the Brooks-Gelman-Rubin (BGR) or $\hat{R}$ ("Rhat") statistic (Brooks & Gelman, 1998; Gelman & Rubin, 1992). This is a convergence test statistic that is related to the F-test in a one-way analysis of variance testing for a chain effect (Quinn & Keough, 2002). At convergence, there is no longer any chain effect and all the variability among the draws is within rather than between the chains, resulting in an Rhat of 1. For better or worse, a common rule of thumb takes 1.1 as a threshold for the acceptance of chain convergence. Rhat is useful for a quick screening of convergence in models with many parameters, for instance by producing a plot of the values of Rhat. However, we should always visually skim traceplots for at least the main parameters in the model, because very rarely (in some pathological cases) the Rhat as computed by `jagsUI` is unable to identify non-convergence.

Only once we have assessed convergence do we look at a summary of the posterior distributions.

```
print(out4.4, 3)    # Produce a summary of the fitted model object

JAGS output for model 'model4.4.txt', generated by jagsUI.
Estimates based on 3 chains of 12000 iterations,
adaptation = 1000 iterations (sufficient),
burn-in = 2000 iterations and thin rate = 1,
yielding 30000 total samples from the joint posterior.
MCMC ran in parallel for 0.255 minutes at time 2023-01-10 09:59:54.
```

	mean	sd	2.5%	50%	97.5%	overlap0	f	Rhat	n.eff
pop.mean	598.848	0.921	597.025	598.852	600.650	FALSE	1	1	18299
pop.sd	29.061	0.647	27.830	29.050	30.367	FALSE	1	1	40000
pop.var	844.959	37.677	774.529	843.880	922.157	FALSE	1	1	40000
deviance	9574.631	1.999	9572.685	9574.010	9580.019	FALSE	1	1	14017

```
Successful convergence based on Rhat values (all < 1.1).
Rhat is the potential scale reduction factor (at convergence, Rhat=1).
For each parameter, n.eff is a crude measure of effective sample size.

overlap0 checks if 0 falls in the parameter's 95% credible interval.
f is the proportion of the posterior with the same sign as the mean;
i.e., our confidence that the parameter is positive or negative.

DIC info: (pD = var(deviance)/2)
pD = 2 and DIC = 9576.629
DIC is an estimate of expected predictive error (lower is better).
```

The summary returns a lot of information, divided into three parts. The header gives some general information about the MCMC settings and about the timing of the run. The latter becomes important when your models take hours to run. Then, you will want to know how many iterations you can run overnight so that you have some results to study the next morning, or over the weekend to have something on Monday morning to chew on.

The middle part is the crucial part: it presents statistics based on the samples from the marginal posterior distribution of each saved parameter. Typically, we take the mean as a point estimate, the posterior standard deviation as a quantity akin to a standard error, and the 2.5% and 97.5% percentiles as a 95% credible interval, or CRI. For skewed posterior distributions we should better report the posterior median or the mode – actually, it is the posterior mode which numerically corresponds to the MLE in an analysis with vague priors. Here, mean and median are almost identical, indicating symmetry of the posterior distribution for the population mean, though a little less so for the population variance. We will see below how we can access the posterior draws and produce plots of the posterior distributions or do our own calculations with them.

The columns with heading `overlap0` and `f` are explained in the footer of the output and can be used to conduct a type of Bayesian significance test. We explained the `Rhat` column previously. The final column is the effective sample size from the posterior, which is an estimate of the size of a hypothetical independent sample that contains the same amount of information about a parameter as the dependent sample produced by our MCMC algorithm. Increasing serial autocorrelation in a chain will reduce the size of `n.eff` relative to the nominal sample size chosen in the MCMC settings. To minimize the contribution of simulation error, also called MC error, to the parameter estimates, we want large effective sample sizes (e.g., in the thousands), especially when we are interested in the credible intervals or other percentiles near the tail of the posterior distributions.

The footer of this output gives some explanations and also includes the value of the Bayesian deviance information criterion or DIC (Spiegelhalter et al., 2002), which is a Bayesian analogue to the frequentist Akaike information criterion (AIC; Akaike, 1972) and can be used for model selection. As the AIC, the DIC is computed as a deviance that is penalized by model complexity, where pD is the estimated number of parameters. For non-hierarchical models such as generalized linear models (GLMs), the DIC is directly applicable, but it should in general not be used for most hierarchical models, especially, when there are discrete parameters as in models like those in chapters 19 and 19B. See Chapter 18 for more on model selection and criteria such as AIC, DIC or the Watanabe-Akaike information criterion (WAIC).

We have now achieved a major feat: we have obtained our first set of parameter estimates from one of the general-purpose model fitting engines, JAGS. We might summarize our inferences in a paper as follows: "*We estimate the mean body mass of male peregrines at 598.8 g (posterior mean), with a 95% Bayesian credible interval of 597.0–600.7 g. The standard deviation of body size was estimated at 29.1 g (95% CRI 27.8–30.4 g)*". In practice, though, you would be much more concise and might just write "*the mean body mass of male peregrines was estimated at 598.8 g (95% CRI 597.0–600.7 g) and the population sd, characterizing among-individual variability, at 29.1 (95% CRI 27.8–30.4 g)*".

The object `out4.4` produced by `jagsUI` contains *a lot* more information besides posterior summary statistics. We can see this by listing all objects contained within `out4.4`.

```
names(out4.4)
```

```
 [1]  "sims.list"   "mean"        "sd"          "q2.5"
 [5]  "q25"         "q50"         "q75"         "q97.5"
 [9]  "overlap0"    "f"           "Rhat"        "n.eff"
[13]  "pD"          "DIC"         "summary"     "samples"
[17]  "modfile"     "model"       "parameters"  "mcmc.info"
[21]  "run.date"    "parallel"    "bugs.format" "calc.DIC"
```

Some of the most useful elements include the following:

- `sims.list` contains the merged output from all chains of the posterior draws for each monitored parameter (i.e., the merged samples from below). This is useful for plotting or for producing the posterior distributions of derived quantities, which are functions of one or more parameters.
- The following 11 elements contain the same information as we just saw in the posterior summary table. These are particularly useful for plotting, where, for instance, you will use the posterior mean and the lower and upper bound of the 95% Bayesian credible interval, which are contained in the elements named `mean`, `q2.5` and `q97.5`. Element `Rhat` may be useful for a quick screening of convergence in models that are very parameter-rich.

- Object `summary` is just the posterior summary that we have seen above.
- `samples` is the raw output from JAGS. This object contains the individual posterior draws for each parameter separated by chain. This information is stored as a `mcmc.list` object, which can be manipulated using tools from various MCMC-related R packages such as `coda` or `MCMCvis` (Youngflesh, 2018).
- One of the elements in the `mcmc.info` is the elapsed time, which is important for your planning of JAGS runs over the night, the week-end or your holidays.

As an example of some things that you can do with this output: for a quick check whether any of the parameters has a BGR diagnostic greater than 1.1, i.e., has Markov chains that have not converged, you can type this:

```
hist(out4.4$summary[,8])            # Rhat values in column 8 of the summary
which(out4.4$summary[,8] > 1.1)     # check for non-convergence: none here
```

Understanding the `samples` component in the output is particularly useful. Many software packages that use similar simulation methods, such as Stan and NIMBLE, can generate `mcmc.list` output in the same form as `samples`. Thus any code we write that operates on the `samples` component here can potentially be applied to the output from these other software packages. For example, if we call the `summary()` function on `samples`, we get posterior means, SDs, and quantiles, as well as time-series SE.

```
summary(out4.4$samples)   # not shown
```

The time-series SE, equivalent to the MC error, is helpful for deciding if we have run the chains for long enough. One rule of thumb says MC error should be at most 1–5% of the posterior SD of a parameter (Lunn et al., 2013).

The `samples` component of the output can be used to create our own trace plots:

```
par(mfrow = c(3,1), mar = c(5,5,4,2))                  # not shown
pars <- colnames(out4.4$samples[[1]])[1:3]
for (i in pars){
  matplot(as.array(out4.4$samples)[,i,], type = 'l', lty = 1,
          col = c("green", "red", "blue", "orange"),
          xlab = "Iteration", ylab = "Value", main = i, frame = FALSE)
}
```

We can also produce other graphical summaries from `samples`, e.g., histograms or density plots of the posterior distributions for each parameter (Fig. 4.4) (note this would also work with content in `sims.list`):

```
par(mfrow = c(3, 2), mar = c(5,5,4,2))                 # Fig. 4.4
for (i in pars){
  samps <- as.matrix(out4.4$samples[,i])
  hist(samps, col = 'grey', breaks = 100, xlab = i, main = "")
  plot(density(samps), type = 'l', lwd = 2, col = 'gray40', main = "",
  xlab = i, frame = FALSE)
}
```

Recall from Chapter 2 that output from MCMC (or HMC) algorithms, such as those shown in Fig. 4.4, represent draws from the posterior distributions that arise from the application of Bayes rule to our model and data. Thus, Fig. 4.4 is a graphical representation of the left-hand side ($p(\theta|y)$) of Bayes rule, repeated here from Chapter 2:

$$p(\theta|y) = \frac{p(y|\theta)p(\theta)}{p(y)}$$

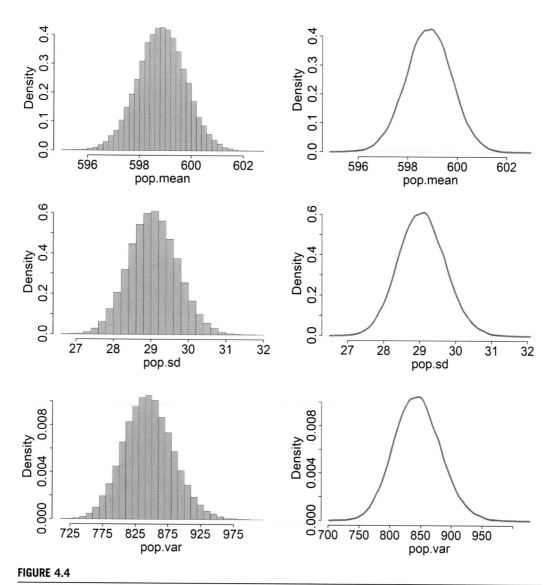

FIGURE 4.4

Two variants of density plots for the marginal posterior distributions of the mean, the standard deviation and the variance of the body mass in the population of male peregrines from which our sample of 1000 birds was drawn. We could also overlay the density estimator (the line on the right) on the histogram, but only if we choose the argument `freq = FALSE` in the histogram.

We can play "probability games" based on the posterior samples produced from a model to estimate the probability that a given parameter has a certain range of values. For instance, we can evaluate the probability that the mean body mass of male peregrines is less than 597 g by conducting a logical test on each draw from the posterior distribution, and then tally up the proportion of draws for which the test evaluates to true (using the mean function for the latter as a shortcut). We find it is near 0.02.

```
sims <- as.matrix(out4.4$samples)
head(sims[,"pop.mean"] < 597)    # show first of MANY logical tests!
mean(sims[,"pop.mean"] < 597)    # Prob(mu < 597)
```

We can also look at the bivariate posterior distribution for two parameters simultaneously; for instance, to check whether the estimates of two parameters are correlated. Here we find that the population mean and population SD are not correlated – there is a circular cloud (Fig. 4.5) and not an elongated one as we saw in Fig. 2.17. We can also play probability games in two dimensions. For example, based on the joint posterior distribution, the probability that the mean mass of male peregrines is less than 600 g, but the standard deviation is more than 30 g, can be estimated at about 0.07.

```
# Fig. 4.5: compute 2d probability game and plot bivariate posterior
test.true <- sims[,"pop.mean"] < 600 & sims[,"pop.sd"] > 30
mean(test.true)
par(mfrow=c(1,1))
plot(sims[,"pop.mean"], sims[,"pop.sd"], pch = 16, col = rgb(0,0,0,0.3),
  cex = 0.8, frame = FALSE)
points(sims[test.true,"pop.mean"], sims[test.true,"pop.sd"], pch = 16,
  col = rgb(1,0,0,0.6), cex = 0.8)
abline(h = 30, col = 'red')
abline(v = 600, col = 'red')

# alternatively do this to see bivariate posterior plots
pairs(sims[,1:3])
```

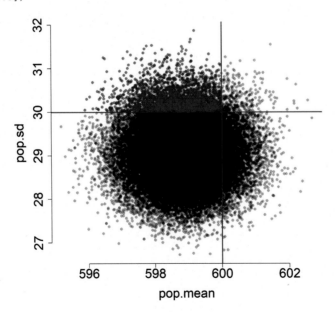

FIGURE 4.5

A bivariate representation of the joint posterior distribution with the red showing those draws for which a two-dimensional condition is met: for them, the mean is less than 600 g and the standard deviation is more than 30 g (this is the top left corner). The probability of this is evaluated at 0.07, which is the proportion of red points among the total number of points.

We can obtain numerical summaries of the posterior distribution, with the standard deviation requested separately:

```
apply(as.matrix(out4.4$samples[,1:2]), 2, summary)

            pop.mean     pop.sd
Min.       595.1759    26.71667
1st Qu.    598.2338    28.61658
Median     598.8524    29.04962
Mean       598.8482    29.06097
3rd Qu.    599.4679    29.49258
Max.       602.6826    31.86843

apply(as.matrix(out4.4$samples[,1:2]), 2, sd)

 pop.mean        pop.sd
0.9210760     0.6472911
```

Comparison of these Bayesian estimates with the results from `lm()` based on least squares in Section 4.3 reveals results that are nearly identical.

```
# Compare likelihood with Bayesian estimates and with truth
jags_est <- out4.4$summary[1:2,1]
print(cbind(truth = truth, lm = lm_est, JAGS = jags_est), 5)

        truth       lm      JAGS
mean      600  598.846   598.848
sd         30   29.017    29.061
```

Of all the general-purpose engines we will apply in this book, we believe that JAGS is the most intuitive and easiest to learn for a beginner. For instance, we can easily specify likelihoods with discrete parameters (see Chapters 19 and 19B) and missing values in the response present no problem (see Chapter 11). As we will see, it is not as easy to handle these two issues in some of the other engines. Thus, the JAGS version of the analysis will take center stage and garner the most details in each chapter. However, as we will see, the other engines have advantages as well.

4.5 NIMBLE

4.5.1 Introduction to NIMBLE

NIMBLE stands for **N**umerical **I**nference for statistical **M**odels for **B**ayesian and **L**ikelihood **E**stimation (de Valpine et al., 2017). It has quickly grown in popularity among ecologists since the first version of the R package was published on CRAN in 2016. Broadly, NIMBLE is a software package providing both a modelling language and an algorithm language for fitting models, and is very similar to JAGS in some ways. Like JAGS, it uses a variant of the BUGS language to describe a model, and it provides tools to fit models with MCMC. Also like JAGS, we can specify models directly with discrete parameters in the likelihood and our response variable may contain missing values. As a result, switching from JAGS to NIMBLE is typically very easy. We can supply model code we originally wrote for JAGS to NIMBLE instead and it will usually just work with only very minor modifications. Unlike JAGS, there's no separate standalone NIMBLE software – it's all included in the R package.

However, to use NIMBLE you need to be able to compile C++ code from R. Setup for this depends on your operating system. On Windows you need to install the Rtools bundle (https://cran.r-project.org/bin/windows/Rtools/), on Mac OS you need Xcode (https://developer.apple.com/xcode/), and on Linux you need `make` and a C++ compiler such as `gcc` (which is probably already installed). See Chapter 4 in the NIMBLE manual for details on setup (https://r-nimble.org/html_manual/cha-installing-nimble.html). We'll need to be able to compile C++ code for two other engines as well (Stan and TMB), for which the same setup applies.

Despite the similarities, NIMBLE is not simply a clone of JAGS or older BUGS software. As the name suggests, NIMBLE provides significant additional flexibility both in writing models and in implementation of sampling algorithms such as (but not limited to) MCMC. For example, it is possible in NIMBLE to write your own custom functions and distributions from within R, and then use them in your BUGS code. In other software packages that implement BUGS, such as JAGS, it is much more complicated to write your own functions. NIMBLE also makes many other welcome modifications to BUGS, such as giving multiple options for parameterizing distributions (e.g. as we will see, you can use the standard deviation instead of precision to parameterize normal distributions).

Finally, NIMBLE allows for very deep customization of model-fitting algorithms, such as modifying the MCMC samplers used, or using Monte Carlo Expectation Maximization (MCEM), Hamiltonian Monte Carlo (HMC; Turek et al. 2024), or Laplace approximation. While these customization tools are incredibly powerful, it is not necessary to apply them in every case, and doing so is beyond the scope of this book. We will stick to default NIMBLE settings and MCMC throughout this book but encourage you to explore the excellent NIMBLE documentation (https://r-nimble.org) if you need these features in your work. Also, the NIMBLE developers provide top-class support in the NIMBLE users email discussion group.

There are two different approaches to a NIMBLE workflow we will demonstrate. The first workflow we will use for NIMBLE is similar to what we did for JAGS. The main differences are that we will use function `nimbleMCMC()` instead of `jags()`, and the output will be in a slightly different (but still familiar) format.

1. Organize the input data in R into a named list of numeric components
2. Write the BUGS code as a `nimbleCode` object
3. Specify MCMC settings, such as initial values, number of chains, iterations, and thinning rate
4. Supply data, model code, and settings to the `nimbleMCMC` function in the `nimble` R package
5. `nimbleMCMC` builds and compiles the model and runs MCMC
6. Resulting draws from the posterior distributions of model parameters are returned to our R environment

Note that unlike with JAGS and `jagsUI`, `nimbleMCMC` does not include a way to automatically run chains in parallel. It is possible to do this manually, for example by using the `parallel` package to set up several cores and then running one chain in each core. Allowing `nimbleMCMC` to do all the work from one function call is convenient, but gives us less control over the configuration. In the second approach, step 5 is broken into a series of separate steps using a sequence of specialized NIMBLE functions, with which we build the model, configure the MCMC, compile the model and MCMC algorithm, and finally run the algorithm. This approach is a little more complicated, but gives us more control.

4.5.2 Fit the model with NIMBLE

The first step is to load the NIMBLE library.

```
# Load NIMBLE
library(nimble)
```

We will fit the model using NIMBLE in a single step using function `nimbleMCMC`. The function has the following key arguments (among others, which you can see with `?nimbleMCMC`):

```
nimbleMCMC(code, constants, inits, monitors,
    nburnin, niter, nthin, nchains, samplesAsCodaMCMC, WAIC)
```

Before we call `nimbleMCMC`, we will need to specify the input for each argument:

- `code` is our BUGS code wrapped in a call to function `nimbleCode`
- `constants` is our R data list (which may contain both constants and data according to NIMBLE terminology; see below)
- `inits` will be our function that generates random starting values for some or all parameters
- `monitors` is a vector with the names of the model parameters for which we want to save samples from the joint posterior distribution
- The next four arguments define what we call the MCMC settings and include: `nburnin`, the number of iterations that are discarded as a burn-in; `niter`, the total number of iterations per chain; `nthin`, the thinning rate; and `nchains`, the number of Markov chains to run
- `samplesAsCodaMCMC` is a switch to indicate if the resulting samples from the posterior distribution should be returned as a `mcmc.list` object (the same format we saw for the `samples` component of the `jagsUI` output in the previous section)
- `WAIC` is a switch to indicate if the Watanabe-Akaike information criterion (Watanabe, 2010) for the model should be returned in the output (see Chapter 18)

We'll start with the data, for which we can use the same list we used for JAGS. In principle, NIMBLE makes a distinction between `constants` and `data`. As the name suggests, `constants` cannot be changed once the model is defined, but `data` can be changed. Indices used in for loops in the model code are a good example of `constants`. On the other hand, any response variable, found on the left-hand size of a stochastic relationship, is typically considered `data`. In practice the distinction is actually a bit more involved and beyond the scope of this book. Luckily, NIMBLE allows us to simply specify everything as `constants`, which is what other software like JAGS is ultimately doing.

```
# Bundle data (same as for JAGS)
str(dataList <- list(mass = y1000, n = length(y1000)))   # not shown
```

Next we write the model code in the BUGS language. With JAGS, we saved this code to a separate text file using the `cat` function. For NIMBLE, this is not necessary: instead, we wrap our BUGS code in a call to the convenient `nimbleCode` function. As with JAGS, we'll show the full model code below and discuss it afterwards. We highlight NIMBLE model code with an olive box.

```
# Write Nimble model file
model4.5 <- nimbleCode({
# Priors and linear models
pop.mean ~ dunif(0, 5000)              # Normal parameterized by precision
precision <- 1 / pop.variance          # Precision = 1/variance
pop.variance <- pop.sd * pop.sd
pop.sd ~ dunif(0, 100)

# Likelihood
for(i in 1:n){
  mass[i] ~ dnorm(pop.mean, precision)
}
})
```

If we compare this to the JAGS code in Section 4.4.2, we can see it is essentially identical, except that we have replaced the enclosing `model{ }` brackets in the JAGS code with `nimbleCode({ })`. In general this will be true in later chapters as well, although the BUGS code used for NIMBLE may require some small modifications. For example, it is possible with NIMBLE to parameterize a normal distribution using the standard deviation instead of the precision, which we will illustrate in the code for future chapters.

Initial values and MCMC settings can be defined in the same way as for JAGS, except that NIMBLE does not have a separate adaptive phase that we need to specify. Thus we can re-use the same code as before.

```
# Can use same function to generate starting values as for JAGS
inits <- function()
   list (pop.mean = rnorm(1, 600), pop.sd = runif(1, 1, 30))

# Parameters monitored: same as before
params <- c("pop.mean", "pop.sd", "pop.variance")

# MCMC settings
ni <- 3000 ; nb <- 1000 ; nc <- 4 ; nt <- 1
```

We're now ready to call `nimbleMCMC`. We wrap the call to `nimbleMCMC` into a call to `system.time()` to time the model run – for the same reasons we have mentioned in Section 4.4. Also, we specify that we want NIMBLE to provide the output as an `mcmc.list` object, as well as a summary of the output.

```
# Call NIMBLE (ART 30 sec)
system.time(
   out4.5 <- nimbleMCMC(code = model4.5, constants = dataList,
      inits = inits, monitors = params, nburnin = nb, niter = ni,
      thin = nt, nchains = nc, samplesAsCodaMCMC = TRUE, summary = TRUE) )
```

This takes a little longer than JAGS, because NIMBLE first has to translate our model to C++ code and then compile the C++ code (i.e., translate into machine language) before sampling. The result is an `mcmc.list`, which as we discussed in the previous section is a special type of list, where each list element represents one MCMC chain. Below we use the `lapply()` function to run the `head()` function on each list element/MCMC chain, in order to view the first few values in each chain.

```
lapply(out4.5, head)   # print first 5 values of each chain (not shown)
```

We can check convergence with the `traceplot` function and get a summary of the posteriors with `summary`.

```
par(mfrow=c(1, 3));
   coda::traceplot(out4.5$samples)     # not shown
out4.5$summary$all.chains              # Posterior summaries from all 4 chains
```

	Mean	Median	St.Dev.	95%CI_low	95%CI_upp
pop.mean	598.86376	598.85848	0.9432491	597.02593	600.66415
pop.sd	29.04457	29.03387	0.6437659	27.78823	30.32504
pop.variance	844.00121	842.96545	37.4167898	772.18590	919.60791

We could also apply the same code from the end of Section 4.4.2 to this output to get similar summary figures and statistics. In order to get summary output that resembles the output from `jagsUI`, we have included the `nimble_summary` function in the companion R package to this book, `ASMbook`.

```
library(ASMbook)
nsum <- nimble_summary(out4.5$samples, params)    # Summary table
round(nsum, 3)
nimble_est <- nsum[1:2,1]                          # save estimates
```

	mean	sd	2.5%	50%	97.5%	Rhat
pop.mean	598.864	0.943	597.026	598.858	600.664	1.002
pop.sd	29.045	0.644	27.788	29.034	30.325	1.002
pop.variance	844.001	37.417	772.186	842.965	919.608	1.002

This one-step approach to running NIMBLE is very practical for a beginner, but later on you may want to separate the different steps that are combined in this single call to `nimbleMCMC`. This will give you better control and save time by avoiding re-compilation of the model each time you run it. To fit our NIMBLE model in this way we will perform several steps: (1) Build a model; (2) configure and build an MCMC algorithm to use when sampling; (3) convert our model and MCMC algorithm from R code to C++ and compile it to increase speed; and (4) generate posterior samples using this compiled code.

First, we will construct a complete NIMBLE model object, which we will call `rawModel`, using the `nimbleModel` function. We need to provide our model code (contained in object `model4.5`), our list of constants (i.e. input data; `dataList`), and a set of initial values for each model parameter. The initial values cannot be supplied as a function as we did with `nimbleMCMC` above. Rather, we must provide a list of actual random initial values. To get this list, we can simply execute our initial value function once.

```
# Create a NIMBLE model from BUGS code
rawModel  <- nimbleModel(code = model4.5, constants = dataList, inits = inits())
```

Next, we will configure our desired MCMC algorithm for our model, and tell NIMBLE which parameters we want to monitor (`params`). We will use the default MCMC settings here and throughout the book, but deep customization is possible in this step. See the NIMBLE manual, Chapter 7, for more details.

```
# Configure the MCMC algorithm: create a default MCMC configuration
# and monitor our selected list of parameters
mcmcConfig <- configureMCMC(rawModel, monitors = params)
```

We then supply this configuration object to the `buildMCMC` function to create our MCMC algorithm with default settings, called `rawMCMC`.

```
# Build the MCMC algorithm function
rawMCMC  <- buildMCMC(mcmcConfig)
```

It is now possible to generate posterior samples with this MCMC function. However, since it is still written in R code, it will be quite slow.

```
system.time(rawMCMC$run(10))   # Take 10 samples, takes about 7 sec
as.matrix(rawMCMC$mvSamples)   # View samples, not shown
```

The solution is to translate our slow R code into C++ code, which is then compiled into fast machine code. NIMBLE can do this for us automatically. We start by compiling our model specification (`rawModel`) from step 1 using the `compileNimble` function.

```
compModel  <- compileNimble(rawModel)
```

Next we compile our MCMC function (`rawMCMC`) as well, using the same `compileNimble` function. We must also provide our original model object (`rawModel`) as an argument.

```
compMCMC  <- compileNimble(rawMCMC, project = rawModel)
```

We can now take samples from the compiled MCMC function, which will be much faster.

```
system.time(compMCMC$run(10))   # not shown, takes about 0.02 sec
```

Finally, we will use `compMCMC` and the `runMCMC` function to generate posterior samples according to the same settings we used previously with `nimbleMCMC`. The output object will be a `mcmc.list`, same as with the `nimbleMCMC` function.

```
system.time(
samples  <- runMCMC(compMCMC, niter = ni, nburnin = nb, thin = nt,
  nchains = nc, samplesAsCodaMCMC = TRUE) )

# Peak at samples
lapply(samples, head)                          # not shown

# Produce marginal posterior summaries
(nsum  <- nimble_summary(samples))             # not shown

# Traceplots
par(mfrow=c(2,2)); coda::traceplot(samples)    # not shown
```

This longer approach gives us a lot more control over configuration, which may be useful when we want to customize our sampling algorithm. This can greatly speed up the algorithms and increase the rates of convergence.

Overall, we find NIMBLE to be a very powerful and flexible engine for model fitting, with the big, added convenience of using the BUGS language for model definition. This includes the ability to directly specify likelihoods with discrete parameters (see Chapters 19 and 19B) and to have missing values in the response (see Chapter 11). Furthermore, as of writing, the software is being developed at a fast rate. As a result, we anticipate continued migration of JAGS users to NIMBLE in the future.

4.6 Stan

4.6.1 Introduction to Stan

Stan is a platform for coding statistical models, first released in 2012 and named after Stanislaw Ulam, co-inventor of modern Monte Carlo methods in the 1940s. Similar to BUGS, Stan can refer to both a series of related software packages (such as CmdStan, PyStan, and RStan) as well as an associated programming language (Carpenter et al., 2017). Stan was inspired by the BUGS family of software, in that it implements a programming language for describing statistical models, as well as algorithms for sampling from posterior distributions (Carpenter et al., 2017). Stan primarily uses a Bayesian approach for estimating model parameters, but also has some support for maximum-likelihood based methods. In this book we'll focus only on the Bayesian methods that Stan supports. Like both BUGS and NIMBLE, Stan can be accessed from R via an R interface package (rstan). Like NIMBLE it is not necessary to install separate Stan software beyond the R package, but R must be able to compile C++ code. See Section 4.5.1 for R/C++ compiler setup instructions, as well as Stan's relevant documentation at https://github.com/stan-dev/rstan/wiki/RStan-Getting-Started.

Stan's default No U-Turn Sampler (NUTS) is a variant of Hamiltonian Monte Carlo (HMC) algorithm (Monnahan et al., 2017) and is a distinguishing feature of the software relative to JAGS (note that NIMBLE also supports HMC; Turek et al., 2024). For some models, HMC/NUTS is a much more efficient sampler than the MCMC algorithms we use when fitting a model with JAGS and often also with NIMBLE's default samplers, requiring fewer iterations, and producing faster chain convergence. An exploration of the detailed differences between these sampling algorithms is beyond the scope of this book. We have found both more traditional MCMC and HMC to work similarly well for the models we present throughout the book, as you will also discover.

The Stan programming language has some superficial similarities to BUGS (and more generally R), but has both additional flexibility and additional constraints as we will see in this and future chapters. As with NIMBLE's variant of BUGS, you can specify custom functions in Stan, and there are many built-in mathematical functions. Furthermore, Stan is arguably a more "complete" programming language than is BUGS in terms of the range of control statements (for and while loops, if/else conditionals) that it supports, which can enable specification of more complex model structures.

However, Stan also imposes additional constraints on model structure, which can be both a benefit and a challenge. For example, unlike BUGS, Stan requires the model file to be divided into specific sections, and the exact order of commands in the Stan model file is sometimes important. This reduces the ways in which a given model can be written, but may also make the model easier to read. In addition to inspiration from BUGS, Stan also has roots in C++. One result of this is that Stan shares the C++ requirement to explicitly define the *type* (e.g. double, integer, scalar, vector, matrix; see Section 4.6.2) and the dimensions of your data and model parameters *a priori*. For those not familiar with C++, this can add to the challenge of writing a working Stan model. On the other hand, this added requirement can help us think through our model structure more carefully and avoid certain errors. Finally, unlike BUGS, Stan does not support sampling discrete parameters, which is a consequence of its reliance on HMC. Latent discrete parameters are particularly useful for many classes of hierarchical models in ecology such as capture-recapture, occupancy and N-mixture models (Royle & Dorazio, 2008; Kéry & Royle, 2016, 2021; Schaub & Kéry, 2022), where having these discrete parameters makes for a much more intuitive way of specifying a model. All of these models can in principle also be fit in Stan, but the discrete parameters have to be marginalized over, i.e., removed by summation, when specifying the likelihood. This can actually be a good thing, since it will often speed up code also in JAGS or NIMBLE (Joseph, 2020a, 2020b; Ponisio et al., 2020; Turek et al., 2016, 2021; Yackulic et al., 2020). However, it makes writing Stan code for these models far more challenging for non-statisticians. We will revisit marginalization in Chapters 12, 19 and 19B. For more on writing Stan models, see the Stan documentation at https://mc-stan.org/users/documentation/.

Our Stan workflow will closely resemble our workflow for both JAGS and NIMBLE:

1. Organize the input data in R into a named list of numeric components
2. Write the Stan code to a separate text file
3. Specify settings for HMC/NUTS sampling, such as initial values, number of chains, iterations, and thinning rate
4. Supply data, model code, and settings to the `stan` function in the `rstan` R package
5. `stan` builds and compiles the model and runs HMC/NUTS
6. Resulting posterior draws of model parameters and summaries are returned to our R environment

4.6.2 Fit the model with Stan

First, we need to load the `rstan` package.

```
# Load Stan R package
library(rstan)
```

Stan models are fit with the `stan` function, which has the following key arguments (among others, which you can see with `?stan`):

```
stan(file, data, pars, chains, iter, warmup, thin, init, cores)
```

Before we call `stan`, we will need to specify the input for each argument:

• `file` is the name of our Stan code file
• `data` is our R data list
• `pars` is a vector with the names of the model parameters for which we want to save samples from the joint posterior distribution
• The next four arguments define what we call the HMC settings and include: `chains`, the number of Markov chains to run; `iter`, the total number of iterations per chain; `warmup`, the number of iterations that are discarded as a burnin; `thin`, the thinning rate
• `init` is how Stan should set initial values for each chain. This can be a function as with JAGS and NIMBLE, among other options
• `cores` specifies how many cores should be used to run chains in parallel, if desired

Our list of data is identical to the one we used for JAGS and NIMBLE. Note, though, that in contrast to JAGS and NIMBLE, Stan does not allow any missing values in our response data. If you have any, then they must be eliminated from the data bundle and the model may have to be adapted accordingly. See Chapter 11 for an example.

```
# Bundle and summarize data
str(dataList <- list(n = length(y1000), mass = y1000))
```

Now we need to write our Stan model to a separate text file (`model4_6.stan`) the same way we did with JAGS, using `cat`. Stan models have a more strictly defined structure than BUGS models. They must be divided into multiple sections called `data`, `parameters`, and `model` (there are other possible sections we will show in later chapters). As the names suggest, in the `data` section we describe the types and dimensions of each element in our data list. In the `parameters` section we describe the types and dimensions of parameters in the model. And finally, the `model` section resembles our BUGS code and is where we specify the priors and the likelihood. While this approach to writing a model is more verbose than with BUGS, it does force us to be more careful and may help to prevent errors.

The syntax of the Stan programming language has many similarities to both R and BUGS. For example, the distribution of data and parameters is specified with the tilde ~. The names of distributions are a little different from BUGS but are usually easily recognizable (e.g., `normal` instead of `dnorm`). Perhaps the trickiest part of the Stan language for those not familiar with C++ is the required declaration of variable *type*. Stan requires us to tell it exactly what each model element (i.e., data and parameters) represents using a specific code: for example, an integer (`int`), a decimal number (`real`), a vector of decimal numbers of length *n* (`vector[n]`), a decimal number with lower bound 0 (`real <lower = 0>`) and so on. Two final considerations are: you should not use any periods (`.`) in your variable names (instead consider using `_`) and each line (besides lines with opening or closing brackets) in a Stan model *must* end in a semi-colon (`;`). Below, we show the complete model and we will describe it in detail afterwards. As throughout the book, we will highlight Stan code with a purple box.

```
# Write text file with model description in Stan language
cat(file = "model4_6.stan",          # This line is still R code
"data {                               // This is the first line of Stan code
  int n;                              // Define the format of all data
  vector[n] mass;                     // ...including the dimension of vectors
}
parameters {                          // Same for the parameters
  real pop_mean;
  real <lower = 0> pop_sd;
}
model {
  // Priors
  pop_mean ~ normal(0, 1000);
  pop_sd ~ cauchy(0, 10);

  // Likelihood
  for (i in 1:n){
    mass[i] ~ normal(pop_mean, pop_sd);
  }
}                                     // This is the last line of Stan code
" )
```

As we noted above, the model is divided into `data`, `parameters`, and `model` sections each wrapped in brackets and with the double-backslash now used for commenting. In the `data{}` section, we must provide detailed information about each element of our data list `dataList`, including both the *type* and *dimensions* of each data element as appropriate. In this case we have just two pieces of data: (1) the number of data points n, which must be an integer and thus has type `int`, and (2) the actual vector of mass values `mass`, which has type `vector` and has one dimension with n values.

Next, the `parameters{}` section contains the type and dimensions of parameters we are estimating in the model: the mean `pop_mean` and the standard deviation `pop_sd`. Both are of type `real`, meaning they are each a single value that can take on any real value (as opposed to `int` which can only be integers). However, we also tell Stan to force `pop_sd` to have a lower bound at 0, since you can't have a negative standard deviation.

Finally, the `model{}` section should look familiar to our JAGS model (with `normal` replacing `dnorm`). We provide priors on both `pop_mean` and `pop_sd` using ~. Note that the normal density is defined in terms of the standard deviation rather than in terms of the precision as in JAGS. We use a half-Cauchy prior for the standard deviation. This is a half-t distribution with 1 d.f., which resembles a half-normal distribution but has fatter tails. The model file is a text file which should have the extension `.stan` and which must not use in its name any periods, as we did for JAGS and NIMBLE; hence we use underlines instead. A final note about the model definition: Stan also supports vectorization in a manner similar to R. So, for example, you could define the likelihood simply as

```
mass ~ normal(pop_mean, pop_sd)
```

instead of using a for loop as we have done. As with R, vectorizing in this way will typically be faster. However in this book we generally write out the complete loop in order to make the code for the different engines more directly comparable and perhaps more intuitive for a beginner.

The configuration options for Stan's HMC/NUTS sampler look similar to those we previously used for MCMC with JAGS and NIMBLE. In general, HMC is usually slower than MCMC per iteration, but is far more efficient in the sense of producing posterior draws with much less serial correlation and thus with a better ratio of effective to nominal sample size per unit time. Hence, we will usually be fine with shorter chains, and sometimes much shorter ones, than in JAGS and NIMBLE with default samplers, and with little or no thinning.

```
# HMC settings
ni <- 1200 ; nb <- 200 ; nc <- 4 ; nt <- 1
```

We will allow Stan to pick random initial values for us and thus will leave argument `init` at its default. We also will not set the `pars` argument, which means Stan will save output for all parameters in the model. Therefore we now have everything we need to run Stan using the `stan` function.

```
# Call STAN (ART 55 / 3 sec)
system.time(
  out4.6 <- stan(file = "model4_6.stan", data = dataList,
    chains = nc, iter = ni, warmup = nb, thin = nt) )
```

Stan will first compile the Stan code in the model file, and then run the sampler. For simpler models, compilation can actually take much longer than sampling. When re-fitting the model in the same R session, the model will run much faster (in about 3 sec), since compilation only needs to be done once.

As with JAGS and NIMBLE, we can check convergence using trace plots:

```
# Check convergence
rstan::traceplot(out4.6)   # not shown
```

The structure of the summary output and the estimates from Stan closely resemble the output from JAGS.

```
# Print posterior summaries
print(out4.6)

# Save estimates
stan_est <- summary(out4.6)$summary[1:2,1]

Inference for Stan model: model4_6.
4 chains, each with iter=1200; warmup=200; thin=1;
post-warmup draws per chain=1000, total post-warmup draws=4000.

              mean  se_mean   sd      2.5%       25%       50%       75%      97.5%
pop_mean    598.85    0.01   0.90    597.06    598.24    598.85    599.48    600.60
pop_sd       29.01    0.02   0.63     27.81     28.58     28.99     29.42     30.31
lp__      -3867.39    0.03   0.96  -3869.85  -3867.78  -3867.10  -3866.70  -3866.45
             n_eff   Rhat
pop_mean      4272      1
pop_sd        1521      1
lp__          1435      1
```

The final row called `lp__` gives the log-predictive density of the data under the model, and not the deviance as with JAGS. This is the sum over all data of the log-density of the data under the model (see Chapter 18). As with JAGS, the output object in R is huge and should be inspected.

```
str(out4.6)    # not shown
```

There are a few options to access the full set of HMC samples from our Stan output. The first option is to use the `extract()` function, which returns a list of samples for each parameter.

```
sims <- extract(out4.6)
lapply(sims, head)    # not shown
```

We can also get the posterior samples in `mcmc.list` format as we have done for the other engines, allowing us to re-use our previous code.

```
sims <- As.mcmc.list(out4.6)                   # Note capital 'A' !
lapply(sims, head)                             # not shown
par(mfrow = c(2,2)); coda::traceplot(sims)     # not shown
```

We find Stan to be a powerful alternative to the BUGS-based packages. While the syntax of the Stan programming language may be a little more challenging to learn than BUGS (especially when you have discrete parameters and missing values in the response), its more organized structure and large number of supported features (e.g., more available built-in mathematical functions and control statements) make it a valuable choice for some more complex models. Furthermore, the HMC/ NUTS sampler Stan uses may work better for some models that have difficulty converging with JAGS. Stan is actively developed and its team are at the forefront of research on Bayesian methods, so new advances are often quickly available in Stan.

4.7 Maximum likelihood in R

4.7.1 Introduction to maximum likelihood in R

The first three engines we have discussed (JAGS, NIMBLE, and Stan) are primarily for fitting a model in a Bayesian framework. In the final two sections of this chapter, we examine maximum likelihood approaches. First, we'll look at doing maximum likelihood estimation from scratch in R. We call this "do-it-yourself MLE" or DIY-MLE. We also refer you to Chapter 2 for more detailed information about maximum likelihood, especially about how to obtain standard errors and confidence intervals.

In some ways this is the simplest "engine" for estimating parameters using our model of the mean. There are really only three steps: (1) write the mathematical formulation of the likelihood for the model (see Chapter 2); (2) convert the math into an R function; and (3) optimize the function for your data set, i.e., identify the parameter values for which the value of the likelihood function is maximized (hence, "maximum likelihood") when the function is applied to your data. Note that rather than maximizing the likelihood function, it is equivalent to find the parameter values that *minimize* the *negative* of the likelihood function. And for computational reasons it is more common to work on the log scale, so the final R function is written to minimize the negative log-likelihood. In practice, step 1 can be very hard, especially for hierarchical or otherwise complex models or new models for which you have to figure out the math yourself (see Chapters 10, 14, 17, 19, and 19B for examples). In some cases it can also be difficult to find reasonably fast-running R code corresponding to a given model.

We must admit that there are probably only a few scenarios in which this DIY maximum likelihood approach would be your best choice for a real analysis. Writing likelihoods in R can be challenging, and estimation of parameters can often be slow even compared with Bayesian approaches, as you will see for instance in Chapter 10. The primary value of our DIY approach is educational. Writing out the likelihood math, and translating it to code, is a very powerful method for ensuring that we fully understand the model we are applying. Furthermore, maximum likelihood is the underlying method for many of the canned functions in R, so having some experience doing it ourselves can help us understand, use, and even contribute to these packages more effectively. And, finally, the likelihood is an integral part of any model fit also in the Bayesian mode of inference, and you will benefit greatly if you have an understanding of likelihood that comes from working with it as we do in the DIY approach in this book.

Here is the workflow for DIY maximum likelihood.

1. Write out the mathematical formula for the likelihood of our model, or (if we're lucky!) find it in the literature
2. Translate this math into an equivalent negative log-likelihood function written in R
3. Organize the input data in R which must correspond to the structure of our R function
4. Supply our data and R function to the built-in `optim()` function in R which performs the optimization
5. Extract maximum likelihood estimates of the parameters, along with standard errors obtained by inverting the Hessian, from the output of `optim()`, and compute Wald-type confidence intervals (CIs) (see Chapter 2)

4.7.2 **Fit the model using maximum likelihood in R**

For our "model of the mean", we're modeling our body mass measurements as coming from a normal distribution. We'll start with the probability density function for the normal distribution, which you can find in any statistics textbook and also in our Chapters 2 and 3:

$$f(y_i|\mu, \sigma) = \frac{1}{\sigma\sqrt{2\pi}} e^{-\frac{1}{2}\left(\frac{y_i-\mu}{\sigma}\right)^2}$$

This equation calculates the probability density of obtaining a certain value of body mass y_i from a normal distribution with mean μ and standard deviation σ. The equivalent R code is simply

```
dnorm(y[i], mu, sigma)
```

Recall from Chapter 2 that on average, the probability density for a given, fixed y_i will be larger the closer the values of μ and σ are chosen to the "true" values of the population parameters (which we know from our simulation to be 600 and 30 g, respectively). We can re-arrange the equation slightly to express this more directly, i.e., to show that the likelihood is a function of the parameters or a given data set:

$$L(\mu, \sigma|y_i) = \frac{1}{\sigma\sqrt{2\pi}} e^{-\frac{1}{2}\left(\frac{y_i-\mu}{\sigma}\right)^2}$$

This equation is algebraically identical to the density, but it calculates the *likelihood* (*L*) that parameters μ and σ equal certain values, given our fixed body mass data. We know that measurement y_i came from a normal distribution with parameter values $\mu = 600$ and $\sigma = 30$ (we simulated it). Thus, arbitrarily choosing our first body mass value y_1, we should expect that the value of this function should be relatively high when we plug in 600 and 30, and relatively low if we pick two values that are further from the known true values, such as 550 and 25:

```
dnorm(y10[1], mean = 600, sd = 30)

[1] 0.01298933

dnorm(y10[1], mean = 550, sd = 25)

[1] 0.00124098
```

We could sit here all day manually plugging in various combinations of μ and σ trying to find the largest possible result (i.e., the largest possible likelihood): that's what we had the computer do when we performed a grid search for the MLEs in Chapter 2. And that's basically also what we'll do later in this section, with help from the computer, except that our function minimization will use information about the gradient of the likelihood when searching for the MLEs, i.e., the local slope at each point of the likelihood surface. Doing this speeds up the search for the extremum relative to a brute-force grid search.

Up to this point we've expressed the likelihood only for a single data point, but of course we actually have multiple data points (1000 in the larger dataset). We want to calculate the likelihood of μ and σ using the entire vector of body masses **y**. If we assume that our measurements of body mass are statistically independent, which is often a reasonable assumption, and independently distributed, then the total likelihood for the entire dataset is simply the product of the likelihoods for each individual body mass measurement

$$L(\mu, \sigma|\mathbf{y}) = \prod_{i=1}^{n} \frac{1}{\sigma\sqrt{2\pi}} e^{-\frac{1}{2}\left(\frac{y_i-\mu}{\sigma}\right)^2}$$

Here, n is our total sample size. We can easily calculate this in R as well using the product function in R:

```
prod(dnorm(y10, mean = 600, sd = 30))
```

```
[1] 2.495448e-22
```

We are multiplying 10 already small values together, so the result is *really* small. However, as before, if we plug in values for the parameters we know are further away from the truth (which here we know, because we simulated these data), we'll get an even smaller number.

```
prod(dnorm(y10, mean = 550, sd = 25))
```

```
[1] 2.229322e-37
```

We now have our likelihood equation for the model of the mean. Before moving on to step 2, however, there is one final consideration. For ease of computation and other reasons beyond the scope of this book, it is common when working with maximum likelihood to express the formula, or at least the corresponding code, in terms of the *log*-likelihood (to be specific, the natural log) rather than likelihood. To get there we just need to take the log of both sides of our likelihood equation, or alternatively specify an extra argument in R. Keeping in mind that the log of a product of values is equal to the sum of the logs of each value, the equation now looks like

$$\log L(\mu, \sigma | \mathbf{y}) = \sum_{i=1}^{n} \log \left(\frac{1}{\sigma\sqrt{2\pi}} e^{-\frac{1}{2}\left(\frac{y_i-\mu}{\sigma}\right)^2} \right)$$

Of course you could further simplify the right-hand-side of the equation, but we'll leave that as exercise for you … or you can look it up in books like Millar (2011) or Pawitan (2013). The equivalent R code is simply this:

```
sum(dnorm(y10, mean = 600, sd = 30, log = TRUE))
```

```
[1] -49.7424
```

Our goal in step 2 of the workflow is to write an R log-likelihood function that corresponds to the equation above. This function needs two arguments: the first should be a vector of model parameters (which in this case will contain only μ and σ) and the second should be our vector of body mass data. Inside the function will be a modified version of the code we just wrote above. We'll call the function LL for log-likelihood, and instead of μ and σ we will call the parameters pop.mean and pop.sd to match the other engines we've worked with so far. Throughout the book, we will highlight the R functions for the negative log-likelihoods in light peach color.

```
# Definition of LL for a normal linear model with intercept only
LL <- function(param, y) {
  pop.mean <- param[1]                          # First parameter is mean (mu)
  pop.sd <- param[2]                            # Second is SD (sigma)
  sum(dnorm(y, mean = pop.mean, sd = pop.sd, log = TRUE))
}
```

If we plug in parameter values and our dataset to the function we should get the same result as above:

```
LL(c(600, 30), y10)
```

```
[1] -49.7424
```

With the likelihood function written, we have completed step 2. In this example, step 3 is easy as our data is a single vector. In step 4 we want to *optimize* our likelihood function for our data. That is, we want to identify the set of parameter values for `pop.mean` and `pop.sd` which result in the largest possible (i.e., the maximum of the) likelihood, when we apply the function to our data set. The result is our maximum likelihood estimates (MLEs) for those parameters. R has a built-in function for optimizing other functions, called `optim()`. Here are the key arguments for `optim`:

```
optim(par, fn, method, lower, upper, hessian, ...)
```

- `par` is a vector of starting values for each parameter in the model
- `fn` is the function to optimize
- `method` is the algorithm to use for searching across possible parameter values – we won't specify it here but we will use it in later chapters. The default is called 'Nelder-Mead' but sometimes we will find that 'BFGS' works better.
- `lower` and `upper` allow us to specify lower and upper bounds on possible values for each parameter. Again, we won't use this here but we will in the future.
- `hessian` specifies if `optim` should return the Hessian matrix. We need this to calculate the standard errors of the MLEs (see Chapter 2 and below).
- ... Any other arguments we provide will be passed along as arguments to our function. In this case we also need to provide our data, `y`.

Picking the starting values can be tricky. It's helpful to make an educated guess, and to not pick values that are mathematically impossible (such as a negative standard deviation). Though we know the real values, we'll pretend we don't and arbitrarily guess 550 for the mean and 25 for the standard deviation. You may need to try a few combinations of values when fitting a complex model but it shouldn't be an issue for a trivial example as ours here.

```
starts <- c(pop.mean = 550, pop.sd = 25)
```

Now for our function to optimize. The documentation for `optim` indicates that by default, it finds the parameter values that *minimize* a function, rather than the ones that maximize it. Not to worry – instead of maximizing our log-likelihood function, we can just minimize the negative of our log-likelihood function, which is equivalent and will yield the same results. We need to make a tiny adjustment to our function `LL`, which we now call `NLL` (for negative log-likelihood):

```
# Definition of NLL for a normal linear model with constant mean
NLL <- function(param, y) {
  pop.mean <- param[1]              # First parameter is mean (mu)
  pop.sd <- param[2]                # Second is SD (sigma)
  -sum(dnorm(y, mean = pop.mean,
    sd = pop.sd, log = TRUE))       # Note minus sign
}
```

We can now call `optim()` using our starting values, NLL function, and data (we'll use the large dataset), specifying that we want to return the Hessian matrix. You may see a warning message but you can ignore it here.

```
out4.7 <- optim(starts, fn = NLL, hessian = TRUE, y = y1000)
```

First we should make sure the optimization worked properly ("converged"). We want a value of 0.

```
out4.7$convergence
```

```
[1] 0
```

The MLEs are contained in $par and should look similar to the results from other engines. They're in the same order they were in our function (pop.mean then pop.sd).

```
out4.7$par
diy_est <- out4.7$par    # save estimates
```

```
pop.mean    pop.sd
598.8533   29.0038
```

To get the standard errors for the MLEs, we first need to invert the Hessian matrix (found in $hessian) which results in the estimated variance-covariance (VC) matrix. Then we take the square root of the diagonal to get the asymptotic standard errors (ASEs). Remember from Section 2.5.3 that the Hessian is the matrix of the 2nd partial derivatives of the log-likelihood with respect to the parameters, and that inverting the negative of the Hessian gives an estimate of the variance-covariance (VC) matrix (Millar, 2011). However, since we work with the negative log-likelihood, what optim() calls 'the Hessian' is in fact the negative of the Hessian matrix of the log-likelihood. Thus, to obtain the VC matrix, we can directly invert the Hessian given by optim().

```
vcov <- solve(out4.7$hessian)    # invert Hessian
ASE <- sqrt(diag(vcov))          # Get asymptotic SEs
ASE
```

```
  pop.mean        pop.sd
0.9171808    0.6485921
```

We've included a function get_MLE in the ASMbook package which does these calculations, computes Wald-type 95% CIs, and organizes the result into a table.

```
get_MLE(out4.7, 5)
```

```
              MLE       ASE    LCL.95    UCL.95
pop.mean  598.853   0.91718   597.056   600.651
pop.sd     29.004   0.64859    27.733    30.275
```

Our estimates are very similar to those from the previous engines, and particularly the results from the canned lm function (although recall lm() uses least squares, not maximum likelihood, and that the residual SE is computed based on n-1 there).

Our first from-scratch maximum likelihood analysis is complete! This is clearly a more challenging approach to fitting a model, even for our simple model of the mean. In return, however, you have hopefully gained a much deeper understanding of this simple model, and especially of how the method of maximum likelihood works more generally. As we noted previously, do-it-yourself maximum likelihood will rarely be the best choice for your actual model fitting, but we think that it is very powerful for your understanding of this foundational method. From a practical standpoint, it can be difficult to write the correct log-likelihood function for complex models, as we will see in later chapters, especially for models with random effects. Also, the method doesn't deal directly with missing values in the response (see Chapter 11). Furthermore, this approach can be very slow given that R needs to call the log-likelihood function hundreds or thousands of times. One solution to the speed issue is to write the log-likelihood function in a programming language that is much faster than R, such as C++. That's what we'll show in the next section with Template Model Builder.

4.8 Maximum likelihood using Template Model Builder (TMB)

4.8.1 Introduction to Template Model Builder

Template Model Builder (TMB) is a general-purpose R package for fitting statistical models using maximum likelihood (Kristensen et al., 2016; Thorson & Kristensen, 2024). The first version of the TMB package was released in 2015. It is inspired by the AD Model Builder (ADMB) software package (Fournier et al., 2012), providing similar functionality, but easier access via R.

Both ADMB and TMB apply automatic differentiation (AD), a computational technique which can obtain derivatives (first, second, third, and so on) of a function such as a likelihood function. For many functions, obtaining these derivatives analytically (i.e., mathematically, "with pen and pencil"; Hooten & Hefley, 2019) would be difficult or impossible. Access to the higher-order derivatives of the likelihood function is rarely necessary for optimizing relatively simple models such as the one in this chapter. However, they are valuable for more complex models that include random effects. For example, TMB uses a method called Laplace approximation to integrate out random effects in a model and then optimizes the resulting marginal likelihood (Royle & Dorazio, 2008: chapter 2; Kéry & Royle, 2016: chapter 2). Laplace approximation requires use of the higher-order derivatives of the likelihood function obtained from AD. The details of AD and the Laplace approximation are far beyond the scope of this book; however in Chapters 10, 14, and 17 we will demonstrate how much easier TMB makes it to handle models with random effects compared to our DIY approach. This lets one fit even complex hierarchical models which were hitherto only possible to fit with Bayesian computational techniques (Thorson & Kristensen, 2024). Note that NIMBLE is also capable of fitting models with Laplace approximation.

As with most of the previous engines we've discussed, fitting a model with TMB involves some work in R and some work in another language. Specifically, the likelihood function must be written in C++. If you are mainly familiar with R, writing code in C++ may be a bit challenging at first. Some of the differences between R and C++ will resemble the differences between R and the Stan programming language (Section 4.6). We'll walk through writing a TMB model in C++ in the next section. Using C++ also means that R will need access to a C++ compiler, as with NIMBLE and Stan (see Section 4.5.1). In exchange, optimization with TMB will typically be much faster than optimizing with an R likelihood function. The part of the TMB workflow conducted in R will resemble the DIY-MLE approach from the previous Section (4.7) – specifically, the use of the `optim` function. Here's our complete workflow for TMB:

1. Organize the input data in an R list
2. Write out the mathematical formula for the likelihood of our model, or find it in the literature
3. Translate this math into an equivalent negative log likelihood function written in C++, according to TMB conventions, and compile it
4. Describe the dimensions of all estimated parameters in the model using an R list structure
5. Create a TMB model object with function `MakeADFun` using the data, parameters, and compiled C++ function
6. Supply TMB model object to R's `optim()` function which performs the optimization
7. Extract the MLEs of the parameters, along with standard errors and confidence intervals, from the TMB model object

4.8.2 **Fit the model using TMB**

As with the other engines, we need to load the TMB library.

```
library(TMB)
```

For this model, we can use the same list of data as the one we used for JAGS, NIMBLE, and Stan.

```
# Bundle data
str(dataList <- list(y = y1000, n = length(y1000)))
```

Our next step is to write out the likelihood for the model of the mean, which we've already done in Section 4.7.1. Therefore, we'll jump right in to translating the likelihood to C++ code for use with TMB. Rather than showing the entire model first, in this case we will explain things line by line.

The first several lines of the model code contain boilerplate, but required, setup code for the function that will be identical for all TMB models we fit in the book. We won't talk about most of this boilerplate code, and you don't need to worry about fully understanding it to use TMB. We will note one thing: recall from our Section 4.6.2 the concept of parameter *types*, which define what kind of value a particular parameter can take on, i.e., integer, real number, etc. When writing a function in C++, we must start by explicitly declaring the type of value that the function will return. In this case (and in all future TMB code), the function type will be, confusingly, a special type called Type (note the capital T). This is because TMB uses a feature of C++ called *templates*, which are useful for writing flexible and re-usable code. A detailed explanation of the use and value of templates is again far beyond the scope of this book.

```
#include <TMB.hpp>
template<class Type>
Type objective_function<Type>::operator() ()
{
```

We're now in the body of our function. In the first section inside the function, we must describe the type of each element of our input data in a manner similar to the data{} section in Stan. We identify n as an integer and y as a vector using DATA_INTEGER and DATA_VECTOR respectively. Also like Stan, we need to end every line without a closing/opening bracket in a semi-colon.

```
DATA_INTEGER(n);
DATA_VECTOR(y);
```

Next, and again like the Stan section parameters{}, we need to tell TMB about the parameters we want TMB to estimate for us. The two key parameters in this model are the population mean pop_mean and population SD pop_sd. The population SD must be a positive value. This can cause problems during optimization, because there is (typically) nothing stopping an optimization algorithm from assigning a negative value to pop_sd instead. One solution to this problem is to specify pop_sd on the log scale instead (i.e., log_pop_sd), and then within our likelihood function calculate pop_sd as exp(log_pop_sd). This will ensure pop_sd is always positive. We did not do this in Section 4.7, and the optimization still worked, but in this section and future chapters we will always work with standard deviations on the log scale.

Both `pop_mean` and `log_pop_sd` are real-valued scalars, so we use the `PARAMETER` function (we will use others such as `PARAMETER_VECTOR` in future chapters).

```
PARAMETER(pop_mean);
PARAMETER(log_pop_sd);
```

After defining the types of the input data and key model parameters, the rest of the function can be structured essentially however we want. For convenience, the first line of this section will calculate `pop_sd` from `log_pop_sd`.

```
Type pop_sd = exp(log_pop_sd);
```

When first declaring or calculating any new parameter (such as `pop_sd`) in our C++ code, we need to define the type of the resulting parameter at the start of the line. For reasons related to TMB's use of templating, essentially all new values we calculate in our model code will have the same basic type as the function itself, i.e., `Type`.

Next, we initialize the total log-likelihood at 0, again specifying `Type`:

```
Type LL = 0.0;
```

Finally we are ready to actually calculate the log-likelihood. As with our DIY likelihood function, we will iterate over each of our n data points, calculate the log-likelihood of that data point using `dnorm`, and add it to our total log likelihood `LL` using a for loop. This loop will look a little different than how it is done in R, but the similarities should nevertheless be evident. One of the most important things to remember is that in C++, and thus in TMB, the index of the first element in a vector is 0, not 1 as it is in R, JAGS, NIMBLE, or even in Stan. Therefore, for loop indexing must start at 0.

```
for (int i=0; i<n; i++){
  LL += dnorm(y(i), pop_mean, pop_sd, true);
}
```

The `LL += something` syntax is shorthand for saying `LL = LL + something`. The main part of this code (the call to `dnorm`) looks almost identical to R, except that instead of brackets `[]` we use `()` to indicate the *i*-th element of vector `y`. As with R, the final `true` argument to the `dnorm` function indicates that we want the log-likelihood to be returned.

Finally, at the end of the function code, we say that we want the negative of total log-likelihood to be returned:

```
return -LL;
```

Here is our complete TMB code for the model of the mean, which we will write to a `.cpp` (for C plus plus) file using `cat` in R. As in Stan models, comments in C++ code are marked with `//` instead of `#` like in R. Throughout the book, we will highlight the TMB model file in peach color.

```
# Write TMB model file
cat(file = "model4_8.cpp",
"#include <TMB.hpp>

template<class Type>
Type objective_function<Type>::operator() ()
{
  // Describe input data
  DATA_INTEGER(n);              // Sample size
  DATA_VECTOR(y);               // observations

  // Describe parameters
  PARAMETER(pop_mean);          // Mean
  PARAMETER(log_pop_sd);        // log(standard deviation)
  Type pop_sd = exp(log_pop_sd);   // Type = match type of function output

  Type LL = 0.0;                // Initialize total log likelihood at 0

  for (int i=0; i<n; i++){      // Note index starts at 0 instead of 1!
                                // Calculate log-likelihood of observation
                                // value of true = function should return log(lik)
    LL += dnorm(y(i), pop_mean, pop_sd, true);
  }

  return -LL;                   // Return negative of total log likelihood
}
")
```

We can now compile our completed model file and load the result into our R session. This takes a little while.

```
# Compile and load TMB function
compile("model4_8.cpp")   # Produces gibberish ... have no fear !
dyn.load(dynlib("model4_8"))
```

Next we need to tell TMB about the dimensions of our parameters, which we do by creating a starting value for each parameter and combining them in a named list. Note that the list names must exactly match the names of the parameters in your model code! We'll set the starting values at 0 for both parameters. Remember that we're working with the log of the population standard deviation, so it is OK to start with a value of 0.

```
# Provide dimensions and starting values for parameters
params <- list(pop_mean = 0, log_pop_sd = 0)
```

Using our input data, compiled model, and parameter information, we create a TMB object with the function `MakeADFun` that is ready to be optimized. The function has the following key arguments:

```
MakeADFun(data, parameters, random, DLL, silent)
```

Here's some basic information about each argument:

- `data` is our named list of data components
- `parameters` is our named list of parameter starting values
- `random` is a vector with the names of the parameters in `parameters` which should be considered random effects. For this model we specify `NULL` because we don't have any random effects
- `DLL` is the name of our compiled C++ model.
- `silent` is a switch which suppresses some console messages

```
# Create TMB object
out4.8 <- MakeADFun(data = dataList, parameters = params, random = NULL,
  DLL = "model4_8", silent = TRUE)
```

Our initial creation of this object simply checks that the function works and the data and parameters have the correct dimensions. We haven't actually optimized the function yet. We do that in the next step, using `optim()` and some components of the TMB object. The TMB object contains starting values (that we previously provided) in `$par`, the likelihood function in `$fn`, and the gradient function (obtained using AD) in `$gr`. Access to the gradient function is a key contribution of TMB – it represents the derivative of the likelihood function and greatly speeds up optimization. We specify an optimization method that makes use of the algorithm called "BFGS".

```
# Optimize TMB object
opt <- optim(out4.8$par, fn = out4.8$fn, gr = out4.8$gr,
  method = "BFGS", hessian = TRUE)
```

As in Section 4.7.2, `opt` contains all the standard output from running the `optim()` function, and we can see the estimated parameter values using

```
opt$par
   pop_mean    log_pop_sd
 598.846095      3.367378
```

Remember that the SD is on the log scale. This information has also been saved into the TMB model object `out4.8` and can be accessed using the `summary` and `sdreport` functions, which return the estimates and their standard errors, which are again obtained by inverting the Hessian given by `optim()` as discussed in the last section.

```
summary(sdreport(out4.8))

             Estimate Std.       Error
pop_mean      598.846095    0.91713599
log_pop_sd      3.367378    0.02236067
```

We've also included a summary function in the ASMbook package to quickly summarize TMB input.

```
tsum <- tmb_summary(out4.8)          # not shown
tmb_est <- tsum[,1]                  # save results
tmb_est[2] <- exp(tmb_est[2])        # convert SD from log scale
```

TMB is probably the most challenging of the engines we present in this book, due to the need to write C++ code. As Stan and our DIY-MLEs, it can't automatically deal with either discrete parameters or missing values in the response. But the good news is that we've already covered the hardest parts of the new language – the boilerplate C++ code, types, and indices starting at 0. Hopefully, based on the example in this chapter you are beginning to see the similarities between C++ and R code. We believe the initial work getting comfortable with C++ will pay big dividends later – as you will see, TMB will do a much better and faster job than R once we start fitting more complex models in later chapters. Additionally, while we wrote this book, a new R package called RTMB was released which allows for specifying TMB models directly with R code instead of C++ (Kristensen, 2023). This should make TMB much more accessible in many cases. We note that there is also an interface between TMB and Stan called `tmbstan` (Monnahan & Kristensen, 2018). Finally, if you like maximum likelihood but still want to fit complex statistical models in population and community ecology, you ought to check out the brandnew book by Thorson and Kristensen (2024).

4.9 Comparison of engines and concluding remarks

In the sections above, we've been saving the parameter estimates (population mean and population SD) from each engine. We'll now build a comparison table for the estimates from every engine starting with the canned `lm()` function all the way to TMB.

```
comp <- cbind(truth = truth, lm = lm_est, JAGS = jags_est,
  NIMBLE = nimble_est, Stan = stan_est, DIY = diy_est, TMB = tmb_est)
print(comp, 5)

         truth      lm     JAGS   NIMBLE      Stan      DIY      TMB
mean       600 598.846  598.848  598.864   598.852  598.853  598.846
sd          30  29.017   29.061   29.045    29.012   29.004   29.002
```

The estimates from all five engines are very similar to each other, to the least-squares estimates from our canned function `lm()`, and to the true values we used in simulating the data set. We want to make two points based on this result. First, there may be many reasons to choose one general-purpose model fitting engine over another (see below), but regardless of which one you choose, you should expect to obtain similar answers if the model is fit correctly, and if you specify vague priors in a Bayesian analysis, and sample size isn't really small. Second, there is value in fitting your model using multiple engines. If you find that two engines give different results, it likely means you've made a mistake somewhere.

The engines have advantages and disadvantages relative to each other and Table 4.1 shows our opinions on these trade-offs. Speed, ease-of-use, and flexibility to handle customizations are all key considerations when choosing an engine to use. For example, if your desired model is relatively standard and widely used, and speed and ease of use are important, then R's various canned functions may be the best choice. If you have a complex custom model with many latent parameters you expect will benefit from a Bayesian analysis, then NIMBLE or Stan may be the best options. In addition, there is the issue of discrete parameters and missing values in the response which can be handled directly by JAGS and NIMBLE (and the canned functions in R), but not by the other engines. We hope that the side-by-side comparisons of the code and workflow for each engine in each chapter of this book will give you the skills to switch between the engines as your needs require.

In terms of statistical modeling, we have covered one of the simplest possible models, which assumes a normal response and has one parameter in the mean and another for the dispersion of the modelled random variables. All other chapters in the book will now add in more complexity, first in the mean (e.g., by adding effects of continuous or discrete explanatory variables) and later also in the variance (by adding additional sources of randomness via random effects).

Table 4.1 Comparison of the general-purpose model fitting engines used in this book.

Engine	Framework	Language	Runtime	Flexibility	Difficulty
Canned R functions	Max likelihood (usually)	R	Fast	Low	Easy
JAGS	Bayesian	BUGS	Slow	Medium	Medium
NIMBLE	Bayesian (primarily)	BUGS	Slow	Medium-high	Medium
Stan	Bayesian (primarily)	Stan	Slow	Medium-high	Medium-hard
DIY maximum likelihood	Maximum likelihood	R	Medium	High	Hard
TMB	Maximum likelihood	C++	Fast	High	Hard

Normal linear regression[⊛]

5.1 Introduction

In Chapter 4, we introduced the "model of the mean," in which a series of measurements y_i is modeled as coming from a normal distribution with constant mean μ and constant variance σ^2:

$$y_i \sim Normal(\mu, \sigma^2)$$

In this simple case, we assumed that a single parameter for the mean (μ) was adequate to model the measurements y_i. But what if we also measured another continuous variable, call it x_i, which we believe affects the value of y_i? Incorporating this variable may improve our model. In addition, it lets us test whether our expectation of an association between x and y is met, and enables us to measure the strength of any such relationship. We can do this by assuming that (1) the value of the mean μ varies by measurement i thus becoming μ_i, and (2) the value of μ_i depends on, that is, is a function of, x_i. Mathematically, this looks like:

$$\mu_i = \alpha + \beta \cdot x_i$$

[⊛]This book has a companion website hosting complementary materials, including all code for download. Visit this URL to access it: https://www.elsevier.com/books-and-journals/book-companion/9780443137150.

Applied Statistical Modelling for Ecologists. DOI: https://doi.org/10.1016/B978-0-443-13715-0.00007-8

which you should recognize as the equation for a straight line with intercept α and slope β. The intercept is the expected response when x is zero, and the slope is the expected change in the response as you change x by 1 unit. If the slope β is positive, the mean μ_i (and therefore y, on average) will increase as x increases (i.e., x has a positive effect on y through its expectation μ), and if β is negative, μ will decrease as x increases (x has a negative effect on y and on μ). If we plug this into our original equation from Chapter 4, we get:

$$y_i \sim Normal(\alpha + \beta \cdot x_i, \sigma^2)$$

By explicitly separating the deterministic part of response y_i (the straight line) from its stochastic part (the normal distribution with variance σ^2), we get an equivalent set of equations:

$$y_i = \alpha + \beta \cdot x_i + \varepsilon_i$$
$$\varepsilon_i \sim Normal(0, \sigma^2)$$

The deterministic part, that is, the linear model $\alpha + \beta \cdot x_i$, is also called the linear predictor of this model, and the stochastic part ε_i is called the residual error, or the error term, which is assumed to be distributed normally with mean 0 and variance σ^2. This model is called a normal linear regression model, or just a linear regression. In its generalization to the generalized linear model (GLM) (see Chapters 3 and 11), the linear regression model is the foundation of basically all analyses we cover in this book. Our main goal with linear regression analysis is to estimate the parameters in the model, which here are: α, β, and σ^2.

As a motivating example for a linear regression analysis we take an imaginary survey of the wallcreeper in Switzerland (Fig. 5.1). This is a spectacular little cliff-inhabiting bird that

FIGURE 5.1

Wallcreeper (*Tichodroma muraria*), Switzerland, 2022 (Photo by Andi Meier).

appears to have declined greatly in Switzerland around the turn of the last century. Assume that we had data on the percentage of sample quadrats in which the species was observed in Switzerland for the years 1990–2005. In addition, let's assume that the random deviations (i.e., the residual errors) about a linear time trend (around the linear predictor) are normally distributed. This is for illustration only: usually, we would use logistic regression (Chapters 15–17) or an occupancy model (see Chapter 19) to make inferences about such data on the distribution of a species that represent a percentage or proportion (i.e., number occupied/number surveyed). However, if observed proportions are well away from the boundary of 0 and 1, a normal model may be adequate.

Importantly, in this chapter, we will also briefly cover model assessment (goodness-of-fit, GOF), that is, whether our model is adequate for our data (see also Chapter 18 for a more in-depth treatment). A key quantity for this is the *residual*, which is a measure for the discrepancy between the observed and the expected response, that is, an estimate of the ε_i above. We review two types of residuals in a classical analysis and also introduce posterior predictive model checking, including computation of the Bayesian p-value (also called the posterior predictive p-value; Gelman et al., 1996; Gelman & Hill, 2007: Chapter 24). This is a versatile method for checking the GOF of a model when using simulation-based Bayesian inference, that is, of making a judgment of how good a model is in absolute terms.

Finally, we also introduce the computation of predictions or fitted values in a regression model. These are estimates of the expected response for specific chosen values of the explanatory variables. Tabulation and plotting of predictions is a key method for presenting the results of fitting a regression model, and understanding what it tells us in the first place.

5.2 Data generation

We simulate a simple wallcreeper data set. First we will set the sample size n and values for the model parameters α, β, and σ^2:

```
set.seed(5)
n <- 16                    # Number of years
a <- 40                    # Intercept
b <- -0.5                  # Slope
sigma2 <- 25               # Residual variance

# Save true values for later comparisons
truth <- c(alpha = a, beta = b, sigma = sqrt(sigma2))
```

Next we set the values for the covariate x, which represent years (which we let run from 1–16 in the actual analysis, but from 1989–2005 in our story and in the plots), and simulate the measurements y_i. Note the similarity between the R code here and the two equations at the end of the previous section. This illustrates our frequent claim that R data simulation code is a powerful manner of defining and explaining a statistical model. We end by plotting the observed percentage occupied over time. Plotting is an integral part of learning from data.

```
# See Fig. 5.4 for a variant of this plot
x <- 1:16                                          # Values of covariate year
eps <- rnorm(n, mean = 0, sd = sqrt(sigma2))       # Residuals
y <- a + b*x + eps                                 # Assemble data set
plot((x + 1989), y, xlab = "Year", las = 1,
  ylab = "Pct occupied (%)", cex = 1.5,
  pch = 16, col = rgb(0,0,0,0.6), frame = FALSE)    # not shown

# Load required libraries
library(ASMbook); library(jagsUI); library(rstan); library(TMB)
```

5.3 Analysis with canned functions in R

In this section and the next, we cover model assessment, prediction, and GOF using residuals in a normal linear model fit with frequentist methods. We introduce both "traditional" residuals (i.e., the simple difference between observed and expected response/fitted values) and the more general *randomized quantile residuals* (Dunn & Smyth, 1996). Quantile residuals have the advantage that they readily generalize to GLMs as well as to mixed and other hierarchical models (e.g., Gelman & Hill, 2007; Warton et al., 2017; Hartig, 2022). Moreover, they can similarly be applied in a classical analysis using maximum likelihood and in a Bayesian analysis with posterior inference. Indeed, the simulation-based assessment of quantile residuals in R package DHARMa (Hartig, 2022; Section 5.3.2) corresponds exactly to the concept of Bayesian posterior predictive checks (PPC) (Section 5.4.2).

5.3.1 Fitting the model

Here is a classical analysis of our wallcreeper data using lm(), the built-in normal linear regression function in R. As a reminder, lm() uses a least-squares approach to parameter estimation, not maximum likelihood. The estimates of the parameters in the mean will be identical to maximum likelihood estimates (MLEs), and the estimate of the residual variance almost identical to the MLE of the variance in moderate to large samples (see Section 5.10). In the output, the estimates for the intercept and slope associated with year x are obvious. The estimate of σ^2 is a little less obvious: R calls it the residual standard error (SE).

```
# Fit model and save lm estimates
summary(out5.3 <- lm(y ~ x))
lm_est <- c(coef(out5.3), sigma = sigma(out5.3))

Coefficients:
             Estimate Std.   Error   t value   Pr(>|t|)
(Intercept)   40.7426  2.4017  16.964   9.89e-11 ***
x             -0.6907   0.2484  -2.781    0.0147 **
---
Signif. codes: 0 '***' 0.001 '**' 0.01 '*' 0.05 '.' 0.1 ' ' 1

Residual standard error: 4.58 on 14 degrees of freedom
Multiple R-squared: 0.3558, Adjusted R-squared: 0.3098
F-statistic: 7.734 on 1 and 14 DF, p-value: 0.01472
```

5.3.2 **Goodness-of-fit assessment using traditional and quantile residuals**

Assessments of the adequacy of regression models typically focus on properties of the residuals. Residuals can be defined in various ways (Cox & Snell, 1968; McCullagh & Nelder, 1989), but they always provide a metric of the difference between the observed response and what the model says the response should be. The latter is the fitted value, or expected response, in a regression model.

The simplest residual definition is that of the raw residual, which for each datum i is simply the difference between observed and fitted value of the response, $y_i - \hat{y}_i$, and equivalently also the difference between the observed and the expected value of the response $y_i - \hat{\mu}_i$.

```
# Print out traditional residuals (=raw residuals)
print(residuals(out5.3), 4)
```

1	2	3	4	5	6	7	8
-4.756	6.561	-6.448	0.371	8.768	-2.613	-1.768	-2.394
9	10	11	12	13	14	15	16
-0.455	1.855	7.494	-2.463	-3.665	1.140	-3.241	1.614

For instance, we see that the first data point is much lower than what we would expect based on the model, the second is much higher, and so forth. These residuals are the estimates of the ε_i in the model, that is, the unexplained part of the response for which we adopt a statistical distribution. For normal linear regression models, the normality and constant-variance assumptions are commonly assessed via histograms or QQ plots for the residuals, and plots of the observed versus predicted values, respectively, which we assume you are familiar with. In R, we can get these and other visual model assessments using the `plot` function on the fitted model object. For very small samples such as ours, these assessments may be of dubious value. We show them here mostly for illustrative purposes, because they have an important place in the workflow of a statistical analysis.

```
# Visually assess adequacy of fitted linear regression model
hist(residuals(out5.3), breaks = 20)      # not shown
plot(out5.3)                              # not shown
```

For GLMs, the expected variance of the residuals often depends on the mean. Therefore a residual definition that is commonly used for nonnormal GLMs is the Pearson residual, which is the raw residual divided by the expected SE of the fitted value (see Chapter 11). While Pearson residuals are preferable over raw residuals, they can often show apparent patterns due to the asymmetry of the GLM distribution or the discrete nature of the observed data. See McCullagh & Nelder (1989) for the deviance residual, another type of residual which can also be used for GLMs.

A more recently developed type of residual is called *randomized quantile residual*. The basic idea of these residuals is to express the discrepancy between observed and expected value of each datum by the percentile, or quantile, of the cumulative distribution function (CDF) of each response. The problem of discrete observations such as counts is solved by adding some noise that distributes residuals homogenously across the CDF, that is, by a sort of jittering. A quantile residual of 0 thus means that an actual observation is smaller than any observation that one might expect for the given covariates under the model, while a value of, say, 0.9 denotes an observation for which 10% of the values expected under the model are greater than the observed value. These quantile residuals are uniformly distributed, but can be transformed into whatever distribution is desired, for example, into a standard normal as in the original formulation by Dunn & Smyth (1996).

Quantile residuals can be easily computed for nonnormal GLMs or any other parametric model, including mixed or other hierarchical models, simply by simulating data from the fitted model. Thus we can simulate, say, 1000 data sets under our model and then compute the quantile residual for each observation as a percentile within the empirical CDF obtained from the 1000 simulations for that data point. That is, the residual for a datum y_i will be given by the proportion of the simulated replicates of that datum that are smaller than or equal to the observed value y_i. This simulation-based approach is implemented in the R package DHARMa (Hartig, 2022), which supports many of R's most widely used regression and mixed-modeling functions, including those used in our book. In addition, the package allows Bayesian posterior draws to be read in and quantile residuals to be computed for them as well (see package vignette *"DHARMa for Bayesians"*). When used in this way, the package does PPCs on the observed data as explained in Section 5.4.2.

Here we give a brief introduction to DHARMa for residual assessments in our normal linear model and in the next section we do the same for the case when we fit our model using Bayesian posterior inference with JAGS. In subsequent chapters, we provide a brief code block to do model assessments using quantile residuals for models fit with canned functions in R and for specific examples also provide code for the same for Bayesian posterior inference with JAGS, as just one illustration for fit assessments with DHARMa for Bayesian simulations.

The main function in DHARMa is called simulateResiduals and simulates n new data sets under the estimated model. The simulated data are then used to estimate the empirical CDF for every observation, and calculate the quantile residual. The default is $n = 250$, but since for all models in this book simulations are swift, we choose $n = 1000$ to obtain a better approximation of the empirical CDFs.

```
# Computation and goodness-of-fit assessment using quantile residuals
library(DHARMa)
simOut <- simulateResiduals(out5.3, n = 1000)
residuals(simOut)          # print out scaled quantile residuals
```

The simOut object we created contains the scaled quantile residuals. For a fitting model, they will be uniformly distributed on (0, 1) both when we produce a histogram as well as against any covariate.

```
[1]    0.161   0.932   0.088   0.555   0.977   0.268   0.358 0.302
[9]    0.438   0.661   0.942   0.298   0.191   0.596   0.234 0.644
```

The format of scaled quantile residuals that DHARMa defaults to is convenient for interpretation, since a uniform distribution can easily be assessed visually, and the scaling between 0 and 1 allows a direct interpretation of a residual as a percentile. However, it is less well suited for assessing extreme residuals, that is, to assess the tails of the distribution of residuals. For that, DHARMa allows transformation of the default 0–1-scaled quantile residuals into equivalent quantile residuals that are scaled as a standard normal distribution, as in the original definition by Dunn & Smyth (1996). Under a fitting model, these should follow a standard normal distribution.

```
# Normal-transformed quantile residuals
print(residuals(simOut, quantileFunction = qnorm), 4)
```

```
[1]    -0.9904   1.4909   -1.3532    0.1383    1.9954   -0.6189   -0.3638   -0.5187
[9]    -0.1560   0.4152    1.5718   -0.5302   -0.8742    0.2430   -0.7257    0.3692
```

There is a variety of graphical options and significance tests that can be applied to quantile residuals in DHARMa. Below we show two diagnostic plots (Fig. 5.2): a QQ plot (with several GOF test results) on the left, and a plot of the scaled residuals versus a predictor of choice on the right. See the package help files including the vignette (Hartig, 2022) for more information about these tests, additional

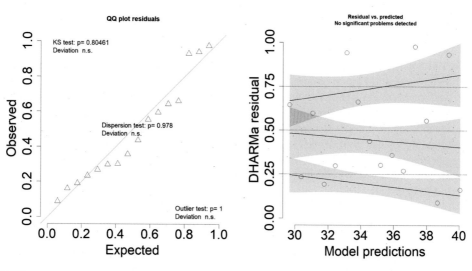

FIGURE 5.2

Model assessment in the normal linear regression model using scaled randomized quantile residuals in package DHARMa.

functions in general, and about options how to diagnose typical failures of model assumptions. (Note that rank = TRUE below produces a series of errors for our small sample. These are not important for the purposes of this book and thus we will not discuss them.)

```
par(mfrow = c(1,2))
plotQQunif(simOut)
plotResiduals(simOut, rank = FALSE)      # See comment above
par(mfrow = c(1,1))
testDispersion(simOut)                   # Test for over- or underdispersion
                                         # (not shown, and note this test really
                                         # makes sense only for nonnormal GLMs)
```

We cover GOF assessment via residuals in our example mostly for illustrative purposes, since the value of residual checks for small data sets is limited. However, this is an important topic of applied statistical modeling. Thus we add a short block of DHARMa functions in most "canned function" sections in the remainder of the book.

5.3.3 Computing predictions

Predictions are the expected values of a response variable or of some modeled parameter at selected values of the explanatory variable(s). For the model in this chapter, the predictions are given by $\hat{\mu}_i = \alpha + \beta x_i$. That is, to get predictions we simply plug into this equation the parameter estimates from the fitted model (i.e., the coefficients just estimated) along with one or more values of choice of the covariate x. When we plug into this equation the observed values of the covariate x, we get what is also called the *fitted values:* $\hat{y} = \alpha + \beta x_i$. More commonly perhaps, we plug into the equation a large number of possible values of the covariate within a meaningful range and compute the expected response, to depict the functional form of the estimated regression relationship.

Predictions are extremely important in applied statistical modeling for two reasons. First, predictions, especially when represented in a graph, are one of the best ways of communicating what can be learned from a model. Second, for more complex models (e.g., with polynomial terms, interactions, and for Poisson or binomial models; see Chapters 11–20), predictions may be the only way to understand what a model is telling us about the functional forms of covariate relationships. For the simple normal straight-line model in this chapter, we could of course simply look at magnitude and sign of the slope estimate to understand what the model is telling us about the population trend in Swiss wallcreepers. However, for models with two or more covariates, this may become less straightforward and it has been our experience that practitioners are often confused by how predictions are obtained once a model has been fit. Therefore we next compute predictions for our simple model and obtain SEs and confidence intervals (CIs) around these predictions. See also Section 9.3, where we compute predictions from a more complicated model with two explanatory variables.

In R there is a method for computing predictions called `predict()`. Its precise functioning depends on the fitted model object on which it is applied. Broadly, it typically does two things: first, it takes the parameter estimates and combines them with selected values of all covariates in the model by essentially plugging estimates and covariate values into the model equation and evaluating the result; and second, it typically also computes SEs and sometimes CIs. For a fitted model object from `lm()`, we must compute CIs ourselves (see Section 2.5.3). We use `predict()` to obtain the estimated regression line (also called "line of best fit"), which is simply a plot of the estimated mean response $\hat{\mu}$, for a large number of possible values of the covariate x. Note that a larger number of covariate values for which the regression equation is evaluated by `predict()` will lead to smoother plots of predictions. Here we choose 100 points for the covariate "year."

```
xpred <- seq(1, 16, length.out = 100)          # Predict for 100 values of x
newdata <- list(x = xpred)
pred <- predict(out5.3, newdata = newdata, se.fit = TRUE)   # also get SEs
LCL <- pred$fit - 1.96 * pred$se.fit           # Lower 95% confidence limit
UCL <- pred$fit + 1.96 * pred$se.fit           # Upper 95% confidence limit
pred.table <- cbind(x = newdata$x, "point_est" =
  pred$fit, "se" = pred$se.fit, "LCL" = LCL, "UCL" = UCL)   # Table with x and predictions
print(pred.table)                              # Shown only partially,
                                               # but see also Fig. 5.4

         x    point_est        se         LCL        UCL
1    1.000000     40.05187   2.186554   35.76623   44.33752
2    1.151515     39.94722   2.154583   35.72424   44.17020
3    1.303030     39.84256   2.122798   35.68188   44.00325
. . . .
98   15.696970    29.90035   2.122798   25.73967   34.06103
99   15.848485    29.79569   2.154583   25.57271   34.01868
100  16.000000    29.69104   2.186554   25.40540   33.97668
```

We will plot the values in this table in Fig. 5.4.

5.4 Bayesian analysis with JAGS

Next, we conduct a Bayesian analysis of the linear regression model with JAGS. The model includes code for two approaches to checking model fit to the data set: (1) residual plots and (2) a PPC plus computation of a Bayesian p-value (Gelman et al., 1996). These approaches are described in more detail in Section 5.4.2. For the residuals, we cover both traditional (here, raw) residuals and quantile residuals using functionality in package DHARMa (Hartig, 2022).

5.4.1 Fitting the model

We need to send three pieces of data to JAGS: the measurements y, the covariate values x, and the sample size n.

```
# Bundle and summarize data
str(dataList <- list(y = y, x = x, n = n))

List of 3
 $ y :  num   [1:16]   35.3  45.9  32.2  38.4  46.1  ...
 $ x :  int   [1:16]   1  2  3  4  5  6  7  8  9  10  ...
 $ n :  num   16
```

We write the JAGS model to a file in the usual way using `cat()`. We show the entire model below and discuss it afterwards.

```
# Write JAGS model file
cat(file = "model5.4.txt", "
model {
# Priors
alpha ~ dnorm(0, 0.0001)
beta ~ dnorm(0, 0.0001)
sigma ~ dunif(0, 100)              # May consider smaller range
tau <- pow(sigma, -2)

# Likelihood
for (i in 1:n) {
  y[i] ~ dnorm(mu[i], tau)         # Response y distributed normally
  mu[i] <- alpha + beta * x[i]     # Linear predictor
}

# Assess model fit using a sums-of-squares-type discrepancy
for (i in 1:n) {
  residual[i] <- y[i]-mu[i]        # Raw residuals for observed data
  predicted[i] <- mu[i]            # Predicted, or fitted, values
  sq[i] <- residual[i]^2           # Squared residuals for observed data

# Posterior predictive check
# based on posterior predictive distribution of the data
# Generate replicate data and compute fit stats for them
  y.new[i] ~ dnorm(mu[i], tau)     # one new data set at each MCMC iteration:
                                   # this is the posterior predictive distribution of y
  sq.new[i] <- (y.new[i]-predicted[i])^2  # Squared residual for new data
}
fit <- sum(sq[])                   # Sum of squared residuals for actual data set
fit.new <- sum(sq.new[])           # Sum of squared residuals for new data set
}
")
```

First we place priors on each parameter in the model: alpha, beta, and sigma. Remember that the normal distribution in JAGS is parameterized with mean and precision. Thus we have to convert sigma into precision ($1/\sigma^2$), which we call tau. The likelihood part of the model looks very similar to our data simulation code and the equations in Section 5.1, after replacing σ^2 with tau.

After the likelihood, we calculate several derived parameters. As the name suggests, these are new variables which are functions of (i.e., are derived from) the three basic parameters in the model. The great thing about analyses in JAGS and, in general, of simulation-based Bayesian analyses is that we can easily obtain posterior distributions for these derived parameters, just as we obtain posteriors for the other parameters. In the first part of the derived parameters section, we calculate the residuals and the squared residuals for our actual data set. Then we evaluate the posterior predictive distribution of the observed data simply by simulating a new data set based on the fitted model at every iteration of the Markov Chain Monte Carlo (MCMC) algorithm. For each, we then calculate the squared residuals for the new data set: this yields the posterior predictive distribution of that discrepancy measure for every datum. Finally, we generate overall fit statistics for the model by summing up the squared residuals for the actual data set (fit) and for the simulated data set (fit.new) at every iteration. We'll talk more about these fit statistics in the next section but note that for use with DHARMa, we need to obtain posterior simulations of the data under the model and thus we add to the saved parameters y.new.

With the model specified, we next define our initial values, parameters to save, and MCMC settings in the same way as in Chapter 4. We then provide all this information to the jags function.

```
# Function to generate starting values
inits <- function(){ list(alpha = rnorm(1),      # Note lognormal init for sigma
  beta = rnorm(1), sigma = rlnorm(1))}           # to avoid values < 0

# Parameters to estimate
params <- c("alpha","beta", "sigma", "fit", "fit.new", "residual", "predicted", "y.new")

# MCMC settings
na <- 1000 ; ni <- 6000 ; nb <- 1000 ; nc <- 4 ; nt <- 1

# Call JAGS (ART <1 min), check convergence and summarize posteriors
out5.4 <- jags(dataList, inits, params, "model5.4.txt", n.iter = ni, n.burnin = nb,
  n.chains = nc, n.thin = nt, n.adapt = na, parallel = TRUE)
jagsUI::traceplot(out5.4)                          # not shown
print(out5.4, 2)
```

	mean	sd	2.5%	50%	97.5%	overlap0	f	Rhat	n.eff
alpha	40.66	2.69	35.33	40.67	45.95	FALSE	1.00	1	20000
beta	-0.68	0.28	-1.23	-0.68	-0.12	FALSE	0.99	1	20000
sigma	5.06	1.09	3.44	4.89	7.66	FALSE	1.00	1	16197
fit	346.56	63.31	294.79	326.35	516.59	FALSE	1.00	1	18574
fit.new	427.88	260.09	124.70	366.82	1093.72	FALSE	1.00	1	14051
residual[1]	-4.68	2.45	-9.51	-4.70	0.17	TRUE	0.97	1	20000
residual[2]	6.63	2.22	2.27	6.61	11.03	FALSE	1.00	1	20000
residual[3]	-6.39	2.00	-10.32	-6.40	-2.41	FALSE	1.00	1	20000

[...]

```
# Save JAGS estimates
jags_est <- out5.4$summary[1:3, 1]

# Compare results of lm() and JAGS
comp <- cbind(truth = truth, lm = lm_est, JAGS = jags_est)
print(comp, 4)
```

```
         truth       lm      JAGS
alpha     40.0  40.7426   40.6590
beta      -0.5  -0.6907   -0.6827
sigma      5.0   4.5798    5.0598
```

We comment more on these estimates in Section 5.9.

5.4.2 Goodness-of-fit assessment in Bayesian analyses

People sometimes ask whether you have to do residual checks for a Bayesian model fit, too. The answer is a resounding "Yes," since the model and its assumptions have nothing to do with whether we fit the model with least-squares, maximum likelihood, or with Bayesian posterior inference. The only practical difference is that with the Bayesian engines used in this book, you will have to code up these assessments yourself for the most part, starting by computation of the residuals, for which we added a line of code. In this section, as an instructive example, we first assess the adequacy of the model using a check of traditional residuals (note that we could code additional assessments as well, such as QQ plots). Second, we will introduce a very general method for GOF assessment of models fit with Bayesian posterior inference based on posterior predictive distributions of the data and of things we compute from them, including Bayesian p-values (Gelman et al., 1996, 2014a; Gelman & Hill, 2007). This latter offers an overall fit assessment for our chosen fit criterion. In addition, Bayesian posterior simulations of the data can be fed into DHARMa, allowing us to obtain a range of useful PPCs at the level of each individual data point.

Plots of traditional residuals: One commonly produced graphical check of the residuals of a linear model is a plot of the residuals against the predicted values. Under the normal linear regression model, residuals are assumed to be an independent random sample from a single normal distribution, that is, the data are assumed to be i.i.d. (Chapter 2). There should be no visible structure in the residuals. In particular, the scatterplot of the residuals should not have the shape of a fan; this would indicate that the variance is not constant but is (usually when this happens) larger for larger responses. We check this first and find no sign of a violation of the so-called homoscedasticity (or "equal-variance") assumption (Fig. 5.3 left). Note that the residuals must be estimated, since they depend in part on the parameters, and thus that they each have a distribution. We here take their posterior means as a point estimate.

```
# Fig. 5.3 (left)
par(mfrow = c(1, 2), mar = c(5,5,2,3))
plot(out5.4$mean$predicted, out5.4$mean$residual, main = "", las = 1,
   xlab = "Predicted values", ylab = "Residuals", frame = FALSE,
   pch = 16, col = rgb(0,0,0,0.5), cex = 1.5)
abline(h = 0)
```

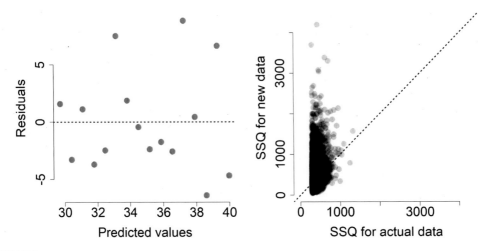

FIGURE 5.3

Model assessment in the normal linear regression model. (Left) One type of a residual plot for the linear regression analysis for trend in the Swiss wallcreeper species distribution. (Right) Graphical posterior predictive check of the model adequacy for the wallcreeper analysis plotting predictive versus realized sums-of-squares discrepancies. The Bayesian *p*-value is equal to the proportion of plot symbols above the 1:1 line and should be near 0.5 for a fitting model.

PPC and Bayesian p-values: Posterior predictive checks (PPCs) are a very general way of assessing model fit when using MCMC techniques (Gelman et al., 1996; Gelman & Hill, 2007). The idea of a PPC is to compare the lack of fit (or discrepancy between observed and expected) of the model for the actual data set with the lack of fit of the model computed with respect to replicated, "ideal" data sets, that is, the posterior predictive distribution of the data given the model and the observed data. These "ideal" data sets conform exactly to the assumptions made in the model. They can easily be generated using the parameter estimates obtained from the analysis of the actual data set. In contrast to a frequentist analysis, where the solution of a model consists of a single value for each parameter, we estimate a whole distribution for anything unknown in a Bayesian analysis. Discrepancy measures are functions of the known data and the unknown parameters and therefore also have posterior distributions.

These "ideal" data sets are also called the predictive distribution of the data. To obtain them, at each MCMC iteration one replicate data set is assembled under the same model that we fit to the actual data set and using the values of all parameters from the current MCMC iteration. A discrepancy measure chosen to quantify a certain kind of lack of fit can then be computed for both that ideal data set and for the actual data set. Therefore at the end of an MCMC run for *n* chains of length *m*, we have $n \cdot m$ draws from the posterior predictive distribution of the discrepancy measure applied to the actual data set, as well as for the discrepancy measure applied to an ideal data set.

What does "discrepancy measure" mean and how is it chosen? The discrepancy measure is any measure chosen to assess a particular feature of the model. Often, some global measure of lack of fit will be selected, for example, a sums of squares (SSQ) discrepancy as we do here, or a Chi-squared-type discrepancy (see Chapter 19B for another example in a more complex hierarchical model). However, entirely different measures may also be chosen, for instance the incidence or magnitude of extreme values to assess the adequacy of the model for outliers; see Gelman et al. (1996) for examples.

One of the best ways to assess model adequacy based on PPCs is graphically in a plot of the lack of fit for the ideal data versus the lack of fit for the actual data. If the model fits the data, then about half of the points should lie above a 1:1 line. Alternatively, a numerical summary, called a Bayesian or posterior predictive *p*-value, can be computed. It gives the proportion of MCMC draws where the discrepancy measure for the ideal data sets is greater than the discrepancy measure computed for the actual data set. A fitting model has a Bayesian *p*-value near 0.5, and values close to 0 or close to 1 suggest poor model fit.

```
# Fig. 5.3 (right)
lim <- c(0, 4300)
plot(out5.4$sims.list$fit, out5.4$sims.list$fit.new, main = "", las = 1,
    xlab = "SSQ for actual data", ylab = "SSQ for new data", xlim = lim,
    ylim = lim, frame = FALSE, pch = 16, col = rgb(0,0,0,0.3), cex = 1.5)
abline(0, 1)

mean(out5.4$sims.list$fit.new > out5.4$sims.list$fit)    # Bayesian p-value

[1] 0.5565
```

The graphical PPC (Fig. 5.3 right) and the numerical Bayesian *p*-value near 0.5 suggest that our fitted model is adequate for the wallcreeper data … something that will hardly come as a surprise since we simulated these data under the very same model. We note that the hard lower boundary for the SSQ for the observed data is represented by the least-squares estimates: the discrepancy can't get any lower than that for the observed data. Thus there is nothing wrong with this feature of the plot. PPCs are often criticized because they use the data twice: first, to generate the replicate data and second to compare them with these replicates. In many cases, PPCs have relatively weak power to detect a poorly fitting model, and PPCs should not be used for model selection. See Ntzoufras (2009: Chapter 10), Conn et al. (2018) and Chapter 18 for alternatives.

Randomized quantile residuals in a Bayesian analysis: Functionality in DHARMa can be accessed for the results from a Bayesian analysis also and to do a PPC at the level of each datum; see the vignette *"DHARMa for Bayesians."* The only difference between this and the application of the package in a non-Bayesian analysis is that the results now incorporate the full posterior uncertainty from the Bayesian analysis. In contrast, the residuals computed from the model fit in a frequentist analysis condition on the MLE, that is, on the point estimates, and ignore the uncertainty associated with the parameter estimates as represented by the SEs. We show this here as an example and note that we give a further example in Chapter 16.

```
# How to use DHARMa for quantile residuals when fitting model in JAGS
y.new <- out5.4$sims.list$y.new
sim <- createDHARMa(simulatedResponse = t(y.new), observedResponse = y,
    fittedPredictedResponse = apply(y.new, 2, median))
par(mfrow=c(1,2))      # Graphical results as in Fig. 5.2 (not shown)
plotQQunif(sim)
plotResiduals(sim, rank = FALSE)
```

Thus, to use `DHARMa` functions on the output from a Bayesian analysis, all we have to do is to define a node in the model for replicate data, or else in R produce a sample of the posterior distribution using posterior samples of all parameters produced by JAGS or another Bayesian engine. This allows us to evaluate the posterior predictive distribution of the data, simulations of which can then be used with `DHARMa` for easy PPCs for the data level. In contrast, when we code up such checks in JAGS directly, we will often add up data-level lack of fit measures such as residuals to some overall measure of lack of fit, for example, Chi-squared, Freeman-Tukey, or sums of squared residuals (Kéry & Royle, 2016: section 2.8). This can give complementary information to output given from `DHARMa`.

5.4.3 Computing predictions

Obtaining predictions with Bayesian inference and JAGS works conceptually in the same way as for the classical (least-squares) case, except that we have to code up everything ourselves. We show how to do this here. Predictions depend on the values of the estimated parameters. They are unknown in this sense and must be estimated, and therefore in a Bayesian analysis, get a posterior distribution. We can then summarize these distributions for a 95% uncertainty interval (called a credible interval, or CRI) around the trend line estimated in the wallcreeper analysis. To obtain the posterior distribution of the prediction, for a given value of x (i.e., a year) we take the first draw of the intercept `alpha` and the slope `beta` from the posterior distribution accumulated by JAGS, plug them along with the value of x into the regression formula shown in Section 5.1, and calculate the predicted value of the response y. This process is repeated for each of the other 19,999 samples from the posterior distribution, so that for each year value, we will have a posterior distribution of the expected response that can be estimated based on our 20,000 samples. The 2.5th and 97.5th percentiles of these posterior distributions form the lower and upper bounds of the 95% CRI around the expected response μ_i.

As an aside, the posterior distribution of a prediction is a completely different thing than a posterior predictive distribution of the data, used above in our Bayesian GOF assessment, and visited again in Section 5.4.5. What we are about to compute here is simply the usual quantification, using probability, of our uncertain knowledge about an estimated derived quantity, which here is a prediction or expected response. In contrast, predictive distributions in a Bayesian context are the distributions expected for future or otherwise unobserved data, given the observed data and the model (Conn et al., 2018). As we saw above, they are useful for GOF assessment of a model.

Continuing, we set up an R data structure to hold the predictions, fill it using the regression formula, then determine the appropriate percentile points and produce a plot for both the frequentist and the Bayesian inference on the line of best fit and its associated 95% uncertainty interval (Fig. 5.4). Once more, we see very similar inferences under the two methods of inference. There is less uncertainty from the frequentist analysis and that is likely due to the asymptotic nature of the SEs and CIs. These may be too optimistic (i.e., too narrow) for our small data set with only 16 data points. In contrast, Bayesian CRIs are exact for any sample size.

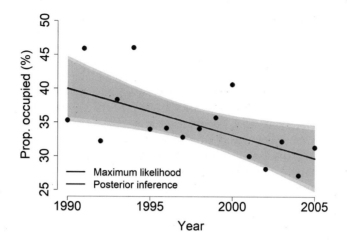

FIGURE 5.4

Observed (points) and predicted (lines) change in the distribution of Swiss wallcreepers (blue—maximum likelihood, red—Bayesian inference). Shaded areas show 95% confidence intervals (CIs) in blue and 95% credible intervals (CRIs) in red. The lines of best fit show the estimates $\hat{\mu}_i = \hat{\alpha} + \hat{\beta}x_i$ for the explanatory variable x (year). Compare this with the prediction interval of the data in Fig. 5.6.

```
# Get predictions for Bayesian model fit
sims <- out5.4$sims.list                    # Grab posterior draws produced by JAGS
predictions <- array(dim = c(length(xpred), length(sims$alpha)))
for(i in 1:length(xpred)){
  predictions[i,] <- sims$alpha + sims$beta * xpred[i]
}
BPE <- rowMeans(predictions)                 # Bayesian point estimates (post. means)
LPB <- apply(predictions, 1, quantile, probs = 0.025) # Lower bound
UPB <- apply(predictions, 1, quantile, probs = 0.975) # Upper bound

# Plot predictions with 95% uncertainty intervals from both model fits
# Fig. 5.4
plot(1990:2005, y, xlab = "Year", las = 1, ylab = "Prop. occupied (%)", pch = 16,
  ylim = c(25, 50), col = rgb(0,0,0,0.5), frame = FALSE, cex = 1.5)

# CIs from frequentist inference with lm() (Section 5.3.3)
lines(xpred+1989, pred.table[,2], lwd = 2, col = 'blue')
polygon(c(newdata$x+1989, rev(newdata$x+1989)), c(pred.table[,4],
  rev(pred.table[,5])), col = rgb(0, 0, 1, 0.1), border = NA)

# CRIs from Bayesian inference (here)
lines(xpred+1989, BPE, lwd = 2, col = 'red')
polygon(c(xpred+1989, rev(xpred+1989)), c(LPB, rev(UPB)), col = rgb(1, 0, 0, 0.2),
  border = NA)
legend('bottomleft', legend = c("Maximum likelihood", "Posterior inference"),
  cex = 1.2, bty = 'n', lty = 1, col = c('blue', 'red'), lwd = 3)
```

5.4.4 **Interpretation of confidence interval versus credible intervals**

Consider the frequentist inference about the slope parameter; −0.6907, SE 0.2484. A Wald-type 95% CI is given by −0.6907 ± 1.96 · 0.2484 = (−1.1776, −0.2038) (a profile interval is (−1.2234, −0.1580)). This means that if we took, for example, 1000 replicate sample observations of 16 annual surveys each in the same Swiss wallcreeper population and estimated 1000 times an annual trend with an associated 95% CI using linear regression, then on average we would expect that 950 intervals would contain the true value of the population trend, and 50 would not. We cannot make any direct probability statement about the trend itself; the true value of the trend is either inside or out of our single interval. There is no probability associated with this, since only Bayesians, not frequentists, make probability expressions directly about parameters. In particular, it would be wrong to say that the population trend of the wallcreeper lies between −1.1776 and −0.2038 with a probability of 95%. The probability statement associated with a 95% CI refers to the reliability of the tool, that is, computation of the CI, and not to the parameter for which a CI is constructed.

In contrast, the posterior probability in a Bayesian analysis measures our degree of belief about the likely value of a parameter, given the model, the observed data, and our priors. Hence, we can make direct probability statements about a parameter based on its posterior distribution. Let's do this here for the slope parameter, which represents the population trend of the wallcreeper in Switzerland (Fig. 5.5).

```
par(mfrow = c(1, 1), mar = c(5,5,2,3))      # Fig. 5.5
hist(out5.4$sims.list$beta, main = "", col = "grey", xlab = "Trend estimate",
  xlim = c(-2, 1), breaks = 100, freq = FALSE)
abline(v = 0, col = "black", lwd = 2, lty = 3)
mean(out5.4$sims.list$beta < 0)             # Probability of population decline

[1] 0.9893
```

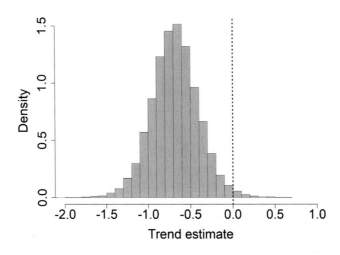

FIGURE 5.5

Posterior distribution of the distributional trend in Swiss wallcreepers (simulated data). The value of zero (representing no trend, or a stable extent of its distribution) is shown as a dashed vertical line.

We see clearly that values representing no decline or even an increase, that is, values of the slope of 0 and from there to the right, have very little mass under this posterior distribution. Based on our posterior draws, we estimate the probability of a population decline at 0.99. We can thus say that the probability of a stable or increasing wallcreeper population is very small at 0.01. Such a statement is exactly what most consumers of statistics such as politicians or journalists would like to have, rather than the somewhat contorted statement about the likely magnitude of a population trend as based on the frequentist CI.

5.4.5 And what about prediction intervals?

You may note in Fig. 5.4 that a substantial part (and surely more than 5%) of the observed data lie outside of the 95% CIs or CRIs. Is there something wrong here?

No, all is fine. But if you were wondering about this, you were thinking about an uncertainty interval for a different thing: for the output of the stochastic process represented by our model, that is, for the data that are a realization of that process, rather than for its mean. The intervals shown there convey our uncertainty about the *expected or mean values* of y and not about the *actual data*. That is, the CIs/CRIs we discuss in Sections 5.4.3 and 5.4.4 are associated with our estimate of the expected response $\hat{\mu} = \hat{\alpha} + \hat{\beta}x$, where the hats denote estimates. There is an uncertainty associated with these estimates, and we must propagate this into any estimate of the expected value μ. The CIs/CRIs do that.

However, perhaps you are interested in an assessment of the likely magnitude of future values of observable data that come *from the same process as the one that produced our data at hand*. That is, you want to produce an estimate of a future or otherwise unobserved datum $\tilde{y}$, which is given by $\tilde{y} \sim \text{Normal}(\hat{\mu}, \hat{\sigma}^2)$, with $\hat{\mu} = \hat{\alpha} + \hat{\beta}x$. Clearly, the prediction interval for new data $\tilde{y}$ will be wider than the CI/CRI for $\hat{\mu}$, since in addition to uncertainty about the intercept and the slope it will include contributions from the sampling variability of the data (σ^2) as well as from the uncertainty associated with this variability, that is, for the fact that we don't know σ^2, but instead work with an estimate of it ($\hat{\sigma}^2$). A prediction interval for the data can be obtained very easily by taking a range of values for the predictor of our model (x) and then combining this with the posterior draws for our three parameters `alpha`, `beta`, and `sigma`.

Next, we do this and plot the prediction intervals alongside the data, in a plot similar to Fig. 5.4. First, we generate a sequence of 1000 values of our covariate x and store them in a new vector `xpred`. Second, for each value in `xpred`, we compute samples from the posterior distribution of the expected value `mu` using our 20,000 draws from the posterior distributions of `alpha` and `beta`. Third, we supply the samples of `mu` along with samples from the posterior of the residual error `sigma` to R's normal RNG function `rnorm()`, generating samples of $\tilde{y}$ for each value in `xpred`. Finally, we use these samples to calculate the 95% prediction interval at each value of `xpred`.

```
# Get posterior predictive distribution of y at fine resolution
sims <- out5.4$sims.list                  # Grab all posterior draws
str(sims)                                 # Remind ourselves of MCMC output format
xpred <- seq(1, 16, length.out = 1000)    # Predict for 1000 values of x
y_tilde <- array(NA, dim = c(1000, 20000))  # Post. predictive dist.
for(k in 1:1000){                         # Loop over all 1000 values of xpred
  mu_k <- sims$alpha + sims$beta * xpred[k]
  y_tilde[k,] <- rnorm(20000, mean = mu_k, sd = sims$sigma)
}
```

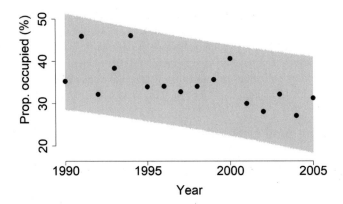

FIGURE 5.6

Observed data (points) and 95% prediction interval for new data $\tilde{y}$ (shaded polygon) for the Swiss wallcreeper data. Compare this with the much tighter 95% CIs/CRIs for the mean of that process in Fig. 5.4. *CIs*, Confidence intervals; *CRIs*, credible intervals.

```
# Get 95% prediction interval for the data y_tilde
LPI <- apply(y_tilde, 1, quantile, probs = 0.025)    # Lower bound
UPI <- apply(y_tilde, 1, quantile, probs = 0.975)    # Upper bound

# Plot 95% prediction interval for the data (Fig. 5.6)
plot(1990:2005, y, xlab = "Year", las = 1, ylab = "Prop. occupied (%)",
  pch = 16, ylim = c(18, 58), col = rgb(0,0,0,0.5), frame = FALSE, cex = 1.5)
polygon(c(xpred+1989, rev(xpred+1989)), c(LPI, rev(UPI)),
  col = rgb(1, 0, 0, 0.2), border = NA)
```

We note that these 95% prediction intervals in Fig. 5.6 are far wider than the CIs/CRIs for the mean of the same process in Fig. 5.4. This makes sense because they contain two additional sources of uncertainty compared to the latter: the sampling variability in the data (σ^2) and the parametric uncertainty about the estimate of that quantity.

5.5 Bayesian analysis with NIMBLE

NIMBLE code for the model is available on the book website.

5.6 Bayesian analysis with Stan

To fit the model in Stan, our data is the same.

```
# Bundle and summarize data (same as before)
str(dataList <- list(y = y, x = x, n = n))    # not shown
```

The model definition looks similar in the Stan language: the main difference is that it is broken up into the special sections required by Stan. Below we show the entire model, and describe it in more detail afterwards.

```
# Write text file with model description in Stan language
cat(file = "model5_6.stan",              # This line is R code
"data {                                  // This is the first line of Stan code
  int<lower = 0> n;                      // Define the format of all data
  vector[n] y;                           //... including the dimension of vectors
  vector[n] x;                           //
}
parameters {                             // Define format for all parameters
  real alpha;
  real beta;
  real<lower = 0> sigma;                 // sigma (sd) cannot be negative
}

transformed parameters {
  vector[n] mu;
  for (i in 1:n){
    mu[i] = alpha + beta * x[i];         // Calculate linear predictor
  }
}

model {
  // Priors
  alpha ~ normal(0, 100);
  beta ~ normal(0, 100);
  sigma ~ cauchy(0, 10);
  // Likelihood (could be vectorized to increase speed)
  for (i in 1:n){
    y[i] ~ normal(mu[i], sigma);
  }
}

generated quantities {
  vector[n] residuals;
  vector[n] sq;
  vector[n] sq_new;
  vector[n] y_new;
  real fit;
  real fit_new;
  for (i in 1:n){
    residuals[i] = y[i] - mu[i];
    sq[i] = residuals[i]^2;
    y_new[i] = normal_rng(mu[i], sigma);
    sq_new[i] = (y_new[i] - mu[i])^2;
  }
  fit = sum(sq);
  fit_new = sum(sq_new);
}                                        // This is the last line of Stan code
" )
```

As explained in Chapter 4, we describe the types and dimensions of our data and parameters in the `data` and `parameters` sections, respectively. We add a new section, transformed parameters, to calculate values which are functions of the parameters in the parameters section. In this case, we only calculate the linear predictors mu. Note also that at the beginning of this section we need to explicitly define the type and dimensions of any new parameters we calculate.

Next is the model section, which includes the priors and likelihood, which look very similar to BUGS code for this model. Finally, we have another new section generated quantities, which is where we calculate the derived parameters. The code here should look similar to the equivalent code for JAGS and NIMBLE. The only difference is that at the beginning of the section, as with transformed parameters, we need to define the type and dimensions of each new parameter we calculate later in the section.

```
# HMC settings
ni <- 3000 ; nb <- 1000 ; nc <- 4 ; nt <- 1

# Call STAN (ART 55/3 sec), assess convergence, save estimates
system.time(
    out5.6 <- stan(file = "model5_6.stan", data = dataList,
      chains = nc, iter = ni, warmup = nb, thin = nt))
rstan::traceplot(out5.6)          # not shown
print(out5.6, dig = 2)            # not shown
stan_est <- summary(out5.6)$summary[1:3,1]
```

5.7 Do-it-yourself maximum likelihood estimates

First, a quick review. We want to estimate three parameters using maximum likelihood: the intercept α, the slope β, and the residual standard deviation (SD) σ. As input data we have the response y (percent occupied) and the predictor variable x (year). We apply our standard approach for MLE. Our first step is to write out the likelihood math. Recall from Chapter 4 the likelihood for what we call the model of the mean:

$$L(\mu, \sigma | y) = \prod_{i=1}^{n} \frac{1}{\sigma\sqrt{2\pi}} e^{-\frac{1}{2}\left(\frac{y_i - \mu}{\sigma}\right)^2}$$

The only change we need to make now is to specify that the mean μ is a function of α, β, and our covariate data x.

$$L(\alpha, \beta, \sigma | y, x) = \prod_{i=1}^{n} \frac{1}{\sigma\sqrt{2\pi}} e^{-\frac{1}{2}\left(\frac{y_i - (\alpha + \beta \cdot x_i)}{\sigma}\right)^2}$$

This change is even simpler when expressed in R code. For a given set of parameter values for α, β, and σ, we first calculate the mean value μ_i for each data point:

```
for (i in 1:n){
  mu[i] <- alpha + beta * x[i]
}
```

Then we just plug the vector of means mu into the dnorm function as we did in Chapter 4 and sum up over all the data points.

```
for (i in 1:n){
   LL[i] <- dnorm(y[i], mean = mu[i], sd = sigma, log = TRUE)
}
sum(LL)
```

Converting this code into an R function that, when applied to some data, evaluates the negative log-likelihood:

```
NLL <- function(params, y, x) {
  alpha <- params[1]
  beta <- params[2]
  sigma <- exp(params[3])               # convert from log scale
  n <- length(y)                        # number of datapoints

  mu <- LL <- numeric(n)                # empty vectors
  # You could vectorize this loop to speed things up
  for (i in 1:n){
    mu[i] <- alpha + beta * x[i]
    LL[i] <- dnorm(y[i], mean = mu[i], sd = sigma, log = TRUE)
  }
  -sum(LL)
}
```

The next step is to provide the NLL function to function optim() to find the set of parameter values which minimizes the negative log-likelihood, which also maximizes the likelihood, for our data set.

```
# Minimize that NLL to find MLEs and also get SEs
inits <- c('alpha' = 0, 'beta' = 0, 'log.sigma' = 0)
out5.7 <- optim(inits, NLL, y = y, x = x, hessian = TRUE, method = 'BFGS')
get_MLE(out5.7, 4)
```

```
                MLE      ASE    LCL.95    UCL.95
alpha       40.7426   2.2466    36.339   45.1459
beta        -0.6907   0.2323    -1.146   -0.2353
log.sigma    1.4549   0.1768     1.108    1.8014
```

```
# Save DIY estimates
diy_est <- c(out5.7$par[1:2], exp(out5.7$par[3]))
```

5.8 Likelihood analysis with TMB

We begin the Template Model Builder (TMB) analysis by bundling together the data into a list, just as we did for JAGS, NIMBLE, and Stan. The list contains a vector of responses y, a vector of predictors x (years), and the number of datapoints n.

```
# Bundle and summarize data (same as before)
str(tmbData <- list(y = y, x = x, n = n))   # not shown
```

As with the DIY maximum likelihood in the previous section, for TMB we also need to write a function that calculates the negative log-likelihood, only we now do this in C++. This function will resemble the one in the previous section in many ways, but has some additional TMB-specific code. As usual, the function is contained in its own .cpp file and contains four parts: (1) the usual boilerplate TMB code, (2) a description of the input data, (3) a description of the parameters to estimate, and (4) the actual likelihood math.

```
# Write TMB model file
cat(file = "model5_8.cpp",
"#include <TMB.hpp>                      // (1) First some boilerplate code

template<class Type>
Type objective_function<Type>::operator() ()
{
  // (2) Describe input data
  DATA_VECTOR(y);                        // response
  DATA_VECTOR(x);                        // covariate
  DATA_INTEGER(n);                       // Number of obs

  // (3) Describe parameters
  PARAMETER(alpha);                      // Intercept
  PARAMETER(beta);                       // Slope
  PARAMETER(log_sigma);                  // log(residual standard deviation)
  Type sigma = exp(log_sigma);           // Type = match type of function output

  // (4) Calculate the log-likelihood
  Type LL = 0.0;                         // Initialize total log likelihood at 0
  Type mu;                               // Initialize the expected value

  for (int i=0; i<n; i++){               // Note index starts at 0 instead of 1!
    mu = alpha + beta * x(i);
    LL += dnorm(y(i), mu, sigma, true);
    //Equivalent to LL = LL + dnorm(...)
  }

  return -LL;                            // Return negative of total log likelihood
}
")
```

As you can see, the likelihood calculation is nearly identical to that in our DIY-MLE function in the previous section. Next, we compile our C++ code so it can be used in R, and load it into the R session.

```
# Compile and load TMB function
compile("model5_8.cpp")
dyn.load(dynlib("model5_8"))
```

The final piece of information TMB needs is a named list of the parameters to estimate (taken from the parameters section in the model above), plus information about parameter dimensions.

Since all our parameters to estimate are scalars (single values), we provide a single value of 0 for each.

```
# Provide dimensions, names and starting values for parameters
params <- list(alpha = 0, beta = 0, log_sigma = 0)
```

We can now create our TMB model object.

```
# Create TMB object
out5.8 <- MakeADFun(data = tmbData, parameters = params,
                    DLL = "model5_8", silent = TRUE)
```

The TMB object is a list containing the negative log-likelihood function (out5.8$fn) and the gradient function (out5.8$gr), which we will pass to optim in order to obtain parameter estimates just as we did with our DIY likelihood function in the previous section.

```
# Optimize TMB object and print results
starts <- rep(0, length(unlist(params)))
opt <- optim(starts, fn = out5.8$fn, gr = out5.8$gr, method = "BFGS", hessian = TRUE)
(tsum <- tmb_summary(out5.8))                    # not shown
```

```
# Save TMB estimates
tmb_est <- c(opt$par[1:2], exp(opt$par[3]))
```

Note that for the maximum likelihood engines, we originally estimated sigma on the log scale (log_sigma). To convert back to the natural scale and make the values comparable with the Bayesian engines, we need to remember to calculate exp(log_sigma). This will also be true for dispersion parameters in other chapters.

5.9 Comparison of the parameter estimates

We see that all engines give similar estimates for the two parameters in the mean of the response, but much less so for the dispersion parameter sigma. Thus lm() gives a slightly different estimate from DIY-MLE and TMB, and the three Bayesian engines give yet slightly different estimates again. However, for the key parameter in this analysis, the estimated population trend represented by beta, all the engines yield essentially identical estimates. Note that successfully running the code below requires that you fit the model in NIMBLE, for which code is available only on the book website. This will also be the case for similar comparisons in all later chapters.

```
# Compare results of all engines with truth
comp <- cbind(truth = truth, lm = lm_est, JAGS = jags_est, NIMBLE = nimble_est,
   Stan = stan_est, DIY = diy_est, TMB = tmb_est)
print(comp, 4)
```

	truth	lm	JAGS	NIMBLE	Stan	DIY	TMB
alpha	40.0	40.7426	40.6590	40.670	40.7383	40.7426	40.7429
beta	-0.5	-0.6907	-0.6827	-0.682	-0.6894	-0.6907	-0.6907
sigma	5.0	4.5798	5.0598	5.037	4.9657	4.2840	4.2840

Hence, there are slight discrepancies and they are interesting. They may have at least two different causes. First, we commented earlier that least-squares and maximum likelihood yield the same estimates for the parameters in the mean in normal linear models. But this is not so for the residual variance, where the least-squares estimate is the ratio of the residual sums of squares and the residual degrees of freedom (d.f.), 14 in our case (Millar, 2011). To see the residual sums of squares and d.f., run `anova` (`out5.3`) and you will see that the least-squares estimate of the residual SE is $\sqrt{293.65/14} = 4.58$. In contrast, the MLE has a denominator equal to the sample size, that is, 16 in our case. Hence, the MLE of the residual SE is $\sqrt{293.65/16} = 4.28$. The MLE variance estimator is known to be biased for small samples, but this bias will diminish rapidly with increasing sample size.

Second, the discrepancy between the three Bayesian estimates of `sigma` and the least-squares estimate is also a result of small sample size. In this case, our Uniform(0, 100) prior in JAGS and NIMBLE and the Cauchy(0, 10) in Stan lead to estimates that are closer to the truth for this particular data set, but which are in fact inadvertently a little informative for this small sample size. You can try that out by doing a prior sensitivity analysis: repeating the analysis with a Uniform(0,10) prior.

5.10 Summary and outlook

In this chapter, we fit a simple linear regression using all five engines, in addition to a canned function. For the Bayesian engines (JAGS, NIMBLE, and Stan) as well as for the canned code in R we also demonstrated calculation of statistics related to model fit: residuals, and PPCs and Bayesian p-values, which are both based on the predictive distribution of the data. PPCs are a very general and flexible way of assessing the GOF of a model analyzed using MCMC. The R package `DHARMa` provides functionality to obtain residual assessments based on predictive distributions of the data, where the latter are obtained by simulating from the model to produce new data. When used on the output from an MCMC algorithm, it produces directly PPCs. In addition, we note that a frequentist analog to such a simulation-based GOF assessment would be a parametric bootstrap (see Section 2.5.4, and Chapter 18 and also Dixon, 2006 and Manly, 2006).

This chapter also clarified the differences between a frequentist CI and a Bayesian CRI. A CI is essentially a variability statement about the data and a method used to calculate an uncertainty interval from the data. In contrast, a CRI is an uncertainty statement about a parameter. In addition, we have also explained the prediction interval, which is a statement about likely magnitudes of future or otherwise unobserved data produced by the same process that generated our data set at hand. All three types of intervals are often confused.

Finally, when comparing the estimates among our engines, we noted some discrepancies that were due to a specific form of small-sample bias of the MLE for the variance in this type of model on the one hand, and to some influence of even a vague prior in a Bayesian analysis with very small sample size. There is nothing wrong with these discrepancies. It is a fact of life that in small samples, MLEs may exhibit bias (Millar, 2011) and that even vague priors in a Bayesian analysis may exert some influence on the posterior when the information in the data is weak.

The extension from a simple to a multiple linear regression would be straightforward: we simply add more terms such as $\beta_2 \cdot x_{2,i}$ to the linear predictor of our model. Here, $x_{2,i}$ is the value of covariate x_2 for the i-th datum, and the subscript 2 is needed for the coefficient name to distinguish it from the other `beta` already in there. The geometrical interpretation of such a model with two continuous covariates is that of a plane. Virtually all considerations of the simple linear regression carry over also to the multiple linear case.

Comparing two groups in a normal model[⊛]

6

Chapter outline

6.1 Introduction

In Chapter 4, we presented a linear regression model with no covariates, or what we called the "model of the mean." We had a sample of n measurements that were assumed to be what statisticians call i.i.d.: independent and identically distributed. We assumed that all measurements came from a distribution with the same mean, and variance or standard deviation. In this chapter, we will make things a little more complex: we present models which compare two groups. The two groups may have different means and possibly also different variances.

In Chapter 5, we described a linear regression model with a single explanatory variable, which in that case was a continuous covariate. One approach to modeling a comparison between two groups involves just a very small variation of the linear regression model of Chapter 5: we simply

---

[⊛]This book has a companion website hosting complementary materials, including all code for download. Visit this URL to access it: https://www.elsevier.com/books-and-journals/book-companion/9780443137150.

Applied Statistical Modelling for Ecologists. DOI: https://doi.org/10.1016/B978-0-443-13715-0.00014-5
171

replace the continuous covariate with a categorical covariate, or factor, with two levels, one for each group. Such a model allows us to estimate a mean for each of the groups and test if they are different from each other. A second, basically equivalent approach is to apply a statistical method called a t-test (Zar, 1998). Historically and in many introductory stats classes, the t-test approach has been emphasized for comparison of two groups, but in this chapter we focus on the more general linear modeling approach. We do this to make clear the connections between the models in this and other chapters, for instance, with models that have one or two classifications (or factors) with more than two levels each, which we cover in Chapters 7 and 8.

One place where t-tests and linear regression fit with `lm()` are not necessarily equivalent is in the treatment of the variance parameter. Both the mean *and* the variance could differ between groups, but the basic linear regression approach we describe above only accounts for differences between the means and assumes the two groups have equal variance. On the other hand, there are variants of the t-test (such as Welch's t-test; Zar, 1998) which do not make this assumption. However, using the power of our general-purpose modeling engines, we can fairly easily allow for unequal group variances in a linear regression modeling framework as well. We will demonstrate two-group models with both equal and unequal variance in this chapter.

6.2   Comparing two groups with equal variances

First, we will examine the simpler case where we assume the two groups have equal variance. We will consider a model represented by exactly the same equation as in Chapter 5 for linear regression:

$$y_i = \alpha + \beta x_i + \varepsilon_i$$

$$\varepsilon_i \sim Normal(0, \sigma^2)$$

In Chapter 5, x was a continuous covariate. Here, x_i codes for the group membership of data point i, that is, a categorical explanatory variable or factor. As there are two groups, x_i can take on only two possible values: 0 if data point i belongs to group 1, and 1 if it belongs to group 2. This type of covariate indexing group membership is also called a *dummy or an indicator variable*. Considering only the linear predictor (i.e., ignoring the residual error) in the equation above and plugging in the two possible values for x, this results in two possible equations:

$$\alpha + \beta \cdot 0 = \alpha, \quad \text{when } x_i = 0 \text{ (group 1)}$$

$$\alpha + \beta \cdot 1 = \alpha + \beta, \quad \text{when } x_i = 1 \text{ (group 2)}$$

Thus the value of β, equivalent to the slope in the linear regression in Chapter 5, now represents the difference between the means of group 1 and of group 2. We call this an *effects parameterization* of the model because β represents the additive effect of being in group 2 relative to being in group 1 (we note it's also called a corner-point parameterization in the GLM literature, or the treatment contrast parameterization in R). In total, as in Chapter 5, we have three parameters to estimate: the intercept α (which as you can see in the equation above, also represents the mean of group 1), the slope β, and the residual variance common to both groups σ^2.

6.2.1 **Data generation**

We first simulate data under this model and for a motivating example return to the magic peregrine falcon. We imagine that we had measured the wingspan of a number of male and female birds and were interested in a sex difference in this measure of size. For Western Europe, Monneret (2006) gives the range of male wingspan as 70–85 cm and that for females as 95–115 cm. Assuming normal distributions for wingspan, this implies means and standard deviations of about 77.5 and 2.5 cm for males, and of 105 and 3 cm for females. For the group membership covariate x, females are assigned a value of 0 and males a value of 1, which means that females will be treated as the reference category represented by the intercept in the model (Fig. 6.1).

```
# Generate a data set
set.seed(61)
n1 <- 60                        # Number of females
n2 <- 40                        # Number of males
mu1 <- 105                      # Population mean of females
mu2 <- 77.5                     # Population mean of males
sigma <- 2.75                   # Average population SD of both

n <- n1+n2                      # Total sample size
y1 <- rnorm(n1, mu1, sigma)     # Data for females
y2 <- rnorm(n2, mu2, sigma)     # Date for males
y <- c(y1, y2)                  # Merge both data sets
x <- rep(c(0,1), c(n1, n2))     # Indicator variable indexing a male

# Make a plot (Fig. 6.1)
par(mfrow = c(1, 1), mar = c(6,6,6,3), cex.lab = 1.5, cex.axis = 1.5, cex.main = 2)
boxplot(y ~ x, col = "grey", xlab = "Male", ylab = "Wingspan (cm)", las = 1, frame = FALSE)
```

The manner in which we just generated this data set (Fig. 6.1) corresponds to a means, as opposed to an effects, parameterization. Here is a different way to generate an identical data set, which more closely matches the equations above.

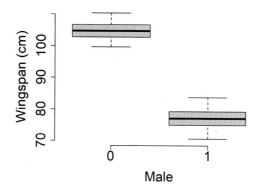

FIGURE 6.1

A boxplot of the generated data set on the wingspan of female and male peregrines when the residual variance is constant between the two groups.

```
set.seed(61)
n <- n1 + n2                              # Total sample size
alpha <- mu1                             # Mean for females serves as the intercept
beta <- mu2 - mu1                        # beta is the difference male-female
E.y <- alpha + beta*x                    # Expectation (linear predictor)
y.obs <- rnorm(n = n, mean = E.y, sd = sigma)  # Add random variation
boxplot(y.obs ~ x, col = "grey", xlab = "Male", ylab = "Wingspan (cm)", las = 1, frame = FALSE)

# Save true values for later comparisons
truth <- c(alpha = mu1, beta = mu2 - mu1, sigma = sigma)
```

An important aside again: To get a feel for the effects of chance, or technically, for sampling variance or sampling error, it is good to repeatedly execute one of the previous sets of commands without setting a seed and observe just how different repeated realizations of one and the same random process can be.

```
# Load required libraries
library(ASMbook); library(DHARMa); library(jagsUI); library(rstan); library(TMB)
```

6.2.2 Likelihood analysis with canned functions in R

First we fit a linear regression in R with `lm()`, which will look identical to the corresponding analysis in Chapter 5. As we noted, `lm()` estimates only one residual variance value and thus assumes equal variance for the two groups.

```
summary(out6.2.2 <- lm(y ~ x))
lm_est <- c(coef(out6.2.2), sigma(out6.2.2))  # save results

Call:
lm(formula = y ~ x)

Residuals:
    Min      1Q   Median      3Q     Max
-6.5222  -2.1287  -0.0261  1.9165  6.5952

Coefficients:
              Estimate Std.    Error   t value   Pr(>|t|)
(Intercept)      104.7706      0.3579    292.7    <2e-16***
x                -27.6184      0.5659    -48.8    <2e-16***

---
Signif. codes:    0 '***'  0.001 '**'  0.01 '*'    0.05 '.'  0.1 ' ' 1

Residual standard error: 2.772 on 98 degrees of freedom
Multiple R-squared: 0.9605,  Adjusted R-squared: 0.9601
F-statistic: 2382 on 1 and 98 DF, p-value: < 2.2e-16
```

Recall that the intercept estimate is the mean for group 1 (females), and the mean for group 2 (males) is given by the intercept plus the slope for x, or $104.77 + (-27.62) = 77.15$. Because the "slope," which really is a group difference here, is significantly different from 0 ($p < 2e-16$), we conclude the two groups have different means.

We can also use R's `t.test` function to get similar results. Note that we have to explicitly tell the function to assume equal variance.

```
t.test(y ~ x, var.equal = TRUE)

        Two Sample t-test

data: y by x
t = 48.805, df = 98, P-value < 2.2e-16
alternative hypothesis: true difference in means between group 0 and group 1 is not equal to 0
95 percent confidence interval:
26.49536  28.74135
sample estimates:
  mean in group 0   mean in group 1
        104.77058         77.15223
```

The output from t.test gives us the actual mean estimates for both groups, rather than the estimate for group 1 and the difference between group 1 and group 2.

For this example data set, we've defined the group indicator covariate as numeric with only values 0 and 1. However, in R it is also possible to treat the group covariate explicitly as a *factor* (i.e., categorical) variable. We can create an equivalent factor variable with the following code:

```
xfac <- factor(rep(c("group1", "group2"), c(n1, n2)))
head(xfac)          # look at the first few values

[1] group1 group1 group1 group1 group1 group1
Levels: group1 group2
```

R has automatically identified the possible levels of our factor variable (group1 and group2). If we plug this new version of the covariate into lm(), we should get identical estimates. Being able to denote factor levels with meaningful labels is really useful when using canned functions in R. Unfortunately, for all our other engines you will see that there will be a much more limited choice for formatting factor levels.

```
summary(lm(y ~ xfac))   # not shown (effects or treatment contrast parameterization)
```

Also, in R you can reparameterize the model in terms of the group means, such that the two parameters directly represent female and male peregrines, as follows. We explain this reparameterization of the linear model for a factor in Chapter 3.

```
summary(lm(y ~ xfac - 1))   # not shown (means parameterization)
```

Here is our code block to assess model fit using quantile residuals.

```
# Diagnostic checks of residuals/model GOF (not shown)
plot(out6.2.2)                # Traditional Pearson residual check
simOut <- simulateResiduals(out6.2.2, n = 1000, plot = TRUE)
plotResiduals(simOut)         # same as right plot ...
testDispersion(out6.2.2)      # Test for over- or underdispersion
```

6.2.3 Bayesian analysis with JAGS

Our setup code for JAGS will look similar to the regression model from Chapter 5. Our list of data includes the response, the covariate, and the total number of data points.

```
# Bundle and summarize data
str(dataList <- list(y = y, x = x, n = n))

List of 3
 $ y: num [1:100] 104 104 100 106 101 ...
 $ x: num [1:100] 0 0 0 0 0 0 0 0 0 0 ...
 $ n: num 100
```

```
# Write JAGS model file
cat(file = "model6.2.3.txt", "
model {
# Priors
alpha ~ dnorm(0, 0.0001)          # Intercept (=female mean)
beta ~ dnorm(0, 0.0001)           # Slope (=diff between females and males)
tau <- pow(sigma, -2)
sigma ~ dunif(0, 10)              # common SD

# Likelihood
for (i in 1:n) {
  y[i] ~ dnorm(mu[i], tau)
  mu[i] <- alpha + beta * x[i]    # linear predictor
}

# Derived parameters
mu1 <- alpha                      # female mean
mu2 <- alpha + beta               # male mean
for (i in 1:n){                   # residuals
  residual[i] <- y[i] - mu[i]
}
}")
```

The model looks nearly identical to the linear regression model in Chapter 5. The only difference is that we calculate a couple of extra derived parameters representing the group means for females and males. The other setup code (initial values, parameters, and MCMC settings) is also similar to what we've seen before. Note that we specify a log-normal initial value for `sigma`, which forces it to be positive as required. Otherwise, if a negative initial value was provided for `sigma`, JAGS would stop with an error.

```
# Function to generate starting values
inits <- function(){list(alpha = rnorm(1), beta = rnorm(1), sigma = rlnorm(1))}

# Parameters to estimate
params <- c("alpha", "beta", "sigma", "mu1", "mu2", "residual")

# MCMC settings
na <- 1000 ; ni <- 3000 ; nb <- 1000 ; nc <- 4 ; nt <- 1

# Call JAGS (ART <1 min), check convergence and summarize posteriors
out6.2.3 <- jags(data = dataList, inits = inits, parameters.to.save = params,
  model.file = "model6.2.3.txt", n.iter = ni, n.burnin = nb, n.chains = nc,
  n.thin = nt, n.adapt = na, parallel = TRUE)
jagsUI::traceplot(out6.2.3)                        # not shown
print(out6.2.3, 3)
```

	mean	sd	2.5%	50%	97.5%	overlap0	f	Rhat	n.eff
alpha	104.767	0.368	104.033	104.774	105.474	FALSE	1.000	1.000	5210
beta	-27.616	0.574	-28.711	-27.609	-26.494	FALSE	1.000	1.000	4351
sigma	2.805	0.202	2.443	2.793	3.234	FALSE	1.000	1.001	2462
mu1	104.767	0.368	104.033	104.774	105.474	FALSE	1.000	1.000	5210
mu2	77.151	0.442	76.292	77.151	78.010	FALSE	1.000	1.000	4664
residual[1]	-0.814	0.368	-1.521	-0.821	-0.080	FALSE	0.985	1.000	5210
residual[2]	-0.798	0.368	-1.505	-0.805	-0.064	FALSE	0.984	1.000	5210
[...]									

```
# Compare likelihood with Bayesian estimates and with truth
jags_est <- out6.2.3$summary[1:3,1]
comp <- cbind(truth = truth, lm = lm_est, JAGS = jags_est)
print(comp, 4)

        truth       lm     JAGS
alpha  105.00  104.771  104.767
beta   -27.50  -27.618  -27.616
sigma    2.75    2.772    2.805
```

Comparing the inference from JAGS with that from lm() (i.e., the least-square estimates), we see that the point estimates are almost identical.

One of the nicest things about a Bayesian analysis is that parameters that are functions of primary parameters and their uncertainty (e.g., standard errors or credible intervals) can very easily be obtained using the MCMC samples. Thus, in the above model code the primary parameters are the female mean (alpha or mu1) and the male-female difference (beta), but we also added a line that computes the mean for males (mu2) at every iteration. We then obtain samples from the posterior distributions of not only the female mean wingspan and the sex difference, but also directly of the mean wingspan of a male. In a frequentist mode of inference, this would require application of the delta method which is more complicated and also makes more assumptions. In the Bayesian analysis, estimation error is automatically propagated into functions of parameters, or derived quantities.

Before making an inference about possible sex differences in the wingspan in this peregrine population, we should really check whether the model is adequate. Of course, the check of model adequacy is somewhat contrived since we use exclusively simulated and therefore, in a sense, perfect data sets. However, it is important to practice, so we will check the residuals here. For illustration, we here plot the residuals against the order in which individuals are present in the data set and then produce a boxplot for male and female residuals to get a feel whether the distributions of residuals for the two groups are similar.

```
par(mfrow = c(1, 2), mar = c(5,5,4,3), cex.lab = 1.5, cex.axis = 1.5)
plot(1:100, out6.2.3$mean$residual)                           # not shown
abline(h = 0)
boxplot(out6.2.3$mean$residual ~ x, col = "grey", xlab = "Male",
  ylab = "Wingspan residuals (cm)", las = 1)                  # not shown
abline(h = 0)
```

No violation of the model assumption of homoscedasticity is apparent from these residual checks.

6.2.4 Bayesian analysis with NIMBLE

Code to fit the model with NIMBLE is extremely similar to that for JAGS. We show the NIMBLE code on the book website. Remember to get and execute the NIMBLE code if you want to make the usual grand comparison among the estimates by all engines at the end of this section.

6.2.5 Bayesian analysis with Stan

Next, the same with Stan. Our data list is identical to JAGS and NIMBLE.

```
# Bundle and summarize data (same as before)
dataList <- list(y = y, x = x, n = n)
```

Our model file looks very similar to the Stan model in Chapter 5. To review, we define the types and dimensions of the data in data, and then the types and dimensions of the parameters in parameters. In the transformed parameters section, we calculate values for the linear predictor mu, which is a function of the other parameters. The priors and likelihood are in model and the derived parameters (in this case the residuals) are in generated quantities. Remember that Stan needs the name of the model file (before the file extension) *to not have any periods*, so we have replaced them with underscores.

```
# Write text file with model description in BUGS language
cat(file = "model6_2_5.stan",                # This line is R code
"data {                                      // This is the first line of Stan code
  int<lower=0> n;                            // Define the format of all data
  vector[n] y;                               //...including the dimension of vectors
  vector[n] x;                               //
}

parameters {                                 // Define format for all parameters
  real alpha;
  real beta;
  real<lower=0> sigma;
}

transformed parameters{
  vector[n] mu;
  for (i in 1:n){
    mu[i] = alpha + beta * x[i];             // Calculate linear predictor
  }
}

model {
  // Priors
  alpha ~ normal(0, 100);
  beta ~ normal(0, 100);
  sigma ~ cauchy(0, 10);

  // Likelihood
  for (i in 1:n){
    y[i] ~ normal(mu[i], sigma);
  }
}

generated quantities {
  vector[n] residuals;
  for (i in 1:n){
    residuals[i] = y[i] - mu[i];
  }
}                                            // This is the last line of Stan code
" )
```

A nice feature of the Stan language is that you can write some loops in a more compact and usually more efficient *vectorized* format. This is analogous to vectorizing code in R. For example, in R, you could generate 10 random normal values using a loop:

```r
rand <- numeric(10)
for (i in 1:10){
  rand[i] <- rnorm(1, mean = 0, sd = 1)
}
```

You could also get the same result in a vectorized form like this:

```r
rand <- rnorm(10, mean = 0, sd = 1)
```

Here's the same Stan model as above, only this time we replace all the loops with their vectorized equivalents. This is both more concise and also generally runs faster in Stan. In future chapters, we may use the vectorized notation.

```
# Write text file with model description in BUGS language
cat(file = "model6_2_5.stan",          # This line is R code
"data {                                 // This is the first line of Stan code
  int<lower=0> n;                       // Define the format of all data
  vector[n] y;                          //...including the dimension of vectors
  vector[n] x;                          //
}

parameters {                            // Define format for all parameters
  real alpha;
  real beta;
  real<lower=0> sigma;
}

transformed parameters{
  vector[n] mu;
  mu = alpha + beta * x;                // Calculate linear predictor
}

model {
  // Priors
  alpha ~ normal(0, 100);
  beta ~ normal(0, 100);
  sigma ~ cauchy(0, 10);

  // Likelihood
  y ~ normal(mu, sigma);
}

generated quantities {
  vector[n] residuals;
  residuals = y - mu;
}                                       // This is the last line of Stan code
" )
```

```
# HMC settings
ni <- 3000 ; nb <- 1000 ; nc <- 4 ; nt <- 1

# Call STAN (ART 57/4 sec), check convergence, summarize posteriors and save estimates
system.time(
  out6.2.5 <- stan(file = "model6_2_5.stan", data = dataList,
    chains = nc, iter = ni, warmup = nb, thin = nt) )
rstan::traceplot(out6.2.5)                    # not shown
print(out6.2.5, dig = 2)                       # not shown
stan_est <- summary(out6.2.5)$summary[1:3,1]
```

6.2.6 Do-it-yourself maximum likelihood estimates

As we discussed in the introduction to Section 6.2, a comparison of two groups with equal variance is equivalent to a normal linear regression with a single covariate x that takes on values of 0 or 1. Thus the equation for the likelihood is the same as in Chapter 5 (remember bold face denotes a vector or a matrix).

$$L(\alpha, \beta, \sigma | \mathbf{y}, \mathbf{x}) = \prod_{i=1}^{n} \frac{1}{\sigma\sqrt{2\pi}} e^{-\frac{1}{2}\left(\frac{y_i - (\alpha + \beta \cdot x_i)}{\sigma}\right)^2}$$

The equivalent negative log-likelihood function is also the same as in Chapter 5. Don't forget that for convenience we estimate $\log(\sigma)$ instead of σ.

```
NLL <- function(params, y, x) {
  alpha <- params[1]
  beta <- params[2]
  sigma <- exp(params[3])                      # convert from log scale
  n <- length(y)                               # number of datapoints

  mu <- LL <- numeric(n)                        # empty vectors
  # You could vectorize this loop to speed things up
  for (i in 1:n){
    mu[i] <- alpha + beta * x[i]
    LL[i] <- dnorm(y[i], mean = mu[i], sd = sigma, log = TRUE)
  }
  -sum(LL)
}
```

```
# Minimize that NLL to find MLEs and also get SEs
inits <- c('alpha' = 50, 'beta' = 10, 'log_sigma' = 0)
out6.2.6 <- optim(inits, NLL, y = y, x = x, hessian=TRUE)
diy_est <- c(out6.2.6$par[1:2], exp(out6.2.6$par[3]))   # save estimates
get_MLE(out6.2.6, 5)
```

	MLE	ASE	LCL.95	UCL.95
alpha	104.7563	0.355651	104.05925	105.4534
beta	-27.6037	0.562328	-28.70589	-26.5016
log_sigma	1.0134	0.070979	0.87423	1.1525

Impeccable!

6.2.7 **Likelihood analysis with TMB**

As with do-it-yourself (DIY) maximum likelihood, the analysis of the two-group model using TMB will look very similar to the equivalent analysis in Chapter 5.

```
# Bundle and summarize data (same as before)
str(dataList <- list(y = y, x = x, n = n))   # not shown
```

We have made one addition to the model file from Chapter 5. TMB allows us to calculate and monitor new parameters that are functions of the estimated parameters in the model (similar to derived parameters in JAGS and NIMBLE, and generated quantities in Stan). Using this functionality, we calculate values for the group 1 (= female) mean (mu1, equal to alpha as we discussed in Section 6.2) and the group 2 (= male) mean (mu2, equal to alpha plus beta). We can tell TMB to add estimates of these derived parameters (and, importantly, corresponding estimates of uncertainty) in the reported output by wrapping the parameter names in the special ADREPORT function.

```
# Write TMB model file
cat(file = "model6_2_7.cpp",
"#include <TMB.hpp>

template<class Type>
Type objective_function<Type>::operator() ()
{
  //Describe input data
  DATA_VECTOR(y);                       //response
  DATA_VECTOR(x);                       //covariate
  DATA_INTEGER(n);                      //Number of obs

  //Describe parameters
  PARAMETER(alpha);                     //Intercept
  PARAMETER(beta);                      //Slope
  PARAMETER(log_sigma);                 //log(residual standard deviation)

  Type sigma = exp(log_sigma);          //Type = match type of function output (double)
  Type LL = 0.0;                        //Initialize total log likelihood at 0
  Type mu;

  for (int i=0; i<n; i++){              //Note index starts at 0 instead of 1!
    mu = alpha + beta * x(i);
    //Calculate log-likelihood of observation and add to total
    LL += dnorm(y(i), mu, sigma, true); //Add log-lik of obs i
  }

  //Derived parameters
  Type mu1 = alpha;
  Type mu2 = alpha + beta;
  ADREPORT(mu1);                        // save mu1 and mu2 to output
  ADREPORT(mu2);

  return -LL;                           //Return negative of total log likelihood
}
")
```

```
# Compile and load TMB function
compile("model6_2_7.cpp")
dyn.load(dynlib("model6_2_7"))

# Provide dimensions and starting values for parameters
params <- list(alpha = 0, beta = 0, log_sigma = 0)

# Create TMB object
out6.2.7 <- MakeADFun(data = dataList, parameters = params,
                      DLL = "model6_2_7", silent = TRUE)

# Optimize TMB object, print results and save estimates
opt <- optim(out6.2.7$par, fn = out6.2.7$fn, gr = out6.2.7$gr, method = "BFGS", hessian = TRUE)
tmb_est <- c(opt$par[1:2], exp(opt$par[3]))         # save estimates
(tsum <- tmb_summary(out6.2.7))                     # look at output
```

	Estimate	Std. Error
alpha	104.770584	0.35430576
beta	-27.618356	0.56020659
log_sigma	1.009577	0.07071073
mu1	104.770584	0.35430576
mu2	77.152229	0.43393416

Note that we also get estimates and standard errors for the group means mu1 and mu2 (for females and males, respectively) as we requested.

6.2.8 Comparison of the parameter estimates

As expected, the engines yield similar results for the parameter estimates (remember to run the NIMBLE code from our website if you want this comparison). Don't forget that for some engines (DIY, TMB) we needed to convert log_sigma back to sigma in order to compare with the Bayesian engines. We note that with a sample of size 100, there are hardly any differences between estimates of sigma from the least-squares and maximum likelihood methods. There are very slight differences between Bayes and non-Bayes, which may be due to the effects of the priors or perhaps due to a slight skew of the posterior. But in most practical applications of modeling, a difference between 2.75 and 2.81 will be completely unimportant.

```
# Compare results with truth and previous estimates
comp <- cbind(truth = truth, lm = lm_est, JAGS = jags_est, NIMBLE = nimble_est,
   Stan = stan_est, DIY = diy_est, TMB = tmb_est)
print(comp, 4)
```

	truth	lm	JAGS	NIMBLE	Stan	DIY	TMB
alpha	105.00	104.771	104.767	104.767	104.765	104.756	104.771
beta	-27.50	-27.618	-27.616	-27.613	-27.612	-27.604	-27.618
sigma	2.75	2.772	2.805	2.812	2.813	2.755	2.744

6.3 Comparing two groups with unequal variances

The previous analysis assumed that inter-individual variation in wingspan is identical for male and female peregrines. This may well not be the case, and it may be better to use a model that can accommodate possibly different variances—a "heterogeneous groups" model. Our model from Section 6.2 then becomes this:

$$y_i = \alpha + \beta x_i + \varepsilon_i$$

$$\varepsilon_i \sim Normal(0, \sigma_1^2) \quad \text{for} \quad x_i = 0 \quad \text{(females)}$$

$$\varepsilon_i \sim Normal(0, \sigma_2^2) \quad \text{for} \quad x_i = 1 \quad \text{(males)}$$

We now have one additional parameter to estimate. Estimating a different variance parameter for males and females will properly accommodate for the heteroskedasticity in the data.

6.3.1 Data generation

We first simulate data under the heterogeneous groups model (Fig. 6.2).

```
set.seed(63)

# Generate data set
n1 <- 60                        # Number of females
n2 <- 40                        # Number of males
mu1 <- 105                      # Population mean for females
mu2 <- 77.5                     # Population mean for males
sigma1 <- 3                     # Population SD for females
sigma2 <- 2.5                   # Population SD for males

n <- n1+n2                      # Total sample size
y1 <- rnorm(n1, mu1, sigma1)    # Data for females
y2 <- rnorm(n2, mu2, sigma2)    # Data for males
y <- c(y1, y2)                  # Combine both data sets
x <- rep(c(0,1), c(n1, n2))     # Indicator for male
```

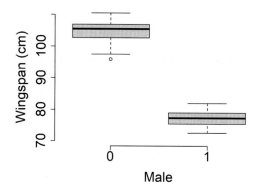

FIGURE 6.2

A boxplot of the generated data set on wingspan of female and male peregrines when the magnitude of the variation of the residuals depends on sex.

```
# Save true values for later comparisons
truth <- c(mu1 = mu1, mu2 = mu2, sigma1 = sigma1, sigma2 = sigma2)

# Make a plot (Fig. 6.2)
par(mfrow = c(1, 1), mar = c(6,6,6,3), cex.lab = 1.5, cex.axis = 1.5, cex.main = 2)
boxplot(y ~ x, col = "grey", xlab = "Male", ylab = "Wingspan (cm)", las = 1, frame = FALSE)
```

6.3.2 Frequentist analysis with canned functions in R

The canned `lm()` function in R, which we used in the previous example, does not support this model. However, we can use the `t.test` function to fit the so-called Welch two-sample *t*-test which allows for unequal variances. We use the same code as in Section 6.2.2 except that we leave the `var.equal` argument at its default value of `FALSE`.

```
(out6.3.2 <- t.test(y ~ x))
tt_est <- c(out6.3.2$est, NA, NA)   # save estimates of means

        Welch Two Sample t-test

data: y by x
t = 50.278, df = 96.17, p-value < 2.2e-16
alternative hypothesis: true difference in means is not equal to 0
95 percent confidence interval:
  26.51684 28.69664
sample estimates:
mean in group 0   mean in group 1
     104.96214           77.35539
```

The `t.test` function does not return estimates of the variance for the two groups, only their means.

Next, we assess goodness-of-fit using quantile residuals using DHARMa. The package doesn't support the `t.test` function, so we will instead fit the model from Section 6.2.2 with homogenous variance and see whether this mis-specification shows up in the residuals ... and we find it does not.

```
# Diagnostic checks of residuals/model
summary(fm <- lm(y ~ as.factor(x)-1) )   # Fit model from 6.2.2
simOut <- simulateResiduals(fm, n = 1000, plot = TRUE)
```

6.3.3 Bayesian analysis with JAGS

Now the first Bayesian analysis of this model. In effect, what we will do is simultaneously fit two separate models of the mean (see Chapter 4), one for females and one for males. Each of the two submodels will have its own mean and variance. This reflects the way our data were simulated in Section 6.3.1—that is, as two separate random processes. Our list of data looks similar to the previous section, except that we have now have two separate y elements for the wingspan values in each group (1 = female and 2 = male), and two separate sample sizes.

```
# Bundle and summarize data
str(dataList <- list(y1 = y1, y2 = y2, n1 = n1, n2 = n2))

List of 4
 $ y1: num [1:60] 109 99.4 106.5 99.3 106.6 ...
 $ y2: num [1:40] 76 79.1 75.4 78.3 75.7 ...
 $ n1: num 60
 $ n2: num 40
```

```
# Write JAGS model file
cat(file = "model6.3.3.txt", "
model {
# Priors
mu1 ~ dnorm(0, 0.0001)
mu2 ~ dnorm(0, 0.0001)
tau1 <- pow(sigma1, -2)
sigma1 ~ dunif(0, 1000)
tau2 <- pow(sigma2, -2)
sigma2 ~ dunif(0, 1000)

# Likelihood
for (i in 1:n1) {            # First sample (females)
  y1[i] ~ dnorm(mu1, tau1)
}

for (i in 1:n2) {            # Second sample (males)
  y2[i] ~ dnorm(mu2, tau2)
}

# Derived quantities
delta <- mu2 - mu1
}
")
```

As we noted previously, the JAGS model looks as if we combined two copies of the model of the mean JAGS code from Chapter 4, one for each group. We have a separate set of priors for each group, and a separate for loop. There's one other addition: we calculate a derived parameter `delta` that represents the difference between females and males. This value is similar in interpretation to the slope estimate (`beta`) from Section 6.2.

Before calling JAGS, we set initial values for the two means and the two standard deviations and tell JAGS to save the posteriors for the means, standard deviations, and `delta`.

```
# Function to generate starting values
inits <- function(){ list(mu1 = rnorm(1), mu2 = rnorm(1), sigma1 = rlnorm(1), sigma2 = rlnorm(1))}

# Parameters to estimate
params <- c("mu1", "mu2", "delta", "sigma1", "sigma2")

# MCMC settings
na <- 1000 ; ni <- 3000 ; nb <- 1000 ; nc <- 4 ; nt <- 1

# Call JAGS (ART <1 min), check convergence, summarize posteriors and save estimates
out6.3.3 <- jags(data = dataList, inits = inits, parameters.to.save = params,
  model.file = "model6.3.3.txt", n.iter = ni, n.burnin = nb, n.chains = nc,
  n.thin = nt, n.adapt = na, parallel = TRUE)
jagsUI::traceplot(out6.3.3)                              # not shown
jags_est <- out6.3.3$summary[c(1, 2, 4, 5), 1]          # save estimates
print(out6.3.3, 3)
```

	mean	sd	2.5%	50%	97.5%	overlap0	f	Rhat	n.eff
mu1	104.963	0.414	104.163	104.963	105.769	FALSE	1	1	8000
mu2	77.345	0.396	76.580	77.343	78.125	FALSE	1	1	5194
delta	-27.618	0.577	-28.740	-27.622	-26.469	FALSE	1	1	8000
sigma1	3.186	0.307	2.638	3.167	3.852	FALSE	1	1	8000
sigma2	2.453	0.290	1.963	2.431	3.087	FALSE	1	1	4350

The posterior for `delta` (the difference between the means) does not cover 0; thus we conclude that the means of the two sexes differ. To formally test whether the two variances also differ, we could reparameterize the model such that the variance for one group is expressed as the variance of the other plus some constant to be estimated. Actually, this could also be done outside of JAGS in R by forming the difference, for each draw in the Markov chain, between `sigma1` and `sigma2`. If the credible interval for that parameter covers zero, that would be taken as lack of evidence for different variances. For example:

```
samps <- as.matrix(out6.3.3$samples)
quantile(samps[,'sigma2'] - samps[,'sigma1'], c(0.025,0.975))

     2.5%      97.5%
-1.5555412  0.1009973
```

This is an important idea; that derived variables with their full posterior uncertainty can also be computed outside of JAGS in R, if Markov chains of all of their components are available. This is often easier than putting the added code into the JAGS model description.

6.3.4 Bayesian analysis with NIMBLE

Code to fit the model with NIMBLE is extremely similar to that for JAGS. We show the NIMBLE code on the book website. Get it and run it if you want the full comparison table in the end.

6.3.5 Bayesian analysis with Stan

The data are the same as with JAGS and NIMBLE.

```
# Bundle and summarize data
str(dataList <- list(y1 = y1, y2 = y2, n1 = n1, n2 = n2))
```

As with JAGS, the model file should look similar to the Stan model of the mean in Chapter 4, except that we have two groups each with their own mean and standard deviation. In this model, we've applied the vectorization approach throughout as described in Section 6.2.5. As with the JAGS and NIMBLE models, we also calculate a derived parameter `delta` (as usual, found in the `generated quantities` section) representing the difference between the two group means.

```
# Write text file with model description in BUGS language
cat(file = "model6_3_5.stan",              # This line is R code
"data {                                     //This is the first line of Stan code
  int n1;                                   //Define the format of all data
  int n2;
  vector[n1] y1;
  vector[n2] y2;
}

parameters {                                //Define format for all parameters
  real mu1;
  real mu2;
  real<lower=0> sigma1;
  real<lower=0> sigma2;
}

model {                                     //Priors and likelihood here
  //Priors
  mu1 ~ normal(0, 100);
  mu2 ~ normal(0, 100);
  sigma1 ~ cauchy(0, 10);
  sigma2 ~ cauchy(0, 10);
  //Likelihood
  y1 ~ normal(mu1, sigma1);
  y2 ~ normal(mu2, sigma2);
}

generated quantities{
  real delta = mu2 - mu1;
}                                           //This is the last line of Stan code
" )

# HMC settings
ni <- 2000  ; nb <- 1000  ; nc <- 4  ; nt <- 1

# Call STAN (ART 55/1 sec)
system.time(
  out6.3.5 <- stan(file = "model6_3_5.stan", data = dataList,
    chains = nc, iter = ni, warmup = nb, thin = nt) )
rstan::traceplot(out6.3.5)                  # not shown
stan_est <- summary(out6.3.5)$summary[1:4, 1] # save estimates
print(out6.3.5, dig = 2)                    # not shown
```

6.3.6 Do-it-yourself maximum likelihood estimates

Similar to JAGS, NIMBLE, and Stan, we will approach this model by considering it a combination of two "models of the mean" from Chapter 4. The likelihood for the model of the mean with just a single group is this:

$$L(\mu, \sigma|\mathbf{y}) = \prod_{i=1}^{n} \frac{1}{\sigma\sqrt{2\pi}} e^{-\frac{1}{2}\left(\frac{y_i-\mu}{\sigma}\right)^2}$$

For two groups of data, each with their own mean and standard deviation parameters, we can multiply together the likelihoods for each group (and we note that we will revisit this general way of combining information from multiple data sets in a single model in Chapter 20).

$$L(\mu_1, \mu_2, \sigma_1, \sigma_2 | \mathbf{y}_1, \mathbf{y}_2) = \prod_{i=1}^{n_1} \frac{1}{\sigma\sqrt{2\pi}} e^{-\frac{1}{2}\left(\frac{y_{1,i}-\mu_1}{\sigma_1}\right)^2} \prod_{j=1}^{n_2} \frac{1}{\sigma\sqrt{2\pi}} e^{-\frac{1}{2}\left(\frac{y_{2,j}-\mu_2}{\sigma_2}\right)^2}$$

The negative log-likelihood function will also look similar to the one in Chapter 4. There are two inputs instead of one (y1 and y2). We'll calculate the sum of the log-likelihood for each group, and then add them together (remember adding the log-likelihood is equivalent to multiplying the likelihood). A consequence of this approach (a means parameterization) is that while we estimate the mean for each population, we can't also explicitly estimate the difference between the two means (as with the *t*-test). You could reformulate the likelihood to use an effects parameterization instead in order to explicitly estimate the difference between the means. As in Section 6.2.6, we'll estimate the sigma parameters on the log scale for numerical convenience.

```
NLL <- function(param, y1, y2){

  mu1 <- param[1]
  mu2 <- param[2]
  sigma1 <- exp(param[3])
  sigma2 <- exp(param[4])

  LL1 <- dnorm(y1, mu1, sigma1, log = TRUE)    # females
  LL2 <- dnorm(y2, mu2, sigma2, log = TRUE)    # males
  LL <- sum(LL1) + sum(LL2)                    # total

  -LL                                          # negative log likelihood
}
```

Surprisingly perhaps, we need to give `optim()` a little help by providing semi-informative initial values. Otherwise, all is as before and we get good estimates of the four parameters.

```
# Minimize the NLL to find MLEs, get SEs and CIs and save estimates
inits <- c('mu1' = 50, 'mu2' = 50, 'log_sigma1' = 1, 'log_sigma2' = 1)
out6.3.6 <- optim(inits, NLL, y1 = y1, y2 = y2, hessian = TRUE)
diy_est <- c(out6.3.6$par[1:2], exp(out6.3.6$par[3:4]))
get_MLE(out6.3.6, 5)
```

	MLE	ASE	LCL.95	UCL.95
mu1	104.94770	0.390281	104.18275	105.7126
mu2	77.35551	0.381175	76.60841	78.1026
log_sigma1	1.10626	0.089498	0.93085	1.2817
log_sigma2	0.87994	0.115208	0.65414	1.1058

6.3.7 Likelihood analysis with TMB

The TMB analysis looks similar to the DIY likelihood analysis: we calculate the log-likelihoods for each group, and add them together. As with the Bayesian engines, we are also able to calculate a derived parameter `delta` that represents the difference between the two means, and add it to our output report with `ADREPORT`.

```r
# Bundle and summarize data (same as before)
str(dataList <- list(y1 = y1, y2 = y2, n1 = n1, n2 = n2))

# Write TMB model file
cat(file = "model6_3_7.cpp",
"#include <TMB.hpp>

template<class Type>
Type objective_function<Type>::operator() ()
{
  //Describe input data
  DATA_VECTOR(y1);                         //response pop1
  DATA_VECTOR(y2);                         //response pop2
  DATA_INTEGER(n1);                        //Number of obs pop1
  DATA_INTEGER(n2);                        //Number of obs pop2

  //Describe parameters
  PARAMETER(mu1);                          //Intercept pop1
  PARAMETER(mu2);                          //Intercept pop2
  PARAMETER(log_sigma1);                   //log(residual standard deviation) pop1
  PARAMETER(log_sigma2);                   //log(residual standard deviation) pop2

  Type sigma1 = exp(log_sigma1);
  Type sigma2 = exp(log_sigma2);
  Type LL = 0.0;                           //Initialize total log likelihood at 0

  for (int i=0; i<n1; i++){                //Note index starts at 0 instead of 1
    LL += dnorm(y1(i), mu1, sigma1, true); //Add log-lik of obs i
  }

  for (int i=0; i<n2; i++){                //Note index starts at 0 instead of 1
    LL += dnorm(y2(i), mu2, sigma2, true); //Add log-lik of obs i
  }

  Type delta = mu2 - mu1;
  ADREPORT(delta);

  return -LL;                              //Return negative of total log likelihood
}
")

# Compile and load TMB function
compile("model6_3_7.cpp")
dyn.load(dynlib("model6_3_7"))

# Provide dimensions and starting values for parameters
params <- list(mu1 = 0, mu2 = 0, log_sigma1 = 0, log_sigma2 = 0)

# Create TMB object
out6.3.7 <- MakeADFun(data = dataList, parameters = params,
                      DLL = "model6_3_7", silent = TRUE)
```

```
# Optimize TMB object, print results and save estimates
opt <- optim(out6.3.7$par, fn = out6.3.7$fn, gr = out6.3.7$gr,
            method = "BFGS", hessian = TRUE)
tmb_est <- c(opt$par[1:2], exp(opt$par[3:4]))   # save estimates
(tsum <- tmb_summary(out6.3.7))
```

6.3.8 Comparison of the parameter estimates

```
# Compare results with truth and previous estimates
comp <- cbind(truth = truth, tt = tt_est, JAGS = jags_est,  NIMBLE = nimble_est,
   Stan = stan_est, DIY = diy_est, TMB = tmb_est)
print(comp, 3)
```

	truth	tt	JAGS	NIMBLE	Stan	DIY	TMB
mu1	105.0	105.0	104.96	104.99	104.96	104.95	104.96
mu2	77.5	77.4	77.35	77.36	77.35	77.36	77.36
sigma1	3.0	NA	3.19	3.19	3.17	3.02	3.08
sigma2	2.5	NA	2.45	2.45	2.44	2.41	2.34

The variance estimates are slightly lower for the maximum likelihood engines, but otherwise estimates are very similar.

6.4 Summary and a comment on the modeling of variances

In this chapter, we compared the means between two groups, both when the group variances were equal and when they were unequal. As we discussed in Section 6.1, the former can be done by fitting a linear regression with a single covariate (see Chapter 5) representing group membership; such a covariate is also called a dummy or indicator variable, or more generally a factor with two levels. The latter can be done by fitting a combination of two models of the mean (see Chapter 4). Traditionally, both of these analyses have also been done using different forms of the t-test (Zar, 1998).

The version of that test with unequal variances is the only place in this book where we explicitly model the variance (except for the modeling of variances by variance components, or in hierarchical models; see Chapters 8, 10, 14, 17, 19, and 19B). This chapter shows that not only the mean but that also the variance may be modeled in the fashion of a linear model. In classical statistics, modeling the variance may be rather hard and fairly obscure in its application to an ecologist. In contrast in JAGS and some of the other engines, the modeling of variances, for example, as a function of some covariate, could be simply undertaken by use of a log link function; see Lee & Nelder (2006) and Lee et al. (2017) for (frequentist) examples of such models. Modeling the variance, either for the residuals or for random effects, may be required in order to adequately characterize the stochastic system components even when inference is focused on the mean structure in a model. Alternatively, one may focus on a relation between an explanatory variable and a variance, for instance, to test a hypothesis that some conditions increase the variance in some trait.

Models with a single categorical covariate with more than two levels[*]

7

Chapter outline

7.1 Introduction: fixed and random effects

In Chapter 6, we modeled the effects of a categorical explanatory variable, or factor, which had two groups, or levels. Here now, we generalize that model such that we consider a single factor with more than two groups. In addition, we will introduce random effects. In random-effects models, a set of effects or parameters (e.g., group means) is constrained to come from some distribution, which is most often a normal, though it may also be a Bernoulli (see Chapter 19) or a Poisson (see Chapter 19B) or yet another distribution. In the first half of the chapter, we will generate and analyze data under a model with a single factor where the parameters associated with the factor levels will be assumed to be fixed effects. Then, in the second half of the chapter, we will repeat this with a model with parameters that will be treated as random effects. The linear models considered in this chapter are those that underlie the frequentist analysis technique known as a fixed- and random-effects one-way analysis of variance (ANOVA; Zar, 1998; Quinn & Keough, 2002).

---

[*]This book has a companion website hosting complementary materials, including all code for download. Visit this URL to access it: https://www.elsevier.com/books-and-journals/book-companion/9780443137150.

Applied Statistical Modelling for Ecologists. DOI: https://doi.org/10.1016/B978-0-443-13715-0.00024-8
191

This chapter is the first of a series that cover mixed-effects models, that is, models that contain both fixed and random effects and where the latter are draws from a normal distribution and appear as additive terms in the linear predictor of a generalized linear model (GLM). In Chapters 10, 14, and 17, we will cover more mixed models, with both continuous explanatory variables and factors, and for normal, Poisson and binomial distributions for the response variable.

As a motivating example for this chapter, we use simulated measurements of snout-vent length (SVL) in five populations of smooth snakes (Fig. 7.1). We are interested in characterizing SVL in each population and comparing its mean among the populations. Another motivation might be the decomposition of the total variation in this size measurement into a component among populations and another among individuals within populations, i.e., variance components decomposition.

The linear model underlying a group comparison within a single factor can be parameterized in various ways (see Chapter 3). We adopt a means parameterization of the linear model for the fixed effects of populations and write this as follows:

$$y_i = \alpha_{j(i)} + \varepsilon_i$$

$$\varepsilon_i \sim Normal(0, \sigma^2)$$

Here, y_i is the observed SVL of smooth snake i in population j, $\alpha_{j(i)}$ is the expected SVL of snake i in population j, and residual ε_i is the random SVL deviation of snake i from its population mean $\alpha_{j(i)}$. This residual is assumed to be normally distributed around zero with a constant variance σ^2.

FIGURE 7.1

Smooth snake (*Coronella austriaca*), Switzerland (Photo by Andreas Meyer).

Absent any further assumptions, the population means $\alpha_{j(i)}$ are simply some unknown constants that are estimated as fixed effects. If, however, we add a distributional assumption about the population means α_j, we obtain the random-effects analog of the same model:

$$y_i = \alpha_{j(i)} + \varepsilon_i$$

$$\varepsilon_i \sim Normal(0, \sigma^2)$$

$$\alpha_j \sim Normal(\mu, \tau^2)$$

It is this single additional line which defines the values of α as random! The interpretation of α_j and ε_i as population-specific mean SVL and as the snake-level residual, respectively, remains unchanged. But now, the α_j parameters are no longer assumed to be independent, but rather we assume that they are draws from a second normal distribution with mean μ and variance τ^2. The latter are also called hyperparameters, since they occur at a higher level than the parameters α_j that they govern. For the same reason, this model can also be called a multilevel or hierarchical model.

Thus, the two variants of the model with fixed and random effects of populations differ only subtly. So, how do we know when to apply each in practice? There are differing views on this decision, see Gelman & Hill (2007, p. 245). The traditional view goes about as follows. When you have a particular interest in the studied factor levels and/or when you have included (nearly) all conceivable levels of a factor in a study, the associated factor should be viewed as having fixed effects. You estimate the effects of each level, but are not interested in the variance among levels. Importantly, you cannot generalize to factor levels that were not included in the study. In contrast, you consider a factor as random when you don't have a particular interest in the levels that actually appear in your study and/or when these levels form a sample from a (much) larger set of possible levels that you *could* have included in your study. Typically, you do want to generalize to this larger population and you are more interested in the variation among the factor levels in that population, though you may still want to estimate the effects of the levels actually observed in your study. Thus, typical fixed-effects factors would be sex or the treatments in a manipulative experiment (Quinn & Keough, 2002). Typical random-effects factors might be time (e.g., year, month, or day), species, individuals, or location, such as experimental blocks or other spatial units on which repeated measurements are taken.

However, another and perhaps more modern view of the distinction between fixed and random effects is simply based on the question of whether these effects could plausibly have come from some distribution of effects. Under this view, random effects are generated by some stochastic process and statisticians also say they are *exchangeable*. In common language you can think of them as similar, but not identical, where the similarity is due to the common stochastic process that generated them, thus creating a stochastic relationship among the effects of the levels of a random-effects factor. In contrast, when factor levels are modeled as fixed, they are completely unrelated.

In Chapter 3, we discussed several reasons for why one may treat a set of parameters as random. Perhaps the five most important reasons are as follows:

- Viewing the studied effects as a random sample from some statistical population enables one to extrapolate to that wider population and obtain information about hyperparameters like μ and τ^2. This can only be achieved by explicitly modeling the stochastic process that generates the realized values of the random effects, for example, by assuming a normal distribution for the α_j above.
- Declaring factor effects as random acknowledges that when repeating our study, we would normally obtain a different set of effects, so the resulting parameter estimates will differ from those in our current study. Random-effects modeling properly accounts for this added uncertainty in our inference about the analyzed system.

- When making a random-effects assumption about a factor, these effects are no longer estimated independently; instead, estimates are influenced by each other and in this sense somewhat dependent. Specifically, the random-effects estimates α_j are "pulled in" toward the common mean μ, that is, they will be somewhat closer to μ than the corresponding fixed-effects estimates of α_j. This is why random-effects estimators are said to be "shrinkage estimators." Individual estimates that are more imprecise, for instance because they are based on a smaller sample size, are shrunk more. Shrinkage results in better estimates (e.g., with smaller prediction error) when effects are indeed exchangeable than estimates obtained from a fixed-effects analysis. This is why one also says that a random-effects analysis "borrows strength." Interestingly, despite this sharing of information there is a sense in which we actually need more information when want to estimate a set of parameter as random effects. Typically, if we have few factor levels (e.g., just 2 or 3 or perhaps even up to 5 or more), the resulting estimates of the hyperparameters will usually turn out to be extremely imprecise and hence we don't gain anything if our interest is focused on them. In these situations it will often be better to revert to the simpler fixed-effects formulations of a model.
- Modeling of correlations among parameters is typically done by treating parameters as random variables on which we place a distribution such as the multivariate normal, which has hyperparameters that govern the magnitude and direction of the correlations. Typical examples include temporal and spatial autocorrelation, while in Chapter 10 we will see an example with correlations among two types of parameters, the intercepts and the slopes in a cluster of regression lines.
- Finally, placing a distribution on a set of parameters also acts as a way of penalization, or constraint, on the resulting estimates. This can be beneficial for multiple reasons; see Section 18.8.

Random-effects modeling can also be viewed as a compromise between the extremes of assuming no effects of the levels of a factor and fully independent effects of the levels of a factor. When assuming a factor has no effect, you pool its effects, while when assuming it has fixed effects, you treat all effects as completely independent instead. When assuming a factor has random effects, you pool effects only partially, and the degree of pooling is based on the amount of information that is available about the effect of each level. According to this view, and in contrast to what we just said about the increased information requirements of random effects, some authors argue that you should always assume all factors as random and let the data determine the degree of pooling (Gelman, 2005; Gelman & Hill, 2007).

During the last 20 years, statisticians seem to prefer the view of random-effects factors as those whose levels result from a common stochastic process, with the resulting benefits of the ability to extrapolate, more honest accounting for uncertainty, and shrinkage estimation. For instance, Sauer & Link (2002) assessed population trends in a large numbers of bird species in North America and showed how imprecise estimates for species with little information borrowed strength from the "ensemble" (i.e., from the group of all the species) and got pulled toward the group mean, and this yielded better predictions. Similarly, Welham et al. (2004) analyzed a huge wheat variety testing experiment and treated variety as random. Again, they found that this gave better predictions of future yield than treating variety as fixed. Kéry & Royle (2021; Chapter 3) report on a simulation study where treating species-level effects as draws from a normal distribution (i.e., as random effects) reduced the estimation error in a multispecies analysis even though in the data simulation a uniform distribution was used to produce the species-level effects.

In the next section, we generate one data set under a fixed-effects design and another under a random-effects design. We do this in a "linear model" fashion, that is, by first specifying a design matrix and choosing parameter values associated with each level in our classification, i.e., for each population. For the fixed-effects analysis, we will arbitrarily select these values, while for the random-effects analysis, we will draw them from a normal distribution with hyperparameters that we pick, and to which we can then compare our estimates. Then we multiply the design matrix with the parameter vector to obtain the linear predictor, to which we add residuals to obtain the actual simulated measurements.

7.2 Fixed-effects models

7.2.1 Data generation

We assume five populations with 10 snakes measured in each, with SVL averages of 50, 40, 45, 55, and 60. This corresponds to a baseline population mean of 50 and effects of populations 2 to 5 of −10, −5, 5, and 10. We choose a residual standard deviation of SVL of 5 and assemble everything (Fig. 7.2).

```
# Simulate a data set
set.seed(72)                              # Initialize RNGs
nPops <- 5                                # Number of populations
nSample <- 10                             # Number of snakes in each
pop.means <- c(50, 40, 45, 55, 60)        # Population mean SVL
sigma <- 5                                # Residual sd

n <- nPops * nSample                      # Total number of data points
eps <- rnorm(n, 0, sigma)                 # Residuals
pop <- factor(rep(1:5, rep(nSample, nPops)))   # Indicator for population
means <- rep(pop.means, rep(nSample, nPops))
X <- as.matrix(model.matrix(~ pop-1))     # Create design matrix
X                                         # Inspect design matrix
y <- as.numeric(X %*% as.matrix(pop.means) + eps)
                                          # %*% denotes matrix multiplication
```

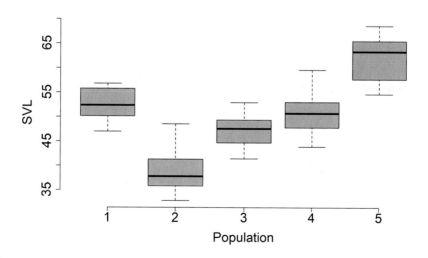

FIGURE 7.2

Snout-vent length (SVL) of smooth snakes in five populations simulated under a fixed-effects model.

```
# Save true values for later comparisons
truth <- c(pop.means, sigma)
names(truth) <- c(paste0("pop", 1:5), "sigma")

# Make plot (Fig. 7.2)
par(mfrow = c(1, 1), mar = c(6,6,6,3), cex.lab = 1.5, cex.axis = 1.5, cex.main = 2)
boxplot(y ~ pop, col = "grey", xlab = "Population", ylab = "SVL", main = "", las = 1, frame = FALSE)

# Load required libraries
library(ASMbook); library(DHARMa); library(jagsUI); library(rstan); library(TMB)
```

7.2.2 Likelihood analysis with canned functions in R

We continue working with function lm() to produce least-squares estimates, which for this model class are identical to the maximum likelihood estimates (MLEs) for the parameters in the mean and very similar to the MLE for the variance. Remember from Chapter 3 how the parameters associated with the levels of a factor can be expressed in a variety of ways. We also show the effects parameterization, but retain estimates from the means parameterization.

```
# Default treatment contrast, or effects, parameterization (not shown)
summary(out <- lm(y ~ pop))
# Means parameterization of factor levels
summary(out72.2 <- lm(y ~ pop - 1))

Coefficients:
        Estimate  Std. Error  t Value  Pr(>|t|)
pop1     52.473      1.359      38.62    <2e-16 ***
pop2     39.306      1.359      28.93    <2e-16 ***
pop3     47.034      1.359      34.62    <2e-16 ***
pop4     50.611      1.359      37.25    <2e-16 ***
pop5     62.257      1.359      45.83    <2e-16 ***
---

Signif. codes:  0 '***' 0.001 '**' 0.01 '*' 0.05 '.' 0.1 ' ' 1

Residual standard error: 4.296 on 45 degrees of freedom
Multiple R-squared: 0.9936,  Adjusted R-squared: 0.9929
F-statistic: 1403 on 5 and 45 DF, p-value: < 2.2e-16

# Save least-squares estimates
lm_est <- c(coef(out72.2), sigma = sigma(out72.2))
```

We next assess both traditional residuals and quantile residuals using DHARMa and not so surprisingly find all in order (except for a couple error messages which we ignore).

```
# Check goodness-of-fit using traditional and quantile residuals
plot(out72.2)      # Traditional residual check for comparison
simOut <- simulateResiduals(out72.2, n = 1000, plot = TRUE)
```

7.2.3 Bayesian analysis with JAGS

We fit a means parameterization of the model again and obtain effects estimates, i.e., differences in the mean SVL among populations, as derived quantities. Note JAGS' elegant double-indexing (alpha[pop[i]]) to specify the expected SVL of snake i according to the i-th value of the population index pop. We also add two lines to show how custom hypotheses can easily be tested as derived quantities. Test 1 examines whether snakes in populations 2 and 3 have the same size on average as those in populations 4 and 5. Test 2 checks whether the size difference between snakes in populations 5 and 1 is on average twice that between populations 4 and 1. Both are fairly arbitrary of course, but they illustrate how easily we can test very specific quantitative hypotheses as a derived quantity in a Bayesian analysis.

```
# Bundle and summarize data
# Note that JAGS requires us to convert pop from a factor to numeric
str(dataList <- list(y = y, pop = as.numeric(pop), n = n, nPops = nPops))

List of 4
 $ y     : num  [1:50] 56.9  55.8  50.2  55.2  53.2 ...
 $ pop   : num  [1:50] 1  1  1  1  1  1  1  1  1  1 ...
 $ n     : num  50
 $ nPops : num  5


# Write JAGS model file
cat(file = "model72.3.txt", "
model {
# Priors
for (i in 1:nPops){                                # Define alpha as a vector
  alpha[i] ~ dnorm(0, 0.0001)
}
tau <- pow(sigma, -2)
sigma ~ dunif(0, 10)

# Likelihood
for (i in 1:n) {
  y[i] ~ dnorm(mean[i], tau)
  mean[i] <- alpha[pop[i]]
}

# Derived quantities (effects estimates)
effe2 <- alpha[2] - alpha[1]
effe3 <- alpha[3] - alpha[1]
effe4 <- alpha[4] - alpha[1]
effe5 <- alpha[5] - alpha[1]

# Custom hypothesis test/Define your own contrasts
test1 <- (effe2+effe3) - (effe4+effe5)      # Equals 0 when 2+3 = 4+5
test2 <- effe5 - 2 * effe4                   # Equals 0 when effe5 = 2*effe4
}
")

# Function to generate starting values
inits <- function(){ list(alpha = rnorm(nPops, mean = mean(y)), sigma = rlnorm(1) )}

# Parameters to estimate
params <- c("alpha", "sigma", "effe2", "effe3", "effe4", "effe5", "test1", "test2")

# MCMC settings
na <- 1000  ;  ni <- 3000  ;  nb <- 1000  ;  nc <- 4  ;  nt <- 1
```

```
# Call JAGS (ART <1 sec), check convergence and summarize posteriors
out72.3 <- jags(dataList, inits, params, "model72.3.txt", n.iter = ni,
  n.burnin = nb, n.chains = nc, n.thin = nt, n.adapt = na, parallel = TRUE)
jagsUI::traceplot(out72.3)                      # not shown
print(out72.3, 3)
```

	mean	sd	2.5%	50%	97.5%	overlap0	f	Rhat	n.eff
alpha[1]	52.457	1.397	49.745	52.452	55.182	FALSE	1.000	1.000	8000
alpha[2]	39.302	1.421	36.544	39.299	42.117	FALSE	1.000	1.000	8000
alpha[3]	47.044	1.391	44.353	47.045	49.757	FALSE	1.000	1.000	5969
alpha[4]	50.610	1.387	47.844	50.614	53.364	FALSE	1.000	1.000	8000
alpha[5]	62.239	1.396	59.480	62.237	65.005	FALSE	1.000	1.001	4547
sigma	4.431	0.493	3.591	4.391	5.516	FALSE	1.000	1.001	2931
effe2	-13.155	1.991	-17.105	-13.149	-9.196	FALSE	1.000	1.000	8000
effe3	-5.412	1.973	-9.265	-5.411	-1.564	FALSE	0.997	1.000	5817
effe4	-1.847	1.981	-5.670	-1.854	2.097	TRUE	0.828	1.000	8000
effe5	9.783	1.978	5.827	9.783	13.714	FALSE	1.000	1.001	3852
test1	-26.503	2.822	-31.944	-26.506	-20.905	FALSE	1.000	1.000	8000
test2	13.476	3.407	6.716	13.499	20.195	FALSE	1.000	1.000	8000

```
# Compare likelihood with Bayesian estimates and with truth
jags_est <- out72.3$summary[1:6, 1]
comp <- cbind(truth = truth, lm = lm_est, JAGS = jags_est)
print(comp, 4)
```

	truth	lm	JAGS
pop1	50	52.473	52.457
pop2	40	39.306	39.302
pop3	45	47.034	47.044
pop4	55	50.611	50.610
pop5	60	62.257	62.239
sigma	5	4.296	4.431

Comparison with the ML solutions shows once more how with vague priors, a Bayesian analysis typically yields very similar inferences as does a frequentist analysis. However, one of the most compelling things about a Bayesian analysis conducted using MCMC methods is the ease with which derived quantities can be estimated and custom tests conducted. In the above JAGS model code, we see how easily such custom contrasts (i.e., focused comparisons) can be estimated with full error propagation from all the involved random quantities. Thus, although we specified the model in the means parameterization, we can easily recover the parameters under an effects parameterization as derived quantities. Of course, for a simple model such as a normal linear model with a single factor, or a one-way ANOVA, this can also be done easily in standard stats packages. However, in JAGS this is *equally simple for any kind of parameter*, for example, for variances (see Chapter 5), *and in any type of model*, for example, mixed models, GLMs, generalized linear mixed models (GLMMs), or indeed any hierarchical models.

7.2.4 Bayesian analysis with NIMBLE

NIMBLE code for the model is available on the book website.

7.2.5 **Bayesian analysis with Stan**

Stan requires the same list of data as JAGS and NIMBLE.

```
# Bundle and summarize data (same as for NIMBLE)
str(dataList <- list(y = y, pop = as.numeric(pop), n = n, nPops = nPops))

# Write text file with model description in BUGS language
cat(file = "model72_5.stan",
"
data {
  int<lower=1> n;                          // Declare all data
  int<lower=1> nPops;
  vector[n] y;
  array[n] int <lower=1> pop;
}

parameters {                               // Define format for all parameters
  vector [nPops] alpha;                    // Mean value for each pop
  real<lower=0> sigma;                     // Standard deviation must be positive
}

model {
  // Priors
  alpha ~ normal(0, 100);
  sigma ~ cauchy(0, 10);
  // Likelihood
  for(i in 1:n) {
    y[i] ~ normal(alpha[pop[i]], sigma);
  }
}

generated quantities {
  // Derived quantities
  real effe2 = alpha[2] - alpha[1];
  real effe3 = alpha[3] - alpha[1];
  real effe4 = alpha[4] - alpha[1];
  real effe5 = alpha[5] - alpha[1];

  // Custom hypothesis test / Define your own contrasts
  // test1 equals 0 when 2+3 = 4+5
  real test1 = (effe2+effe3) - (effe4+effe5);
  real test2 = effe5 - 2 * effe4;
}
" )

# HMC settings
ni <- 3000   ;   nb <- 1000   ;   nc <- 4   ;   nt <- 1

# Call STAN (ART 74/2 sec)
system.time(
  out72.5 <- stan(file = "model72_5.stan", data = dataList,
    chains = nc, iter = ni, warmup = nb, thin = nt) )
rstan::traceplot(out72.5)              # not shown
print(out72.5, dig = 2)               # not shown
stan_est <- summary(out72.5)$summary[1:6,1]
```

7.2.6 Do-it-yourself maximum likelihood estimates

The likelihood equation for this model looks similar to that in Chapters 5 and 6. We just need to modify how we calculate the vector of expected values for each data point, μ. We'll calculate these values by matrix-multiplying the model matrix X with the vector of population means α, the same way we did when we generated the data in Section 7.2.1.

$$L(\alpha, \sigma | y, X) = \prod_{i=1}^{n} \frac{1}{\sigma \sqrt{2\pi}} e^{-\frac{1}{2}\left(\frac{y_i - \mu_i}{\sigma}\right)^2}$$

$$\mu = X\alpha$$

Given a set of input population means `alpha` and a model matrix `X`, we calculate the expected mean value `mu` for each datum based on its population membership. For each observed `y`, we then calculate the log of the probability (i.e., the log-likelihood) that `y` came from a normal distribution with mean `alpha` and standard deviation `sigma`. The closer the provided values of `alpha` and `sigma` are to the values we used to generate the dataset, the greater should be the log-likelihood. The greater the log-likelihoods for each datum, the smaller the negative of their total sum (the negative log-likelihood). Function `optim()` attempts to find parameters that get us to the minimum possible value.

```
# Definition of NLL for a one-factor linear model with Gaussian errors
NLL <- function(param, y, X) {
  alpha <- param[1:5]                      # Population means
  sigma <- param[6]                        # Residual SD
  mu <- X %*% alpha                        # Get pop mean for each datum
  LL <- dnorm(y, mu, sigma, log = TRUE)    # log-likelihood each datum
  NLL <- -sum(LL)                          # NLL for all data points
  return(NLL)
}
```

```
# Get desired design matrix (means parameterization)
X <- model.matrix(~ pop - 1)
```

```
# Minimize that NLL to find MLEs, get SEs and CIs and save estimates
inits <- c('mu1' = 50, 'mu2' = 50, 'mu3' = 50, 'mu4' = 50, 'mu5' = 50, 'sigma' = 10)
out72.6 <- optim(inits, NLL, y = y, X = X, hessian = TRUE)
get_MLE(out72.6, 4)
diy_est <- out72.6$par
```

```
         MLE      ASE    LCL.95   UCL.95
mu1    52.466   1.2873   49.943   54.989
mu2    39.304   1.2873   36.780   41.827
mu3    47.040   1.2873   44.517   49.563
mu4    50.616   1.2873   48.093   53.139
mu5    62.262   1.2873   59.739   64.785
sigma   4.071   0.4064    3.274    4.867
```

7.2.7 **Likelihood analysis with TMB**

The data list for Template Model Builder (TMB) is similar to that used for the other engines. We must make one small change: for JAGS, NIMBLE, and Stan, indices start with 1, but for TMB they start with 0. Thus, we must subtract 1 from our vector of population indices pop so that the smallest value is 0.

```
# Bundle and summarize data (similar as before, except for pop)
str(tmbData <- list(y = y, pop = as.numeric(pop) - 1, n = n))
```

The TMB model begins with the usual boilerplate code, followed by a description of the data and the parameters. The model description itself should look very similar to our DIY likelihood and to Stan, except with the slightly more verbose TMB/C++ method of describing a for loop. Finally, note that TMB requires you to explicitly tell it to save output for derived parameters using the function ADREPORT.

```
# Write TMB model file
cat(file = "model72_7.cpp",
"#include <TMB.hpp>

template<class Type>
Type objective_function<Type>::operator() ()
{
  //Describe input data
  DATA_VECTOR(y);                       //response
  DATA_IVECTOR(pop);                    //Population index: note IVECTOR, not VECTOR
  DATA_INTEGER(n);                      //Number of obs

  //Describe parameters
  PARAMETER_VECTOR(alpha);              //Population means
  PARAMETER(log_sigma);                 //log(residual standard deviation)

  Type sigma = exp(log_sigma);          //Standard deviation

  Type LL = 0;                          //Initialize total log likelihood at 0

  // Likelihood of each datapoint
  for (int i=0; i<n; i++){              //Note index starts at 0 instead of 1!
    LL += dnorm(y(i), alpha(pop(i)), sigma, true);
  }

  // Derived effects (note index starts at 0!)
  Type effe2 = alpha(1) - alpha(0);
  Type effe3 = alpha(2) - alpha(0);
  Type effe4 = alpha(3) - alpha(0);
  Type effe5 = alpha(4) - alpha(0);

  // Custom hypothesis test
  Type test1 = (effe2+effe3) - (effe4+effe5);
  Type test2 = effe5 - 2 * effe4;

  // Tell TMB to save derived parameter values
  ADREPORT(effe2);
  ADREPORT(effe3);
  ADREPORT(effe4);
  ADREPORT(effe5);
  ADREPORT(test1);
  ADREPORT(test2);

  return -LL;                           //Return negative of total log likelihood
}
")
```

```
# Compile and load TMB function
compile("model72_7.cpp")    # gibberish....
dyn.load(dynlib("model72_7"))

# Provide dimensions and starting values for parameters
params <- list(alpha = rep(0, nPops), log_sigma = 0)

# Create TMB object
out72.7 <- MakeADFun(data = tmbData, parameters = params,
                     DLL = "model72_7", silent = TRUE)

# Optimize TMB object, print results and save estimates
opt <- optim(out72.7$par, fn = out72.7$fn, gr = out72.7$gr, method = "BFGS", hessian = TRUE)
(tsum <- tmb_summary(out72.7))                # not shown
tmb_est <- c(opt$par[1:5], exp(opt$par[6]))
```

7.2.8 Comparison of the parameter estimates

As always, we compare among the engines the estimates of all parameters.

```
# Compare results with truth
comp <- cbind(truth = truth, lm = lm_est, JAGS = jags_est,
  NIMBLE = nimble_est, Stan = stan_est, DIY = diy_est, TMB = tmb_est)
print(comp, 4)
```

	truth	lm	JAGS	NIMBLE	Stan	DIY	TMB
pop1	50	52.473	52.457	52.442	52.461	52.466	52.473
pop2	40	39.306	39.302	39.276	39.319	39.304	39.306
pop3	45	47.034	47.044	47.050	47.010	47.040	47.034
pop4	55	50.611	50.610	50.605	50.597	50.616	50.611
pop5	60	62.257	62.239	62.232	62.243	62.262	62.257
sigma	5	4.296	4.431	4.398	4.395	4.071	4.076

We note very similar estimates in the mean part of the model, but slight differences in the dispersion part. For this particular data set, all estimates of `sigma` are somewhat off, with the Bayesian estimates a little less so than the least-squares estimate. This difference is likely a consequence of our priors chosen, which even if vague, do provide a little information in this moderately sized data set. In addition, we see again that the MLE of `sigma` is biased for small samples; see comments in Section 5.9.

7.3 Random-effects models

7.3.1 Data generation

For our second data set we assume that population SVL means come from a normal distribution with hyperparameters that we choose. The code for data simulation is only slightly different from the previous data-generating code. First, we choose the two sample sizes: the number of populations and that of snakes examined in each.

```
# Simulate a data set
set.seed(73)
nPops <- 10              # Number of populations: choose 10 rather than 5
nSample <- 12            # Number of snakes in each
n <- nPops * nSample     # Total number of data points
```

We choose the hyperparameters of the normal distribution from which the random population means are thought to come from and use `rnorm()` to draw one realization from that distribution for each population. Then, we select the residual standard deviation and draw residuals.

```
pop.grand.mean <- 50                    # Grand mean SVL
pop.sd <- 3                             # sd of population effects about mean
pop.means <- rnorm(n = nPops, mean = pop.grand.mean, sd = pop.sd)
sigma <- 5                              # Residual sd
eps <- rnorm(n, 0, sigma)               # Draw residuals
```

We build the design matrix, expand the population effects to the chosen (larger) sample size, use matrix multiplication to assemble our data set, and have a look at what we've created (Fig. 7.3):

```
pop <- factor(rep(1:nPops, rep(nSample, nPops)))
Xmat <- as.matrix(model.matrix(~ pop - 1))
y <- as.numeric(Xmat %*% as.matrix(pop.means) + eps)

# Save true values for later comparisons
truth <- c(pop.grand.mean = pop.grand.mean, pop.sd = pop.sd, residual.sd = sigma)

# Make a plot (Fig. 7.3)
par(mfrow = c(1, 1), mar = c(6,6,6,3), cex.lab = 1.5, cex.axis = 1.5, cex.main = 2)
boxplot(y ~ pop, col = "grey", xlab = "Population", ylab = "SVL", main = "", las = 1, frame = FALSE)
abline(h = pop.grand.mean)
```

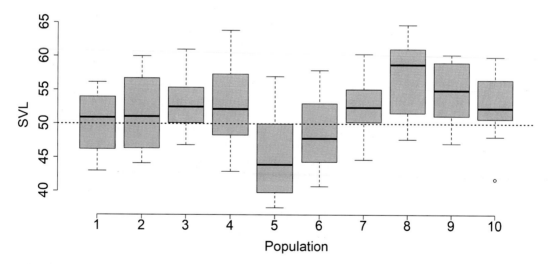

FIGURE 7.3

Simulated snout-vent length (SVL) of smooth snakes (*Coronella austriaca*) in 10 populations. Note that there is no way to tell from this graph that the population effects were simulated as random now rather than as fixed effects as in Fig. 7.2. The horizontal line shows the hyperparameter for the grand mean.

7.3.2 **Likelihood analysis with canned functions in R**

A variety of packages allow us to fit mixed models in R using likelihood methods. These include the "mother" of all such packages in R, lme4 (Bates et al., 2014), and more recently also a wrapper package for TMB called glmmTMB (Brooks et al., 2017). Both packages let us fit a wide range of linear and generalized linear mixed models (LMMs and GLMMs), with either maximum likelihood (ML) or restricted maximum likelihood (REML). REML is a variant of ML known to yield better variance estimates for mixed models than straight ML (Bolker et al., 2009), because with ML we usually underestimate the group-level variances. Here and throughout the book we use glmmTMB to fit mixed models, but note that the syntax for fitting them with lme4 is virtually identical. Both packages use a variation on the typical R formula language. Random effects are specified in the formula using the | (bar) character. For example, to specify a model with random intercepts for each population, we include + (1|pop) in our formula.

```
library(glmmTMB)    # Load glmmTMB

# Fit model and inspect results
summary(out73.2ML <- glmmTMB(y ~ 1 + 1 | pop, REML = FALSE))      # ML
summary(out73.2REML <- glmmTMB(y ~ 1 + 1 | pop, REML = TRUE))     # REML
ranef(out73.2REML)                                                # Look: THESE are the
                                                                  # estimated random effects !


> summary(out73.2ML <- glmmTMB(y ~ 1 + 1 | pop, REML = FALSE))    # ML
Family: gaussian    (identity)
Formula:          y ~ 1 + 1 | pop

....

Random effects:

Conditional model:
 Groups    Name          Variance  Std.Dev.
 pop       (Intercept)    7.172     2.678
 Residual                28.100     5.301
Number of obs: 120, groups: pop,10

Dispersion estimate for gaussian family (sigma^2): 28.1

Conditional model:
            Estimate Std. Error  z value  Pr(>|z|)
(Intercept)  51.9202    0.9754    53.23    <2e-16  ***
....


> summary(out73.2REML <- glmmTMB(y ~ 1 + 1 | pop, REML = TRUE))   # REML
Family: gaussian (identity)
Formula:          y ~ 1 + 1 | pop

....
```

```
Random effects:

Conditional model:
 Groups    Name              Variance  Std.Dev.
 pop       (Intercept)        8.229     2.869
 Residual                    28.100     5.301
Number of obs: 120, groups: pop, 10

Dispersion estimate for gaussian family (sigma^2): 28.1

Conditional model:
             Estimate  Std. Error  z value  Pr(>|z|)
(Intercept)   51.920      1.028      50.5    <2e-16 ***
....

> ranef(out73.2REML)  # Look: THESE are the estimated random effects!
$pop
      (Intercept)
1     -1.2155429
2     -0.3084802
3      0.7353944
4      1.0092069
5     -5.1650833
6     -2.5510673
7      0.6936551
8      3.9226811
9      2.2541683
10     0.6250679

# Save estimates from glmmTMB and make interim comparison
gtmb_est <- c(fixef(out73.2ML)$cond, sqrt(VarCorr(out73.2ML)$cond$pop[1]),
  sigma(out73.2ML))
gtmb_REMLest <- c(fixef(out73.2REML)$cond, sqrt(VarCorr(out73.2REML)$cond$pop[1]),
  sigma(out73.2REML))
comp <- cbind(truth = truth, ML = gtmb_est, REML = gtmb_REMLest)
print(comp, 4)            # not shown, but note REML estimate of pop.sd greater
```

Next is the assessment of GOF using quantile residuals. In mixed models such as our random-effects model, simulated quantile residuals can either be computed conditional on (i.e., given, or based on) the realized value of the random effects (here, the population means) or in an unconditional way, in which new draws of the random effects are produced for each simulated data set. As Hartig (2022) explains, both may be useful in different circumstances. However, at the time of writing, use of DHARMa with glmmTMB fitted model objects only supports unconditional simulations so this is what we do.

```
# Check goodness-of-fit using quantile residuals
simOut <- simulateResiduals(out73.2ML, n = 1000, plot = TRUE)
```

7.3.3 **Bayesian analysis with JAGS … and a quick fixed-random comparison**

Now, JAGS. Remember that random effects are a set of parameters on which we place a statistical distribution, which itself has parameters that are shared among the random effects, and which we estimate from the data. You can check this out in the BUGS code here below, where adopting a common prior distribution for the 10 population means represents the random-effects assumption in our model.

```
# Bundle and summarize data
str(dataList <- list(y = y, pop = as.numeric(pop), nPops = nPops, n = n))

List of 4
 $ y     : num [1:120] 45.5  54.1  56.2  52.1  47 ...
 $ pop   : num [1:120] 1  1  1  1  1  1  1  1  1  1 ...
 $ nPops : num 10
 $ n     : num 120

# Write JAGS model file
cat(file = "model73.3.txt", "
model {
# Priors (and also some derived quantities)
for (i in 1:nPops){
  # Prior for population means
  pop.mean[i] ~ dnorm(pop.grand.mean, pop.tau) # Random effects !
  # Pop. effects as derived quantities
  effe[i] <- pop.mean[i] - pop.grand.mean      # Ranef as in lme4
}
pop.grand.mean ~ dnorm(0, 0.0001)              # Hyperprior for grand mean svl
pop.sd ~ dt(0, 0.01, 1)T(0,)                   # Hyperprior for sd of population effects
residual.sd ~ dunif(0, 10)                     # Prior for residual sd

pop.tau <- pow(pop.sd, -2)
residual.tau <- pow(residual.sd, -2)

# 'Likelihood'
for (i in 1:n) {
  y[i] ~ dnorm(mu[i], residual.tau)
  mu[i] <- pop.mean[pop[i]]
}
}
")
```

We make a couple of comments here. First, derived quantity `effe` expresses the random effects as parameterized in `lme4` and `glmmTMB`, namely, as draws from a zero-mean normal distribution. We compute these quantities in case you want to compare them with the solutions from these engines. As a prior for the population standard deviation `pop.sd`, we give a half-Cauchy distribution with suitable width. Gelman (2006) found this prior was often better than alternative priors for variance parameters. JAGS does not explicitly support the Cauchy distribution, but it is equivalent to a t-distribution (which JAGS does support) with 1 degree of freedom. To constrain the distribution to only positive values (i.e., to make a Cauchy distribution into half-Cauchy), we add the `T(0,)` truncation notation after the distribution.

```
# Function to generate starting values
inits <- function(){ list(pop.grand.mean = runif(1, 1, 100),
  pop.sd = runif(1), residual.sd = runif(1)) }

# Parameters to estimate
params <- c("pop.grand.mean", "pop.sd", "residual.sd", "pop.mean", "effe")

# MCMC settings
na <- 1000   ;   ni <- 3000   ;   nb <- 1000   ;   nc <- 4   ;   nt <- 1

# Call JAGS (ART 1 sec), check convergence, summarize posteriors and save estimates
out73.3 <- jags(dataList, inits, params, "model73.3.txt", n.iter = ni,
  n.burnin = nb, n.chains = nc, n.thin = nt, n.adapt = na, parallel = TRUE)
jagsUI:: traceplot(out73.3)                         # not shown
print(out73.3, 3)
jags_est <- out73.3$summary[1:3,1]
```

	mean	sd	2.5%	50%	97.5%	overlap0	f	Rhat	n.eff
pop.grand.mean	51.887	1.237	49.417	51.890	54.380	FALSE	1.000	1.001	4105
pop.sd	3.376	1.174	1.649	3.189	6.324	FALSE	1.000	1.003	1917
residual.sd	5.381	0.378	4.710	5.364	6.182	FALSE	1.000	1.001	2067
pop.mean[1]	50.693	1.407	47.912	50.702	53.430	FALSE	1.000	1.001	3160
pop.mean[2]	51.615	1.395	48.804	51.629	54.331	FALSE	1.000	1.000	3613
pop.mean[3]	52.628	1.417	49.851	52.621	55.419	FALSE	1.000	1.000	8000
pop.mean[4]	52.948	1.402	50.199	52.953	55.673	FALSE	1.000	1.000	8000
pop.mean[5]	46.651	1.566	43.497	46.644	49.737	FALSE	1.000	1.001	2220
pop.mean[6]	49.324	1.428	46.507	49.327	52.091	FALSE	1.000	1.000	5885
pop.mean[7]	52.626	1.395	49.865	52.638	55.390	FALSE	1.000	1.000	8000
pop.mean[8]	55.898	1.495	52.989	55.894	58.845	FALSE	1.000	1.000	5399
pop.mean[9]	54.195	1.428	51.404	54.175	56.984	FALSE	1.000	1.000	8000
pop.mean[10]	52.557	1.403	49.813	52.526	55.324	FALSE	1.000	1.000	5470
effe[1]	-1.194	1.719	-4.652	-1.182	2.163	TRUE	0.764	1.001	4966
effe[2]	-0.272	1.731	-3.760	-0.279	3.185	TRUE	0.565	1.001	2328
....									

```
# Compare likelihood with Bayesian estimates and with truth
comp <- cbind(truth = truth, ML = gtmb_est, REML = gtmb_REMLest, JAGS = jags_est)
print(comp, 4)
```

	truth	ML	REML	JAGS
pop.grand.mean	50	51.920	51.920	51.887
pop.sd	3	2.678	2.869	3.376
residual.sd	5	5.301	5.301	5.381

Comparing parameter estimates using ML and REML in `glmmTMB()` with our Bayesian analysis, we find some agreement between the three methods. While the grand mean and the residual standard deviation are estimated at near identical values, the estimates of the among-population

SVL variation (pop.sd) are more variable among methods than what we have come to expect. In this specific case, the REML estimate is closer to the true value than is the Bayesian posterior mean; however, we would have to run a simulation with, say, 1000 analysed replicate data sets to test the generality of this finding (this would be a good exercise).

What we see when we plot the posterior distribution of this parameter from the analysis in JAGS (Fig. 7.4) is a very drawn out and, especially, right-skewed distribution. This shows that it is difficult to estimate a variance component with small samples—and what we do here is estimate the variance of 10 values (i.e., the population means), which are not even observed, but must be estimated. Presumably, the estimates would become better with larger sample size, especially with more populations (and this would be another good exercise).

```
# Check shape of posterior distribution for pop.sd (Fig. 7.4)
hist(out73.3$sims.list$pop.sd, col = 'grey', breaks = 30, main = '', freq = F)
```

Also, the skew of the posterior suggests that the posterior mode may be a better point estimator than the mean. And indeed, the mode is much closer to the MLEs.

```
# Compute the posterior mode of pop.sd as a better point estimate
library(MCMCglmm)
posterior.mode(mcmc(out73.3$sims.list$pop.sd))

      var1
2.833314
```

It is really instructive to compare fixed- and random-effects estimates for the same data set. Therefore, we next repeat the fixed-effects analysis from the last section and compare the two sets of estimates for the population means. We recycle everything we can; in particular, the data are identical. We also drop some things from the model which are tangential for this comparison.

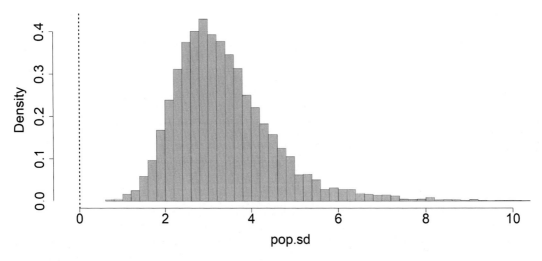

FIGURE 7.4

Posterior distribution of the among-population SD of snout-vent length in simulated smooth snakes.

```
# Write JAGS model file
cat(file = "model73.3fix.txt", "
model {
# Priors (and also some derived quantities)
for (i in 1:nPops){
  # Prior for population means: no estimated hyperparameters now !
  pop.mean[i] ~ dnorm(0, 0.0001)
}
residual.sd ~ dunif(0, 10)                      # Prior for residual sd
residual.tau <- pow(residual.sd, -2)

# 'Likelihood'
for (i in 1:n) {
  y[i] ~ dnorm(mu[i], residual.tau)
  mu[i] <- pop.mean[pop[i]]
}
}
")

# Function to generate starting values
inits <- function(){ list(pop.mean = runif(nPops, 30, 70), residual.sd = runif(1)) }

# Parameters to estimate
params <- c("pop.mean", "residual.sd")

# Call JAGS (ART <1 min), check convergence and summarize posteriors
out73.3fix <- jags(dataList, inits, params, "model73.3fix.txt", n.iter = ni,
  n.burnin = nb, n.chains = nc, n.thin = nt, n.adapt = na, parallel = TRUE)
jagsUI::traceplot(out73.3fix)                   # not shown
print(out73.3fix, 3)                            # also not shown

# Compare fixed- and the random-effects population means (Fig. 7.5)
off <- 0.15
plot((1:nPops)-off, pop.means, pch = 16, cex = 2, col = 'red', frame = FALSE,
  xlab = 'Population', ylab = 'Mean SVL in population', las = 1, ylim = c(38, 62))
abline(h = out73.3$mean$pop.grand.mean, lty = 2, lwd = 3, col = 'blue')
points(1:nPops, out73.3$mean$pop.mean, pch = 16, cex = 2, col = 'blue')
segments(1:nPops, out73.3$q2.5$pop.mean, 1:nPops, out73.3$q97.5$pop.mean, lwd = 2, col = 'blue')
points((1:nPops)+off, out73.3fix$mean$pop.mean, pch = 16, cex = 2, col = 'black')
segments((1:nPops)+off, out73.3fix$q2.5$pop.mean, (1:nPops)+off,
  out73.3fix$q97.5$pop.mean, lwd = 2, col = 'black')
legend('topleft', pch = 16, cex = 1.5, col = c("red", "blue", "black"), legend = c
("Truth", "Random effects", "Fixed effects"), bty = 'n')
```

In Fig. 7.5, we see shrinkage of the population mean estimates toward the estimated grand mean, especially for the more extreme population means. Overall, there appears to be a tendency for the random-effects estimates to be closer to the true values (red) than the fixed-effects estimates, although a proper simulation would be necessary to confirm this impression. In addition, the random-effects estimates are slightly more precise, as shown by the shorter 95% credible intervals, illustrating what is meant by "borrowing strength." Both effects would become stronger with smaller sample sizes.

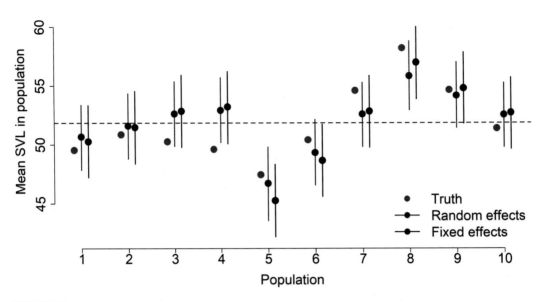

FIGURE 7.5

Comparison of the fixed- and the random-effects estimates for the same data set with the true population means in the data simulation. Dashed blue line is the estimated population grand mean in the random-effects model, to which the random-effects population means are shrunken to. Note how one parameter is outside of the 95% credible interval... yes, this happens about 5% of the times also in a Bayesian analysis. Note that due to printing limitations, the blue color for the random-effects estimates can hardly be distinguished from the black fixed-effects estimates. The order of points in each cluster, from left to right, is truth, random effects estimate, fixed effect estimate.

7.3.4 Bayesian analysis with NIMBLE

NIMBLE code for the model is available on the book website.

7.3.5 Bayesian analysis with Stan

Our data for Stan is the same as for JAGS and NIMBLE, and once again we must divide up our code into `data`, `parameters`, `model`, and `generated quantities` sections. The specification of the half-Cauchy prior on `pop_sd` is more straightforward in Stan, as the distribution is explicitly supported with `cauchy`. To make the distribution a half-Cauchy, we specify in `parameters` that `pop_sd` must be positive (`<lower = 0>`). In the Stan model, we use the half-Cauchy distribution for the residual standard deviation as well.

```r
# Bundle and summarize data (same as before)
str(dataList <- list(y = y, pop = as.numeric(pop), nPops = nPops, n = n))

# Write text file with model description in BUGS language
cat(file = "model73_5.stan",
"data {                                 // Describe input data
  int n;
  int nPops;
  vector[n] y;
  array[n] int pop;                     // A vector of integers of length n
}

parameters {                            // Define parameters
  real<lower=0> pop_grand_mean;
  real<lower=0> pop_sd;
  real<lower=0> residual_sd;
  vector<lower=0>[nPops] pop_mean;
}

model {
  // Vector to hold expected value for each datapoint
  vector[n] mu;

  // Priors
  pop_grand_mean ~ normal(0, 100);
  pop_sd ~ cauchy(0, 10);
  residual_sd ~ cauchy(0, 10);

  for (i in 1:nPops) {
    pop_mean[i] ~ normal(pop_grand_mean, pop_sd);
  }

  // 'Likelihood'
  for(i in 1:n) {
    mu[i] = pop_mean[pop[i]];
    y[i] ~ normal(mu[i], residual_sd);
  }
}

generated quantities {
  vector[nPops] effe;
  for (i in 1:nPops){
    effe[i] = pop_mean[i] - pop_grand_mean;
  }
}
" )

# HMC settings
ni <- 2000   ;   nb <- 1000   ;   nc <- 4   ;   nt <- 1

# Call STAN (ART 61/3 sec), assess convergence, print and save estimates
system.time(
  out73.5 <- stan(file = "model73_5.stan", data = dataList,
    chains = nc, iter = ni, warmup = nb, thin = nt) )
rstan::traceplot(out73.5)                    # not shown
print(out73.5, dig = 2)                       # not shown
stan_est <- summary(out73.5)$summary[1:3,1]
```

7.3.6 Do-it-yourself maximum likelihood estimates

Writing our own ML function is substantially more complicated for the random-effects model than for the fixed-effects version in the previous section. Our fixed-effects ML function took as input the vector of population means, plus the residual standard deviation. Instead, we now have to obtain MLEs for two hyperparameters (population mean and standard deviation) along with the residual standard deviation. We can't obtain estimates of the random effects directly by including them in our set of input parameters; instead they must be "integrated out" of the likelihood by integrating over all possible values the random effect could take for each population (Chapter 3).

Within our likelihood function, we start by dividing up the dataset by population. The contribution to the likelihood for one population i containing n snakes with measurements y looks like this:

$$L_i = \int\limits_{-\infty}^{\infty} \prod_{j=1}^{n} \left[f\left(y_j | \mu + r_i, \sigma_\varepsilon\right) \right] \cdot f(r_i | 0, \sigma_{pop}) dr_i$$

where r_i represents the random effect for population i, f is the normal probability density function, μ is the grand mean SVL, σ_ε is the residual standard deviation, and σ_{pop} is the population standard deviation.

This equation can be represented in R code as follows, using the built-in `integrate` function (note this is not executable code yet, but a component of the NLL function further down):

```
L = integrate(function(r){
    tot <- 1        # Starting value for product over J
    # Calculate product
    # Iterate over each snake j (1-12) in pop i
    for (j in 1:nSample){
        tot = tot * dnorm(ysub[j], pop.grand.mean + r, sigma)
    }
    tot <- tot * dnorm(r, 0, pop.sd)
    tot
}, lower=-Inf, upper=Inf)$value
```

We need to repeat this calculation for each population in our dataset, sum up the log-likelihoods, and then return the negative of that sum, so that we can use the function with `optim()`. The likelihood function below does this.

```
# Definition of NLL for a model with one random-effects factor and Gaussian errors
NLL <- function(pars) {
pop.grand.mean <- pars[1]
  pop.sd <- exp(pars[2])
  sigma <- exp(pars[3])

  LL <- 0

  for (i in 1:nPops){
    # Subset data to just pop i
    ysub <- y[pop == i]

    L <- integrate(function(r){
      tot <- 1 # Starting value for product over J
      # Iterate over each snake j (1-12) in pop i
      for (j in 1:nSample){
        tot = tot * dnorm(ysub[j], pop.grand.mean + r, sigma)
      }
      tot <- tot * dnorm(r, 0, pop.sd)
      tot
    }, lower = -Inf, upper = Inf, subdivisions = 20)$value
    LL <- LL + log(L)
  }
  return(-LL)
}
```

For our optimization to work properly, we need to initialize the parameters at reasonable values; if we set all the inits to 0, we will have problems. One reasonable approach is to initialize the population grand mean at the overall mean of our SVL dataset, and the residual standard deviation as the overall standard deviation of the dataset. Don't forget we need to provide initials for standard deviations on the log scale.

```
# Minimize NLL to find MLEs, also get SEs and CIs and save estimates
inits <- c('pop.grand.mean' = mean(y), 'log(pop.sd)' = 0, 'log(residual.sd)' = log(sd(y)))
out73.6 <- optim(inits, NLL, hessian = TRUE, method = 'BFGS')
get_MLE(out73.6, 4)
exp(out73.6$par[2:3])          # Backtransform the SD's (ignore 'log' in name)
diy_est <- c(out73.6$par[1],   # Save estimates
  exp(out73.6$par[2:3]))
```

	MLE	ASE	LCL.95	UCL.95
pop.grand.mean	51.8966	0.95653	50.0219	53.771
log(pop.sd)	0.9786	0.29282	0.4047	1.553
log(residual.sd)	1.6678	0.06743	1.5357	1.800

log(pop.sd)	log(residual.sd)
2.660759	5.300746

Aren't these DIY-MLEs for the random-effects model cool?

7.3.7 **Likelihood analysis with TMB**

Input data for TMB is the same as in other sections, except as usual we need to adjust the population index `pop` so that it starts at 0 instead of 1.

```
# Bundle and summarize data; adjust pop so it starts at 0 instead of 1
tmbData <- list(y = y, pop = as.numeric(pop)-1, nPops = nPops, n = n)
```

As TMB uses ML, our approach will in some ways be similar to the previous DIY section. However, thankfully, specifying the random effects in TMB is much easier. In fact, the ability of TMB to easily specify random effects is one of the most powerful and useful features of this engine. Because this is the first time we use TMB to specify a random effect, we will go into a bit more detail describing each piece of the model file.

First, the parts you've seen before: the boilerplate heading, plus a description of the input data:

```
#include <TMB.hpp>
template<class Type>
Type objective_function<Type>::operator() ()
{
   //Describe input data
   DATA_VECTOR(y);         //response
   DATA_IVECTOR(x);        //Population index: note IVECTOR, not VECTOR
   DATA_INTEGER(nPops);    //Number of populations
   DATA_INTEGER(n);        //Number of obs
```

We also must describe the parameters. Notice that we have included hyperparameters (`pop_grand_mean` and `log_pop_sd`), the residual standard deviation, *and* the vector of random population intercepts, called `pop_mean`.

```
   //Describe parameters
   PARAMETER(pop_grand_mean);
   PARAMETER(log_pop_sd)
   PARAMETER(log_residual_sd);
   PARAMETER_VECTOR(pop_mean);

   Type pop_sd = exp(log_pop_sd);
   Type residual_sd = exp(log_residual_sd);
```

Next comes the key part: the description of the random effects. We will specify the model using a means parameterization. We begin by initializing our log-likelihood at 0. Then, we loop over each population, and specify that each population mean comes from a normal distribution with mean `pop_grand_-mean` and standard deviation `pop_sd`. We add this to our total log-likelihood. You should see the similarity in this step to the DIY function, but we don't have to worry about integrating out the random effects—TMB will do that for us. We must explicitly tell TMB that `pop_mean` is meant to be a random effect, which we will do in a later step. We also calculate the difference between each population mean and the grand mean for comparison to results from `glmmTMB`.

```
   Type LL = 0;               //Initialize total log likelihood at 0

   // Random intercepts
   vector<Type> effe(nPops);  //Pop effects
   for (int i = 0; i<nPops; i++){
     LL += dnorm(pop_mean(i), pop_grand_mean, pop_sd, true);
     effe(i) = pop_mean(i) - pop_grand_mean;
   }
```

Finally, we calculate the log-likelihood for each of our data points and add it to our total log-likelihood, tell TMB to report our derived parameter `effe` in the output, and return the negative log-likelihood.

```
// Likelihood of obs
for (int i = 0; i<n; i++){   //Note index starts at 0 instead of 1!
  LL += dnorm(y(i), pop_mean(x(i)), residual_sd, true);
}

ADREPORT(effe);
return -LL;                    //Return negative of total log likelihood
```

All together our model file looks like this:

```
# Write TMB model file
cat(file = "model73_7.cpp",
"#include <TMB.hpp>

template<class Type>
Type objective_function<Type>::operator() ()
{
  //Describe input data
  DATA_VECTOR(y);                     //response
  DATA_IVECTOR(pop);                  //Population index: note IVECTOR, not VECTOR
  DATA_INTEGER(nPops);                //Number of populations
  DATA_INTEGER(n);                    //Number of obs

  //Describe parameters
  PARAMETER(pop_grand_mean);
  PARAMETER(log_pop_sd);
  PARAMETER(log_residual_sd);
  PARAMETER_VECTOR(pop_mean);

  Type pop_sd = exp(log_pop_sd);
  Type residual_sd = exp(log_residual_sd);

  Type LL = 0;                        //Initialize total log likelihood at 0

  // Random intercepts
  vector<Type> effe(nPops);           //Pop effects
  for (int i = 0; i<nPops; i++){
    LL +=dnorm(pop_mean(i), pop_grand_mean, pop_sd, true);
    effe(i) = pop_mean(i) - pop_grand_mean;
  }

  // Likelihood of obs
  for (int i = 0; i<n; i++){                //Note index starts at 0 instead of 1!
    LL += dnorm(y(i), pop_mean(pop(i)), residual_sd, true);
  }

  ADREPORT(effe);

  return -LL;                         //Return negative of total log likelihood
}
")
```

```
# Compile and load TMB function
compile("model73_7.cpp")
dyn.load(dynlib("model73_7"))
```

We must provide starting values and the correct dimensions for all parameters, including random effects.

```
# Provide dimensions and starting values for parameters
params <- list(pop_grand_mean = 0, log_pop_sd = 0, log_residual_sd = 0,
               pop_mean = rep(0, tmbData$nPops))
```

Another key change: when constructing our TMB model object, we must tell TMB that `pop_mean` is a random effect so that it is handled properly. We don't do this for hyperparameters, just the actual vector of random effects.

```
# Create TMB object
out73.7 <- MakeADFun(data = tmbData, parameters = params,
   random = "pop_mean", DLL = "model73_7", silent = TRUE)

# Optimize TMB object and print results (including random effects)
opt <- optim(out73.7$par, fn = out73.7$fn, gr = out73.7$gr, method = "BFGS", hessian = TRUE)
(tsum <- tmb_summary(out73.7))                         # not shown

# Save estimates from TMB
tmb_est <- c(opt$par[1], exp(opt$par[2:3]))
```

7.3.8 Comparison of the parameter estimates

As always, we make our grand comparison table for all the engines used to fit the random-effects normal model with a single factor.

```
# Compare results with truth and previous estimates
comp <- cbind(truth = truth, ML = gtmb_est, REML = gtmb_REMLest, JAGS = jags_est,
   NIMBLE = nimble_est, Stan = stan_est, DIY = diy_est, TMB = tmb_est)
print(comp, 4)
```

	truth	ML	REML	JAGS	NIMBLE	Stan	DIY	TMB
pop.grand.mean	50	51.920	51.920	51.887	51.930	51.91	51.897	51.920
pop.sd	3	2.678	2.869	3.376	3.329	3.27	2.661	2.678
residual.sd	5	5.301	5.301	5.381	5.363	5.36	5.301	5.301

We see virtually identical estimates of the grand mean and of the residual standard deviation for all engines and methods (i.e., ML and REML). In contrast, for `pop.sd` we see slightly different estimates when using ML (ML, DIY, and TMB) than when using REML, and again slightly different estimates when using Bayesian posterior inference and reporting the posterior mean. However, in practice neither of these differences would probably matter.

7.4 **Summary and outlook**

We have introduced fixed- and random-effects models and fitted the models using all our model fitting engines, and using both ML and REML. We have compared the fixed- and the random-effects estimates for the same data set and observed (visually in Fig. 7.5) those two hallmarks of random-effects estimates: shrinkage and increased precision compared to their fixed-effects counterparts.

You will use random effects extremely often in your statistical modeling. Hence, it is absolutely crucial that you obtain a sound practical understanding of what they imply, when we use them, and what the consequences are of treating a set of parameters as random rather than fixed. We believe that the model description in algebra, as well as in the BUGS language, is the most lucid explanation what random effects mean: that we place a statistical distribution on a set of parameters and estimate some shared (hyper-)parameters for them.

In the next chapter, we move to yet another important feature of the modeling of grouped data. While in this chapter we only considered the groups (here, populations) in a single classification, i.e., a single factor, we will next see how we can model the effects of two classifications or factors by considering main effects and interactions.

Comparisons along two classifications in a model with two factors

Chapter outline

8.1 Introduction: main and interaction effects

We now extend the linear model for a single factor in the previous chapter by adding another factor and arrive at the linear model that underlies a least-squares analysis technique known as two-way analysis of variance or two-way ANOVA (Zar, 1998; Quinn & Keough, 2002). We only consider fixed effects here. We will describe two manners in which the effects of two factors A and B can be combined, and the associated models are called main-effects and interaction-effects models (for a third way called nesting, see Section 8.11). In the main-effects model, the effects of A and B are additive, that is, the effect of one level of factor A, say a_1, does not depend on whether it is assessed at one level of B,

---

⊛This book has a companion website hosting complementary materials, including all code for download. Visit this URL to access it: https://www.elsevier.com/books-and-journals/book-companion/9780443137150.

Applied Statistical Modelling for Ecologists. DOI: https://doi.org/10.1016/B978-0-443-13715-0.00021-2

say b_1, or at another, say b_2. In contrast, in the presence of an interaction between factors A and B, some or all effects depend on some or all of each other; hence, the effect of a_1 may not be identical when assessed at b_1 or at b_2. Interaction is symmetric, so the effect of b_1 will then also not be the same whether assessed at a_1 or at a_2. However, the interaction model is still linear, since effects are simply added together, only with an additional set of effects: those for the combination of each level of two or more factors. These are called the interaction effects in the linear model. For some models with fixed-effects factors, not all effects will in general be estimable, see Section 8.5.1. In contrast, in a random-effects model with interaction, all effects will typically be estimable.

In Chapter 3, we already saw linear models with two crossed factors, i.e., with both their main effects and their interaction. In algebra, we can represent the model with two factors A and B as follows. When a data point y_i has a particular level a of factor A and level b of factor B, then the effects parameterization of the model is

$$y_{i(a,b)} = \mu + \alpha_a + \beta_b + \gamma_{a,b} + \varepsilon_i$$

Here μ is the expected value at reference levels of A and B, α_a is the effect of factor level a relative to the reference level of A, β_b is the effect of factor level b relative the reference level of B, and $\gamma_{a,b}$ is the interaction effect of the particular levels a and b. Note that when a is the reference level of A, then $\alpha_a = 0$ (and same with b and B). Alternatively, the means parameterization of this model can be written like this:

$$y_{i(a,b)} = \mu_{a,b} + \varepsilon_i$$

FIGURE 8.1

Mourning cloak (*Nymphalis antiopa*), Switzerland, 2006 (Photo by Thomas Marent).

Table 8.1 **The 15 parameters estimated in our mourning cloak example, using the effects parameterization of the linear model for two crossed factors with interaction. The intercept represents the mean of butterflies in `pop1` and `hab1`. See also the end of Section 8.2 for a numerical translation of this scheme into the actual $5 \times 3 = 15$ group means in terms of the actual data in our simulated data set. Note that in our simulated data set we will have four butterflies for each cell of this cross-classification.**

Intercept	hab2	hab3
pop2	pop2.hab2	pop2.hab3
pop3	pop3.hab2	pop3.hab3
pop4	pop4.hab2	pop4.hab3
pop5	pop5.hab2	pop5.hab3

where $\mu_{a,b}$ is the expected value for the particular combination of factor levels a and b of factors A and B, respectively. In both cases, we also have $\varepsilon_i \sim Normal(0, \sigma^2)$.

We will use the beautiful mourning cloak (Fig. 8.1) as an illustration for this chapter and assume that we had measured wing length of four butterflies in each combination of three habitat types and five different populations and that the effects of these factors interact. Table 8.1 shows the meaning of the coefficients in the linear model for fully crossed effects of population and habitat in the effects parameterization.

If the population factor has `n.pop` levels and the habitat factor `n.hab` levels, we have one intercept, `n.pop-1 = 4` effects for the population factor, `n.hab-1 = 2` effects for the habitat factor, and `(n.pop-1) * (n.hab-1)` = 8 effects for the interaction between population and habitat. This adds up to the 15 degrees of freedom that it will cost us to fit this model to a data set that contains observations in every cell (i.e., combination of levels) in the cross-classification of population and habitat in our example.

8.2 Data generation

In this chapter, we will generate a single data set first, which has both main and interaction effects, and then analyze this data set with our model-fitting engines first with a main-effects model (which will of course be mis-specified for our data set) and second with the data-generating interaction-effects model. We assume 4 butterflies measured in every combination of five populations and three habitat types, where you can think of the latter as low, medium, and high vegetation cover. The relationship between wing length and habitat type is not homogeneous among the five studied populations, so there is a population-habitat interaction. We choose the residual wing-length standard deviation (SD) to be 3.

```
set.seed(8)

# Choose sample size
nPops <- 5
nHab <- 3
nSample <- 4
n <- nPops * nHab * nSample

# Create factor levels
pop <- gl(n = nPops, k = nHab * nSample, length = n)
hab <- gl(n = nHab, k = nSample, length = n)
```

```
# Choose values for effects
baseline <- 40                               # Intercept
pop.eff <- c(-10, -5, 5, 10)                 # Population effects
hab.eff <- c(5, 10)                          # Hab effects
interaction.eff <- c(-2, 3, 0, 4, 4, 0, 3, -2)   # Interaction effects
all.eff <- c(baseline, pop.eff, hab.eff, interaction.eff)

# Residual standard deviation
sigma <- 3
eps <- rnorm(n, 0, sigma)                    # Residuals

Xmat <- as.matrix(model.matrix(~ pop * hab) )   # Create design matrix
Xmat                                         # Have a look at design matrix,
                                             # make sure you understand it!
str(Xmat)                                    # not shown
```

We use matrix multiplication to assemble all components for the final wing length measurements which we inspect in a grouped boxplot (Fig. 8.2).

```
wing <- as.vector(Xmat %*% all.eff + eps)
boxplot(wing ~ hab*pop, col = "grey", xlab = "Habitat-by-Population combination",
  ylab = "Wing length", main = "Simulated data set", las = 1,
  ylim = c(20, 70), frame = FALSE)                       # Plot of generated data
abline(h = 40)
```

We have generated data for which the wing length–habitat relationship varies considerably among the five populations. This can also be seen nicely in a conditioning plot, which can be drawn using the function xyplot() in the lattice package. The data can be viewed in two ways, with either plot showing that the effects of population and habitat are not independent (Fig. 8.3).

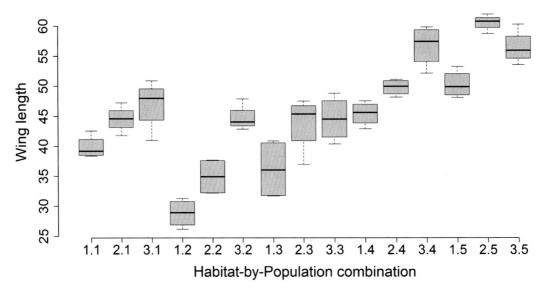

FIGURE 8.2

Mean wing length of mourning cloaks at every combination of three habitat types and five populations. Boxplots are ordered first by habitat type and second by population. Hence, the box for 3.2 shows the mean wing length of butterflies in population 2 at habitat type 3.

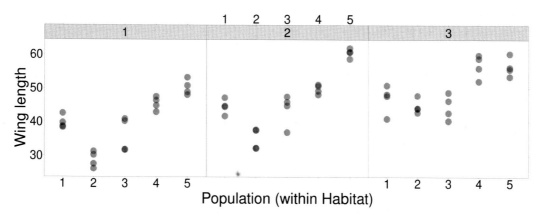

FIGURE 8.3

Conditioning plots showing interaction of the effects on wing length of habitat and population.

```
library(lattice)                    # Load the lattice library
xyplot(wing ~ hab | pop, ylab = "Wing length", xlab = "Habitat",
  main = "Population-specific relationship between wing and habitat type", pch = 16,
  cex = 2, col = rgb(0,0,0,0.4))

xyplot(wing ~ pop | hab, ylab = "Wing length", xlab = "Population",
  main = "Habitat-specific relationship between wing and population",
  pch = 16, cex = 2, col = rgb(0,0,0,0.6))                     # Fig. 8.3

# Save true values for later comparisons
truth <- c(all.eff, sigma)
```

We simulated the data in the effects or treatment contrast parameterization of the model. However, below we will also fit models in the means parameterization. For that, it will be useful to also re-express the "truth" vector to be in the same format. So here we construct the vector of true group means. To understand this, you can refer to Table 8.1, and vice versa.

```
( true.group.means <- c(baseline, baseline + pop.eff, baseline + hab.eff[1],
  baseline + pop.eff + hab.eff[1] + interaction.eff[1:4], baseline + hab.eff[2],
  baseline + pop.eff + hab.eff[2] + interaction.eff[5:8]) )

[1]  40  30  35  45  50  45  33  43  50  59  50  44  45  58  58

# Required libraries
library(ASMbook); library(jagsUI); library(rstan); library(TMB); library(DHARMa)
```

8.3 Likelihood analysis with canned functions in R

We begin by using R to fit the main-effects-only model, where we stipulate additive effects for habitat and population. We fit the factor levels in the effects, or treatment contrast, parameterization. The means parameterization is not possible for a main-effects model and that's one of the reasons for why the treatment contrast parameterization is typically the default in statistics packages.

```
# Fit main-effects model and save estimates
summary( out83.main <- lm(wing ~ hab + pop) )
lm_est_me <- c(coef(out83.main), sigma = sigma(out83.main))
```

```
Coefficients:
              Estimate   Std. Error   t value   Pr(>|t|)
(Intercept)     38.294        1.235    30.997    < 2e-16 ***
hab2             6.629        1.144     5.796   3.82e-07 ***
hab3             9.769        1.144     8.541   1.55e-11 ***
pop2            -7.633        1.477    -5.170   3.64e-06 ***
pop3            -2.275        1.477    -1.540      0.129
pop4             6.876        1.477     4.657   2.20e-05 ***
pop5            11.973        1.477     8.108   7.51e-11 ***
---
Signif. codes:  0 `***' 0.001 `**' 0.01 `*' 0.05 `.' 0.1 ` ' 1

Residual standard error: 3.617 on 53 degrees of freedom
Multiple R-squared: 0.8475,  Adjusted R-squared: 0.8302
F-statistic: 49.07 on 6 and 53 DF, p-value: < 2.2e-16
```

The estimate of the residual SD (or standard error [SE]) is larger than the value we used to generate the data set, which makes sense since the missing interaction effects in the model get absorbed into the residuals.

Next, we fit the interaction-effects model in both the mean and the treatment contrast parameterizations. We don't show the former here, but will need its estimates further down in a comparison.

```
# Interaction-effects model (means parameterization)
summary(out83.intX <- lm(wing ~ pop:hab-1) )    # not shown
```

```
# Interaction-effects model (treatment contrast parameterization)
summary(out83.int <- lm(wing ~ pop*hab) )
```

```
Coefficients:
              Estimate   Std. Error   t value   Pr(>|t|)
(Intercept)    39.8061       1.5346    25.938    < 2e-16 ***
pop2          -11.0188       2.1703    -5.077   7.12e-06 ***
pop3           -3.6614       2.1703    -1.687    0.09852 .
pop4            5.6075       2.1703     2.584    0.01309 *
pop5           10.4514       2.1703     4.816   1.70e-05 **
hab2            4.7210       2.1703     2.175    0.03490 *
hab3            7.1399       2.1703     3.290    0.00195 **
pop2:hab2       1.4117       3.0693     0.460    0.64777
pop3:hab2       2.9236       3.0693     0.953    0.34591
pop4:hab2      -0.3478       3.0693    -0.113    0.91028
pop5:hab2       5.5538       3.0693     1.809    0.07706 .
pop2:hab3       8.7444       3.0693     2.849    0.00659 **
pop3:hab3       1.2367       3.0693     0.403    0.68892
pop4:hab3       4.1529       3.0693     1.353    0.18280
pop5:hab3      -0.9898       3.0693    -0.322    0.74857
---
Signif. codes:  0 `***' 0.001 `**' 0.01 `*' 0.05 `.' 0.1 ` ' 1

Residual standard error: 3.069 on 45 degrees of freedom
Multiple R-squared: 0.9067,  Adjusted R-squared: 0.8777
F-statistic: 31.25 on 14 and 45 DF, p-value: < 2.2e-16
```

Note how the estimated residual SE is much closer to the value used to generate the data set (3), now that we have correctly specified the model.

As a reminder of the important place of model assessment in the workflow of a statistical analysis, we next assess quantile residuals using DHARMa for both the main- and the interaction-effects models. Neither detects lack of fit based on residual tests implemented in the main function.

```
# Check goodness-of-fit using quantile residuals
# Main-effects model
plot(out83.main)                    # Traditional residual check for comparison
simOut <- simulateResiduals(out83.main,
   n = 1000, plot = TRUE)           # DHARMa

# Interaction-effects model
plot(out83.int)                     # Traditional residual check for comparison
simOut <- simulateResiduals(out83.int,
   n = 1000, plot = TRUE)           # DHARMa
```

8.4 An aside: using simulation to assess bias and precision of an estimator ... and to understand what a standard error is

Let's quickly compare our parameter estimates with the parameter values we used when simulating the data:

```
# Compare estimates with truth
lm_est <- c(coef(out83.int), sigma = sigma(out83.int))
comp <- cbind(truth = truth, lm = lm_est)
print(comp, 4)
```

	truth	lm
(Intercept)	40	39.8061
pop2	-10	-11.0188
pop3	-5	-3.6614
pop4	5	5.6075
pop5	10	10.4514
hab2	5	4.7210
hab3	10	7.1399
pop2:hab2	-2	1.4117
pop3:hab2	3	2.9236
pop4:hab2	0	-0.3478
pop5:hab2	4	5.5538
pop2:hab3	4	8.7444
pop3:hab3	0	1.2367
pop4:hab3	3	4.1529
pop5:hab3	-2	-0.9898
sigma	3	3.0693

We note several fairly substantial disagreements, especially for some interaction terms. So, is this bad? Would this mean that these estimates are biased?

The answer is that we can't judge bias from the analysis of a single data set, since bias is a property of an estimator (i.e., an estimation method), and not of the actual estimate (which is just one estimate produced by that method). In our case, the coefficient estimates don't necessarily resemble very much the parameters from which we simulated these data; after all, our sample size is rather small. So, to check whether these differences are simply due to sampling variation, as they should be, we repeat this data generation-analysis cycle 10,000 times and then average over the random sampling variation to convince ourselves that our estimators from the linear model are indeed unbiased. Simulations of this kind can be done easily in R.

```
simrep <- 10^4                         # Desired number of iterations
estimates <- array(dim = c(simrep,
  length(all.eff)))                    # Data structure to hold results

for(i in 1:simrep) {                   # Run simulation simrep times
  if(i %% 100 == 0) cat(paste('\n iter', i))  # Counter
  eps <- rnorm(n, 0, sigma)            # Residuals
  y <- as.vector(Xmat %*% all.eff + eps)      # Assemble data
  fit.model <- lm(y ~ pop*hab)         # Break down data
  estimates[i,] <- fit.model$coef      # Save values of coefficients
}
```

We can now compare the input (i.e., the chosen values of the effects) and the output when averaged over the sampling variation, that is, over the variation among repeated data sets. We see that these averages are much closer, and we can get a feel for the unbiasedness of this estimator. Indeed, depending on the number of iterations, we can get arbitrarily close to the input. Alternatively, we could increase the sample size in the data sets, for example, from 4 in each group to 40,000, and we would again get estimates that are much closer to the input values.

```
data.frame(params = names(fit.model$coefficients), truth = all.eff,
           mean.of.estimates = round(apply(estimates, 2, mean), 3))
```

	params	truth	mean.of.estimates
1	(Intercept)	40	40.009
2	pop2	-10	-10.015
3	pop3	-5	-4.990
4	pop4	5	5.004
5	pop5	10	9.987
6	hab2	5	4.972
7	hab3	10	9.987
8	pop2:hab2	-2	-1.944
9	pop3:hab2	3	2.972
10	pop4:hab2	0	0.002
11	pop5:hab2	4	4.044
12	pop2:hab3	4	4.033
13	pop3:hab3	0	-0.020
14	pop4:hab3	3	3.005
15	pop5:hab3	-2	-1.987

This comparison suggests unbiased estimators ... and indeed we know from theory that this is the case for reasonable sample sizes (Chapter 2). Such simulations can be very powerful to create or train your intuition for many fundamental statistical concepts. So, let's drive home the meaning of the SE of an estimator. We can do this using the very same simulation.

In frequentist statistics, the SE is a measure of how uncertain an estimate is owing to sampling variation, in terms of the SD of hypothetical replicates. In our simulation, we can study sampling variation directly: it is the variation (SD) of the estimates of a parameter among replicated data sets. This can also be called a bootstrapped estimate of the SE for the parameter estimates (see Section 2.5.4). In addition, we can also obtain the SE from statistical theory: this is the SE of a parameter given by lm() when we fit the model to our single, actual data set. Let's compare the analytical and the bootstrapped SEs.

```
SE.lm <- summary(out83.int)$coef[,2]
data.frame(SE.lm = round(SE.lm, 3), bootstrapped.SE = round(apply(estimates, 2, sd), 3))
```

```
               SE.lm    bootstrapped.SE
(Intercept)    1.535             1.487
pop2           2.170             2.104
pop3           2.170             2.112
pop4           2.170             2.119
pop5           2.170             2.104
hab2           2.170             2.139
hab3           2.170             2.105
pop2:hab2      3.069             2.996
pop3:hab2      3.069             3.019
pop4:hab2      3.069             3.017
pop5:hab2      3.069             3.014
pop2:hab3      3.069             2.998
pop3:hab3      3.069             2.992
pop4:hab3      3.069             3.004
pop5:hab3      3.069             2.973
```

Here, the theory-based SEs turn out to be really good. Arguably, this is due to the fact that the data set is not too small relative to the complexity of the fitted model (this may be different in other analyses) and that the data-generation and the data-analysis models match (which is never the case in the real world).

It can also be useful for your statistical intuition to actually plot the bootstrapped sampling distribution for each estimator, that is, for the collection of simulation replicates of the estimates of each parameter. For illustration, we plot the sampling distributions of the intercept, pop2, hab2, and pop2:hab2 parameters (Fig. 8.4). We see nice Gaussian-looking sampling distributions for the bootstrap replicates of the estimates of all four parameters, which is what is expected from theory (see Chapter 2). Thus, a precisely estimated parameter in the frequentist view of uncertainty is one which will be estimated with greater repeatability, that is, with less variation across replicates, which translates into a more concentrated sampling distribution. In contrast, a parameter that is

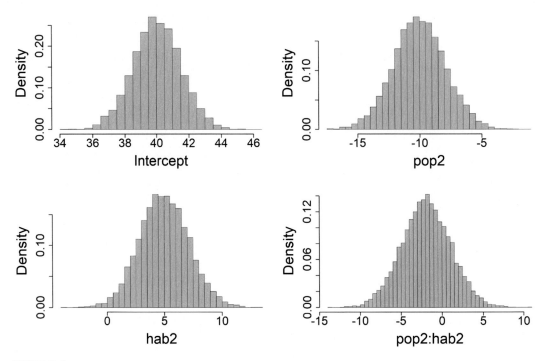

FIGURE 8.4

The bootstrapped sampling distributions of the least-square estimators for four parameters in the interaction-effects linear model for two factors. *The SE of a parameter estimate is the SD of these distributions.* Under the assumptions of the parametric model it can be obtained from theory; this is the SE given by function lm(). However, we can also assess the SE by resampling techniques, such as a parametric bootstrap as we do here.

estimated with less precision will have a more drawn out sampling distribution and hence wider SEs or confidence intervals.

```
# Plot bootstrapped sampling distributions of four estimators (Fig. 8.4)
par(mfrow = c(2,2), mar = c(5,6,5,2), cex.lab = 1.5, cex.axis = 1.5, cex.main = 1.5)
hist(estimates[,1], xlab = '', freq = F, main = 'Intercept',  col = 'grey',  breaks = 40)
hist(estimates[,2], xlab = '', freq = F, main = 'pop2',        col = 'grey',  breaks = 40)
hist(estimates[,6], xlab = '', freq = F, main = 'hab2',        col = 'grey',  breaks = 40)
hist(estimates[,8], xlab = '', freq = F, main = ' pop2:hab2', col = 'grey',  breaks = 40)
```

8.5 Bayesian analysis with JAGS

8.5.1 Main-effects model

We fit the main-effects model in the effects parameterization, where we set to zero the parameters associated with the first levels of each factor, to avoid overparameterization. In the posterior summary output below, we note that these parameters fixed at zero still appear, but since JAGS produces for them only draws of zero, there is no variation in the output from the MCMC for them, n.eff = 1 and Rhat cannot be computed.

```
# Bundle and summarize data
str(dataList <- list(wing = wing, hab = as.numeric(hab), pop = as.numeric(pop), n = length(wing)))

List of 4
 $ wing: num [1:60] 39.7  42.5  38.6  38.3  47.2 ...
 $ hab : num [1:60] 1  1  1  1  2  2  2  2  3  3 ...
 $ pop : num [1:60] 1  1  1  1  1  1  1  1  1  1 ...
 $ n   : int 60

# Write JAGS model file
cat(file = "model85.main.txt", "
model {
# Priors
alpha ~ dnorm(0, 0.0001)                 # Intercept
beta.pop[1] <- 0                         # set to zero effect of 1st level
for(j in 2:5){                           # Loop over remaining factor levels
  beta.pop[j] ~ dnorm(0, 0.0001)
}
beta.hab[1] <- 0                         # ditto
for(j in 2:3){                           # Loop over remaining factor levels
  beta.hab[j] ~ dnorm(0, 0.0001)
}
sigma ~ dunif(0, 100)
tau <- pow(sigma, -2)

# Likelihood
for (i in 1:n) {
  wing[i] ~ dnorm(mean[i], tau)
  mean[i] <- alpha + beta.pop[pop[i]] + beta.hab[hab[i]]
}
}
")

# Function to generate starting values
inits <- function(){ list(alpha = rnorm(1), sigma = rlnorm(1) )}

# Parameters to estimate
params <- c("alpha", "beta.pop", "beta.hab", "sigma")

# MCMC settings
na <- 1000  ;  ni <- 3000  ;  nb <- 1000  ;  nc <- 4  ;  nt <- 1

# Call JAGS (ART <1 min), check convergence, summarize posteriors and save estimates
out85me <- jags(dataList, inits, params, "model85.main.txt", n.iter = ni,
  n.burnin = nb, n.chains = nc, n.thin = nt, n.adapt = na, parallel = TRUE)
jagsUI::traceplot(out85me)                          # not shown
print(out85me, 3)
jags_est_me <- out85me$summary[c(1, 8:9, 3:6, 10), 1]
```

	mean	sd	2.5%	50%	97.5%	overlap0	f	Rhat	n.eff
alpha	38.297	1.285	35.771	38.306	40.859	FALSE	1.000	1.000	8000
beta.pop[1]	0.000	0.000	0.000	0.000	0.000	FALSE	1.000	NA	1
beta.pop[2]	-7.635	1.548	-10.731	-7.620	-4.624	FALSE	1.000	1.000	8000
beta.pop[3]	-2.281	1.534	-5.295	-2.257	0.710	TRUE	0.935	1.000	5990
beta.pop[4]	6.870	1.514	3.861	6.883	9.838	FALSE	1.000	1.000	8000
beta.pop[5]	11.951	1.548	8.905	11.959	14.998	FALSE	1.000	1.000	8000
beta.hab[1]	0.000	0.000	0.000	0.000	0.000	FALSE	1.000	NA	1
beta.hab[2]	6.636	1.176	4.342	6.628	8.966	FALSE	1.000	1.000	8000
beta.hab[3]	9.761	1.177	7.477	9.748	12.107	FALSE	1.000	1.000	8000
sigma	3.703	0.371	3.065	3.674	4.527	FALSE	1.000	1.001	5639

```
# Compare likelihood and Bayesian estimates
comp <- cbind(lm = lm_est_me, JAGS = jags_est_me)
print(comp, 4)
```

	lm	JAGS
(Intercept)	38.294	38.297
hab2	6.629	6.636
hab3	9.769	9.761
pop2	-7.633	-7.635
pop3	-2.275	-2.281
pop4	6.876	6.870
pop5	11.973	11.951
sigma	3.617	3.703

We can't compare these estimates with the truth, since the data-generation model has the interaction terms and the data-analysis model does not, but we see that the MLEs and the Bayesian estimates agree very well.

8.5.2 Interaction-effects model

We will specify the means parameterization for ease of coding and illustrate how parameters in JAGS can be arrays with two (or more) dimensions. This is handy when organizing an analysis. In our code, we refer by "group" to a combination of a specific level of the habitat and population factors.

```
# Bundle and summarize data
str(dataList <- list(wing = wing, hab = as.numeric(hab), pop = as.numeric(pop),
    n = length(wing), nHab = length(unique(hab)), nPops = length(unique(pop))) )

List of 6
 $ wing : num [1:60] 39.7 42.5 38.6 38.3 47.2 ...
 $ hab  : num [1:60] 1 1 1 1 2 2 2 2 3 3 ...
 $ pop  : num [1:60] 1 1 1 1 1 1 1 1 1 1 ...
 $ n    : int 60
 $ nHab : int 3
 $ nPops : int 5
```

```
# Write JAGS model file
cat(file = "model85.int.txt", "
model {
# Priors
for (i in 1:nPops){
  for(j in 1:nHab) {
    group.mean[i,j] ~ dnorm(0, 0.0001)
    }
  }
tau <- pow(sigma, -2)
sigma ~ dunif(0, 100)

# Likelihood
for (i in 1:n) {
  wing[i] ~ dnorm(mean[i], tau)
  mean[i] <- group.mean[pop[i], hab[i]]
  }
}
")

# Function to generate starting values
inits <- function(){list(sigma = rlnorm(1) )}

# Parameters to estimate
params <- c("group.mean", "sigma")

# MCMC settings
na <- 1000   ;   ni <- 3000   ;   nb <- 1000   ;   nc <- 4   ;   nt <- 1

# Call JAGS (ART <1 min), check convergence, summarize posteriors and save estimates
out85ie <- jags(dataList, inits, params, "model85.int.txt", n.iter = ni,
  n.burnin = nb, n.chains = nc, n.thin = nt, n.adapt = na, parallel = TRUE)
jagsUI::traceplot(out85ie)                        # not shown
print(out85ie, 3)
jags_est_ie <- out85ie$summary[1:16,1]
```

	mean	sd	2.5%	50%	97.5%	overlap0	f	Rhat	n.eff
group.mean[1,1]	39.823	1.575	36.690	39.823	42.940	FALSE	1	1.001	3453
group.mean[2,1]	28.771	1.573	25.636	28.770	31.833	FALSE	1	1.001	3268
group.mean[3,1]	36.126	1.581	33.030	36.131	39.212	FALSE	1	1.000	8000
group.mean[4,1]	45.388	1.589	42.244	45.390	48.558	FALSE	1	1.000	7893
group.mean[5,1]	50.225	1.597	47.077	50.239	53.289	FALSE	1	1.000	6831
group.mean[1,2]	44.554	1.567	41.427	44.570	47.623	FALSE	1	1.000	8000
group.mean[2,2]	34.902	1.584	31.768	34.915	38.033	FALSE	1	1.000	8000
group.mean[3,2]	43.766	1.588	40.614	43.771	46.857	FALSE	1	1.000	4728
group.mean[4,2]	49.791	1.580	46.670	49.806	52.874	FALSE	1	1.000	8000
group.mean[5,2]	60.519	1.598	57.332	60.524	63.643	FALSE	1	1.000	5524
group.mean[1,3]	46.928	1.586	43.817	46.923	50.074	FALSE	1	1.000	7628
group.mean[2,3]	44.650	1.575	41.542	44.659	47.825	FALSE	1	1.000	6006
group.mean[3,3]	44.511	1.591	41.432	44.507	47.703	FALSE	1	1.001	3015
group.mean[4,3]	56.656	1.580	53.508	56.666	59.804	FALSE	1	1.000	8000
group.mean[5,3]	56.397	1.579	53.330	56.403	59.556	FALSE	1	1.000	8000
sigma	3.156	0.346	2.568	3.128	3.920	FALSE	1	1.002	2278

For comparison of the MLEs and the Bayesian estimates, we use the model fit by `lm()` in the means parameterization from above and create another version of the vector of truth.

```
truthMP <- c(true.group.means, sigma)    # Truth for means parameterization

# Compare likelihood with Bayesian estimates and with truth
lm_est_ie <- c(coef(out83.intX), sigma = sigma(out83.intX))
comp <- cbind(truth = truthMP, lm = lm_est_ie, JAGS = jags_est_ie)
print(comp, 4)
```

```
              truth      lm     JAGS
pop1:hab1        40  39.806   39.823
pop2:hab1        30  28.787   28.771
pop3:hab1        35  36.145   36.126
pop4:hab1        45  45.414   45.388
pop5:hab1        50  50.258   50.225
pop1:hab2        45  44.527   44.554
pop2:hab2        33  34.920   34.902
pop3:hab2        43  43.789   43.766
pop4:hab2        50  49.787   49.791
pop5:hab2        59  60.532   60.519
pop1:hab3        50  46.946   46.928
pop2:hab3        44  44.672   44.650
pop3:hab3        45  44.521   44.511
pop4:hab3        58  56.706   56.656
pop5:hab3        58  56.408   56.397
sigma             3   3.069    3.156
```

We again see the limits in the agreement between the truth and some of the estimates that we had observed and discussed in Section 8.4, but note the overall excellent match between the likelihood and the Bayesian estimates of all parameters.

8.5.3 Forming predictions

Let's present the Bayesian inference for the interaction-effects model in a graph showing the predicted response, analogous to least-square means in a classical analysis (Zar, 1998), for each combination of habitat and population (Fig. 8.5). This plot corresponds to the boxplot of the data set (Fig. 8.2), and selects the order of the predictions to match that in Fig. 8.2.

```
par(mfrow = c(1,1), mar = c(5,5,4,3), cex.lab = 1.5, cex.axis = 1.5)
ord <- c(1,4,7,10,13,2,5,8,11,14,3,6,9,12,15)                          # select order
plot(ord, out85ie$mean$group.mean, xlab = "Hab-by-Population", las = 1,
  ylab = "Predicted wing length", cex = 2, ylim = c(20, 70), frame = FALSE,
  pch = 16, col = rgb(0,0,0,0.5))
segments(ord, out85ie$q2.5$group.mean, ord, out85ie$q97.5$group.mean,
  col = rgb(0,0,0,0.5), lwd = 2)
abline(h = 40)
```

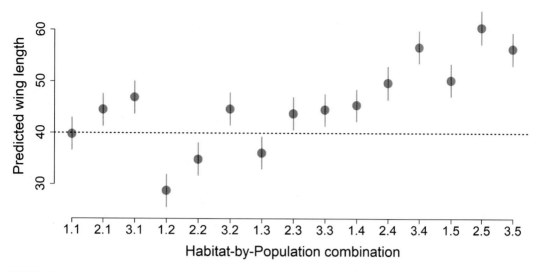

FIGURE 8.5

Predicted wing length of mourning cloak butterflies for each habitat–population combination. The order of the 15 groups is the same as in Fig. 8.2. Error bars are 95% CRIs.

8.6 Bayesian analysis with NIMBLE

8.6.1 Main-effects model

NIMBLE code for the model is available on the book website.

8.6.2 Interaction-effects model

NIMBLE code for the model is available on the book website.

8.7 Bayesian analysis with Stan

8.7.1 Main-effects model

We take a similar approach to fitting the main-effects model in Stan. It is a little more difficult in Stan to fix a parameter value at 0. We must do this in two stages: first, in the `parameters` block, we define a parameter vector `beta_raw` on which we set the prior. Second, we make a copy of this parameter vector called `beta` in the `transformed parameters` block, and set the first element equal to 0. We must do this for both the population and habitat parameter vectors. We then use these modified vectors in the `model` block. As with JAGS, Stan will give us a warning about Rhat values for the parameters fixed at 0 in this model, which we can ignore.

```
# Bundle and summarize data (same as for JAGS)
dataList <- list(wing = wing, hab = as.numeric(hab), pop = as.numeric(pop),
  n = length(wing))
```

```
# Write text file with model description in Stan
cat(file = "model87_main.stan",   # This line is R code
"
data {
  int n;
  vector[n] wing;
  array[n] int pop;
  array[n] int hab;
}

parameters {
  real alpha;
  vector[5] beta_pop_raw;
  vector[3] beta_hab_raw;
  real<lower=0> sigma;
}

transformed parameters {
  vector[5] beta_pop = beta_pop_raw;
  vector[3] beta_hab = beta_hab_raw;
  beta_pop[1] = 0;
  beta_hab[1] = 0;
}

model {
  vector [n] mu;

  beta_pop_raw ~ normal(0, 100);
  beta_hab_raw ~ normal(0, 100);
  sigma ~ cauchy(0, 10);

  for (i in 1:n){
    mu[i] = alpha + beta_pop[pop[i]] + beta_hab[hab[i]];
    wing[i] ~ normal(mu[i], sigma);
  }
}                                 // This is the last line of Stan code
" )

# HMC settings
ni <- 2000  ;  nb <- 1000  ;  nc <- 4  ;  nt <- 1

# Parameters to save
params <- c("alpha", "beta_pop", "beta_hab", "sigma")

# Call STAN (ART 50/3 sec)
system.time(
  out87me <- stan(file = "model87_main.stan", data = dataList,
    pars = params, chains = nc, iter = ni, warmup = nb, thin = nt) )
rstan::traceplot(out87me)                    # not shown
print(out87me, dig = 2)                       # not shown
stan_est_me <- summary(out87me)$summary[c(1, 8, 9, 3, 4, 5, 6, 10), 1]
```

8.7.2 Interaction-effects model

The interaction-effects model is actually simpler to code in Stan than the main-effects model. The code looks very similar to the BUGS code. One exception is that Stan can't put a prior directly on a matrix parameter—we must convert it to a vector first. We do this with the `to_vector` function.

```
# Bundle and summarize data
str(dataList <- list(wing = wing, hab = as.numeric(hab), pop = as.numeric(pop),
  n = length(wing), nHab = nHab, nPops = nPops))

# Write text file with model description in Stan
cat(file = "model87_int.stan",    # This line is R code
"
data {
  int n;
  int nPops;
  int nHab;
  vector[n] wing;
  array[n] int hab;
  array[n] int pop;
}

parameters {
  matrix[nPops, nHab] group_mean;
  real<lower=0> sigma;
}

model {
  to_vector(group_mean) ~ normal(0, 100);
  sigma ~ cauchy(0, 10);

  for (i in 1:n) {
    wing[i] ~ normal(group_mean[pop[i], hab[i]], sigma);
  }
}                                  // This is the last line of Stan code
" )

# HMC settings
ni <- 2000   ;   nb <- 1000   ;   nc <- 4   ;   nt <- 1

# Call STAN (ART 48/3 sec)
system.time(
  out87ie <- stan(file = "model87_int.stan", data = dataList, chains = nc, iter = ni,
    warmup = nb, thin = nt) )
rstan::traceplot(out87ie)                            # not shown
print(out87ie, dig = 2)                              # not shown

# Save estimates
# Stan sorts summary by row, then by column, opposite of JAGS
stan_est_ie <- summary(out87ie)$summary[c(1,4,7,10,13,2,5,8,11,14,3,6,9,12,15,16),1]
```

8.8 Do-it-yourself maximum likelihood estimates

8.8.1 Main-effects model

First, the main-effects-only model. The likelihood equation looks nearly identical to the one in Section 7.2.6. We calculate the vector of expected values μ by matrix-multiplying the appropriate model matrix X with our vector of parameters β, the same way we did when we generated the data in Section 8.2.

$$L(\beta, \sigma | y, X) = \prod_{i=1}^{n} \frac{1}{\sigma\sqrt{2\pi}} e^{-\frac{1}{2}\left(\frac{y_i-\mu_i}{\sigma}\right)^2}$$

$$\mu = X\beta$$

To translate this equation into an R function, we first need to calculate the model matrix again:

```
head(Xmat <- model.matrix(~ hab + pop))   # Reminder of what it looks like
```

Our two factor variables (hab and pop) have now been converted to a series of dummy variables. Notice that there is no variable for the first habitat level (hab1) or the first pop level (pop1), because the intercept represents these base levels of each factor.

For the likelihood function, we first separate the parameters into a vector beta and sigma. Then we matrix-multiply Xmat and beta, calculate the log-likelihood for each data point, sum them up, and return the negative.

```
# NLL for main-effects normal linear model with two factors
NLL <- function(param, y, Xmat) {
  beta <- param[1:7]                  # Intercept and slope coefficients
  sigma <- exp(param[8])              # Residual SD
  mu <- Xmat %*% beta                 # Multiply each row of Xmat by beta
  LL <- dnorm(y, mu, sigma, log = TRUE)   # Log-likelihood for each obs
  NLL <- -sum(LL)                     # NLL for all observations (whole data set)
  return(NLL)
}
```

```
# Minimize that NLL to find MLEs, get SEs and CIs and save estimates
inits <- c(mean(wing), rep(0, 6), log(sd(wing)))
names(inits) <- c(names(coef(out83.main)), 'log-sigma')
out88me <- optim(inits, NLL, y = wing, Xmat = Xmat, hessian = TRUE, method = 'BFGS')
get_MLE(out88me, 4)
diy_est_me <- c(out88me$par[-8], exp(out88me$par[8]))
```

	MLE	ASE	LCL.95	UCL.95
(Intercept)	38.294	1.16112	36.018	40.5695
hab2	6.629	1.07498	4.522	8.7363
hab3	9.769	1.07498	7.662	11.8757
pop2	-7.633	1.38780	-10.353	-4.9132
pop3	-2.275	1.38780	-4.995	0.4455
pop4	6.876	1.38780	4.156	9.5960
pop5	11.973	1.38780	9.253	14.6929
log-sigma	1.224	0.09129	1.045	1.4025

8.8.2 Interaction-effects model

Next, the version of the model with interactions. The likelihood equation is the same as in Section 8.8.1. All we're changing is the contents of the model matrix and the (corresponding) number of parameters in the β vector, as we now have new interaction parameters to estimate. To match previous implementations of this model, we estimate a parameter for each combination of hab and pop (hab:pop in the formula) and then drop the intercept term (-1 in the formula):

```
head( Xmat <- model.matrix(~ pop:hab-1))   # not shown
```

Here's the likelihood function again (note that beta has more parameters now).

```
# NLL for interaction-effects normal linear model with two factors
NLL <- function(param, y, Xmat) {
  beta <- param[1:15]                    # Interaction parameters
  sigma <- exp(param[16])                # Residual SD
  mu <- Xmat %*% beta
  LL <- dnorm(y, mu, sigma, log = TRUE)  # Log-lik for each observation
  NLL <- -sum(LL)                        # NLL for all observations (whole data set)
  return(NLL)
}
```

```
# Minimize that NLL to find MLEs, get SEs and CIs and save estimates
inits <- c(rep(mean(wing), 15), log(sd(wing)))
names(inits) <- c(colnames(Xmat), 'log-sigma')
out88.int <- optim(inits, NLL, y = wing, Xmat = Xmat, hessian = TRUE, method = 'BFGS')
get_MLE(out88.int, 4)
diy_est_ie <- c(out88.int$par[-16], exp(out88.int$par[16]))
```

	MLE	ASE	LCL.95	UCL.95
pop1:hab1	39.8061	1.32904	37.2012	42.411
pop2:hab1	28.7874	1.32904	26.1824	31.392
pop3:hab1	36.1447	1.32904	33.5398	38.750
pop4:hab1	45.4137	1.32904	42.8087	48.019
pop5:hab1	50.2576	1.32904	47.6526	52.862
pop1:hab2	44.5272	1.32904	41.9222	47.132
pop2:hab2	34.9201	1.32904	32.3152	37.525
pop3:hab2	43.7894	1.32904	41.1844	46.394
pop4:hab2	49.7869	1.32904	47.1820	52.392
pop5:hab2	60.5324	1.32904	57.9275	63.137
pop1:hab3	46.9460	1.32904	44.3411	49.551
pop2:hab3	44.6717	1.32904	42.0667	47.277
pop3:hab3	44.5213	1.32904	41.9163	47.126
pop4:hab3	56.7064	1.32904	54.1015	59.311
pop5:hab3	56.4076	1.32904	53.8027	59.013
log-sigma	0.9776	0.09129	0.7987	1.157

8.9 Likelihood analysis with TMB

8.9.1 Main-effects model

We start with the main-effects-only model. Our TMB model will look almost identical to the do-it-yourself (DIY) model, so again we start with building a model matrix and including it in our input data for TMB.

```
# Bundle and summarize data
Xmat <- model.matrix(~ hab + pop)
str(tmbData <- list(wing = wing, Xmat = Xmat, n = length(wing)))
```

You should be able to see the similarity between the DIY likelihood function and the TMB code below. The only major addition in the TMB code is the preamble in which we describe the structure of the input data and parameters.

```
# Write TMB model file
cat(file = "model89_main.cpp",
"#include <TMB.hpp>

template<class Type>
Type objective_function<Type>::operator() ()
{
  //Describe input data
  DATA_VECTOR(wing);                       //response
  DATA_MATRIX(Xmat);                       //Model matrix
  DATA_INTEGER(n);                         //Number of obs

  //Describe parameters
  PARAMETER_VECTOR(beta);                  //Intercept and slopes
  PARAMETER(log_sigma);                    //log(residual standard deviation)

  Type sigma = exp(log_sigma);
  Type LL = 0;                             //Initialize total log likelihood at 0

  matrix<Type> mu = Xmat * beta;           //Matrix multiply Xmat and beta

  for (int i=0; i<n; i++){                 //Note index starts at 0 instead of 1!
    LL += dnorm(wing(i), mu(i), sigma, true);
  }

  return -LL;                              //Return negative of total log likelihood
}
")
# Compile and load TMB function
compile("model89_main.cpp")
dyn.load(dynlib("model89_main"))

# Provide dimensions and starting values for parameters
params <- list(beta = rep(0, ncol(Xmat)), log_sigma = 0)

# Create TMB object
out89me <- MakeADFun(data = tmbData, parameters = params,
                     DLL = "model89_main", silent = TRUE)
```

```
# Optimize TMB object, print results and save estimates
opt1 <- optim(out89me$par, fn = out89me$fn, gr = out89me$gr, method="BFGS", hessian = TRUE)
(tsum <- tmb_summary(out89me))
tmb_est <- c(opt1$par[1:7], exp(opt1$par[8]))
```

8.9.2 Interaction-effects model

Finally, the interaction-effects model. For this model, the TMB code looks more like our JAGS and Stan code. Rather than using a model matrix as we did in Section 8.9.1, we have a matrix of group means `group_mean` which we index using our `pop` and `hab` vectors. Remember in TMB indices start at 0, so we have to subtract 1 from both `pop` and `hab`.

```
# Bundle and summarize data
str(tmbData <- list(wing = wing,
    pop = as.numeric(pop)-1,   # TMB indices start at 0
    hab = as.numeric(hab)-1,   # TMB indices start at 0
    n = length(wing)))

# Write TMB model file
cat(file = "model89_int.cpp",
"#include <TMB.hpp>
template<class Type>
Type objective_function<Type>::operator() ()
{
  //Describe input data
  DATA_VECTOR(wing);                 //response
  DATA_IVECTOR(pop);                 //Population of each obs
  DATA_IVECTOR(hab);                 //Habitat group of each obs
  DATA_INTEGER(n);                   //Number of obs

  //Describe parameters
  PARAMETER_MATRIX(group_mean);
  PARAMETER(log_sigma);              //log(residual standard deviation)

  Type sigma = exp(log_sigma);       //Type = match type of function output

  Type LL = 0.0;                     //Initialize total log likelihood at 0

  for (int i=0; i<n; i++){           //Note index starts at 0 instead of 1!
    LL += dnorm(wing(i), group_mean(pop(i), hab(i)), sigma, true);
  }

  return -LL;                        //Return negative of total log likelihood
}
")

# Compile and load TMB function
compile("model89_int.cpp")
dyn.load(dynlib("model89_int"))

# Provide dimensions and starting values for parameters
params <- list(group_mean = matrix(0, nPops, nHab), log_sigma=0)
```

```
# Create TMB object
out89ie <- MakeADFun(data = tmbData, parameters = params,
                     DLL = "model89_int", silent = TRUE)

# Optimize TMB object, print results and save estimates
starts <- rep(0, nPops*nHab+1)
opt2 <- optim(starts, fn = out89ie$fn, gr = out89ie$gr, method = "BFGS", hessian = TRUE)
(tsum <- tmb_summary(out89ie))                        # not shown
tmb_est_ie <- c(opt2$par[1:15], exp(opt2$par[16]))
```

8.10 Comparison of the parameter estimates

And, our final comparison in the end (which we show for the interaction-effects model only)...

```
# Compare interaction-effects model estimates with truth
comp <- cbind(truth = truthMP, lm = lm_est_ie, JAGS = jags_est_ie, NIMBLE = nimble_est_ie,
   Stan = stan_est_ie, DIY = diy_est_ie, TMB = tmb_est_ie)
print(comp, 4)
```

	truth	lm	JAGS	NIMBLE	Stan	DIY	TMB
pop1:hab1	40	39.806	39.823	39.790	39.771	39.806	39.806
pop2:hab1	30	28.787	28.771	28.778	28.756	28.787	28.787
pop3:hab1	35	36.145	36.126	36.111	36.139	36.145	36.145
pop4:hab1	45	45.414	45.388	45.410	45.410	45.414	45.414
pop5:hab1	50	50.258	50.225	50.242	50.269	50.258	50.258
pop1:hab2	45	44.527	44.554	44.517	44.499	44.527	44.527
pop2:hab2	33	34.920	34.902	34.910	34.922	34.920	34.920
pop3:hab2	43	43.789	43.766	43.803	43.765	43.789	43.789
pop4:hab2	50	49.787	49.791	49.784	49.779	49.787	49.787
pop5:hab2	59	60.532	60.519	60.507	60.501	60.532	60.533
pop1:hab3	50	46.946	46.928	46.939	46.897	46.946	46.946
pop2:hab3	44	44.672	44.650	44.668	44.709	44.672	44.672
pop3:hab3	45	44.521	44.511	44.495	44.512	44.521	44.521
pop4:hab3	58	56.706	56.656	56.700	56.709	56.706	56.707
pop5:hab3	58	56.408	56.397	56.395	56.390	56.408	56.408
sigma	3	3.069	3.156	3.154	3.143	2.658	2.658

8.11 Summary and outlook

We have introduced the concepts of main effects and interaction effects in a linear model with two factors. In an aside, we have illustrated R's flexibility to conduct simulations to verify the unbiasedness of the estimators in the model of this chapter as well as the effects of sampling variation on the parameter estimates. In so doing, we have given one more illustration of the use of a parametric bootstrap for uncertainty assessment (Section 2.5.4).

In a main-effects model, the effects of one factor are estimated without regard for the levels of the other, while in an interaction-effects model, the effects of one factor may depend on the effects of another factor. When there is interaction, the effects of one factor are not independent from those of another. The same concept of interaction also carries over to more than two factors, except that you have to do many more mental pushups to understand the meaning of the resulting parameters. For this, plotting predictions for groups defined by the levels of the factors involved in an interaction becomes really important.

In this chapter, we have dealt with the effects of two factors that are completely crossed. In the Wilkinson-Rogers (1973) formula language used in R, this is what we write by $A + B + A{:}B$ for two factors A and B, or in shorthand $A{*}B$. This denotes the main effects of both plus the interaction. There is another type of combination of the effects of two factors that is known as "nesting" (Quinn & Keough, 2002). An example would be $A + A{:}B$, where we fit the main effects of A only and then for every combination of A and B a second set of effects. Thus, there are no main effects of B, and every level of B is only ever expressed inside of a specific level of A. Being able to specify the effects of nested factors in a linear model becomes important when you have hierarchical orderings in your analysis, such as species that are nested within genera and families, or in certain experimental designs known as split-plots (Quinn & Keough, 2002). Examples of BUGS code for nested factor effects are given in many places, including Qian & Shen (2007), Li et al. (2017), and Pizarro Muñoz et al. (2018).

The last four chapters have focused on the modelling of effects of factors and here we have covered the combination of two factors in the same model. In the next chapter, we will extend the model such that we still consider two explanatory variables, but where only one will be a factor and the other continuous. We will also study interaction between the two variables.

General linear model for a normal response with continuous and categorical explanatory variables[*]

9

Chapter outline

9.1 Introduction

In the previous four chapters, we have covered normal-response models that had either one continuous explanatory variable (a "covariate", i.e., in the linear regression model in Chapter 5), or one or two categorical explanatory variables (factors) in Chapters 6–8. Here, we combine the two types of explanatory variables in a single model in what is also known as the *general linear model* ($\neq$ GLM!; see Chapter 12). This model expresses a continuous response as a linear combination of the effects of both categorical and continuous explanatory variables, plus a single random contribution from a normal distribution, whose variance is estimated along with the coefficients of all explanatory variables. In the context of least-squares analyses, the model in this chapter is also called an ANCOVA, or analysis of covariance, model (Steel et al., 1996; Zar, 1998).

In many practical applications, we will have both categorical and continuous explanatory variables. In addition, we will want to fit both main effects of these covariates as well as some or all of their pairwise or even higher-order interactions. As a simple example, we will consider a linear model with an interaction between a categorical and a continuous covariate. We saw how to fit interactions between two factors (i.e., categorical covariates) in Chapter 8. The inferential setting considered in this chapter is the relationship between body mass and body length of the asp viper (Fig. 9.1) in three different populations: Pyrenees, Massif Central, and the Jura mountains. We are particularly interested in population-specific differences of the mass-length relationship, that is, in

---

[*]This book has a companion website hosting complementary materials, including all code for download. Visit this URL to access it: https://www.elsevier.com/books-and-journals/book-companion/9780443137150.

Applied Statistical Modelling for Ecologists. DOI: https://doi.org/10.1016/B978-0-443-13715-0.00017-0

FIGURE 9.1

Male Asp viper (*Vipera aspis*), French Jura mountains, 2020 (Photo by Thomas Ott).

an interaction between length and population. The means parameterization of the model we will fit can be written as (see Section 3.4.6)

$$y_i = \alpha_{j(i)} + \beta_{j(i)} \cdot x_i + \varepsilon_i, \text{ and}$$
$$\varepsilon_i \sim Normal(0, \sigma^2),$$

where y_i is body mass of individual i, $\alpha_{j(i)}$ and $\beta_{j(i)}$ are the intercept and the slope, respectively, of the mass–length relationship in population j to which individual i belongs, and x_i is the body length of snake i. In addition, and as usual, ε_i describes the combined effects of all unmeasured influences on the body mass of snake i as well as measurement error, and is assumed to behave like a zero-mean normal random variable with variance σ^2.

The effects parameterization of the same model is this:

$$y_i = \alpha_{Pyr} + \beta_1 \cdot x_{MC(i)} + \beta_2 \cdot x_{Jura(i)} + \beta_3 \cdot x_{body(i)} + \beta_4 \cdot x_{body(i)} \cdot x_{MC(i)} + \beta_5 \cdot x_{body(i)} \cdot x_{Jura(i)} + \varepsilon_i$$

In addition to y_i and ε_i that are as before, α_{Pyr} is the expected mass of snakes in the Pyrenees at length 0, β_1 is the difference between the expected mass of snakes in the Massif Central and those in the Pyrenees, and $x_{MC(i)}$ is an indicator variable for snakes caught in the Massif Central. β_2 is the difference between the expected mass in the Jura to that in the Pyrenees, $x_{Jura(i)}$ is an indicator variable for snakes in the Jura, β_3 is the slope of the regression of body mass on body length x_{body} in the Pyrenees, β_4 is the difference in that slope between the Massif Central and the Pyrenees, and β_5 the difference of slopes between Jura and the Pyrenees. Thus, snakes in the Pyrenees act as baseline with which snakes from the Massif Central and the Jura are compared, but as usual, this choice of reference category has no effect on inference. It's just a different parameterization for the same model as above.

9.2 **Data generation**

For simplicity, we assume equal numbers of snakes studied in all three mountain ranges, but a balanced design is not required in principle.

```
set.seed(9)
nPops <- 3
nSample <- 10
n <- nPops * nSample          # Total number of data points
x <- rep(1:nPops, rep(nSample, nPops))  # Indicator for population
pop <- factor(x, labels = c("Pyrenees", "Massif Central", "Jura"))
length <- runif(n, 45, 70)    # Adult body length rarely <45 cm
lengthC <- length-mean(length)  # Use centered length
```

First, we build the design matrix of an interactive combination of length and population, inspect that, and select the parameter values, that is, pick values for α_{Pyr}, β_1, β_2, β_3, β_4 and β_5. Note that we chose these by trial and error such that the resulting mass data looked good.

```
Xmat <- model.matrix(~ pop * lengthC)
print(Xmat, dig = 2)   # not shown, but make sure to understand this!
beta.vec <- c(80, -30, -20, 6, -3, -4)
```

Next, we build up the body mass measurements y_i by adding the residual to the value of the linear predictor, with residuals drawn from a zero-mean normal distribution with a standard deviation of our choice (we take a value of 10 here). The value of the linear predictor is obtained by matrix multiplication of the design matrix (Xmat) and the parameter vector (beta.vec). Many of our vipers are probably overweight, but that doesn't really matter for our purposes (Fig. 9.2).

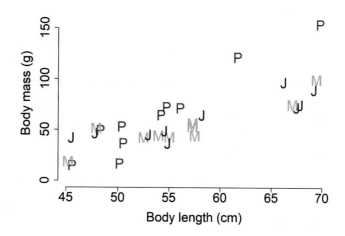

FIGURE 9.2

Simulated data set showing body mass versus length of 10 asp vipers in each of three populations (P—Pyrenees, M—Massif Central, and J—Jura).

```
sigma <- 10                               # Choose residual SD
lin.pred <- Xmat[,] %*% beta.vec          # Value of lin.predictor
eps <- rnorm(n = n, mean = 0, sd = sigma) # residuals
mass <- as.numeric(lin.pred + eps)        # response = lin.pred + residual
hist(mass)                                # Inspect what we've created (not shown)

par(mar = c(5,5,4,2), cex.lab = 1.5, cex.axis = 1.5)
matplot(cbind(length[1:10], length[11:20], length[21:30]), cbind(mass[1:10],
  mass[11:20], mass[21:30]), ylim = c(0, max(mass)), ylab = "Body mass (g)",
  xlab = "Body length (cm)", col = c("Red","Green","Blue"), pch = c("P", "M", "J"),
  las = 1, cex = 1.6, cex.lab = 1.5, frame = FALSE)          # Fig. 9.2

# Save chosen values for later comparisons
truth <- c(beta.vec, sigma)
```

We have created a data set in which vipers from the Pyrenees have the steepest slope between mass and length, followed by those from the Massif Central and finally those in the Jura mountains (Fig. 9.2). Now let's analyze these data.

```
# Load required libraries
library(ASMbook); library(DHARMa); library(jagsUI); library(rstan); library(TMB)
```

9.3 Likelihood analysis with canned functions in R

As always, the code for an analysis using a canned function in R is very parsimonious.

```
summary(out9.3 <- lm(mass ~ pop * lengthC))

[...]
Coefficients:
                          Estimate  Std. Error  t value  Pr(>|t|)
(Intercept)                78.3370      3.7846   20.699   < 2e-16 ***
popMassif Central         -25.3304      5.2271   -4.846  6.14e-05 ***
popJura                   -20.9909      5.3102   -3.953  0.000593 ***
lengthC                     5.9006      0.5339   11.051  6.74e-11 ***
popMassif Central:lengthC  -3.3176      0.7361   -4.507  0.000146 ***
popJura:length             -3.8032      0.6895   -5.516  1.13e-05 ***
---
Signif. codes:  0 '***' 0.001 '**' 0.01 '*' 0.05 '.' 0.1 ' ' 1

Residual standard error: 11.4 on 24 degrees of freedom
Multiple R-squared: 0.881, Adjusted R-squared: 0.8562
F-statistic: 35.53 on 5 and 24 DF, p-value: 2.462e-10
```

The coefficients can directly be compared with the beta vector, since we simulated the data exactly in the default (effects, or treatment contrast) parameterization of a linear model specified in R. The residual standard deviation is called residual standard error by R. Up to sampling error, we get back what we input into the data.

```
# Save least-squares estimates
lm_est <- c(coef(out9.3), sigma = summary(out9.3)$sigma)
```

Next we assess traditional and then quantile residuals using DHARMa. Both illustrate well the fact that for small samples, very strange-looking residual patterns may show up even for a fitting model.

```
# Check goodness-of-fit using quantile residuals
plot(out9.3)    # Traditional residual check for comparison
simOut <- simulateResiduals(out9.3, n = 1000, plot = TRUE)
```

Finally, we illustrate one approach to plotting the effects of the two covariates in the model. We want to generate two plots: one plot showing the effect of length on mass, with population held constant, and one plot showing the effect of population on mass, with length held constant. We'll use the built in `predict` function to generate predictions from the model.

First, we'll plot the effect of length on mass. We need to create a `data.frame` containing each value of length we want to generate a mass prediction for. To create a smooth line, we'll generate a sequence of 100 equidispersed length values from the minimum observed length to the maximum. Our data frame must also contain a value for the population in order for the function to generate a prediction. Here, we only want to vary one covariate at a time, so we will hold population fixed at the reference level (Pyrenees). We also plot the raw data for comparison (Fig. 9.3).

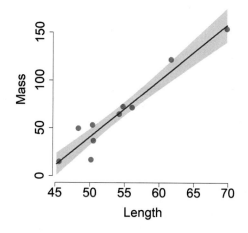 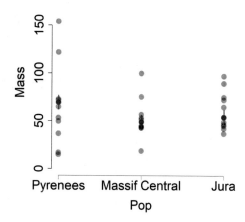

FIGURE 9.3

Model predictions (in red) for effect of length on mass (left plot) and effect of population on mass (right plot). Raw data points are shown in black, and red shaded area or lines give the Wald-type 95% CIs.

```
par(mfrow = c(1,2))
# Plot effect of length, holding pop constant (left plot)
# make sequence of 100 length values
length.seq <- seq(min(length), max(length), length.out = 100)
# Center the length sequence using the original mean
lengthC.seq <- length.seq - mean(length)
# make newdata data.frame, predict and compute 95% CI around prediction
newdata <- data.frame(lengthC = lengthC.seq, pop = "Pyrenees")
pr.length <- predict(out9.3, newdata = newdata, se.fit = TRUE)
LCL <- pr.length$fit - 1.96 * pr.length$se.fit        # Wald 95% LCL
UCL <- pr.length$fit + 1.96 * pr.length$se.fit        # Wald 95% UCL

# Draw the plot (Fig. 9.3 left)
plot(length[pop == "Pyrenees"], mass[pop == "Pyrenees"], pch = 16,
  ylab = "Mass", xlab = "Length", main = "Effect of length on mass for Pyrenees",
  frame = FALSE, ylim = c(0, 180), cex = 1.5, col = rgb(0,0,0,0.5))
lines(length.seq, pr.length$fit, col = 'red', lwd = 3)
polygon(c(length.seq, rev(length.seq)), c(LCL, rev(UCL)), col = rgb(1,0,0,0.2), border = NA)
```

Second, we want to plot the effect of population on mass. This plot will look different because population is a categorical or factor covariate with three distinct values, rather than a continuous covariate as with length. Thus, our data frame will only have three rows instead of 100, with each row corresponding to one population value. We also need a value for length in each row, which we want to hold constant. The choice of value is arbitrary, but it usually makes sense to pick the median or mean value.

```
# Plot effect of pop, holding length constant at its median (right plot)
newdata <- data.frame(lengthC = median(lengthC), pop = levels(pop))
pr.pop <- predict(out9.3, newdata = newdata, se.fit = TRUE)
LCL <- pr.pop$fit - 1.96 * pr.pop$se.fit
UCL <- pr.pop$fit + 1.96 * pr.pop$se.fit

plot(as.numeric(pop), mass, pch = 16, xaxt='n', xlab = "Pop",
  ylab = "Mass", main = "Effect of pop on mass at median length",
  frame = FALSE, cex = 1.5, col = rgb(0,0,0,0.4))   # Draw the plot (Fig. 9.3 right)
axis(1, at = 1:3, labels = levels(pop))
points(1:3, pr.pop$fit, pch = 16, col = "red", cex = 1.5)
segments(1:3, LCL, 1:3, UCL, lwd = 2, col = "red")
par(mfrow = c(1,1))
```

9.4 Bayesian analysis with JAGS

In JAGS, we find it easier to fit the means parameterization of the model, that is, to specify three separate linear regressions, one for each mountain range. The effects, that is, the differences of the intercepts or slopes with reference to the intercept or the slope in the Pyrenees, are trivially easy to recover as derived parameters in just a few lines of JAGS code. This allows for better comparison between the parameters we used to simulate the data and the JAGS output.

```
# Bundle and summarize data
str(dataList <- list(mass = as.numeric(mass), pop = as.numeric(pop), lengthC = lengthC,
  nPops = nPops, n = n) )

List of 5
 $ mass    : num [1:30] 36.9  14.9  17  53.3  71.5 ...
 $ pop     : num [1:30] 1  1  1  1  1  1  1  1  1  1 ...
 $ lengthC : num [1:30] -5.78  -10.72  -6.14  -5.93  -0.23 ...
 $ nPops   : num 3
 $ n       : num 30

# Write JAGS model file
cat(file = "model9.4.txt", "
model {
# Priors
for (i in 1:nPops){
  alpha[i] ~ dnorm(0, 0.0001)          # Intercepts
  beta[i] ~ dnorm(0, 0.0001)           # Slopes
}
sigma ~ dunif(0, 100)                  # Residual standard deviation
tau <- pow(sigma, -2)

# Likelihood
for (i in 1:n) {
  mass[i] ~ dnorm(mu[i], tau)
  mu[i] <- alpha[pop[i]] + beta[pop[i]] * lengthC[i]
}
# Define effects relative to baseline level
a.effe2 <- alpha[2] - alpha[1]         # Intercept Massif Central vs. Pyr.
a.effe3 <- alpha[3] - alpha[1]         # Intercept Jura vs. Pyr.
b.effe2 <- beta[2] - beta[1]           # Slope Massif Central vs. Pyr.
b.effe3 <- beta[3] - beta[1]           # Slope Jura vs. Pyr.

# Custom comparison
test1 <- beta[3] - beta[2]             # Slope Jura vs. Massif Central
}
")

# Function to generate starting values
inits <- function(){ list(alpha = rnorm(nPops, 0, 2), beta = rnorm(nPops, 1, 1), sigma = runif(1))}

# Parameters to estimate
params <- c("alpha", "beta", "sigma", "a.effe2", "a.effe3", "b.effe2", "b.effe3", "test1")

# MCMC settings
na <- 1000  ;  ni <- 3000  ;  nb <- 1000  ;  nc <- 4  ;  nt <- 1
```

```
# Call JAGS (ART 2 sec), check convergence, summarize posteriors and save estimates
out9.4 <- jags(dataList, inits, params, "model9.4.txt", n.iter = ni, n.burnin = nb,
  n.chains = nc, n.thin = nt, n.adapt = na, parallel = TRUE)
jagsUI::traceplot(out9.4)                              # not shown
print(out9.4, 3)
jags_est <- out9.4$summary[c(1, 8, 9, 4, 10, 11, 7), 1]
```

This is a simple model that converges very rapidly. We inspect the results and compare them with the truth in the data-generating random process as well as with the inference from lm():

	mean	sd	2.5%	50%	97.5%	overlap0	f	Rhat	n.eff
alpha[1]	78.138	4.016	70.287	78.148	85.744	FALSE	1.000	1.000	8000
alpha[2]	52.907	3.815	45.319	52.949	60.540	FALSE	1.000	1.000	3815
alpha[3]	57.255	3.963	49.378	57.258	65.169	FALSE	1.000	1.000	8000
beta[1]	5.891	0.571	4.760	5.887	7.040	FALSE	1.000	1.000	8000
beta[2]	2.586	0.542	1.505	2.597	3.627	FALSE	1.000	1.000	8000
beta[3]	2.102	0.469	1.187	2.103	3.013	FALSE	1.000	1.000	7603
sigma	11.991	1.803	9.071	11.795	16.025	FALSE	1.000	1.001	8000
a.effe2	-25.230	5.517	-36.005	-25.294	-14.387	FALSE	1.000	1.000	8000
a.effe3	-20.883	5.683	-32.138	-20.987	-9.612	FALSE	0.999	1.000	5761
b.effe2	-3.305	0.790	-4.815	-3.306	-1.771	FALSE	1.000	1.000	8000
b.effe3	-3.789	0.739	-5.265	-3.780	-2.350	FALSE	1.000	1.000	8000
test1	-0.484	0.719	-1.923	-0.485	0.922	TRUE	0.751	1.000	8000

```
# Compare likelihood with Bayesian estimates and with truth
comp <- cbind(truth = truth, lm = lm_est, JAGS = jags_est)
print(comp, 4)
```

	truth	lm	JAGS
(Intercept)	80	78.337	78.138
popMassif Central	-30	-25.330	-25.230
popJura	-20	-20.991	-20.883
length	6	5.901	5.891
popMassif Central:length	-3	-3.318	-3.305
popJura:length	-4	-3.803	-3.789
sigma	10	11.402	11.991

We see the typical good numerical agreement between the maximum likelihood estimates (MLEs) and Bayesian posterior means from an analysis with vague priors.

9.5 Bayesian analysis with NIMBLE

NIMBLE code for the model is available on the book website.

9.6 Bayesian analysis with Stan

With Stan, we again find it easier to fit a means parameterization of the model. Thus, the Stan code below looks very similar to the JAGS and NIMBLE code. Of course, we must also define our data and parameters explicitly in the appropriate model blocks as usual for Stan. Calculation of the effects relative to the baseline level (for comparison) must be done in the generated quantities block.

```
# Bundle and summarize data
str(dataList <- list(mass = as.numeric(mass), pop = as.numeric(pop), lengthC = lengthC,
  nPops = nPops, n = n) )

# Write Stan model
cat(file = "model9_6.stan", "

data {
  int n;                              //sample size
  int nPops;                          //number of populations
  vector[n] mass;                     //response
  vector[n] lengthC;                  //covariate
  array[n] int pop;                   //population of each observation
}

parameters {
  vector[nPops] alpha;                //intercepts
  vector[nPops] beta;                 //slopes
  real<lower=0> sigma;                //residual standard deviation
}

model {
  vector[n] mu;                       //expected value of observations

  //Priors
  for (i in 1:nPops){
    alpha[i] ~ normal(0, 100);
    beta[i] ~ normal(0, 100);
  }
  sigma ~ uniform(0, 100);

  //Likelihood
  for (i in 1:n){
    mu[i] = alpha[pop[i]] + beta[pop[i]] * lengthC[i];
    mass[i] ~ normal(mu[i], sigma);
  }
}

generated quantities {
  real a_effe2;
  real a_effe3;
  real b_effe2;
  real b_effe3;
  real test1;

  a_effe2 = alpha[2] - alpha[1];      //Intercept Massif Central vs. Pyr.
  a_effe3 = alpha[3] - alpha[1];      //Intercept Jura vs. Pyr.
  b_effe2 = beta[2] - beta[1];        //Slope Massif Central vs. Pyr.
  b_effe3 = beta[3] - beta[1];        //Slope Jura vs. Pyr.

  test1 = beta[3] - beta[2];          //Slope Jura vs. Massif Central
}
")
```

```
# Parameters to estimate
params <- c("alpha", "beta", "sigma", "a.effe2", "a.effe3", "b.effe2", "b.effe3", "test1")

# HMC settings
ni <- 1000  ;  nb <- 500  ;  nc <- 4  ;  nt <- 1

# Call STAN (ART 40/2 sec), print output and save estimates
system.time(
out9.6 <- stan(file = "model9_6.stan", data = dataList,
          warmup = nb, iter = ni, chains = nc, thin = nt) )
rstan::traceplot(out9.6)                    # not shown
print(out9.6, dig = 3)                       # not shown
stan_est <- summary(out9.6)$summary[c(1, 8, 9, 4, 10, 11, 7), 1]
```

9.7 Do-it-yourself maximum likelihood estimates

Now for our homegrown MLEs. In contrast to JAGS, NIMBLE, and Stan, our do-it-yourself (DIY) MLE function will use an effects parameterization, as with our analysis using lm(). The resulting model will be very similar to the regression models in Chapters 5–8. Instead of just continuous covariates or categorical (factor) covariates, we now have an interaction between a continuous covariate and a factor. Thus, we simply modify the likelihood equation from Chapters 5–8 by adding additional parameters (combined in vector β) to the linear predictor for the expected mass of snake i, μ_i. See Section 9.1 for a more detailed description of the equation for the linear predictor in this model.

$$L\left(\beta, \sigma | y, x_{MC}, x_{Jura}, x_{body}\right) = \prod_{i=1}^{n} \frac{1}{\sigma\sqrt{2\pi}} e^{-\frac{1}{2}\left(\frac{y_i - \mu_i}{\sigma}\right)^2}$$

$$\mu_i = \beta_{0(Pyr)} + \beta_1 \cdot x_{MC(i)} + \beta_2 \cdot x_{Jura(i)} + \beta_3 \cdot x_{body(i)} + \beta_4 \cdot x_{body(i)} \cdot x_{MC(i)} + \beta_5 \cdot x_{body(i)} \cdot x_{Jura(i)}$$

We have made a slight modification to the linear predictor here compared to the one in Section 9.1. We now include the intercept in our vector of parameters β as β_0 instead of α. Thus, we now have six parameters in the vector β: the intercept, three main-effects parameters, and two interaction-effects parameters. Given the complexity of our model, we find it easier to calculate the vector of expected values μ by matrix-multiplying the model matrix **X** we generated in Section 9.2 (which contains all the covariate data) with β, as we did in Chapters 7 and 8. This greatly simplifies the second equation:

$$\mu = X\beta$$

Here's the likelihood translated into an R function. We first separate the complete parameter vector into β and σ. Next we calculate the vector μ by matrix-multiplying the model matrix **X** and β. We then calculate the log-likelihood for each observation using the vector of mass values **y**, μ, and σ, and finally sum up the log-likelihoods and return the negative of the sum.

```
NLL <- function(param, y, Xmat) {
  beta <- param[1:6]                          # Intercept/slopes matching columns of Xmat
  sigma <- exp(param[7])                       # Residual SD
  mu <- Xmat %*% beta                          # Matrix-multiply beta and design matrix
  LL <- dnorm(y, mu, sigma, log=TRUE)          # Log-lik for each obs
  -sum(LL)                                      # NLL for all observations (whole data set)
}
```

Interestingly, we need to provide semi-informative starting values for the intercept β_0 and σ to run the optimization successfully (we encourage you to experiment with other sets of inits). We set the initial values for these two values to the overall mean mass and the log of the overall standard deviation of the mass values.

```
# Minimize NLL to find MLEs, get SEs and 95% CIs and save estimates
inits <- c(mean(mass), rep(0, 5), log(sd(mass)))
names(inits) <- c(colnames(Xmat), 'log-sigma')
out9.7 <- optim(inits, NLL, y = mass, Xmat = Xmat, hessian=TRUE, method = "BFGS")
get_MLE(out9.7, 4)
diy_est <- c(out9.7$par[1:6], exp(out9.7$par[7]))
```

	MLE	ASE	LCL.95	UCL.95
(Intercept)	78.343	3.3849	71.708	84.977
popMassif Central	-25.340	4.6751	-34.504	-16.177
popJura	-20.998	4.7494	-30.307	-11.689
lengthC	5.901	0.4776	4.965	6.837
popMassif Central:lengthC	-3.318	0.6584	-4.609	-2.028
popJura:lengthC	-3.804	0.6167	-5.012	-2.595
log-sigma	2.322	0.1291	2.069	2.575

That looks good!

9.8 Likelihood analysis with TMB

Our input data for TMB is similar to JAGS and Stan, but we need to make the data conform to the C++ custom of starting indices at 0. Thus, we must adjust pop by subtracting 1.

```
# Bundle and summarize data
tmbData <- dataList
tmbData$pop <- tmbData$pop-1
str(tmbData)
```

Our model file looks very similar to our Stan code—we just need to change how to describe the input data and parameters, and remember to start our indices at 0.

```
# Write TMB model file
cat(file = "model9_8.cpp",
"#include <TMB.hpp>

template<class Type>
Type objective_function<Type>::operator() ()
{
  //Describe input data
  DATA_INTEGER(n);                       //number of obs
  DATA_VECTOR(mass);                     //response
  DATA_VECTOR(lengthC);                  //covariate
  DATA_IVECTOR(pop);                     //population index

  //Describe parameters
  PARAMETER_VECTOR(alpha);               //Intercepts
  PARAMETER_VECTOR(beta);                //Slopes
  PARAMETER(log_sigma);                  //Residual sd on log scale

  Type sigma = exp(log_sigma);           //Residual SD

  vector<Type> mu(n);                    //Expected value of observations

  Type LL = 0;                           // Initialize log-likelihood at 0

  //Iterate over observations
  for (int i=0; i<n; i++){               //Note index starts at 0 instead of 1
    mu(i) = alpha(pop(i)) + beta(pop(i)) * lengthC(i);

    //Calculate log-likelihood of observation and add to total loglik
    LL += dnorm(mass(i), mu(i), sigma, true);
  }

  //Calculate derived parameters
  Type a_effe2 = alpha(1) - alpha(0);    //Intercept M. Central vs. Pyr.
  Type a_effe3 = alpha(2) - alpha(0);    //Intercept Jura vs. Pyr.
  Type b_effe2 = beta(1) - beta(0);      //Slope Massif Central vs. Pyr.
  Type b_effe3 = beta(2) - beta(0);      //Slope Jura vs. Pyr.
  Type test1 = beta(2) - beta(1);        //Slope Jura vs. Massif Central

  //Tell TMB to report these derived parameters
  ADREPORT(a_effe2);
  ADREPORT(a_effe3);
  ADREPORT(b_effe2);
  ADREPORT(b_effe3);
  ADREPORT(test1);

  return -LL;                            //Return negative of total log likelihood
}
")
```

Interestingly, when fitting the model, we again need to provide semi-informative starting values, otherwise the optimization won't converge to the MLEs. We use the same ones as we did for the DIY likelihood.

```
# Compile and load TMB function
compile("model9_8.cpp")
dyn.load(dynlib("model9_8"))

# Provide dimensions and starting values for parameters
# Starting values need to be semi-informative (use same ones as DIY)
params <- list(alpha = c(mean(mass), 0, 0), beta = rep(0,3),
          log_sigma = log(sd(mass)))

# Create TMB object
out9.8 <- MakeADFun(data = tmbData, parameters = params,
          DLL = "model9_8", silent = TRUE)

# Optimize TMB object and print and save results
opt <- optim(out9.8$par, fn = out9.8$fn, gr = out9.8$gr, method = "BFGS", hessian = TRUE)
(tsum <- tmb_summary(out9.8))
tmb_est <- c(tsum[c(1, 8, 9, 4, 10, 11), 1], exp(tsum[7, 1]))
```

9.9 Comparison of the parameter estimates

Here's our usual comparison of parameter estimates from all methods:

```
# Compare results with truth and previous estimates
comp <- cbind(cbind(truth = truth, lm = lm_est, JAGS = jags_est,
   NIMBLE = nimble_est, Stan = stan_est, DIY = diy_est, TMB = tmb_est))
print(comp, 3)
```

	truth	lm	JAGS	NIMBLE	Stan	DIY	TMB
(Intercept)	80	78.34	78.14	78.28	78.32	78.34	78.34
popMassif Central	-30	-25.33	-25.23	-25.38	-25.33	-25.34	-25.34
popJura	-20	-20.99	-20.88	-21.09	-21.07	-21.00	-20.99
Length	6	5.90	5.89	5.90	5.91	5.90	5.90
popMassif Central:lengthC	-3	-3.32	-3.30	-3.31	-3.33	-3.32	-3.32
popJura:lengthC	-4	-3.80	-3.79	-3.80	-3.80	-3.80	-3.80
sigma	10	11.40	11.99	12.16	12.04	10.20	10.20

You will see that there is hardly any variation among methods, except for the estimate of the residual standard deviation, `sigma`. As discussed in earlier chapters, this is the result of the relatively small sample size and a variety of modeling approaches used (least-squares vs. Bayesian vs. maximum likelihood). Interestingly, even though the variance estimate of maximum likelihood is known to be biased in small sample sizes, for this particular data set, by chance the MLE of `sigma` comes closest to the truth. As always, to better understand the behavior of these estimation methods, we encourage you to run simulations, for instance, to find out whether this pattern prevails ... if you do this, you will see that it is not generally the case.

9.10 Summary and outlook

In this chapter, we have focused on a model with one discrete and one continuous predictor and which includes their interaction. Understanding this version of a general linear model is an important intermediate step to understanding the linear mixed model in the next chapter.

Of course, the model in this chapter is just a toy example of the kinds of general linear models that you may frequently encounter in your research, where you may have multiple continuous covariates as well as factors, and different kinds of interactions, such as factor/factor (as in Chapter 8), factor/continuous (as in this chapter), or also continuous/continuous. We don't show this third type of an interaction in a linear model, between two (or possibly more) continuous covariates, since they are very easy to specify: we simply take the product of the two covariates and add their effect to our linear predictor. Although particularly easy to specify, these interactions can be challenging to visualize and understand. While a linear model with the main effects of two continuous covariates is represented geometrically as a plane, adding an interaction between them will twist that plane. To understand such an interaction, it is best to form predictions for a grid of values for both covariates and then plot them as an image; for example, see Kéry & Royle (2016: Section 10.9) or Kéry & Royle (2021: Section 4.9), or, in a different context, also the image plot in Fig. 2.12.

Linear mixed-effects model⊛

Chapter outline

⊛This book has a companion website hosting complementary materials, including all code for download. Visit this URL to access it: https://www.elsevier.com/books-and-journals/book-companion/9780443137150.

Applied Statistical Modelling for Ecologists. DOI: https://doi.org/10.1016/B978-0-443-13715-0.00003-0

10.1 Introduction

Mixed-effects or mixed models contain both fixed parameters and parameters that are random, in the sense that we assume they are draws from some distribution with shared hyperparameters that we estimate (Chapter 3). During the last 25 years or so, the use of mixed models has greatly increased in statistical applications in ecology and related disciplines (Pinheiro & Bates, 2000; McCulloch & Searle, 2001; Bolker et al., 2009; Lee et al., 2017). As we discussed in Chapters 3 and 7, there may be multiple motivations for assuming a set of parameters constitutes a random sample from some distribution and estimating the associated hyperparameters as the main structural parameters of a model. These include increased scope of inference, more honest accounting for all sources of uncertainty, efficiency of estimation, and the ability to model patterns in parameters, including correlations between them.

In Chapter 7, we saw a model with a random-effects factor that, apart from an overall mean and two variances, contained only random effects and could be called a variance–components model. That model is in fact also a mixed model, since both the overall mean and the two variances are fixed effects, but this terminology is not standard. Here, we consider a classic mixed model that arises as a direct generalization of the general linear, or ANCOVA, model with a cluster of regression lines in Chapter 9. We modify our Asp viper (Fig. 10.1) data set from that chapter just a little bit and assume

FIGURE 10.1

Female Asp viper (*Vipera aspis*), French Jura mountains (Photo by Thomas Ott).

that we now have measurements from a much larger number of populations, say, 56. A random-effects factor need not possess that many levels (some statisticians suggest considering factors with any number of levels as random; see Gelman, 2005), but in practice one rarely sees fewer than, say, 5–10 parameters treated as random effects. Estimating a variance with so few values, which are moreover unobserved, may result in imprecise and possibly even biased estimates (see also Lambert et al., 2005).

We will re-simulate some Asp viper data using R code fairly similar to that in the previous chapter. However, we will now *constrain the values for at least one set of effects* (the intercepts and/or the slopes) to come from a normal distribution: this is what the random-effects assumption in a mixed model means. There are at least three sets of assumptions that we could make about the random effects for the intercept or the slope of regression lines fitted to grouped (here, population-specific) data:

1. Only intercepts are random, but slopes are identical for all groups,
2. both intercepts and slopes are random, but they are independent, and
3. both intercepts and slopes are random and there is a correlation between them.

(An additional case, where slopes are random and intercepts fixed, is rarely used in practice.) Model No. 1 is known as a *random-intercepts model*, and both models No. 2 and 3 as *random-coefficients models*. As we will see, model No. 3 is the default in the R function `glmmTMB()` when fitting a random-coefficients model.

We will first simulate a random-coefficients data set under model No. 2, where both intercepts and slopes are uncorrelated random effects. Then, we will fit both a random-intercepts (No. 1) and a random-coefficients model without correlation (No. 2) to these data (see Sections 10.2–10.4). After that, we generate another data set that includes a correlation between intercepts and slopes, and in the analysis adopt the random-coefficients model which estimates the correlation as a parameter (No. 3; Chapter 10.5).

This is a key chapter for your understanding of mixed models and we expect its contents to be helpful for the general understanding of mixed models to many ecologists. A close examination of how such data can be assembled (i.e., simulated in R) will be an invaluable help for your understanding of how analogous data sets are broken down (i.e., analyzed) when using any type of mixed model. Indeed, we believe that very few strategies can be more effective to understanding this type of mixed model than the combination of simulating data sets and describing the models fitted in the BUGS language.

In addition, this is the only place in the book where we estimate a correlation between two sets of parameters as a separate parameter. There are many examples in more advanced ecological models where you may need this. Examples include the study of correlations between temporal variation among sets of parameters, such as juvenile and adult survival in a Cormack-Jolly Seber model (e.g., Section 7.6.2 in Kéry & Schaub, 2012) or abundances of interacting species (e.g., Mutshinda et al., 2011).

Here is one way in which to write the random-coefficients model without correlation between the random effects for mass y_i of snake i in population j:

$$y_i = \alpha_{j(i)} + \beta_{j(i)} \cdot x_i + \varepsilon_i$$
$$\alpha_j \sim Normal(\mu_\alpha, \sigma_\alpha^2) \qquad \text{\# Random effects for intercepts}$$
$$\beta_j \sim Normal(\mu_\beta, \sigma_\beta^2) \qquad \text{\# Random effects for slopes}$$
$$\varepsilon_i \sim Normal(0, \sigma^2) \qquad \text{\# Residual "random" effects}$$

Exactly as in the model in Chapter 9, mass y_i is assumed to be related to body length x_i of snake i in population j by a straight-line relationship with population-specific values for intercept α_j and slope β_j. (These regression parameters vary by individual i through their membership to population j.) However, both α_j and β_j are now assumed to be draws from independent normal distributions, with means μ_α and μ_β and variances of σ_α^2 and σ_β^2, respectively. The residuals ε_i for snake i in population j are assumed to come from another independent normal distribution with variance σ^2 and which is centered on zero. As a result, we could also say that when fitting this model, we estimate the parameters of three normal distributions simultaneously, but where the mean of one of them is fixed at zero.

10.2 **Data generation**

As always, we assume a balanced design for simple convenience; for imbalanced data sets and dealing with missing values, see Chapter 11.

```
set.seed(10)
nPops <- 56                        # Number of populations
nSample <- 10                      # Number of vipers in each pop
n <- nPops * nSample               # Total number of data points
pop <- gl(n = nPops, k = nSample)  # Indicator for population
```

We directly normalize covariate length to avoid any numerical trouble with our fitting engines.

```
# Body length (cm)
orig.length <- runif(n, 45, 70)
mn <- mean(orig.length)
sd <- sd(orig.length)
cat("Mean and sd used to normalise original length:", mn, sd, "\n\n")
lengthN <- (orig.length - mn)/sd         # N for 'normalized'
hist(lengthN, col = "grey")              # Not shown
```

We build a design matrix without intercept.

```
Xmat <- model.matrix(~ pop * lengthN - 1 - lengthN)
print(Xmat[1:21,], dig = 2)   # Print top 21 rows (not shown)
```

Next, we choose parameter values, but this time constrain them in the sense that we draw both the values for the vector of intercepts (α) and for the vector of slopes (β) from two normal distributions, for which we will specify four hyperparameters: two means, corresponding to μ_α and μ_β, and two standard deviations, corresponding to the square root of σ_α^2 and σ_β^2. For the residual variation we will use a mean-zero normal distribution with a standard deviation of 30.

```
mu.alpha <- 260
sigma.alpha <- 20
mu.beta <- 60
sigma.beta <- 30

alpha <- rnorm(n = nPops, mean = mu.alpha, sd = sigma.alpha)
beta <- rnorm(n = nPops, mean = mu.beta, sd = sigma.beta)
all.pars <- c(alpha, beta)        # Put them all together
```

We assemble the measurements y_i as in Section 9.2.

```
sigma <- 30                         # Residual standard deviation
lin.pred <- Xmat[,] %*% all.pars    # Value of lin.predictor
eps <- rnorm(n = n, mean = 0, sd = sigma)   # residuals
mass <- lin.pred + eps              # response = lin.pred + residual
mass <- drop(mass)                  # Turn n×1 matrix into vector of length n
hist(mass, col = "grey")            # Inspect what we've created (not shown)


# Save true values for comparisons later
truth <- c(mu.alpha = mu.alpha, mu.beta = mu.beta,
           sigma.alpha = sigma.alpha, sigma.beta = sigma.beta, residual.sd = sigma)
```

We produce a trellis graph of the relationships in all `nPops` populations (Fig. 10.2). Depending on the particular realization of the simulated stochastic system, we generally have

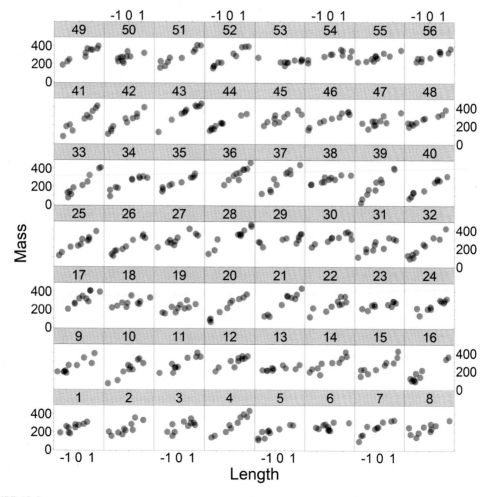

FIGURE 10.2

Trellis plot of the mass–length relationships in 56 asp viper populations. We see heterogeneity among populations in both the intercepts and the slopes.

quite a few large animals and may even have a few negative masses, but this doesn't really matter for our analysis.

```
library(lattice)
xyplot(mass ~ lengthN | pop, xlab = 'Length', ylab = 'Mass', main = 'Realized mass-length relationships',
    pch = 16, cex = 1.2, col = rgb(0, 0, 0, 0.4))
```

We can detect straight-line relationships between mass and length that differ among the 56 populations. What we can't see is the random-effects assumption that we employed to build our data set. That is, we are unable to distinguish a data set under a general linear model as in Chapter 9 from a mixed model data set as in this chapter. Whether to treat a set of parameters as random is a modeling choice, not something fixed for a given data set or even something we could test.

```
# Required libraries
library(ASMbook); library(jagsUI); library(rstan); library(TMB)
```

10.3 Analysis under a random-intercepts model

10.3.1 Likelihood analysis using canned functions in R

We first assume that the slope of the mass–length relationship is identical in all populations and that only the intercepts differ randomly from one population to another. As in Chapter 7, we continue to use mixed-modeling functions with both maximum likelihood (ML) and restricted maximum likelihood (REML) in package glmmTMB, but note that fitting functions in the package lme4 have nearly identical syntax and their use should lead to identical estimates as glmmTMB. Here and throughout this chapter, we had some difficulty fitting models with glmmTMB() and do-it-yourself maximum likelihood. One solution is to change the optimization algorithm, as we do below and in Section 10.4.1.

```
library(glmmTMB)
gtmbData <- data.frame(mass = mass, lengthN = lengthN, pop = pop)
out10.3.ML <- glmmTMB(mass ~ lengthN + (1 | pop), data = gtmbData, REML = FALSE,
                    control = glmmTMBControl(optimizer = optim,
                                         optArgs = list(method = "L-BFGS-B"))) # ML
out10.3.REML <- glmmTMB(mass ~ lengthN + (1 | pop), data = gtmbData, REML = TRUE)  # REML
summary(out10.3.ML)
summary(out10.3.REML)

> summary(out10.3.ML)
 Family: gaussian ( identity )
Formula:    mass ~ length + (1 | pop)

....

Random effects:

Conditional model:
 Groups     Name         Variance   Std.Dev.
 pop        (Intercept)  360.1      18.98
 Residual                1812.3     42.57
Number of obs: 560, groups: pop, 56

Dispersion estimate for gaussian family (sigma^2): 1.81e+03

Conditional model:
             Estimate   Std. Error   z value   Pr(>|z|)
(Intercept)   265.117       3.109     85.27     <2e-16 ***
lengthN        61.194       1.868     32.76     <2e-16 ***
```

```
> summary(out10.3.REML)
 Family: gaussian ( identity )
Formula:    mass ~ length + (1 | pop)

...

Random effects:

Conditional model:
 Groups      Name          Variance   Std.Dev.
 pop         (Intercept)   370        19.23
 Residual                  1816       42.61
Number of obs: 560, groups: pop, 56

Dispersion estimate for gaussian family (sigma^2): 1.82e+03

Conditional model:
              Estimate   Std. Error   z value   Pr(>|z|)
(Intercept)   265.117    3.138        84.48     <2e-16 ***
lengthN       61.188     1.870        32.71     <2e-16 ***


# Save estimates from ML and from REML fit of random-intercepts model
gtmbML_est <- c(fixef(out10.3.ML)$cond,
                attr(VarCorr(out10.3.ML)$cond$pop, "stddev"),
                sigma(out10.3.ML))
gtmbREML_est <- c(fixef(out10.3.REML)$cond,
                attr(VarCorr(out10.3.REML)$cond$pop, "stddev"),
                sigma(out10.3.REML))
```

There are different manners of assessing the goodness-of-fit of a mixed model using simulation. The approaches differ in how new data are simulated using the model. For the first method, data are simulated *conditional* on the estimated random effects. In the second, *unconditional* approach, we re-simulate the random effects before simulating a new data set (Hartig, 2022). We show the two approaches in more detail in the next section with JAGS. Currently, for fitted model objects from `glmmTMB`, DHARMa functionality only allows unconditional simulations. This will test the whole model structure at once.

```
# Assess goodness-of-fit of the model (not shown)
hist(coef(out10.3.ML)$cond$pop[,1], main = 'alpha', breaks = 12)
library(DHARMa)
simOut <- simulateResiduals(out10.3.ML, n = 1000, plot = TRUE)
```

10.3.2 Bayesian analysis with JAGS

Our JAGS code for the random-intercepts model looks similar to the JAGS code in Chapter 9. The main difference is that the intercepts `alpha` now come from a normal distribution with parameters `mu.alpha` and `tau.alpha`. We use a half-Cauchy distribution as a prior for the standard deviation of the random-effects distribution `sd.alpha` (Gelman, 2006). We also add code to calculate new, simulated data under the model using both conditional (i.e., conditional on the estimated random intercepts `alpha`) and unconditional (i.e., `alpha` itself is also re-simulated) approaches. These data sets are used for goodness-of-fit assessments using DHARMa. Using DHARMa on the unconditionally simulated data will assess the entire model structure, while using the conditional simulations will only assess the structure of the "data model."

```
# Bundle and summarize data
str(dataList <- list(mass = mass, pop = as.numeric(pop), lengthN = lengthN, nPops = nPops, n = n) )

List of 5
 $ mass    : Named num [1:560]  278  193  254  306  211 ...
  ..- attr(*, "names") = chr [1:560] "1" "2" "3" "4" ...
 $ pop     : num [1:560]  1 1 1 1 1 1 1 1 1 1  ...
 $ lengthN : num [1:560]  0.0216  -0.6689  -0.2556  0.6603  -1.4315 ...
 $ nPops   : num 56
 $ n       : num 560

# Write JAGS model file
cat(file = "model10.3.2.txt", "
model {
# Priors
for (i in 1:nPops){
  alpha[i] ~ dnorm(mu.alpha, tau.alpha)            # Random intercepts
}

mu.alpha ~ dnorm(0, 0.000001)                      # Mean for random intercepts
tau.alpha <- pow(sigma.alpha, -2)
sigma.alpha ~ dt(0, 1, 0.0001)T(0,)                # SD for random intercepts

mu.beta ~ dnorm(0, 0.000001)                       # Common slope
tau.residual <- pow(sigma.residual, -2)            # Residual precision
sigma.residual ~ dt(0, 1, 0.0001)T(0,)             # Residual standard deviation

# 'Likelihood'
for (i in 1:n) {
  mass[i] ~ dnorm(mu[i], tau.residual)             # The observed random variables
  mu[i] <- alpha[pop[i]] + mu.beta * lengthN[i]    # Expectation
}

# Simulation of new datas for GOF assessment using DHARMa
for (i in 1:nPops){
  alpha.new[i] ~ dnorm(mu.alpha, tau.alpha)
}
for (i in 1:n) {
  mass.new.cond[i] ~ dnorm(mu[i], tau.residual)     # Conditional
  mass.new.un[i] ~ dnorm(mu.new[i], tau.residual)   # Unconditional
  mu.new[i] <- alpha.new[pop[i]] + mu.beta * lengthN[i]
}
}
")

# Function to generate starting values
inits <- function(){list(mu.alpha = rnorm(1, 0, 1), mu.beta = rnorm(1, 0, 1),
  sigma.alpha = rlnorm(1), sigma.residual = rlnorm(1)) }

# Parameters to estimate
params <- c("mu.alpha", "mu.beta", "sigma.alpha", "sigma.residual", "alpha",
  "mass.new.cond", "mass.new.un")
```

```
# MCMC settings
na <- 1000   ;   ni <- 3000   ;   nb <- 1000   ;   nc <- 4   ;   nt <- 1

# Call JAGS (ART <1 min), check convergence, summarize posteriors and save estimates
out10.3.2 <- jags(dataList, inits, params, "model10.3.2.txt", n.iter = ni, n.burnin = nb,
    n.chains = nc, n.thin = nt, n.adapt = na, parallel = TRUE)
par(mfrow = c(2, 2)); jagsUI::traceplot(out10.3.2)              # not shown
print(out10.3.2, 2)
jags_est <- out10.3.2$summary[c(1:4),1]
```

	mean	sd	2.5%	50%	97.5%	overlap0	f	Rhat	n.eff
mu.alpha	265.07	3.15	258.78	265.09	271.21	FALSE	1	1	1514
mu.beta	61.20	1.88	57.53	61.17	64.87	FALSE	1	1	4692
sigma.alpha	19.22	2.89	14.03	19.08	25.28	FALSE	1	1	1194
sigma.residual	42.73	1.37	40.14	42.70	45.52	FALSE	1	1	4740
alpha[1]	275.09	11.16	253.41	275.03	296.92	FALSE	1	1	3898
alpha[2]	257.46	11.01	235.88	257.47	279.14	FALSE	1	1	6853
alpha[3]	245.85	11.29	223.68	246.03	268.46	FALSE	1	1	7112

......

```
# Compare likelihood with Bayesian estimates and with truth
comp <- cbind(truth = truth[-4], gtmbML = gtmbML_est, JAGS = jags_est)
print(comp, 4)
```

	truth	gtmbML	JAGS
mu.alpha	260	265.12	265.07
mu.beta	60	61.19	61.20
sigma.alpha	20	18.98	19.22
residual.sd	30	42.57	42.73

As usual with vague priors, the two analyses yield similar results. Interestingly, the residual standard deviation in both is estimated too high. This is because we simulated the data to contain random variation among the slopes, but we did not fit this model. Therefore, this variation is unaccounted for and gets absorbed into the residual of the model.

We next check the goodness-of-fit of the model by plotting in a histogram the random-effects estimates and by assessment of quantile residuals using DHARMa. By simulating new mass data based on random effects alpha that are also re-simulated, we obtain an assessment that is not conditional on the random effects, but assesses the whole model structure at once.

```
# Distribution of random effects estimates (alpha)
hist(out10.3.2$mean$alpha, main = 'alpha')

# Quantile residual assessments (not shown)
mass.new.un <- out10.3.2$sims.list$mass.new.un        # Unconditional
mass.new.cond <- out10.3.2$sims.list$mass.new.cond   # Conditional

sim.un <- createDHARMa(simulatedResponse = t(mass.new.un), observedResponse = mass,
    fittedPredictedResponse = apply(mass.new.un, 2, median), integerResponse = FALSE)
plot(sim.un)                # Unconditional new data
sim.cond <- createDHARMa(simulatedResponse = t(mass.new.cond), observedResponse = mass,
    fittedPredictedResponse = apply(mass.new.cond, 2, median), integerResponse = FALSE)
plot(sim.cond)                  # New data conditional on estimated random effects
```

10.3.3 **Bayesian analysis with NIMBLE**

NIMBLE code for the model is available on the book website.

10.3.4 **Bayesian analysis with Stan**

The Stan version of the model looks very similar to the BUGS code from the previous two sections. One change is that we can more explicitly specify a half-Cauchy distribution as the prior for the standard deviation parameters. Also, we provide Stan with initial values, unlike previous chapters.

```
# Summarize data set again
str(dataList)   # not shown

# Write Stan model
cat(file = "model10_3_4.stan", "

data {
  int n;                               //Number of observations
  int nPops;                           //Number of populations
  vector[n] mass;                      //Response variable
  vector[n] lengthN;                   //Covariate
  array[n] int pop;                    //Population assignment of each obs
}

parameters {
  real mu_alpha;
  real mu_beta;
  real<lower = 0> sigma_alpha;
  real<lower = 0> sigma_residual;
  vector[nPops] alpha;
}

model {

  vector[n] mu;                        //Expected value

  mu_alpha ~ normal(0, 1000);
  mu_beta ~ normal(0, 1000);
  sigma_alpha ~ cauchy(0, 100);
  sigma_residual ~ cauchy(0, 100);

  for (i in 1:nPops){
    alpha[i] ~ normal(mu_alpha, sigma_alpha);
  }

  //'Likelihood'
  for (i in 1:n){
    mu[i] = alpha[pop[i]] + mu_beta * lengthN[i];
    mass[i] ~ normal(mu[i], sigma_residual);
  }
}
")

# Function to generate starting values
inits <- function(){list(mu_alpha = rnorm(1, 0, 1),
  mu_beta = rnorm(1, 0, 1), sigma_alpha = rlnorm(1),
  sigma_residual = rlnorm(1)) }
```

```
# HMC settings
ni <- 2000  ;  nb <- 1000  ;  nc <- 4  ;  nt <- 1

# Call STAN (ART 50 sec/22 sec) and save results
system.time(
out10.3.4 <- stan(file = "model10_3_4.stan", data = dataList, init = inits,
  warmup = nb, iter = ni, chains = nc, thin = nt) )
rstan::traceplot(out10.3.4)                    # not shown
print(out10.3.4, dig = 3)                      # not shown
stan_est <- summary(out10.3.4)$summary[1:4,1]
```

10.3.5 Do-it-yourself maximum likelihood estimates

We will use the same list of data in this section.

```
str(dataList)  # not shown
```

Now that we have a random-effects factor involved, our likelihood math becomes more complicated. We'll set that extra complexity aside for a moment, and consider the likelihood for a single population j without random intercepts. This looks similar to our likelihood from Section 9.7. The total likelihood for the data in a single population is the product of the K likelihoods for each individual measurement k. Note that the input data *mass* and *lengthN* are both vectors of length K.

$$L(\alpha, \beta, \sigma_{res}|\textbf{mass}, \textbf{lengthN}) = \prod_{k=1}^{K} \text{dnorm}(mass_k|\mu_k, \sigma_{res})$$

$$\mu_k = \alpha + \beta \cdot lengthN_k$$

Now we add back the population-level random intercept, and write the likelihood for just the K data points from a single population j. The intercept α_j is now indexed by j, and is assumed to come from a normal distribution with hyperparameters μ_α and σ_α. To incorporate the random intercept into the likelihood, we multiply our equation above by the likelihood of the random intercept given particular hyperparameter values, equal to $\text{dnorm}(\alpha_j|\mu_\alpha, \sigma_\alpha)$.

$$L(\alpha_j, \mu_\alpha, \sigma_\alpha, \beta, \sigma_{res}|\textbf{mass}_j, \textbf{lengthN}_j) = \prod_{k=1}^{K} \left[\text{dnorm}(mass_{j,k}|\mu_{j,k}, \sigma_{res}) \right] \cdot \text{dnorm}(\alpha_j|\mu_\alpha, \sigma_\alpha)$$

$$\mu_{j,k} = \alpha_j + \beta \cdot lengthN_{j,k}$$

Writing this as an R function (called `f()`) we get:

```
f <- function(alpha.j, mu.alpha, sigma.alpha, mu.beta, sigma.residual,
              mass.j, lengthN.j){
  mu <- alpha.j + mu.beta * lengthN.j
  prod(dnorm(mass.j, mu, sigma.residual)) *
      dnorm(alpha.j, mu.alpha, sigma.alpha)
}
```

As we will see later, we need this function `f()` to be *vectorized*; that is, we need to be able to take a vector of possible `alpha.j` values in its first argument, and return a corresponding vector of likelihood values. To enable this we make some small adjustments to the function:

```
f <- function(alpha.j, mu.alpha, sigma.alpha, mu.beta, sigma.residual,
               mass.j, lengthN.j){
  out <- numeric(length(alpha.j))
  for (i in 1:length(alpha.j)){
    mu <- alpha.j[i] + mu.beta * lengthN.j
    out[i] <- prod(dnorm(mass.j, mu, sigma.residual)) *
      dnorm(alpha.j[i], mu.alpha, sigma.alpha)
  }
  out
}
```

Next, we need to integrate our function f() over the random intercepts α_j, leaving us with the *integrated or marginal likelihood* which depends only on μ_α, β, σ_α, and σ_{res}. We'll call the resulting function g().

$$g\left(\mu_\alpha, \sigma_\alpha, \beta, \sigma_{res}, \textbf{\textit{mass}}_j, \textbf{\textit{lengthN}}_j\right) = \int f(\alpha_j, \mu_\alpha, \sigma_\alpha, \beta, \sigma_{res}, \textbf{\textit{mass}}_j, \textbf{\textit{lengthN}}_j)d\alpha_j$$

We can use the built-in integrate() function in R to do this integration for us. This function assumes that our input function f() can operate on a vector of possible values for α_j, which is why we added that capability earlier. In theory, we should integrate over every possible value for α_j from $-\infty$ to ∞, but in practice this would make optimization very hard. Instead we'll integrate only over a reasonable range of possible population-level intercept values by chosing a range from 100 to 400. Here's the resulting R function g(). As a reminder, this represents the likelihood contribution for all K_j snakes in a single population j.

```
g <- function(mu.alpha, mu.beta, sigma.alpha, sigma.residual,
              mass.j, lengthN.j){
  integrate(f, lower = 100, upper = 400, mu.alpha = mu.alpha,
            sigma.alpha = sigma.alpha,
            mu.beta = mu.beta, sigma.residual = sigma.residual,
            mass.j = mass.j, lengthN.j = lengthN.j)$value
}
```

Finally, we need to write our negative log-likelihood function for all J populations, which is just a sum over the log of function g() evaluated for each population.

```
# Definition of NLL for random-intercepts model with Gaussian errors
NLL <- function(pars, data){
  mu.alpha <- pars[1]
  mu.beta <- pars[2]
  sigma.alpha <- exp(pars[3])
  sigma.residual <- exp(pars[4])
  LL <- 0                                        # initialize LL at 0

  for (j in 1:data$nPops){
    mass.j <- data$mass[data$pop == j]
    lengthN.j <- data$lengthN[data$pop == j]
    L <- g(mu.alpha, mu.beta, sigma.alpha, sigma.residual,
           mass.j, lengthN.j)                    # likelihood for pop j
    LL <- LL + log(L)
  }

  return(-LL)
}
```

This function takes as arguments a vector of parameter values (which includes μ_α, β, σ_α, and σ_{res}) as well as our data list. For each of the J populations, it runs our function g() on the subset of data corresponding to population j, takes the log of the result, and adds it to the total, yielding the total log-likelihood for the entire data set. At the end, we return the negative of the total log-likelihood.

We can now optimize this function for our data set to get our parameter estimates. We need to provide some reasonable initial values as well.

```
# Minimize that NLL to find MLEs and get SEs and CIs
inits <- c(mu.alpha = mean(dataList$mass), mu.beta = 0, log.sigma.alpha = 1,
          log.sigma.residual = log(sd(dataList$mass)))
out10.3.5 <- optim(inits, NLL, hessian = TRUE, data = dataList)
get_MLE(out10.3.5, 4)
```

	MLE	ASE	LCL.95	UCL.95
mu.alpha	265.119	3.1093	259.024	271.213
mu.beta	61.196	1.8680	57.535	64.857
log.sigma.alpha	2.943	0.1430	2.663	3.224
log.sigma.residual	3.751	0.0315	3.689	3.813

This looks good (and don't forget that the dispersion parameters are estimated on the log SD scale).

```
# Save MLEs
diy_est <- out10.3.5$par
diy_est[3:4] <- exp(diy_est[3:4])   # Get SD params on ilog scale
```

10.3.6 Likelihood analysis with TMB

To fit the model in TMB, we need to make the data conform to the C++ custom of starting indices at 0.

```
# Bundle and summarize data
tmbData <- dataList
tmbData$pop <- tmbData$pop-1   # convert pop index to 0-based
str(tmbData)

List of 5
 $ mass   : Named num [1:560] 278 193 254 306 211 ...
   ..- attr(*, "names") = chr [1:560] "1" "2" "3" "4" ...
 $ pop    : num [1:560] 0 0 0 0 0 0 0 0 0 0 ...
 $ lengthN: num [1:560] 0.0216 -0.6689 -0.2556 0.6603 -1.4315 ...
 $ nPops  : num 56
 $ n      : num 560
```

An advantage of TMB is that we do not have to explicitly integrate out the random effects, that is, the population-level intercepts in our model. Instead, we can define the normal distribution of these random intercepts in a way that looks similar to our BUGS or Stan code. Later, we will tell TMB that `alpha` is a random effect and TMB will handle the integration automatically.

```
# Write TMB model file
cat(file = "model10_3_6.cpp",
"#include <TMB.hpp>

template<class Type>
Type objective_function<Type>::operator() ()
{
  //Describe input data
  DATA_VECTOR(mass);                    //response
  DATA_VECTOR(lengthN);                 //covariate
  DATA_IVECTOR(pop);                    //population index
  DATA_INTEGER(nPops);                  //Number of populations
  DATA_INTEGER(n);                      //Number of observations

  //Describe parameters
  PARAMETER(mu_alpha);
  PARAMETER(mu_beta);
  PARAMETER(log_sigma_alpha);
  PARAMETER(log_sigma_residual);
  PARAMETER_VECTOR(alpha);

  Type sigma_alpha = exp(log_sigma_alpha);
  Type sigma_residual = exp(log_sigma_residual);

  Type LL = 0.0;                        //Initialize log-likelihood at 0

  //Distribution of random intercepts - add to log-likelihood
  for (int i=0; i<nPops; i++){          //Note index starts at 0 instead of 1!
    LL += dnorm(alpha(i), mu_alpha, sigma_alpha, true);
  }

  for (int i = 0; i<n; i++){
    Type mu = alpha(pop(i)) + mu_beta * lengthN(i);

    //Calculate log-likelihood of observation and add to total
    LL += dnorm(mass(i), mu, sigma_residual, true);
  }

  return -LL;                           //Return negative log likelihood
}
")

# Compile and load TMB function
compile("model10_3_6.cpp")
dyn.load(dynlib("model10_3_6"))

# Provide dimensions and starting values for parameters
params <- list(mu_alpha = 0, mu_beta = 0,
               log_sigma_alpha = 0, log_sigma_residual = 0,
               alpha = rep(0, tmbData$nPops))

# Create TMB object and tell TMB that alpha is random
out10.3.6 <- MakeADFun(data = tmbData, parameters = params,
             random = "alpha", #Identify which params(s) are random
             DLL = "model10_3_6", silent = TRUE)
```

```
# Optimize TMB object and print results
opt <- optim(out10.3.6$par, fn = out10.3.6$fn, gr = out10.3.6$gr,
             method = "BFGS", hessian = TRUE)
(tsum <- tmb_summary(out10.3.6))             # not shown


# Save TMB estimates
tmb_est <- c(opt$par[1:2], exp(opt$par[3:4]))
```

10.3.7 Comparison of the parameter estimates

Here's our grand comparison table for the random-intercepts model.

```
# Compare all results with truth
comp <- cbind(truth = truth[-4], gtmbML = gtmbML_est, JAGS = jags_est,
  NIMBLE = nimble_est, Stan = stan_est, DIY = diy_est, TMB = tmb_est)
print(comp, 4)
```

	truth	gtmbML	JAGS	NIMBLE	Stan	DIY	TMB
mu.alpha	260	265.12	265.07	265.05	265.22	265.12	265.12
mu.beta	60	61.19	61.20	61.22	61.14	61.20	61.21
sigma.alpha	20	18.98	19.22	19.21	19.57	18.98	18.97
residual.sd	30	42.57	42.73	42.71	42.70	42.57	42.57

The estimates numerically agree nicely across all engines.

10.4 Analysis under a random-coefficients model without correlation between intercept and slope

Next, we want to make the assumption that both slopes and intercepts of the mass–length relationship differ among populations in the fashion of two independent normal random variables. That is, we will declare both as random, but we won't estimate as a parameter a correlation between them. Thus, we will analyze the data under the same model that we used to generate our data set. For a variant of this model but with a correlation added as a separate parameter, see Section 10.5.

10.4.1 Likelihood analysis using canned functions in R

Here is how we specify the random-slopes model without a correlation between the intercepts and the slopes.

```
# the || notation means no correlation between random effects
out10.4.ML <- glmmTMB(mass ~ lengthN + (1 + lengthN || pop),
                      data = gtmbData, REML = FALSE)
out10.4.REML <- glmmTMB(mass ~ lengthN + (1+ lengthN || pop),
                        data = gtmbData, REML = TRUE,
                        control = glmmTMBControl(optimizer = optim,
                                      optArgs = list(method = "BFGS")))
summary(out10.4.ML)
summary(out10.4.REML)
```

```
> summary(out10.4.ML)
 Family: gaussian (identity)
Formula:     mass ~ length + (1 + length || pop)
....

Random effects:

Conditional model:
 Groups     Name            Variance   Std.Dev.   Corr
 pop        (Intercept)     336.6      18.35
            lengthN         930.6      30.51      0.00
 Residual                   905.6      30.09
Number of obs: 560, groups: pop, 56

Dispersion estimate for gaussian family (sigma^2): 906

Conditional model:
              Estimate   Std. Error   z value   Pr(>|z|)
(Intercept)   265.638        2.796     95.02     <2e-16 ***
lengthN        62.873        4.306     14.60     <2e-16 ***


> summary(out10.4.REML)
 Family: gaussian (identity)
Formula:     mass ~ length + (1 + length || pop)
....

Random effects:

Conditional model:
 Groups     Name            Variance   Std.Dev.   Corr
 pop        (Intercept)     344.8      18.57
            lengthN         949.5      30.81      0.00
 Residual                   905.5      30.09
Number of obs: 560, groups: pop, 56

Dispersion estimate for gaussian family (sigma^2): 906

Conditional model:
              Estimate   Std. Error   z value   Pr(>|z|)
(Intercept)   265.640        2.822     94.15     <2e-16 ***
lengthN        62.869        4.345     14.47     <2e-16 ***
```

With balanced data we will get very similar estimates from ML and REML.

```
# Save estimates
gtmbML_est <- c(fixef(out10.4.ML)$cond,
                attr(VarCorr(out10.4.ML)$cond$pop, "stddev"),
                sigma(out10.4.ML))
gtmbREML_est <- c(fixef(out10.4.REML)$cond,
                attr(VarCorr(out10.4.REML)$cond$pop, "stddev"),
                sigma(out10.4.REML))

# Assess goodness-of-fit of the model (not shown)
par(mfrow = c(1, 2))   # Distribution of estimates of alpha and beta
hist(coef(out10.4.ML)$cond$pop[,1], main = 'alpha', breaks = 12)
hist(coef(out10.4.ML)$cond$pop[,2], main = 'beta', breaks = 12)
simOut <- simulateResiduals(out10.4.ML, n = 1000, plot = TRUE)
```

10.4.2 **Bayesian analysis with JAGS**

Next, the Bayesian analysis of the random-coefficients model without intercept-slope correlation.

```
# Bundle and summarize data
str(dataList <- list(mass = mass, pop = as.numeric(pop), lengthN = lengthN,
  nPops = nPops, n = n) )

List of 5
 $ mass   : Named num [1:560] 278 193 254 306 211 ...
   ..- attr(*, "names") = chr [1:560] "1" "2" "3" "4" ...
 $ pop    : num [1:560] 1 1 1 1 1 1 1 1 1 1 ...
 $ lengthN: num [1:560] 0.0216 -0.6689 -0.2556 0.6603 -1.4315 ...
 $ nPops  : num 56
 $ n      : num 560


# Write JAGS model file
cat(file = "model10.4.2.txt", "
model {
# Priors
for (i in 1:nPops){
  alpha[i] ~ dnorm(mu.alpha, tau.alpha)        # Random intercepts
  beta[i] ~ dnorm(mu.beta, tau.beta)           # Random slopes
}

mu.alpha ~ dnorm(0, 0.000001)                  # Mean for random intercepts
tau.alpha <- pow(sigma.alpha, -2)
sigma.alpha ~ dt(0, 1, 0.0001)I(0,)            # SD for random intercepts

mu.beta ~ dnorm(0, 0.000001)                   # Mean for random slopes
tau.beta <- pow(sigma.beta, -2)
sigma.beta ~ dt(0, 1, 0.0001)I(0,)             # SD for slopes

tau.residual <- pow(sigma.residual, -2)        # Residual precision
sigma.residual ~ dt(0, 1, 0.0001)I(0,)         # Residual standard deviation

# Likelihood
for (i in 1:n) {
  mass[i] ~ dnorm(mu[i], tau.residual)
  mu[i] <- alpha[pop[i]] + beta[pop[i]] * lengthN[i]
}}
")

# Function to generate starting values
inits <- function(){ list(mu.alpha = rnorm(1, 0, 1),
  sigma.alpha = rlnorm(1), mu.beta = rnorm(1, 0, 1),
  sigma.beta = rlnorm(1), sigma.residual = rlnorm(1)) }

# Parameters to estimate
params <- c("mu.alpha", "mu.beta", "sigma.alpha", "sigma.beta",
            "sigma.residual", "alpha", "beta")
```

```
# MCMC settings
na <- 1000  ;  ni <- 3000  ;  nb <- 1000  ;  nc <- 4  ;  nt <- 1

# Call JAGS (ART <1 min), check convergence, summarize posteriors and save estimates
out10.4.2 <- jags(dataList, inits, params, "model10.4.2.txt", n.iter = ni, n.burnin = nb,
    n.chains = nc, n.thin = nt, n.adapt = na, parallel = TRUE)
jagsUI::traceplot(out10.4.2)              # not shown
print(out10.4.2, 2)
jags_est <- out10.4.2$summary[c(1:5),1]
```

	mean	sd	2.5%	50%	97.5%	overlap0	f	Rhat	n.eff
mu.alpha	265.60	2.87	259.87	265.61	271.35	FALSE	1.00	1	8000
mu.beta	62.88	4.40	54.05	62.91	71.56	FALSE	1.00	1	6718
sigma.alpha	18.69	2.40	14.45	18.54	23.80	FALSE	1.00	1	3173
sigma.beta	31.25	3.48	25.26	31.01	38.73	FALSE	1.00	1	1581
sigma.residual	30.18	1.00	28.28	30.17	32.19	FALSE	1.00	1	4224
alpha[1]	272.31	9.68	253.36	272.36	291.05	FALSE	1.00	1	6833
alpha[2]	257.00	8.67	239.92	257.06	273.78	FALSE	1.00	1	4636
alpha[3]	235.45	9.16	217.77	235.47	253.66	FALSE	1.00	1	8000
alpha[56]	266.44	8.66	249.58	266.47	283.40	FALSE	1.00	1	8000
beta[1]	46.63	13.62	19.70	46.79	73.01	FALSE	1.00	1	2278
beta[2]	66.96	11.12	45.17	66.80	88.96	FALSE	1.00	1	6528
beta[3]	88.48	11.61	65.45	88.51	111.05	FALSE	1.00	1	8000

......

```
# Compare likelihood with Bayesian estimates and with truth
comp <- cbind(truth = truth, gtmbML = gtmbML_est, JAGS = jags_est)
print(comp, 4)
```

	truth	gtmbML	JAGS
mu.alpha	260	265.64	265.60
mu.beta	60	62.87	62.88
sigma.alpha	20	18.35	18.69
sigma.beta	30	30.51	31.25
residual.sd	30	30.09	30.18

The two sets of estimates agree fairly nicely, as do the solutions obtained by `glmmTMB()` and the input values. We emphasize again that using simulated data and successfully recovering the input values may give one the confidence that the analysis in JAGS has been specified correctly. Finally, we note that we can compare the estimates of the random intercepts (`alpha`) and slopes (`beta`) from JAGS found in the output above to equivalent estimates from `glmmTMB`, which defines the random effects as draws from a zero-mean normal distribution (see Section 3.6.3). In `glmmTMB`, the random intercepts and slopes can be obtained with the `coef` function, for example, `coef(out10.4.ML)`.

10.4.3 Bayesian analysis with NIMBLE

NIMBLE code for the model is available on the book website.

10.4.4 **Bayesian analysis with Stan**

```
# Summarize data set again
str(dataList)   # not shown

# Write Stan model
cat(file = "model10_4_4.stan", "

data {
  int n;                              //Number of observations
  int nPops;                          //Number of populations
  vector[n] mass;                     //Response variable
  vector[n] lengthN;                  //Covariate
  array[n] int pop;                   //Population assignment of each obs
}

parameters {
  real mu_alpha;
  real mu_beta;
  real<lower=0> sigma_alpha;
  real<lower=0> sigma_beta;
  real<lower=0> sigma_residual;
  vector[nPops] alpha;
  vector[nPops] beta;
}

model {

  vector[n] mu;                       //Expected value

  //Priors
  mu_alpha ~ normal(0, 1000);
  mu_beta ~ normal(0, 1000);
  sigma_alpha ~ cauchy(0, 100);
  sigma_beta ~ cauchy(0, 100);

  for (i in 1:nPops){
    alpha[i] ~ normal(mu_alpha, sigma_alpha);
    beta[i] ~ normal(mu_beta, sigma_beta);
  }

  //Likelihood
  for (i in 1:n){
    mu[i] = alpha[pop[i]] + beta[pop[i]] * lengthN[i];
    mass[i] ~ normal(mu[i], sigma_residual);
  }
}
")

# Initial values
inits <- function(){ list(mu_alpha = rnorm(1, 0, 1),
  sigma_alpha = rlnorm(1), mu_beta = rnorm(1, 0, 1),
  sigma_beta = rlnorm(1), sigma_residual = rlnorm(1)) }

# HMC settings
ni <- 2000   ;   nb <- 1000   ;   nc <- 4   ;   nt <- 1
```

```
# Call STAN (ART 50 sec/26 sec)
system.time(
out10.4.4 <- stan(file = "model10_4_4.stan", data = dataList, init = inits,
   warmup = nb, iter = ni, chains = nc, thin = nt)
)
rstan::traceplot(out10.4.4)               # not shown
print(out10.4.4, dig = 3)                 # not shown
stan_est <- summary(out10.4.4)$summary[1:5,1]
```

10.4.5 Do-it-yourself maximum likelihood estimates

Our approach here will be very similar to that in Section 10.3.5, with the addition of random slopes β_j and associated hyperparameters μ_β and σ_β. The math for function $f()$ looks like the following, now also multiplying by the likelihood for the random slopes β_j:

$$f\left(\alpha_j, \beta_j, \mu_\alpha, \sigma_\alpha, \mu_\beta, \sigma_\beta, \sigma_{res}, \textbf{\textit{mass}}_j, \textbf{\textit{lengthN}}_j\right) =$$

$$\prod_{k=1}^{K}\left[\text{dnorm}\left(mass_{j,k}|\mu_{j,k}, \sigma_{res}\right)\right] \cdot \text{dnorm}(\alpha_j|\mu_\alpha, \sigma_\alpha) \cdot \text{dnorm}(\beta_j|\mu_\beta, \sigma_\beta)$$

$$\mu_{j,k} = \alpha_j + \beta_j \cdot lengthN_{j,k}$$

Remember that $f()$ calculates the likelihood for a single population j. The corresponding R function could be written as follows:

```
f <- function(alpha.j, beta.j, mu.alpha, sigma.alpha, mu.beta,
              sigma.beta, sigma.residual, mass.j, lengthN.j){
  mu <- alpha.j + beta.j * lengthN.j
  prod(dnorm(mass.j, mu, sigma.residual)) *
      dnorm(alpha.j, mu.alpha, sigma.alpha) *
      dnorm(beta.j, mu.beta, sigma.beta)
}
```

However, this function is not vectorized, and thus is very slow during optimization. Here's the vectorized version, which allows alpha.j and beta.j to be provided as matrices:

```
f <- function(alpha.j, beta.j, mu.alpha, sigma.alpha, mu.beta,
    sigma.beta, sigma.residual, mass.j, lengthN.j){

  matdim <- dim(alpha.j)
  alpha.j <- as.vector(alpha.j)
  beta.j <- as.vector(beta.j)

  par.mat <- cbind(alpha.j, beta.j)
  Xmat <- cbind(1, lengthN.j)
  mu <- Xmat %*% t(par.mat)
  mass.j.mat <- matrix(mass.j, nrow=nrow(mu), ncol=ncol(mu))
  L <- dnorm(mass.j.mat, mu, sigma.residual)
  L <- apply(L, 2, prod)
  L <- L * dnorm(alpha.j, mu.alpha, sigma.alpha) *
    dnorm(beta.j, mu.beta, sigma.beta)

  matrix(L, nrow = matdim[1], ncol = matdim[2])
}
```

As we now have two sets of random effects, we need to do a double integration in order to integrate them both out.

$$g\left(\mu_\alpha, \sigma_\alpha, \mu_\beta, \sigma_\beta, \sigma_{res}, \boldsymbol{mass_j}, \boldsymbol{lengthN_j}\right) = \iint f(\alpha_j, \beta_j, \ldots) d\alpha_j d\beta_j$$

In the R version of function g(), we'll use the integral2 function in the pracma library to do the double integration. We need to once again provide reasonable limits on both integrations. We also add some code (tryCatch) to gracefully handle errors.

```
library(pracma)
g <- function(mu.alpha, mu.beta, sigma.alpha, sigma.beta, sigma.residual, mass.j, lengthN.j){
  tryCatch({
  integral2(fun = f, xmin = 150, xmax = 400, ymin = 0, ymax = 150,
      mu.alpha = mu.alpha, sigma.alpha = sigma.alpha, mu.beta = mu.beta,
      sigma.beta = sigma.beta, sigma.residual = sigma.residual,
      mass.j = mass.j, lengthN.j = lengthN.j)$Q
  }, error = function(e) return(Inf))
}
```

Finally, we write our complete negative log-likelihood function, which is nearly identical to the function in Section 10.3.5. The only difference is we now estimate a parameter sigma.beta, which is the standard deviation for the random slopes.

```
# Definition of NLL for random-slopes model with Gaussian errors
NLL <- function(pars, data){
  mu.alpha <- pars[1]
  mu.beta <- pars[2]
  sigma.alpha <- exp(pars[3])
  sigma.beta <- exp(pars[4])
  sigma.residual <- exp(pars[5])

  nll <- 0                                          # initialize nll at 0

  for (j in 1:data$nPops){
    mass.j <- data$mass[data$pop == j]
    lengthN.j <- data$lengthN[data$pop == j]
    L <- g(mu.alpha, mu.beta, sigma.alpha, sigma.beta, sigma.residual,
          mass.j, lengthN.j)                        # likelihood for pop j
    nll <- nll - log(L)
  }

  return(nll)

}
```

As in Section 10.3.5, we must provide a set of reasonable starting values to the optimization.

```
inits <- c(mu.alpha = mean(dataList$mass), mu.beta = 0,
          log.sigma.alpha = 1, log.sigma.beta = 1,
          log.sigma.residual = log(sd(dataList$mass)))
```

```
# Optimization (ART 206 sec) and calculation of ASEs and CIs
system.time(
out10.4.5 <- optim(inits, NLL, method = "BFGS", hessian = TRUE, data = dataList,
  control = list(trace = 1, REPORT = 5)) )
get_MLE(out10.4.5, 4)
```

	MLE	ASE	LCL.95	UCL.95
mu.alpha	265.453	2.78694	259.991	270.915
mu.beta	62.772	4.25622	54.430	71.114
log.sigma.alpha	2.906	0.12584	2.659	3.152
log.sigma.beta	3.407	0.10491	3.201	3.612
log.sigma.residual	3.405	0.03975	3.327	3.483

```
# Save the MLEs
diy_est <- out10.4.5$par
diy_est[3:5] <- exp(diy_est[3:5])   # Get SD params on ilog scale
```

We are happy about these homegrown MLEs for another mixed model, but we also start to feel the computational costs of the likelihood maximization, which takes over 3 minutes.

10.4.6 Likelihood analysis with TMB

To fit the model in TMB, we again need to make the data conform to the C++ custom of starting indices at 0.

```
# Bundle and summarize data
tmbData <- dataList
tmbData$pop <- tmbData$pop-1   # convert pop index to 0-based
```

The model looks very similar to the one in Section 10.3.6. We just need to add a new specification of the distribution for the random slopes. As before TMB handles the integrations for us.

```
# Write TMB model file
cat(file = "model10_4_6.cpp",
"#include <TMB.hpp>

template<class Type>
Type objective_function<Type>::operator() ()
{
  //Describe input data
  DATA_VECTOR(mass);                        //response
  DATA_VECTOR(lengthN);                     //covariate
  DATA_IVECTOR(pop);                        //population index
  DATA_INTEGER(nPops);                      //Number of populations
  DATA_INTEGER(n);                          //Number of observations

  //Describe parameters
  PARAMETER(mu_alpha);
  PARAMETER(mu_beta);
  PARAMETER(log_sigma_alpha);
  PARAMETER(log_sigma_beta);
  PARAMETER(log_sigma_residual);
  PARAMETER_VECTOR(alpha);
  PARAMETER_VECTOR(beta);

  Type sigma_alpha = exp(log_sigma_alpha);
  Type sigma_beta = exp(log_sigma_beta);
  Type sigma_residual = exp(log_sigma_residual);

  Type LL = 0.0;                            //Initialize log-likelihood at 0

  //Distribution of random intercepts and random slopes
  for (int i=0; i<nPops; i++){              //Note index starts at 0 instead of 1!
    LL += dnorm(alpha(i), mu_alpha, sigma_alpha, true);
    LL += dnorm(beta(i), mu_beta, sigma_beta, true);
  }

  for (int i=0; i<n; i++){
    Type mu = alpha(pop(i)) + beta(pop(i)) * lengthN(i);
    //Calculate log-likelihood of observation and add to total
    LL += dnorm(mass(i), mu, sigma_residual, true);
  }

  return -LL;                               //Return negative log likelihood
}
")

# Compile and load TMB function
compile("model10_4_6.cpp")
dyn.load(dynlib("model10_4_6"))

# Provide dimensions and starting values for parameters
params <- list(mu_alpha = 0, mu_beta = 0, log_sigma_alpha = 0,
               log_sigma_beta = 0, log_sigma_residual = 0,
               alpha = rep(0, tmbData$nPops),
               beta = rep(0, tmbData$nPops))

# Create TMB object
out10.4.6 <- MakeADFun(data = tmbData, parameters = params,
  random = c("alpha", "beta"),     #Identify which params(s) are random
  DLL = "model10_4_6", silent = TRUE)
```

```
# Optimize TMB object and print and save results
opt <- optim(out10.4.6$par, fn = out10.4.6$fn, gr = out10.4.6$gr,
             method = "L-BFGS-B", hessian = TRUE)
(tsum <- tmb_summary(out10.4.6))    # not shown
tmb_est <- c(opt$par[1:2], exp(opt$par[3:5]))
```

10.4.7 Comparison of the parameter estimates

Here's our usual comparison table for the model that has both random intercepts and random slopes, but no correlation between these parameters.

```
# Compare all results with truth
comp <- cbind(truth = truth, gtmbML = gtmbML_est, JAGS = jags_est,
   NIMBLE = nimble_est, Stan = stan_est, DIY = diy_est, TMB = tmb_est)
print(comp, 4)
```

	truth	gtmbML	JAGS	NIMBLE	Stan	DIY	TMB
mu.alpha	260	265.64	265.60	265.65	265.53	265.45	265.64
mu.beta	60	62.87	62.88	62.82	62.85	62.77	62.88
sigma.alpha	20	18.35	18.69	18.63	19.00	18.27	18.35
sigma.beta	30	30.51	31.25	31.21	31.50	30.16	30.51
residual.sd	30	30.09	30.18	30.15	30.20	30.12	30.09

As so often, we see very similar estimates, although there are small differences in estimates of the SD parameter for the population-specific slopes sigma.beta.

10.5 The random-coefficients model with correlation between intercept and slope

10.5.1 Introduction

In a sense, the random-coefficients model with correlation is a simple extension of the previous model. The mass $y_{j,k}$ of snake j in population k is assumed to be described by the following equation:

$$y_{j,k} = \alpha_j + \beta_j \cdot length_{j,k} + \varepsilon_{j,k}$$

The population random intercepts α_j and random slopes β_j are now modeled as coming from a multivariate normal distribution with mean parameter vector μ and variance–covariance (VC) matrix Σ:

$$\mathbf{B}_j \sim MVN(\mu, \Sigma)$$

Here $\mathbf{B}_j$ is a vector containing (α_j, β_j), μ is a vector containing (μ_α, μ_β), and the VC matrix

$$\Sigma = \begin{pmatrix} \sigma_\alpha^2 & \sigma_{\alpha\beta} \\ \sigma_{\alpha\beta} & \sigma_\beta^2 \end{pmatrix}$$

where σ_α^2 and σ_β^2 are the variances of the intercepts α and slopes β, respectively, and $\sigma_{\alpha\beta}$ is the covariance between α and β. The interpretation of the covariance is such that positive values indicate a steeper mass–length relationship for snakes with a greater mass, while negative values of the covariance are associated with a shallower mass–length relationship for heavier snakes. Finally, the residual error is modeled as a normal distribution with mean 0 and variance σ_{res}^2:

$$\varepsilon_{j,k} \sim Normal(0, \sigma_{res}^2)$$

10.5.2 **Data generation**

We generate data under the random-coefficients model and assume a negative covariance between intercepts and slopes.

```
set.seed(10)
nPops <- 56
nSample <- 10
n <- nPops * nSample
pop <- gl(n = nPops, k = nSample)
```

We generate the covariate `lengthN`, which is normalized length.

```
orig.length <- runif(n, 45, 70)              # Body length (cm)
mn <- mean(orig.length)
sd <- sd(orig.length)
cat("Mean and sd used to normalise original length:", mn, sd, "\n\n")
lengthN <- (orig.length - mn)/sd
hist(lengthN, col = "grey")
```

We build the same design matrix as before.

```
Xmat <- model.matrix(~ pop * lengthN - 1 - lengthN)
print(Xmat[1:21,], dig = 2)   # Print top 21 rows (not shown)
```

We choose the parameter values, i.e., the population-specific intercepts and slopes from a bivariate normal distribution (available in the R package MASS), whose hyperparameters (two means and the four cells of the VC matrix) we need to specify. We use again as residual variation a mean-zero normal distribution with standard deviation of 30.

```
library(MASS)                # Load MASS

mu.alpha <- 260              # Values for five hyperparameters
sigma.alpha <- 20
mu.beta <- 60
sigma.beta <- 30
cov.alpha.beta <- -50        # Covariance
sigma.residual <- 30

mu.vector <- c(mu.alpha, mu.beta)
VC.matrix <- matrix(c(sigma.alpha^2, cov.alpha.beta, cov.alpha.beta, sigma.beta^2),2,2)

ranef.matrix <- mvrnorm(n = nPops, mu = mu.vector, Sigma = VC.matrix)
colnames(ranef.matrix) <- c("alpha", "beta")
head(ranef.matrix)           # Look at what we've created:
# pairs of intercepts and slopes .... THESE are the random effects !
apply(ranef.matrix, 2, mean) # Compare observed statistics with mu.vector above
var(ranef.matrix)            # Compare observed statistics with VC.matrix above

> apply(ranef.matrix, 2, mean)    # Compare with mu.vector above
    alpha      beta
256.85475 67.06664

> var(ranef.matrix)          # Compare with VC.matrix above
          alpha        beta
alpha 461.16402  -91.92475
beta  -91.92475  794.77732
```

Compared with the values of the hyperparamaters that we had picked above, we have some random variation in the realized values of the mean, sd and covariance of `alpha` and `beta`. We assemble the snake measurements $y_{j,k}$.

```
lin.pred <- Xmat[,] %*% as.vector(ranef.matrix)    # Value of lin.predictor
eps <- rnorm(n = n, mean = 0, sd = sigma.residual)  # residuals
mass <- lin.pred + eps                              # response = lin.pred + residual
mass <- drop(mass)                                  # Turn matrix into a vector
hist(mass, col = "grey")                            # Inspect what we've created

# Save true values for comparisons later
truth <- c(mu.alpha = mu.alpha, mu.beta = mu.beta, sigma.alpha = sigma.alpha,
   sigma.beta = sigma.beta, cov.alpha.beta = cov.alpha.beta, sigma.residual = sigma.residual)
```

Again, negative masses are possible for some realizations of the data set (i.e., with a different seed), but this doesn't matter for the statistical demonstration of the model. We look at the simulated data set:

```
library(lattice)                         # not shown
xyplot(mass ~ lengthN | pop, pch = 16, cex = 1, col = rgb(0,0,0,0.5),
   main = 'Mass-length relationships of asp vipers by population')
```

Now, we analyze this second data set allowing for a nonzero covariance between intercept and slope effects and we estimate a parameter for the covariance. That is, in our model we will estimate one additional parameter, which is the correlation between intercepts and slopes.

10.5.3 Likelihood analysis using canned functions in R

In order to estimate the correlation between the random intercepts and slopes in `glmmTMB`, we must make one small change to the formula from Section 10.4.1: instead of two bars (||) in the random effect component of the formula, we use a single bar (|). We again fit the model both using maximum likelihood (ML) and restricted maximum likelihood (REML).

```
gtmbData <- data.frame(mass = mass, pop = pop, lengthN = lengthN)
out10.5.ML <- glmmTMB(mass ~ lengthN + (lengthN | pop),
                    data = gtmbData, REML = FALSE)    # with ML
out10.5.REML <- glmmTMB(mass ~ lengthN + (lengthN | pop),
                    data = gtmbData, REML = TRUE)   # with REML
summary(out10.5.ML)
summary(out10.5.REML)

> summary(out10.5.ML)
 Family:  gaussian (identity)
Formula:   mass ~ length + (length | pop)

...

Random effects:

Conditional model:
 Groups     Name          Variance  Std.Dev.  Corr
 pop        (Intercept)   486.4     22.05
            lengthN       689.0     26.25     -0.16
 Residual                 900.5     30.01
Number of obs: 560, groups: pop, 56

Dispersion estimate for gaussian family (sigma^2): 901

Conditional model:
              Estimate   Std. Error   z value   Pr(>|z|)
(Intercept)   257.855       3.236      79.68    <2e-16 ***
lengthN        66.180       3.771      17.55    <2e-16 ***

> summary(out10.5.REML)
 Family:  gaussian (identity)
Formula:   mass ~ length + (length | pop)

....
```

```
Random effects:

Conditional model:
 Groups    Name          Variance  Std.Dev.  Corr
 pop       (Intercept)   496.8     22.29
           lengthN       703.3     26.52     -0.16
 Residual                900.6     30.01
Number of obs: 560, groups: pop, 56

Dispersion estimate for gaussian family (sigma^2): 901

Conditional model:
            Estimate  Std. Error  z value  Pr(>|z|)
(Intercept)  257.854       3.265    78.97  <2e-16 ***
lengthN       66.176       3.805    17.39  <2e-16 ***

# Save the estimates from ML and REML
vcML <- summary(out10.5.ML)$varcor$cond$pop
gtmbML_est <- c(fixef(out10.5.ML)$cond, attr(vcML, "stddev"), vcML[2], sigma(out10.5.ML))

vcREML <- summary(out10.5.REML)$varcor$cond$pop
gtmbREML_est <- c(fixef(out10.5.REML)$cond, attr(vcREML, "stddev"), vcREML[2], sigma(out10.5.REML))

# Show covariance
gtmbML_est[5]

-94.03955
```

The estimate of the covariance looks a little off, but we ought to assess this discrepancy in the light of our estimation uncertainty associated with that parameter. However, we are not given this information by glmmTMB, so we wait until we get one by Bayesian MCMC, which is next.

10.5.4 Bayesian analysis with JAGS

Here is one way in which to specify a Bayesian analysis of the random-coefficients model with correlation. There is ongoing research about how to best specify the VC matrix; see for instance Gelman & Hill (2007), Kéry & Schaub (2012), and Riecke et al. (2019). In our analysis, we adopt an approach similar to that used by Gelman & Hill (2007, Sections 13.3 and 17.1), who put an inverse Wishart prior on the VC matrix (or equivalently, a Wishart prior on the precision matrix). This approach has the advantage of being easy to specify in JAGS. Alternative, and perhaps better, approaches to specifying the prior on the VC matrix are possible as we will note later.

The Wishart distribution has two parameters: the scale matrix (V) and the degrees of freedom. We use the conventional approach of specifying V in the data as an identity matrix (i.e., a matrix with 1's on the diagonal and 0's for all the off-diagonal elements) with a number of rows and columns equal to the number of correlated parameters (in our case 2). The Wishart degrees of freedom are set to the number of correlated parameters plus 1 (therefore, 3).

```
# Bundle and summarize data
str(dataList <- list(mass = mass, pop = as.numeric(pop), lengthN = lengthN, nPops = nPops,
  n = n, V = diag(2), Wdf = 3) )

List of 7
 $ mass   :  Named num [1:560]  272  179  244  307  188 ...
  ..- attr(*, "names") = chr [1:560]  "1"  "2"  "3"  "4" ...
 $ pop    :  num [1:560]  1 1 1 1 1 1 1 1 1 1 ...
 $ lengthN:  num [1:560]  0.0216  -0.6689  -0.2556  0.6603  -1.4315 ...
 $ nPops  :  num 56
 $ n      :  num 560
 $ V      :  num [1:2, 1:2]  1 0 0 1
 $ Wdf    :  num 3
```

```
# Write JAGS model file
cat(file = "model10.5.4.txt", "
model {
# Priors
mu.alpha ~ dnorm(0, 0.00001)
mu.beta ~ dnorm(0,0.00001)

# Put them into a mean vector
Mu[1] <- mu.alpha
Mu[2] <- mu.beta

# Inverse variance-covariance matrix
Tau[1:2, 1:2] ~ dwish(V[1:2, 1:2], Wdf)

# Get variance-covariance matrix
Sigma <- inverse(Tau)

sigma.alpha <- sqrt(Sigma[1,1])
sigma.beta <- sqrt(Sigma[2,2])
cov.alpha.beta <- Sigma[1,2]

sigma.residual ~ dt(0, 1, 0.0001)I(0, )
tau.residual <- pow(sigma.residual, -2)

for (i in 1:nPops){
  B[i,1:2] ~ dmnorm(Mu[], Tau[,])
  alpha[i] <- B[i,1]
  beta[i] <- B[i,2]
}

# 'Likelihood'
for (i in 1:n){
  mass[i] ~ dnorm(mu[i], tau.residual)
  mu[i] <- alpha[pop[i]] + beta[pop[i]] * lengthN[i]      # Expected value
}}
")

# Function to generate starting values
inits <- function(){ list(mu.alpha = rnorm(1, 0, 1),
  mu.beta = rnorm(1, 0, 1), sigma.residual = rlnorm(1)) }

# Parameters to estimate
params <- c("mu.alpha", "mu.beta", "sigma.alpha", "sigma.beta",
    "cov.alpha.beta", "sigma.residual", "alpha", "beta")

# MCMC settings
na <- 5000  ;  ni <- 10000  ;  nb <- 5000  ;  nc <- 4  ;  nt <- 1

# Call JAGS from R (ART 1 min), check convergence, summarize posteriors and save estimates
out10.5.4 <- jags(dataList, inits, params, "model10.5.4.txt", n.iter = ni, n.burnin = nb,
  n.chains = nc, n.thin = nt, n.adapt = na, parallel = TRUE)
jagsUI::traceplot(out10.5.4)                          # not shown
print(out10.5.4, 2)
jags_est <- out10.5.4$summary[c(1:6),1]
```

	mean	sd	2.5%	50%	97.5%	overlap0	f	Rhat	n.eff
mu.alpha	257.84	3.23	251.57	257.82	264.24	FALSE	1.00	1	20000
mu.beta	66.18	3.80	58.61	66.21	73.69	FALSE	1.00	1	8529
sigma.alpha	21.89	2.54	17.46	21.72	27.35	FALSE	1.00	1	3934
sigma.beta	26.13	2.90	21.02	25.95	32.40	FALSE	1.00	1	20000
cov.alpha.beta	-93.78	92.98	-287.74	-90.96	83.50	TRUE	0.86	1	20000
sigma.residual	30.12	1.02	28.23	30.10	32.17	FALSE	1.00	1	14900
alpha[1]	265.87	9.54	246.81	266.04	284.37	FALSE	1.00	1	18670
alpha[2]	252.45	8.97	234.94	252.48	269.82	FALSE	1.00	1	4661
alpha[3]	247.45	9.26	229.13	247.48	265.54	FALSE	1.00	1	20000
....									

We inspect the results and compare them with the frequentist analysis in Section 10.5.3. We find a fair agreement between the two approaches. We note the very substantial uncertainty in the estimate of the covariance and correspondingly, an extremely wide CRI for this parameter.

```
# Compare likelihood with Bayesian estimates and with truth
comp <- cbind(truth = truth, gtmbML = gtmbML_est, JAGS = jags_est)
print(comp, 4)
```

	truth	gtmbML	JAGS
mu.alpha	260	257.85	257.84
mu.beta	60	66.18	66.18
sigma.alpha	20	22.05	21.89
sigma.beta	30	26.25	26.13
cov.alpha.beta	-50	-94.04	-93.78
sigma.residual	30	30.01	30.12

We see that both estimates of the covariance appear to be quite far off from the target (-50), but we note that they are in fact very close to the empirical covariance between the simulated intercepts and slopes (-91.92; see above). So presumably the discrepancy is mere sampling variability, as we would expect for MLEs, and as we could easily verify by simulation.

10.5.5 Bayesian analysis with NIMBLE

NIMBLE code for the model is available on the book website.

10.5.6 Bayesian analysis with Stan

```
# Summarize data set again
str(dataList)
```

The model code in Stan looks similar to JAGS and NIMBLE, but there are a few important changes to note. First, instead of putting a Wishart prior on the precision matrix, we put an inverse Wishart prior directly on the VC matrix. Second, we define the Wishart scale matrix parameter V and the VC matrix Sigma using a special type: cov_matrix. Finally, we have added a new code block: transformed parameters. In this block we calculate parameters which are functions of other parameters. Our mean vector Mu and the components of the VC matrix fit this definition and so the code to calculate these parameters goes in transformed parameters. We also add code in this section to separate our matrix of random effects B into a vector of intercepts alpha and a vector of slopes beta, to make the output comparable to the other engines.

We used an inverse Wishart prior for the VC matrix in order to make our Stan code directly comparable to JAGS. There are other, and perhaps better, ways of setting a prior on a VC matrix in Stan. See, for example, section 1.13 of the Stan User's Guide.

```
# Write Stan model
cat(file = "model10_5_6.stan", "
data {
  int n;                              //Number of observations
  int nPops;                          //Number of populations
  vector[n] mass;                     //Response variable
  vector[n] lengthN;                  //Covariate
  array[n] int pop;                   //Population assignment of each obs
  cov_matrix[2] V;                    //Wishart scale matrix
  int Wdf;                            //Wishart degrees of freedom
}

parameters {
  real mu_alpha;
  real mu_beta;
  cov_matrix[2] Sigma;
  real<lower = 0> sigma_residual;
  matrix[nPops, 2] B;                 //Matrix of random effects
}

transformed parameters {
  vector[2] Mu;
  real sigma_alpha;
  real sigma_beta;
  real cov_alpha_beta;
  vector[nPops] alpha;
  vector[nPops] beta;

  Mu[1] = mu_alpha;
  Mu[2] = mu_beta;
  sigma_alpha = sqrt(Sigma[1,1]);
  sigma_beta = sqrt(Sigma[2,2]);
  cov_alpha_beta = Sigma[1,2];
  alpha = B[1:nPops,1];
  beta = B[1:nPops,2];
}

model {
  vector[n] mu;

  mu_alpha ~ normal(0, 1000);
  mu_beta ~ normal(0, 1000);
  Sigma ~ inv_wishart(Wdf, V);
  sigma_residual ~ cauchy(0, 100);

  for (i in 1:nPops){
    B[i,1:2] ~ multi_normal(Mu, Sigma);
  }
  for (i in 1:n){
    mu[i] = alpha[pop[i]] + beta[pop[i]] * lengthN[i];
    mass[i] ~ normal(mu[i], sigma_residual);
  }
}
")
```

```
# Function to generate starting values
inits <- function(){ list(mu_alpha = rnorm(1, 0, 1),
  mu_beta = rnorm(1, 0, 1), sigma_residual = rlnorm(1))}

# Parameters monitored: we use hyphens as separators
params <- c("mu_alpha", "mu_beta", "sigma_alpha", "sigma_beta",
            "cov_alpha_beta", "sigma_residual", "alpha", "beta")

# HMC settings
ni <- 2000   ;   nb <- 1000   ;   nc <- 4   ;   nt <- 1

# Call STAN (ART 220 sec/130 sec)
system.time(
out10.5.6 <- stan(file = "model10_5_6.stan", data = dataList,
  pars = params, warmup = nb, iter = ni, chains = nc, thin = nt) )
rstan::traceplot(out10.5.6)            # not shown
print(out10.5.6, dig = 3)              # not shown
stan_est <- summary(out10.5.6)$summary[1:6,1]
```

10.5.7 Do-it-yourself maximum likelihood estimates[1]

For the correlated intercepts and slopes version of the model, we can build off of our math and code from Section 10.4.5. The math for our function f() has the same inputs and now looks like this:

$$f\left(\alpha_j, \beta_j, \mu, \Sigma, \sigma_{res} | mass_j, lengthN_j\right) =$$

$$\prod_{k=1}^{K} \left[\text{dnorm}\left(mass_{j,k} | \mu_{j,k}, \sigma_{res}\right) \right] \cdot \text{dmvnorm}(\mathbf{B}_j | \mu, \Sigma)$$

$$\mu_{j,k} = \alpha_j + \beta_j \cdot lengthN_{j,k}$$

where $\mathbf{B}_j$ is a vector of length 2 containing α_j and β_j, μ is a mean vector of length 2 containing μ_α and μ_β, and Σ is the VC matrix. We have replaced the two separate normal distributions for the random intercepts and slopes with a single bivariate normal distribution. Our corresponding function f() in R uses the dmvnorm function from the mvtnorm package and looks like:

```
library(mvtnorm)
f <- function(alpha.j, beta.j, Mu, Sigma,
              sigma.residual, mass.j, lengthN.j){
  mu <- alpha.j + beta.j * lengthN.j
  B.j <- c(alpha.j, beta.j)
  prod(dnorm(mass.j, mu, sigma.residual)) *
      dmvnorm(B.j, Mu, Sigma)
}
```

However, as in Section 10.3.5, we need to vectorize f() so that alpha.j and beta.j can be provided as matrices of possible values, with matrix output of the same dimensions. Otherwise, our integration, and thus optimization, of this model will be extremely slow. Here's the vectorized version of the function:

---

[1]This section is fairly heavy stuff and can easily be skipped.

```
f <- function(alpha.j, beta.j, Mu, Sigma, sigma.residual, mass.j, lengthN.j){
  matdim <- dim(alpha.j)
  alpha.j <- as.vector(alpha.j)
  beta.j <- as.vector(beta.j)

  par.mat <- cbind(alpha.j, beta.j)
  Xmat <- cbind(1, lengthN.j)
  mu <- Xmat %*% t(par.mat)
  mass.j.mat <- matrix(mass.j, nrow = nrow(mu), ncol = ncol(mu))
  L <- dnorm(mass.j.mat, mu, sigma.residual)
  L <- apply(L, 2, prod)
  L <- L * dmvnorm(par.mat, Mu, Sigma)

  matrix(L, nrow = matdim[1], ncol = matdim[2])
}
```

As in Section 10.4.5, we need to do a double integration to integrate out both the random intercepts and slopes.

$$g\left(\mu, \Sigma, \sigma_{res}, \textit{mass}_j, \textit{lengthN}_j\right) = \iint f(\alpha_j, \beta_j, \ldots)d\alpha_j d\beta_j$$

And in R, we again use `integral2` from library `pracma`:

```
library(pracma)
g <- function(Mu, Sigma, sigma.residual, mass.j, lengthN.j){
  tryCatch({
  integral2(fun = f, xmin = 150, xmax = 400, ymin = 0, ymax = 150,
      Mu = Mu, Sigma = Sigma, sigma.residual = sigma.residual,
      mass.j = mass.j, lengthN.j = lengthN.j)$Q
  }, error = function(e) return(Inf))
}
```

Finally, the negative log-likelihood function. This is similar to the one in Section 10.4.5, except that we have to calculate the parameters μ (the mean vector) and Σ (the VC matrix) from the set of input parameters. The calculation of Σ (`Sigma`) in particular is a little tricky, because we need to make sure the result is a valid VC matrix before passing it to our function `g()`. To convert the input parameter values (which are unconstrained, that is, can take on any value) into a valid VC matrix, we use the log-Cholesky parameterization described by Pinheiro & Bates (1996). This parameterization takes as input a vector of three unconstrained parameter values (θ) and converts them into a valid 2×2 VC matrix. The details of this calculation are beyond the scope of the book, but see Pinheiro & Bates (1996), Section 2, if you are interested.

```
# Definition of NLL
NLL <- function(pars, data){
  mu.alpha <- pars[1]
  mu.beta <- pars[2]
  Beta <- c(mu.alpha, mu.beta)                    # mean vector

  theta <- pars[3:5]
  # Make valid covariance matrix from unconstrained theta vector
  # Log-Cholesky parameterization
  # Sigma = M'M, where M is upper tri matrix with positive diagonal
  M <- diag(exp(theta[1:2]))             # diagonal elements must be positive
  M[1,2] <- theta[3]
  Sigma <- t(M)%*%M

  sigma.residual <- exp(pars[6])

  LL <- 0                                          # initialize LL at 0

  for (j in 1:data$nPops){
    mass.j <- data$mass[data$pop == j]
    lengthN.j <- data$lengthN[data$pop == j]
    L <- g(Beta, Sigma, sigma.residual,
           mass.j, lengthN.j)                       # likelihood for pop j
    LL <- LL + log(L)
  }
  return(-LL)

}
```

As usual, we need reasonable initial values for optimization. We initialize the three θ values used in the log-Cholesky parameterization at values of 1, 1, and 0. In addition, we choose the default Nelder-Mead optimization method of `optim()` rather than BFGS, because it is faster and, with our choice of initial values, obtains what appear to be the correct MLEs, unlike BFGS.

```
# Minimize that NLL to find MLEs and get SEs (ART 420 sec)
inits <- c(mu_alpha = mean(dataList$mass), mu_beta = 0, theta1 = 1, theta2 = 1,
  theta3 = 0, log_sigma_residual = log(sd(dataList$mass)))

system.time(
out10.5.7 <- optim(inits, NLL, hessian = TRUE, data = dataList,
               control = list(trace = 1, REPORT = 5, maxit = 1000)) )
get_MLE(out10.5.7, 4)
```

	MLE	ASE	LCL.95	UCL.95
mu_alpha	257.790	3.24720	251.425	264.154
mu_beta	66.229	3.72578	58.927	73.532
theta1	3.098	0.10241	2.897	3.298
theta2	3.241	0.09914	3.047	3.435
theta3	-4.206	4.01392	-12.073	3.661
log_sigma_residual	3.402	0.03287	3.338	3.467

The optimization returns estimates of θ. In order to convert these to elements of the VC matrix, we need to re-apply the log-Cholesky parameterization described above.

```
# Recover Sigma values from theta via log-Cholesky parameterization
theta <- out10.5.7$par[3:5]
M <- matrix(0, nrow = 2, ncol = 2)
M <- diag(exp(theta[1:2]))
M[1,2] <- theta[3]
(Sigma <- t(M)%*%M)
```

```
            [,1]        [,2]
[1,]  490.52413   -93.15704
[2,]  -93.15704   671.07578
```

We assume that this was pretty rough going in this section for many of you (it was for us). This illustrates the fact that mixed models, especially those with correlations between parameters, are far harder to fit with likelihood methods than are simpler models.

10.5.8 Likelihood analysis with TMB

We make the same adjustments to the data as we did in previous sections to prepare the data for TMB.

```
# Bundle and summarize data
tmbData <- dataList
tmbData$pop <- tmbData$pop-1    # convert pop index to 0-based
str(tmbData)                    # not shown
```

The TMB model code is similar to that in previous engines, with one major difference. For the previous engines, we assumed that the correlated random effects for population j were distributed with mean vector μ and VC matrix Σ; that is, $B_j \sim \text{dmvnorm}(\mu, \Sigma)$. However, in TMB, the available multivariate normal distribution (MVNORM) requires that the mean vector must contain only 0's. Thus, we must change our parameterization so that the random effects are centered on 0 and thus now represent differences from the mean vector μ. In other words,

$$B_j = \mu + E_j$$

$$E_j \sim \text{dmvnorm}([0, 0], \Sigma)$$

This is consistent with the way in which the random effects are parameterized in glmmTMB and lme4. Also, as with our do-it-yourself (DIY) code, we need to apply a log-Cholesky parameterization to the vector of parameters θ to yield a valid VC matrix.

```
# Write TMB model file
cat(file = "model10_5_8.cpp",
"#include <TMB.hpp>

using namespace density;

template<class Type>
Type objective_function<Type>::operator() ()
{
  //Describe input data
  DATA_VECTOR(mass);                    //response
  DATA_VECTOR(lengthN);                 //covariate
  DATA_IVECTOR(pop);                    //population index
  DATA_INTEGER(nPops);                  //Number of populations
  DATA_INTEGER(n);                      //Number of observations

  //Describe parameters
  PARAMETER(mu_alpha);
  PARAMETER(mu_beta);
  PARAMETER_VECTOR(theta);              //Parameters associated with VC matrix
  PARAMETER(log_sigma_residual);
  PARAMETER_MATRIX(B);                  //Random effects matrix

  //Make valid covariance matrix from theta
  //using log-Cholesky parameterization
  matrix<Type> Sigma(2,2);
  matrix<Type> M(2,2);
  M(0,0) = exp(theta(0));
  M(1,1) = exp(theta(1));
  M(0,1) = theta(2);
  M(1,0) = 0;
  Sigma = M.transpose() * M;

  Type sigma_residual = exp(log_sigma_residual);
  Type nll = 0.0;                       //Initialize negative log-likelihood at 0

  //Multivariate normal prior on correlated random slopes and intercepts
  for (int i=0; i<nPops; i++){
    nll += MVNORM(Sigma)(B.row(i));     //MVNORM returns -loglik
  }
  //Add random effects to mean values to get complete intercepts/slopes
  vector<Type> alpha = B.col(0);
  alpha += mu_alpha;
  vector<Type> beta = B.col(1);
  beta += mu_beta;
  for (int i=0; i<n; i++){
    Type mu = alpha(pop(i)) + beta(pop(i)) * lengthN(i);
    //Calculate log-likelihood of observation and add to total
    nll -= dnorm(mass(i), mu, sigma_residual, true);
  }
  ADREPORT(Sigma);
  ADREPORT(alpha); ADREPORT(beta);
  return nll;                           //Return negative log likelihood
}
")
```

```
# Compile and load TMB function
compile("model10_5_8.cpp")
dyn.load(dynlib("model10_5_8"))

# Provide dimensions and starting values for parameters

params <- list(mu_alpha = mean(tmbData$mass), mu_beta = 0,
               theta = c(1, 1, 0),
               log_sigma_residual = log(sd(tmbData$mass)),
               B = matrix(0, nPops, 2))

# Create TMB object
out10.5.8 <- MakeADFun(data = tmbData, parameters = params,
              random = "B", #Identify which params(s) are random
              DLL = "model10_5_8", silent = TRUE)

# Optimize TMB object and print results
opt <- optim(out10.5.8$par, fn = out10.5.8$fn, gr = out10.5.8$gr,
  method = "L-BFGS-B", hessian = TRUE)
(tsum <- tmb_summary(out10.5.8))
```

Finally, we recover the estimated VC matrix from the estimate of θ.

```
# Recover VC matrix and save results
# Recover Sigma values via same parameterization as in the DIY-MLEs
theta <- opt$par[3:5]
M <- matrix(0, nrow=2, ncol = 2)
M <- diag(exp(theta[1:2]))
M[1,2] <- theta[3]
Sigma <- t(M)%*%M

tmb_est <- c(opt$par[1:2], sqrt(diag(Sigma)), Sigma[1,2], exp(opt$par[6]))
```

10.5.9 Comparison of the parameter estimates

Phew .... we're through with one of the most challenging models in the book, both statistically and perhaps even more computationally. Let's look at how good our estimates are and how the engines compare.

```
# Compare all results with truth
comp <- cbind(truth = truth, gtmbML = gtmbML_est, JAGS = jags_est,
  NIMBLE = nimble_est, Stan = stan_est, DIY = diy_est, TMB = tmb_est)
print(comp, 4)
```

	truth	gtmbML	JAGS	NIMBLE	Stan	DIY	TMB
mu.alpha	260	257.85	257.84	257.80	257.86	257.79	257.85
mu.beta	60	66.18	66.18	66.28	66.24	66.23	66.18
sigma.alpha	20	22.05	21.89	21.80	21.83	22.15	22.06
sigma.beta	30	26.25	26.13	26.24	26.08	25.91	26.25
cov.alpha.beta	-50	-94.04	-93.78	-97.34	-96.69	-93.16	-93.97
sigma.residual	30	30.01	30.12	30.12	30.16	30.04	30.01

Overall, we'd say they are pretty good and in addition, all engines give remarkably similar estimates. For this particular data set, the estimate of the covariance is much further off the truth than are the other parameter estimates, but this is nothing to be worried about: covariances are notoriously harder to estimate than parameters in the mean of a model. In addition, we have seen above that the covariance estimate is in fact very close to the sample covariance between the realized intercepts and slopes in our data set.

10.6 Summary and outlook

We have introduced the classic mixed ANCOVA model with random intercepts, random slopes, and the possibility of an intercept-slope covariance. Understanding the material presented in this chapter is essential for a thorough understanding of much of the current mixed modeling in ecology. The ideas presented here appear over and over again, in later chapters of this book, as well as in the applied work of many quantitative ecologists. For instance, you may need the estimation of a covariance matrix in a multivariate normal distribution in widely different settings, including the study of correlations between the temporal variation of juvenile and adult survival rate or the temporal or spatial variation in abundance of multiple and potentially interacting species. However, if you need this "for real," then we would suggest for you to read up some on other parameterizations for covariance matrices (e.g., Gelman & Hill, 2007; Riecke et al., 2019). These may be better, but we chose not to enter in this topic in this introductory book, where getting a conceptual understanding of random effects, and of correlated random effects, is at a premium.

Introduction to the generalized linear model (GLM): comparing two groups in a Poisson regression[⊛]

11

Chapter outline

11.1 Introduction

The unification of a large number of statistical methods for normally distributed data, such as regression, ANOVA, and ANCOVA, under the umbrella of the *general linear model* was a big advancement for applied statistics (Chapter 9). However, even more significant was the unification of an even wider range of statistical methods within the class of the *generalized linear model* or GLM in 1972 by Nelder and Wedderburn (also see McCullagh & Nelder, 1989; Dobson & Barnett, 2018). They showed that a large number of techniques previously thought of as representing quite separate types of analyses, including logistic regression, multinomial regression, log-linear models, as well as the general linear model, were all just special cases of a generalized version of a linear model. In that way, much of what was well understood for the linear model could be carried over to that much larger class of models.

We have introduced GLMs in Chapter 3 already and gave a sneak preview in Chapter 2. There, we have argued that beyond and above their intrinsic value for statistical inferences in many simple analyses, GLMs have an even greater importance as the elementary building blocks for most

---

[⊛]This book has a companion website hosting complementary materials, including all code for download. Visit this URL to access it: https://www.elsevier.com/books-and-journals/book-companion/9780443137150.

Applied Statistical Modelling for Ecologists. DOI: https://doi.org/10.1016/B978-0-443-13715-0.00018-2
295

hierarchical models. Thus, due to their central place in applied statistical modeling, we briefly repeat the salient features of a GLM.

The two main ideas of the GLM are that, first, a *transformation of the expectation* of the response $E(y)$ is expressed as a linear combination of covariate effects, rather than the expected, or mean, response itself. And second, for the random part of the model, *distributions other than the normal* can be chosen, for example, a Poisson or binomial.

Formally, a GLM is described by three components:

1. a *statistical distribution* is used to describe the random variation in the response y; this is the stochastic part of the system description,
2. a so-called *link function g*, which is the transformation applied to the expectation of the response $E(y)$, and
3. a *linear predictor η* ("eta"), which is a linear combination of covariate effects that are thought to make up $g(E(y))$; this is the systematic or deterministic part of the system description.

Poisson, Bernoulli/binomial, and normal are probably the three most widely used statistical distributions in a GLM, and that's why we covered them in detail in Chapter 3. The first two are distributions for non-negative integer responses and therefore suitable for counts, including binary data. The normal is the most widely used distribution for continuous responses such as measurements. The three most widely used link functions are the identity, the log, and the logit (= log(odds) = $\log(x/(1-x))$). For various reasons one link function is typically advantageous, though not obligate, for each of these distributions. For instance, the normal distribution combined with an identity link yields the normal linear model, the Poisson with a log link yields a log-linear model, and the binomial or Bernoulli with a logit link yields a logistic regression. Hence, all the normal linear models we covered in Chapters 4–9 are simply special cases of a GLM.

In the next seven chapters, we will go through a progression from simple to more complex models for Poisson and binomial responses. We skip the "model of the mean" (although we did in fact encounter it in Section 2.5) and instead begin with a comparison of two groups. To better see the analogy with the normal linear model, we start by writing the model for the normal two-group comparison (see Chapter 5) in the classical GLM format:

1. Distribution: $\quad\quad\quad\quad\quad\quad\quad\quad\quad y_i \sim Normal(\mu_i, \sigma^2)$
2. Link function: identity, that is, $\quad \eta_i = \mu_i = E(y_i)$
3. Linear predictor (η): $\quad\quad\quad\quad \eta_i = \alpha + \beta x_i$

Next, we generalize this model to count data. We consider counts of Brown hares (Fig. 11.1) in a sample of 30 arable and 30 grassland study areas. We wonder whether hare density depends on land use, which is either arable or grassland (this example is inspired by Zellweger-Fischer et al., 2011).

The typical distribution assumed for animal counts is a Poisson, which applies when counted things are distributed independently and randomly, and samples of equal size are taken. Then, the number of hares counted per study area (y) will be described by a Poisson. The Poisson has a single parameter, the expected count λ, that is often called the intensity and here represents the mean hare density. In contrast to the normal distribution, the Poisson variance is not a free parameter but is equal to the mean λ. For a Poisson-distributed random variable y, we write $y \sim Poisson(\lambda)$.

FIGURE 11.1

Brown hare (*Lepus europaeus*), Germany, 2008 (Photo by Niklaus Zbinden).

If hare density depends on land use, that is, is different in arable and grassland areas, the assumption of a constant mean density across all 60 study areas does not make sense. And presumably we are specifically interested in whether hare density differs between grassland and arable areas, when we chose 30 of each in our study. Therefore, here is a model for hare count y_i in area i, written in the usual way in which a GLM is defined:

1. Distribution: $y_i \sim Poisson(\lambda_i)$
2. Link function: log, that is, $\eta_i = \log(\lambda_i) = \log(E(y_i))$
3. Linear predictor (η): $\eta_i = \alpha + \beta x_i$

In words, hare count y_i in area i is treated as a Poisson random variable with mean $E(y_i) = \lambda_i$. The log-transformation of λ_i is assumed to be a linear function $\alpha + \beta x_i$, where α and β are unknown constants and x_i is the value of an area-specific covariate. If x_i is an indicator for arable areas, then α becomes the mean hare density on the log-scale in grassland areas and β, again on the log-scale, is the difference in mean density between the two land-use types. Thus, x is a factor with two levels, and we refer you to Chapter 3 for different parameterizations of linear models involving a factor.

11.2 An important but often forgotten issue with count data

Whenever we interpret λ as the mean hare *density*, we make the important assumption that *every individual hare is indeed detected*, that is, that detection probability (p) is equal to 1. In addition, we assume that no hare is counted twice. The former assumption is not very likely for hares nor indeed for any wild animal: typically some individuals are overlooked and in some situations, such as when counting animals in flocks, double counting may easily occur (Yoccoz et al., 2001, 2002; Kellner & Swihart, 2014). Alternatively, we may assume that the *proportion of hares overlooked per area is on average the same* in both land-use types (and similarly for rates of double counting). That is, that detection probability is identical across our dimension of comparison, which here is a habitat classification. In that case, counts are considered just an index to absolute density, that is, a measure for *relative density*. What we model as the Poisson parameter λ_i then is in reality the product between absolute hare density and the proportion p of hares seen on a survey. Only by making the assumption that p is identical on average in both land-use types may we really interpret a mean difference between counts in arable and grassland areas as an indication of a true difference in hare density.

There is a huge body of literature and statistical methods that deals with explicit estimation of, and therefore correction for, the specific type of counting measurement errors leading to imperfect detection and false-positive errors in animals and plants. These go under the rubric of capture-recapture, distance-sampling, and occupancy models and are covered in many textbooks, including Buckland et al. (2001, 2004, 2015), Borchers et al. (2002), Williams et al. (2002), Royle & Dorazio (2008), King et al. (2009), Kéry & Schaub (2012), Royle et al. (2014), Kéry & Royle (2016, 2021), MacKenzie et al. (2017) and Seber & Schofield (2019, 2024). For the most part, these methods are outside of the scope of this book, but in Chapters 19 and 19B we do give a brief introduction to a simple occupancy model and a simple *N*-mixture model. They are, respectively, a model for occurrence and for abundance that allow for a formal correction for imperfect detection. There, we will have more to say on these important topics, the distinction between the imperfectly observed true state (presence/absence or abundance) and the observed data, or, between the ecological and the observation processes underlying all ecological field data.

11.3 How to deal with missing values in our data

In this book, we work exclusively with simulated data (except for Sections 2.6 and 2.8), since this has big advantages for learning new statistical methods. However, our simulated data are not very realistic in every respect, for example, they don't have any missing values. In contrast, real data sets will almost always have missing values (called "NAs" by R, for "not available"). Thus, to apply these new methods that you learn here to your real data, you need to know how to deal with NAs. In this chapter, we briefly deal with this for each model-fitting engine considered.

We have to distinguish between NAs in the explanatory variables and in the response. For NAs in the explanatory variables, the simplest but not necessarily best approach is to simply discard cases with at least one NA anywhere. This is the approach taken for instance by `lm()` or `glm()`. However, this may toss out a lot of information, especially in large data sets with many covariates and where NAs may occur in different cases for different covariates. A better approach may then be to estimate the values of the missing covariates and thus to avoid discarding them. This is called *missing-value imputation* and is a large topic in statistics that is beyond the scope of this book (Rubin, 1976; Little & Rubin, 2002). In short, you will typically specify a model for the NAs to estimate their unobserved values, and then propagate the additional uncertainty arising from this estimation into all the other estimated quantities in the model. See Kéry & Schaub (2012: Section 6.4) and Kéry & Royle (2016: Section 5.5) for simple examples of missing value imputation in a Bayesian analysis.

In this chapter, we deal with the easier case where we have missing values in the response. We will see how with our canned functions in R, and with JAGS and NIMBLE, we can just ignore them, since their presence does not affect the inferences from the model. It may slow down the MCMC algorithm for the Bayesian engines, and therefore you may want to instead discard the cases with NA responses (see Kéry & Royle, 2021: Section 4.10.2). In contrast, for Stan, TMB, and our DIY-MLE method, we must always explicitly accommodate NAs in the response. In this chapter, you will learn how to do this.

11.4 Data generation

For now, we simulate and analyze hare counts under the assumption that detectability is perfect. First we need an indicator for land use:

```
set.seed(11)
nSites <- 30
x <- gl(n = 2, k = nSites, labels = c("grassland", "arable"))
n <- 2 * nSites
```

Let the mean hare density in grassland and arable areas be 2 and 5 hares, respectively. Then, $\alpha = \log(2) = 0.69$ and $\log(5) = \alpha + \beta$, thus, $\beta = \log(5) - \log(2) = 0.92$. We build the expected count for each site, λ_i, accordingly:

```
lambda <- exp(0.69 + 0.92*(as.numeric(x)-1))   # x has levels 1 and 2, not 0 and 1

# Save true values of parameters for later comparison
truth <- c(intercept = 0.69, arable = 0.92)
```

We simulate our count data, adding the noise that comes from a Poisson distribution. Then we make three of the resulting count values NA, to simulate the missing data common to most real data sets.

```
y <- rpois(n = n, lambda = lambda)    # Add Poisson noise
y[c(1, 10, 35)] <- NA                 # Make some observations NA
y                                      # Print the counts (not shown)
```

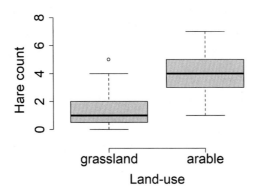

FIGURE 11.2

Relationship between hare count and land-use.

Now we can inspect the hare counts we've generated (Fig. 11.2). Note that when calculating the mean count by habitat using `aggregate()`, we need to tell R to exclude the missing values.

```
aggregate(y, by = list(x), FUN = mean, na.rm = TRUE)   # Observed means
boxplot(y ~ x, col = "grey", xlab = "Land-use",
   ylab = "Hare count", las = 1, frame = FALSE)          # Fig. 11.2
```

Again, you can get a feel for the strong effects of chance (i.e., sampling variation) in this fairly small data set, by repeatedly generating hare counts (without setting a seed) and observing by how much they vary from one sample of 60 counts to another sample of 60 counts (or, 57, to be precise).

```
# Load required libraries
library(ASMbook); library(jagsUI); library(rstan); library(TMB); library(DHARMa)
```

11.5 Likelihood analysis with canned functions in R

We fit the Poisson two-group comparison using the R function `glm(..., family=poisson)`. To test whether mean density in grassland differs from that in arable areas, we can use a test provided by the function `summary()` or a likelihood-ratio test, or analysis of deviance, using `anova()`. There is no big difference here in terms of the inferences. The `glm` function does not technically use the method of maximum likelihood but a variant of least-squares known as iteratively reweighted least squares (McCullagh & Nelder, 1989), which for GLMs produces MLEs. The function handles the missing values in our dataset by automatically removing them and says so in its output. In contrast, many other canned functions such as `lm()` do not print a warning message when missing values are found and removed from the input dataset.

```
out11.5 <- glm(y ~ x, family = poisson)   # Fit the model
summary(out11.5)                          # Two-group comparison
anova(out11.5, test = "Chisq")            # Likelihood-ratio test (LRT)
```

```
Coefficients:
                Estimate   Std. Error   z value   Pr(>|z|)
(Intercept)       0.4290       0.1525     2.813    0.00491 **
xarable           1.0320       0.1768     5.837    5.3e-09 ***
---

...
    Null deviance: 93.119 on 56 degrees of freedom
Residual deviance: 54.157 on 55 degrees of freedom
  (3 observations deleted due to missingness)
[...]

Analysis of Deviance Table
.....
         Df   Deviance Resid.   Df   Resid. Dev   Pr(>Chi)
NULL                             56      93.119
x         1     38.961          55      54.157   4.323e-10 ***

# Save estimates
glm_est <- coef(out11.5)
```

The estimates appear to be far from the truth. Does this mean that there is an error somewhere? No! This is simply the effect of sampling variation, or of chance, which is relatively greater in a small than in a large sample. You could convince yourself that all is fine by repeating data simulation and the analysis with, say, 10,000 sites in each group. Alternatively, you can average out this sampling variation by running a simulation with some large number of replicates; see Section 8.4 for an example.

To assess the adequacy of a GLM, you can again use the plotting function for the fitted model object, which for a GLM works in an analogous way as for a normal linear model (see Chapter 5), except that a different variant of residuals known as Pearson residuals are computed; see the next section. In addition, we can compute quantile residuals using predictive simulations in DHARMa.

```
# Check goodness-of-fit using traditional and quantile residuals
plot(out11.5)              # Traditional residual check for comparison
simOut <- simulateResiduals(out11.5, n = 1000, plot = TRUE)      # From DHARMa
```

11.6 Bayesian analysis with JAGS

Let's now fit the same model in JAGS. To do this, we can take the code from the normal two-group comparison (Chapter 5) and adapt it to the Poisson GLM case. In addition, we will do two more things in the JAGS program below:

1. compute Pearson residuals to assess model fit,
2. conduct a posterior predictive check including calculation of a Bayesian *p*-value (see also Chapters 6, 18 and 19B).

```
# Bundle and summarize data
str(dataList <- list(y = y, x = as.numeric(x)-1, n = length(x)))

List of 3
 $ y : int [1:60] NA 0 2 0 0 5 0 1 4 NA ...
 $ x : num [1:60] 0 0 0 0 0 0 0 0 0 0 ...
 $ n : int 60
```

```
# Write JAGS model file
cat(file = "model11.6.txt", "
model {
# Priors
alpha ~ dnorm(0, 0.0001)
beta ~ dnorm(0, 0.0001)

# Likelihood
for (i in 1:n) {
  y[i] ~ dpois(lambda[i])
  log(lambda[i]) <- alpha + beta *x[i]
  # lambda[i]) <- exp(alpha + beta *x[i])                # Same
}

# Fit assessments
for (i in 1:n) {
  Presi[i] <- (y[i]-lambda[i]) / sqrt(lambda[i])         # Pearson residuals
  y.new[i] ~ dpois(lambda[i])                            # Replicate data set
  Presi.new[i] <- (y.new[i]-lambda[i]) / sqrt(lambda[i]) # Pearson resi
  D[i] <- pow(Presi[i], 2)
  D.new[i] <- pow(Presi.new[i], 2)
}

# Add up discrepancy measures
fit <- sum(D[])
fit.new <- sum(D.new[])
}
")
```

There are a few things to note about this model. First, we are using a log-link function, which is found on the left-hand side of the line of the model that calculates `lambda`. Second, we have not added any code to the JAGS model that specifically handles the missing values in the response `y`. JAGS handles these missing values automatically. When there are missing values in data on the left-hand side of a tilde ~ (such as with `y`), JAGS will simulate possible values for us for that data point based on the model and the available data. A helpful way to think about how this works is to visualize the direction that information is flowing for nonmissing vs. missing data points on the key line `y[i] ~ dpois(lambda[i])`. For example, for `y[2]` (which is not missing), information is flowing from left (the data) to the right (`y[2] -> lambda[2] -> alpha` and `beta`). For `y[1]` (which is missing), JAGS is pushing information from the right (`alpha` and `beta -> lambda[1]-> y[1]`) in order to simulate a value for the missing `y[1]`. However, we don't estimate any new parameters for these missing responses, that is, don't lose any degrees of freedom in the analysis. Instead, we simply simulate from the posterior predictive distribution of the data, which is exactly what we use in a posterior predictive check based on `y.new`.

To emphasize that such missing responses are automatically estimated as part of the model fitting, we will add the response variable `y` to the list of parameters to be saved and thus will be able to see these estimates.

```
# Function to generate starting values
inits <- function(){ list(alpha = rnorm(1), beta = rnorm(1))}

# Parameters to save
params <- c("alpha", "beta", "lambda", "Presi", "fit", "fit.new", "y")
```

```
# MCMC settings
na <- 1000  ;  ni <- 3000  ;  nb <- 1000  ;  nc <- 4  ;  nt <- 1

# Call JAGS (ART <1 min), check convergence and summarize posteriors
out11.6 <- jags(dataList, inits, params, "model11.6.txt", n.iter = ni, n.burnin = nb,
   n.chains = nc, n.thin = nt, n.adapt = na, parallel = TRUE)
jagsUI::traceplot(out11.6)                    # not shown (yet)
print(out11.6, 3)                             # not shown
```

11.6.1 Assessment of model adequacy

For a goodness-of-fit (GOF) assessment of our model, we do something similar to what we did in the normal linear regression example in Chapter 5. That is, we plot the residuals first and then plot the two fit statistics (for the actual data set and for the perfect, or simulated, data sets) against each other and compute the Bayesian p-value as a numerical summary of overall lack of fit. The fit statistic for the new data sets represents the reference distribution for the chosen test statistic, here, the sum of squared Pearson residuals (see also Chapter 18).

For GLMs other than the normal linear model, the variability of the response depends on the mean response. To get residuals with approximately constant variance, Pearson residuals are often computed (for another option, deviance residuals, see McCullagh & Nelder, 1989; Dobson & Barnett, 2018). These are obtained by dividing the raw residuals $(y_i - \bar{y})$ by the standard deviation of y_i; see JAGS code above. Provided that the expected values are not too small, these residuals are approximately standard-normal distributed, although not independent (Dobson & Barnett, 2018: chapter 2). We plot the residuals (Fig. 11.3 left) and conduct a posterior predictive check (PPC) which we plot alongside (Fig. 11.3 right):

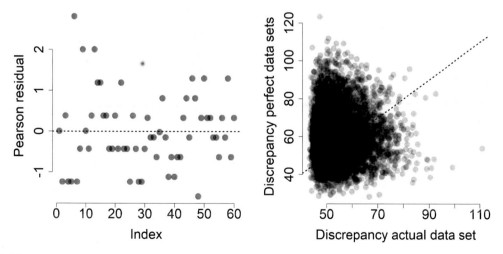

FIGURE 11.3

(left) Pearson residuals for the hare counts. Note that the discrete nature of the integer counts shines through. (right) Graphical posterior predictive check (PPC) based on the sum of squared Pearson residuals.

```
# Fig. 11.3
par(mfrow = c(1, 2), mar = c(5,5,4,2), cex.lab = 1.5, cex.axis = 1.5)
plot(out11.6$mean$Presi, ylab = "Pearson residual", las = 1, pch = 16,
  cex = 1.5, col = rgb(0,0,0,0.5), frame = FALSE)
abline(h = 0)
plot(out11.6$sims.list$fit, out11.6$sims.list$fit.new, main = "",
  xlab = "Discrepancy actual data set", ylab = "Discrepancy perfect data sets",
  pch = 16, cex = 1.5, col = rgb(0,0,0,0.2), frame = FALSE)
abline(0, 1, lwd = 2, col = "black")
```

There is no obvious sign of lack of fit for any particular data point (Fig. 11.3 left) and the graphical posterior predictive check (Fig. 11.3 right) looks good as well, with a similar spread of the points on either side of the 1:1 line. Note that the boundary on the left is caused by the MLE, that is, the sum of the squared Pearson residuals is minimal near the MLE. There is nothing wrong with this, and when doing PPCs for models with more covariates, the posterior cloud will typically be more roundish.

The Bayesian *p*-value (0.650) is the proportion of points above the line. This suggests that the observed data are not unusual in comparison with the "typical" type of data that our model might produce.

```
mean(out11.6$sims.list$fit.new > out11.6$sims.list$fit)
```

```
[1] 0.649875
```

We caution that GOF assessment is a large topic and here we only illustrate two simple techniques which may not always be insightful (see also Conn et al., 2018, and Section 18.5 in this book). In particular, the small size of our data set makes meaningful GOF assessment hard anyway, since it is difficult in such small samples to distinguish structural deficiencies of a model from extreme patterns that arise simply by chance.

11.6.2 Inference under the model

Now that we are convinced that the model is adequate for these data we inspect the estimates and compare them with what we put into the data set, as well as what the frequentist analysis in R tells us.

```
print(out11.6, 2)        # shown partially only
```

	mean	sd	2.5%	50%	97.5%	overlap0	f	Rhat	n.eff
alpha	0.42	0.16	0.10	0.42	0.72	FALSE	0.99	1.00	742
beta	1.04	0.18	0.69	1.04	1.40	FALSE	1.00	1.00	943
lambda[1]	1.53	0.24	1.11	1.52	2.05	FALSE	1.00	1.01	619
lambda[2]	1.53	0.24	1.11	1.52	2.05	FALSE	1.00	1.01	619
[...]									
y[1]	1.54	1.26	0.00	1.00	4.00	TRUE	1.00	1.00	2814
y[2]	0.00	0.00	0.00	0.00	0.00	FALSE	1.00	NA	1
[...]									
y[9]	4.00	0.00	4.00	4.00	4.00	FALSE	1.00	NA	1
y[10]	1.56	1.27	0.00	1.00	4.00	TRUE	1.00	1.00	8000
y[11]	1.00	0.00	1.00	1.00	1.00	FALSE	1.00	NA	1
[...]									
y[34]	3.00	0.00	3.00	3.00	3.00	FALSE	1.00	NA	1
y[35]	4.32	2.17	1.00	4.00	9.00	FALSE	1.00	1.00	8000
y[36]	6.00	0.00	6.00	6.00	6.00	FALSE	1.00	NA	1
[...]									

Remember that the *y* values for areas 1, 10, and 35 were missing from our data set. For these values of *y* in our JAGS output, we get estimates that are similar to the mean values for the group the missing values came from (i.e., grassland or arable). These estimates are given by the posterior predictive distribution of the data (see also Sections 5.4.5 and 18.5). There is quite a bit of uncertainty around these estimates, much more than for the model parameters like `alpha` and `beta`. Notice also that for values of `y` that were *not* missing in the data (such as `y[2]`), JAGS simply returns the value of `y`, and there is no uncertainty, because this value is known.

A comparison of the Bayesian solution with the true values that were used for generating the data set and the solution given by `glm()` shows only a moderate agreement between the two sets of estimates and the truth. However, this is not surprising in view of the small sample size.

```
# Compare likelihood with Bayesian estimates and with truth
jags_est <- unlist(out11.6$mean[1:2])
comp <- cbind(truth = truth, glm = glm_est, JAGS = jags_est)
print(comp, 4)
```

```
           truth    glm     JAGS
intercept   0.69   0.429   0.4159
arable      0.92   1.032   1.0445
```

So, is there a difference in hare density according to land use? Let's look at the posterior distribution of the coefficient for arable (Fig. 11.4 left). The posterior distribution does not overlap zero, so arable sites really do appear to have a different hare density than grassland sites. The same conclusion is arrived at when looking at the 95% credible interval of β in the summary of the analysis above: (0.69–1.40).

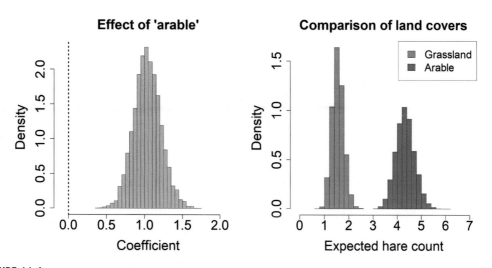

FIGURE 11.4

Posterior distribution of the coefficient of arable (left). Posterior distributions of the expected hare count in grassland and arable areas (right).

Finally, we will form predictions for presentation. Predictions are the expected values of the response under certain conditions, such as for particular covariate values. We have seen earlier that predictions are a valuable means for synthesizing the information that a model extracts from a data set. In a Bayesian analysis, we can base inference about predictions on their posterior distributions. To summarize what we have learned about the differences in hare densities in grassland and arable study areas, we plot the posterior distributions of the effect of arable (Fig. 11.4 left) and of the expected hare counts (λ) for both habitat types (Fig. 11.4 right). We obtain the expected hare counts by exponentiating α and $\alpha + \beta$, respectively.

```
# Fig. 11.4
par(mfrow = c(1, 2), mar = c(5,5,4,3), cex.lab = 1.5,
    cex.axis = 1.5, cex.main = 1.5)
hist(out11.6$sims.list$beta, col = "grey", las = 1,
    xlab = "Coefficient", main = "Effect of 'arable'",
    breaks = 50, freq = FALSE, xlim=c(0,2.2))
abline(v=0, col='red', lty=2, lwd = 3)

hist(exp(out11.6$sims.list$alpha), main= "Comparison of land covers",
    col = "red", xlab = "Expected hare count", xlim = c(0,7),
    breaks = 20, freq = FALSE)
hist(exp(out11.6$sims.list$alpha + out11.6$sims.list$beta),
    col = "blue", breaks = 20, freq = FALSE, add = TRUE)
legend('topright', legend=c('Grassland', 'Arable'),
    fill = c("red", "blue"), cex = 1.5)
```

11.7 Bayesian analysis with NIMBLE

NIMBLE code for the model is available on the book website.

11.8 Bayesian analysis with Stan

Unlike JAGS and NIMBLE, Stan cannot automatically handle missing values in the data. In order to get the model to run, we will need to remove the missing values from y manually before passing them to Stan. We'll also need to remove the corresponding values from the covariate x, and make sure our sample size n refers to the non-missing values only.

```
# Bundle and summarize data
not_na <- which(!is.na(y))
x_stan <- as.numeric(x) - 1
x_stan <- x_stan[not_na]
str(dataList <- list(y = y[not_na], x = x_stan, n = length(x_stan)))

List of 3
 $ y: int [1:57] 0 2 0 0 5 0 1 4 1 2 ...
 $ x: num [1:57] 0 0 0 0 0 0 0 0 0 0 ...
 $ n: int 57
```

The Stan model is otherwise similar to our JAGS and NIMBLE models, with one exception. Instead of having the link function (`log`) on the left-hand side of the `lambda` calculation, as with BUGS, we put the inverse link function (`exp`) on the right-hand side of the calculation. These two approaches are mathematically equivalent, but Stan requires the latter format.

```
# Write Stan model
cat(file = "model11_8.stan", "
data {
  int n;
  array[n] int y;
  vector[n] x;
}

parameters {
  real alpha;
  real beta;
}

transformed parameters {
  vector[n] lambda;
  for (i in 1:n) {
    lambda[i] = exp(alpha + beta * x[i]);
  }
}

model {

  //Priors
  alpha ~ normal(0, 100);
  beta ~ normal(0, 100);

  //Likelihood
  for (i in 1:n) {
    y[i] ~ poisson(lambda[i]);
  }
}

generated quantities {
  vector[n] Presi;
  vector[n] D;
  array[n] int y_new;
  vector[n] Presi_new;
  vector[n] D_new;
  real fit;
  real fit_new;

  for (i in 1:n) {
    Presi[i] = (y[i] - lambda[i])/sqrt(lambda[i]);
    D[i] = pow(Presi[i], 2);
    y_new[i] = poisson_rng(lambda[i]);
    Presi_new[i] = (y_new[i] - lambda[i])/sqrt(lambda[i]);
    D_new[i] = pow(Presi_new[i], 2);
  }

  fit = sum(D);
  fit_new = sum(D_new);
}
")
```

```
# Parameters monitored: same as before
params <- c("alpha", "beta", "lambda", "Presi", "fit", "fit_new")

# HMC settings
ni <- 2000  ;  nb <- 1000  ;  nc <- 4  ;  nt <- 1

# Call STAN (ART 32/3 sec), assess convergence, print estimates and save point estimates
system.time(
out11.8 <- stan(file = "model11_8.stan", data = dataList, pars = params,
            warmup = nb, iter = ni, chains = nc, thin = nt) )
rstan::traceplot(out11.8)                          # not shown
print(out11.8, dig = 3)                            # not shown
stan_est <- summary(out11.8)$summary[1:2,1]
```

11.9 Do-it-yourself maximum likelihood estimates

The likelihood function for a Poisson GLM is conceptually similar to the likelihood for the normal linear model we presented in Section 6.2.6—we just apply the Poisson density instead of the normal density. The equation for the Poisson probability density for a single data point is

$$f(y_i|\lambda) = \frac{\lambda^{y_i} e^{-\lambda}}{y_i!}$$

Converted to a likelihood function, and putting covariates on λ with an exponential inverse-link function, the equation becomes

$$L(\alpha, \beta|\boldsymbol{y},\boldsymbol{x}) = \prod_{i=1}^{n} \frac{\lambda_i^{y_i} e^{-\lambda_i}}{y_i!}$$

$$\lambda_i = \exp(\alpha + \beta x_i)$$

Our DIY likelihood function must perform three tasks: (1) calculate the expected abundance lambda for each of the 60 sites (using the log link function), (2) calculate the log-likelihood of the 60 observed counts using the values of lambda and the dpois() R function, and (3) return the negative of the sum of the log-likelihoods over all counts. For any y that is NA, R will return a corresponding value of NA for the log-likelihood LL. Therefore, we can handle the missing values in y by simply telling R to ignore any NAs when summing up the log-likelihood values over the whole data set (with na.rm = TRUE). Of course, we could also just remove the missing values first as we did with Stan.

```
# Definition of NLL for a Poisson GLM with 1 covariate
NLL <- function(param, y, x) {
  alpha <- param[1]
  beta <- param[2]
  lambda <- exp(alpha + beta * x)         # Calculate lambda for each datum
  LL <- dpois(y, lambda, log = TRUE)      # Log-likelihood for each datum
  -sum(LL, na.rm = TRUE)                   # NLL for all observations
}
```

We now optimize this function as usual using our data.

```
# Minimize that NLL to find MLEs, get SEs and CIs and save estimates
inits <- c('Intercept' = 0, 'beta(arable)' = 0)
out11.9 <- optim(inits, NLL, y = y, x = as.numeric(x)-1, hessian = TRUE)
get_MLE(out11.9, 5)
diy_est <- out11.9$par
```

	MLE	ASE	LCL.95	UCL.95
Intercept	0.42925	0.15248	0.13040	0.72811
beta(arable)	1.03181	0.17677	0.68534	1.37829

11.10 Likelihood analysis with TMB

We use the same list of data as JAGS and NIMBLE, but accommodate the different indexing in C++.

```
# Bundle and summarize data
str(dataList <- list(y = y, x = as.numeric(x)-1, n = length(x)))
```

Our TMB code looks very similar to our DIY likelihood function. As usual we must describe our input data and parameters at the start of the function. In order to handle missing values, we have added a line inside the for loop that tells TMB to skip calculating the log-likelihood (i.e., continue) for values of y that are NA. Alternatively, we could also just remove these values from the input data as we did with Stan.

```cpp
# Write TMB model file
cat(file = "model11_10.cpp",
"#include <TMB.hpp>

template<class Type>
Type objective_function<Type>::operator() ()
{
  //Describe input data
  DATA_VECTOR(y);                              //response
  DATA_VECTOR(x);                              //covariate
  DATA_INTEGER(n);                             //Number of observations

  //Describe parameters
  PARAMETER(alpha);
  PARAMETER(beta);

  Type LL = 0.0;                               //Initialize log-likelihood at 0
  Type lambda;                                 //Initialize lambda

  for (int i=0; i<n; i++){
    if(R_IsNA(asDouble(y(i)))) continue;       // Here accommodate NAs
    lambda = exp(alpha + beta * x(i));         //Expected count

    //Calculate log-likelihood of observation and add to total
    LL += dpois(y(i), lambda, true);
  }

  return -LL;                                  //Return negative log likelihood
}
")
```

```
# Compile and load TMB function
compile("model11_10.cpp")
dyn.load(dynlib("model11_10"))

# Provide dimensions and starting values for parameters
params <- list(alpha = 0, beta = 0)

# Create TMB object
out11.10 <- MakeADFun(data = dataList,
                      parameters = params,
                      DLL = "model11_10", silent = TRUE)

# Optimize TMB object and print results
opt <- optim(out11.10$par, fn = out11.10$fn, gr = out11.10$gr, method = "BFGS")
(tsum <- tmb_summary(out11.10))            # not shown
tmb_est <- c(opt$par[1:2])                 # save estimates
```

11.11 Comparison of the parameter estimates

We compare the parameter estimates produced by our six methods and engines of fitting the same model and find the usual good agreement among the estimates. In contrast, the estimates match the true values only moderately well, as a consequence of the small sample size.

```
# Compare estimates with truth
comp <- cbind(truth = truth, glm = glm_est, JAGS = jags_est,
   NIMBLE = nimble_est, Stan = stan_est, DIY = diy_est, TMB = tmb_est)
print(comp, 4)
```

	truth	glm	JAGS	NIMBLE	Stan	DIY	TMB
intercept	0.69	0.429	0.4159	0.420	0.4147	0.4293	0.429
arable	0.92	1.032	1.0445	1.034	1.0380	1.0318	1.032

11.12 Summary and outlook

We have introduced the generalized linear model, or GLM, for a Poisson response. In a GLM, the covariates are connected to the expectation of a response via a transformation called the link function. The GLM is a key concept that is widely useful in modern applied statistics and in empirical sciences such as ecology. It is useful both in its own right and as a basic building block for hierarchical models, such as most of the models in specialized types of analyses for species distribution, abundance, and demography. In fact, most models covered in books such as Royle & Dorazio (2008), King et al. (2009), Kéry & Schaub (2012), Royle et al. (2014), Kéry & Royle (2016, 2021), or Schaub & Kéry (2022) can be represented simply as a combination of typically two to four GLMs! Thus, a really good grasp of the GLM will be essential if you want to use these models for your data either now or later in your career. We will deepen our understanding of this essential model class in the next six chapters. Furthermore, we will combine the GLM and the concept of random effects to arrive at the most complex models considered in this book, the GLMM in Chapters 14 and 17 and two examples of a general hierarchical model in Chapters 19 and 19B.

We have also examined the practically important topic of how to deal with missing values in the response variable. We have seen that in canned R functions such as `glm()`, but also in the Bayesian engines JAGS and NIMBLE, we can simply ignore them, since they are dealt with in an automatic way (perhaps at some computational costs in JAGS and NIMBLE). However, for our other engines, we have to do some extra work to accommodate missing values. We won't show this in any other chapter in the book, but instead, when you want to adapt a model from the book to your own data which contain NAs, please look up what to do in this chapter.

Overdispersion, zero inflation, and offsets in a Poisson GLM[★]

12

Chapter outline

[★]This book has a companion website hosting complementary materials, including all code for download. Visit this URL to access it: https://www.elsevier.com/books-and-journals/book-companion/9780443137150.

Applied Statistical Modelling for Ecologists. DOI: https://doi.org/10.1016/B978-0-443-13715-0.00011-X

12.1 Introduction

In this chapter, we cover overdispersion (OD), zero inflation, and offsets, three topics that are most typical in Poisson GLMs, although some may also occur in other types of GLMs. OD and offsets are specific to non-normal GLMs. Zero inflation may also occur in normal GLMs and can be viewed as a specific form of OD: there are more zeroes than expected. Here we briefly explain all three in the context of the hare counts example from Chapter 11. To make things simple at the start, we deal with these issues separately. However, in real data analyses they may occur in combination (see Section 12.5).

12.2 Overdispersion

In both of our distributions commonly used to model counts (Poisson and binomial), the dispersion (the variability in the response) is not a free parameter, but rather its magnitude is prescribed and is a function of the mean response. The variance is equal to the mean (λ) for the Poisson and equal to the mean (Np) times $1 - p$ for the binomial distribution (see Chapters 15–17). This means that for a Poisson or binomial random variable the models for the counts come with a "built-in" variability and the magnitude of that variability is assumed to be known. In an analysis of deviance conducted in a classical statistical analysis of the model, the residual deviance of a fitting model will be about the same magnitude as the residual degrees of freedom, that is, the mean deviance ratio (=residual deviance/residual d.f.) is about 1 (McCullagh & Nelder, 1989; Dobson & Barnett, 2018).

However, in real life count data are almost always more variable than expected under the Poisson or binomial models. This is called OD, or extra-Poisson or extrabinomial dispersion. It means that the residual variation is larger than prescribed by a Poisson or binomial distribution. OD often occurs because there are hidden correlations in the data that have not been included in the model, for example, when individuals in family groups are assumed to be independent or when important covariates affecting a response have not been included in a model. When OD is not modeled, tests and confidence intervals will be overconfident (though means won't normally be biased). Therefore, OD should be tested and corrected for when necessary.

Perhaps the simplest way to correct for OD in a classical analysis is by so-called quasi-likelihood (McCullagh & Nelder, 1989; Dobson & Barnett, 2018) and by using the argument `family = quasipoisson` (or `quasibinomial`) in the R function `glm()`. This approach is not explicitly model-based, but simply assumes that the variance is larger by a factor that is estimated, thus making standard errors and confidence intervals appropriately larger. The point estimates are not affected by the choice of quasi-Poisson over a Poisson in `glm()`, as we will see below.

Using our general model-fitting engines such as JAGS, there are several ways in which we can account for OD. One is to specify a distribution that is overdispersed relative to the Poisson, such as the negative binomial (this can be done with a canned variant of the `glm()` function in R package `MASS`). Another solution, and the one we illustrate here, is to add into the linear predictor for the Poisson intensity a normally distributed random effect. Technically, this model is then a Poisson generalized linear mixed model, or Poisson GLMM; see Chapter 14 for a more formal introduction. It is sometimes called a Poisson-lognormal (PLN) model (Millar, 2009).

12.2.1 **Data generation**

We generate a slightly modified hare count data set, where in addition to the land-use difference in mean density, there is also a normally distributed site-specific effect in the linear predictor. For illustrative purposes, we also generate a sister data set without OD.

```
set.seed(122)
nSites <- 50
n <- 2 * nSites
x <- gl(n = 2, k = nSites, labels = c("grassland", "arable"))
eps <- rnorm(2*nSites, mean = 0, sd = 0.5)            # Normal random effect
lambda.OD <- exp(0.69 + (0.92*(as.numeric(x)-1) + eps) )
lambda.Poisson <- exp(0.69 + (0.92*(as.numeric(x)-1)) )   # For comparison

# Save true parameter values
truth <- c(Intercept = 0.69, arable = 0.92, eps_sd = 0.5)
```

We add the noise that comes from a Poisson and inspect the hare counts we've generated (Fig. 12.1):

```
C_OD <- rpois(n = n, lambda = lambda.OD)               # Counts with OD
C_Poisson <- rpois(n = n, lambda = lambda.Poisson)     # Counts without OD

par(mfrow = c(1,2))                                    # Fig. 12.1
boxplot(C_OD ~ x, col = "grey", xlab = "Land-use", main = "With overdispersion",
  ylab = "Hare count", las = 1, ylim = c(0, max(C_OD)), frame = FALSE)
boxplot(C_Poisson ~ x, col = "grey", xlab = "Land-use", main = "Without overdispersion",
  ylab = "Hare count", las = 1, ylim = c(0, max(C_OD)), frame = FALSE )

# Load required libraries
library(ASMbook); library(jagsUI); library(rstan); library(TMB)
```

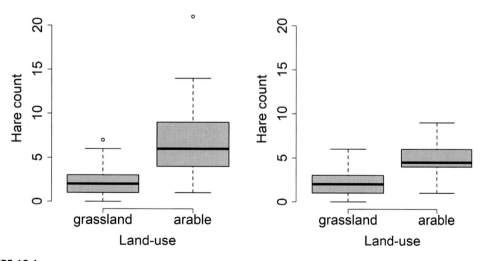

FIGURE 12.1

Two samples of hare counts by land-use with (left) and without (right) overdispersion. Here, we simulated overdispersion by site-specific differences in hare density.

12.2.2 Likelihood analysis with canned functions in R

We start by conducting a classical analysis of the overdispersed data once without and then with correction for OD using `glm(, family = quasi...)`. Note that an alternative might be to fit a linear model with negative binomial distribution for the counts or adoption of a PLN model. We illustrate the latter at the end of this section, since this is also the approach to OD modeling in the other engines.

```
glm.fit.no.OD <- glm(C_OD ~ x, family = poisson)
glm.fit.with.OD <- glm(C_OD ~ x, family = quasipoisson)
summary(glm.fit.no.OD)
glm_est <- c(coef(glm.fit.no.OD), NA)   # Save point estimates


Call:
glm(formula = C.OD ~ x, family = poisson)
...
Coefficients:
              Estimate   Std. Error   z value   Pr(>|z|)
(Intercept)    0.79751      0.09492     8.402    <2e-16 ***
xarable        1.04304      0.11038     9.450    <2e-16 ***
---
Signif. codes: 0 '***' 0.001 '**' 0.01 '*' 0.05 '.' 0.1 ' ' 1

(Dispersion parameter for poisson family taken to be 1)

summary(glm.fit.with.OD)

Coefficients:
              Estimate   Std. Error    t value   Pr(>|t|)
(Intercept)     0.7975       0.1404      5.680    1.38e-07 ***
xarable         1.0430       0.1633      6.388    5.66e-09 ***
---
Signif. codes: 0 '***' 0.001 '**' 0.01 '*' 0.05 '.' 0.1 ' ' 1

(Dispersion parameter for quasipoisson family taken to be 2.188226)
```

We see that the parameter estimates don't change when accounting for OD using the quasi-Poisson, but tests and standard errors do: SEs and p-values become larger when properly accounting for the extra-Poisson dispersion in the data.

For a more explicitly parametric model that accommodates OD, we can fit the PLN model using the R package `PLNmodels` (Chiquet et al., 2018), which is also our approach in our other engines below. For further comparison we save only one of the `glm()` fits, since point estimates are identical.

```
# Canned function for Poisson lognormal
library(PLNmodels)
PLN.fit <- PLN(matrix(C_OD, ncol = 1) ~ x)
PLN_est <- c(coef(PLN.fit), sqrt(sigma(PLN.fit)))
```

In our normal workflow, we now would assess model fit using quantile residual checks with the DHARMa package, but at the time of writing this is not possible with either fitted model object in this section.

12.2.3 Bayesian analysis with JAGS

In JAGS it is easy to move from the simple Poisson model for two groups (with homogeneous variance) in the last chapter to the overdispersed Poisson two-groups model as represented by the Poisson-lognormal model.

```
# Bundle and summarize data
str(dataList <- list(C_OD = C_OD, x = as.numeric(x)-1, n = length(x)))

List of 3
$ C_OD : int [1:100] 4 3 1 2 1 6 1 1 4 2 ...
$ x    : num [1:100] 0 0 0 0 0 0 0 0 0 0 ...
$ n    : int 100

# Write JAGS model file
cat(file = "model12.2.3.txt", "
model {
# Priors
alpha ~ dnorm(0, 0.0001)
beta ~ dnorm(0, 0.0001)
tau <- pow(sigma, -2)
sigma ~ dt(0, 0.1, 1)T(0, )                              # Half-Cauchy

# Likelihood
for (i in 1:n) {
  C_OD[i] ~ dpois(lambda[i])
  log(lambda[i]) <- alpha + beta *x[i] + eps[i]
  eps[i] ~ dnorm(0, tau)
}
}
")

# Function to generate starting values
inits <- function(){ list(alpha = rlnorm(1), beta = rlnorm(1), sigma = rlnorm(1, 0.1))}

# Parameters to estimate
params <- c("alpha", "beta", "sigma")
```

Note that as soon as we start estimating variances (here, of the OD effects eps), we need longer chains.

```
# MCMC settings
na <- 2000  ;   ni <- 12000  ;   nb <- 2000  ;   nc <- 4  ;   nt <- 10

# Call JAGS (ART <1 min), check convergence, summarize and save results
out12.2.3 <- jags(dataList, inits, params, "model12.2.3.txt",
  n.iter = ni, n.burnin = nb, n.chains = nc, n.thin = nt, n.adapt = na, parallel = TRUE)
jagsUI::traceplot(out12.2.3)                             # not shown
print(out12.2.3, 3)
jags_est <- as.numeric(out12.2.3$mean[1:3])             # save estimates
```

	mean	sd	2.5%	50%	97.5%	overlap0	f	Rhat	n.eff
alpha	0.676	0.126	0.421	0.677	0.913	FALSE	1	1.001	1342
beta	1.024	0.156	0.718	1.023	1.330	FALSE	1	1.001	2625
sigma	0.518	0.075	0.379	0.514	0.671	FALSE	1	1.002	1308

```
# Compare likelihood with Bayesian estimates and with truth
comp <- cbind(truth = truth, glm = glm_est, PLN = PLN_est, JAGS = jags_est)
print(comp, 4)
```

	truth	glm	PLN	JAGS
Intercept	0.69	0.7975	0.6897	0.6757
arable	0.92	1.0430	1.0162	1.0239
eps_sd	0.50	NA	0.4914	0.5180

As so often, we get very similar estimates when using maximum likelihood and Bayesian posterior inference. Note that the solution by `glm()` is not based on exactly the same parametric model and thus the estimate of the intercept appears to be a bit off.

12.2.4 Bayesian analysis with NIMBLE

NIMBLE code for the model is available on the book website.

12.2.5 Bayesian analysis with Stan

We can use the same data set.

```
# Bundle and summarize data
str(dataList <- list(C_OD = C_OD, x = as.numeric(x)-1, n = length(x)))
```

The Stan model code looks similar to JAGS and NIMBLE.

```
# Write Stan model
cat(file = "model12_2_5.stan", "
data{
  int n;
  array[n] int C_OD;
  array[n] int x;
}

parameters{
  real alpha;
  real beta;
  real <lower = 0> sigma;
  vector[n] eps;
}

model{
  vector[n] lambda;

  alpha ~ normal(0, 100);
  beta ~ normal(0, 100);
  sigma ~ cauchy(0, 10);

  for (i in 1:n){
    eps[i] ~ normal(0, sigma);
    lambda[i] = exp(alpha + beta * x[i] + eps[i]);
    C_OD[i] ~ poisson(lambda[i]);
  }
}
")
```

We specify our MCMC settings as usual and run Stan. Note that we did not specify a list of parameters; if you don't provide one, Stan will simply save all parameters in the model parameters block, which works fine for this example.

```
# HMC settings
ni <- 2000   ;   nb <- 1000   ;   nc <- 4   ;   nt <- 1

# Call STAN (ART 60/6 sec), assess convergence, print results table and save estimates
system.time(
out12.2.5 <- stan(file = "model12_2_5.stan", data = dataList,
            warmup = nb, iter = ni, chains = nc, thin = nt) )
rstan::traceplot(out12.2.5)                          # not shown
print(out12.2.5, dig = 3)                            # not shown
stan_est <- summary(out12.2.5)$summary[1:3,1]
```

12.2.6 Do-it-yourself maximum likelihood estimates

We can again use the same data list.

```
# Bundle data
str(dataList <- list(C_OD = C_OD, x = as.numeric(x)-1, n = length(x)))
```

Technically, the PLN is a simple case of a Poisson GLMM (see Chapter 14), that is, a Poisson GLM with random effects ε. For a given data point i and a given value of the random effect ε_i, the contribution to the likelihood is the product of the Poisson density for the count C_i and the normal density for ε_i:

$$dpois\,(C_i|\lambda_i) \cdot dnorm\,(\varepsilon_i|0,\sigma)$$

$$\lambda_i = \exp\,(\alpha + \beta x_i + \varepsilon_i)$$

As with other models where we deal with random effects (Chapters 10, 14, 17), we must integrate out the random effects ε_i by integrating over all possible values of ε_i, in which case we are then left with the parameter σ in the integrated likelihood. Finally, we take the product of these integrated likelihoods for all n counts to get our total likelihood.

$$\prod_{i=1}^{n} \int_{-\infty}^{\infty} dpois\,(C_i|\lambda_i) \cdot dnorm\,(\varepsilon_i|0,\sigma)\,d\varepsilon_i$$

Here's the same equation written in R code, except that instead of multiplying the likelihoods for each data point i together, we add the log-likelihoods and in the end negate the result, to obtain the negative log-likelihood.

```
# Definition of NLL for overdispersed Poisson GLM with one covariate
# Note this is technically a Poisson GLMM, as in Chapter 14
NLL <- function(param, data) {
  alpha <- param[1]
  beta <- param[2]
  sigma <- exp(param[3])

  L <- numeric(n)
  for(i in 1:data$n){
    L[i] <- integrate(function(eps){
      lambda <- exp(alpha + beta * data$x[i] + eps)
      dpois(data$C_OD[i], lambda) * dnorm(eps, 0, sigma) },
      lower = -Inf, upper = Inf)$value
  }
  NLL <- -sum(log(L))
  NLL
}
```

We need to provide slightly informative initial values.

```
# Minimize that NLL to find MLEs, get SEs and CIs and save estimates
inits <- c('alpha' = log(mean(C_OD)), 'beta(arable)' = 0, 'log.sigma' = 0)
out12.2.6 <- optim(inits, NLL, data = dataList, hessian = TRUE)
get_MLE(out12.2.6, 5)
diy_est <- c(out12.2.6$par[1:2], exp(out12.2.6$par[3]))
```

```
                   MLE       ASE     LCL.95     UCL.95
alpha          0.68867   0.12269    0.44819    0.92914
beta(arable)   1.01486   0.15155    0.71782    1.31190
log.sigma     -0.69782   0.14412   -0.98029   -0.41535
```

12.2.7 Likelihood analysis with TMB

To fit the PLN model with TMB, we can again use the same data object as before.

```
# Bundle and summarize data
str(dataList <- list(C_OD = C_OD, x = as.numeric(x)-1, n = length(x)))

# Write TMB model file
cat(file = "model12_2_7.cpp",
"#include <TMB.hpp>

template<class Type>
Type objective_function<Type>::operator() ()
{
 //Describe input data
 DATA_VECTOR(C_OD);                                //response
 DATA_VECTOR(x);                                   //covariate
 DATA_INTEGER(n);                                  //Number of observations

 //Describe parameters
 PARAMETER(alpha);
 PARAMETER(beta);
 PARAMETER(log_sigma);
 PARAMETER_VECTOR(eps);

 Type sigma = exp(log_sigma);

 Type LL = 0.0;                                    //Initialize log-likelihood at 0

 for (int i=0; i<n; i++){
 LL += dnorm(eps(i), Type(0.0), sigma, true);

 Type lambda = exp(alpha + beta * x(i) + eps(i));

 //Calculate log-likelihood of observation and add to total
 LL += dpois(C_OD(i), lambda, true);
 }

 return -LL;                                       //Return negative log likelihood
}
")

# Compile and load TMB function
compile("model12_2_7.cpp")
dyn.load(dynlib("model12_2_7"))

# Provide dimensions and starting values for parameters (incl. the random effects 'eps')
params <- list(alpha = 0, beta = 0, log_sigma = 0, eps = rep(0, dataList$n))
```

Note also the declaration of eps as random in the following statement.

```
# Create TMB object
out12.2.7 <- MakeADFun(data = dataList,
                       parameters = params,
                       random = "eps",
                       DLL = "model12_2_7", silent = TRUE)


# Optimize TMB object, print and save results
starts <- rep(0, 3)
opt <- optim(starts, fn = out12.2.7$fn, gr = out12.2.7$gr, method = "BFGS")
(tsum <- tmb_summary(out12.2.7))              # not shown
tmb_est <- c(tsum[1:2,1], exp(tsum[3,1]))
```

12.2.8 Comparison of the parameter estimates

We compare the parameter estimates produced by our seven methods and engines of fitting (almost) the same model and find the usual good agreement.

```
# Compare truth with estimates from all engines and methods
comp <- cbind(truth = truth, glm = glm_est, PLN = PLN_est, JAGS = jags_est,
   NIMBLE = nimble_est, Stan = stan_est, DIY = diy_est, TMB = tmb_est)
print(comp, 4)
```

	truth	glm	PLN	JAGS	NIMBLE	Stan	DIY	TMB
Intercept	0.69	0.7975	0.6897	0.6757	0.680	0.6784	0.6887	0.6882
arable	0.92	1.0430	1.0162	1.0239	1.015	1.0180	1.0149	1.0152
eps_sd	0.50	NA	0.4914	0.5180	0.517	0.5169	0.4977	0.4967

12.3 Zero inflation

Zero inflation can be viewed as a specific form of OD and is frequently found in count data. It means that there are more zeroes than expected under the assumed (e.g., Poisson, binomial or negative binomial) distribution. In the context of the hare counts that form our example in this and the previous chapter, a typical explanation for excess zeroes is that some sites are simply not suitable for hares, such as paved parking lots, roof tops or lakes; hence, counting hares at such sites must result in a zero observation. In the remaining, suitable sites, counts vary according to the assumed distribution. Thus we may imagine a sequential genesis of zero-inflated counts: first, Nature determines whether a site is suitable in principle and may be occupied at all, and second, she selects the counts for those sites that are habitable in principle. Regression models that account for this kind of OD are often called ZIP (zero-inflated Poisson) or ZIB (zero-inflated binomial) models.

One way to write a ZIP model for count C_i at site i algebraically is like this:

$$w_i \sim Bernoulli(\psi_i) \quad \text{1. Suitability of a site}$$
$$C_i \sim Poisson(w_i \lambda_i) \quad \text{2. Observed counts}$$

For each site i, Nature flips a coin that lands heads (i.e., $w_i = 1$) with probability ψ_i. We can't observe w_i perfectly, that is, it is a latent variable or a random effect. Only for sites with $w_i = 1$, Nature then rolls her Poisson(λ_i) die to determine the count C_i at that site. For sites with $w_i = 0$, the Poisson mean is $0 \cdot \lambda_i = 0$ and the corresponding Poisson die produces zero counts only.

We see that a ZIP model simply represents a set of two coupled GLMs: the logistic regression describes the suitability in principle of a site, while the Poisson regression describes the variation of counts among suitable sites, that is, those with $w_i = 1$. All the usual GLM features apply, and in particular, both the Bernoulli and the Poisson parameter can in principle be expressed as a function of covariates on the link scale. These covariates may or may not be the same for both submodels.

We make five comments on the ZIP model in the context of our hare counts. First, the model allows for two entirely different kinds of zero counts: those coming from the Bernoulli and those from the Poisson process. The former are zero counts at unsuitable sites, while the latter are due to Poisson chance, that is, for them, Nature's Poisson die just happened to yield a zero, even for a site that is suitable in principle (i.e., with $w_i = 1$). The actual distribution of an organism, that is, the proportion of sites that is occupied and has non-zero counts, is a result of both processes. Hence, it would be wrong to say that Equation 1 describes the distribution and Equation 2 the abundance (see below for the hurdle model).

Second, the ZIP model is a hierarchical, or random-effects, model with binary instead of normal random effects. It is a simple example of the kind of hierarchical model that is featured extensively in Chapter 19 and in books like Royle & Dorazio (2008), Kéry & Schaub (2012), Kéry & Royle (2016, 2021), or Hooten & Hefley (2019). There, we will encounter the site-occupancy species distribution model, which can be described as another kind of zero-inflated GLM, but one where a Bernoulli or binomial distribution is zero-inflated with another Bernoulli to arrive at a ZIB model.

Third, some authors advocate ZIP models widely for inference about count data (Martin et al., 2005; Joseph et al., 2009). However, on ecological grounds, they appear most adequate in situations where *unknown* environmental covariates determine the suitability of a site. If covariates are known and have been measured, they are probably best added to the model for the Poisson mean. Distribution, or occurrence, is fundamentally a deterministic function of abundance. That is, a species occurs at all sites where abundance is greater than zero. It appears contrived to us to model "distribution" as something separate from abundance. To us, this is the same thing.

Fourth, there is a variant of a ZIP model called a hurdle model (Zeileis et al., 2008), where the first step in the hierarchical genesis of the counts is assumed to be the same as in a ZIP model, that is, $w_i \sim Bernoulli(\psi_i)$. But then, counts at suitable sites (i.e., with $w_i = 1$) are modeled as coming from a zero-truncated Poisson distribution, that is, a Poisson for values excluding zero (see Chapter 20). Hurdles, or thresholds, other than zero are also possible, simply by varying the degree of truncation in the second distribution. Superficially perhaps, this model may appear "better" than a ZIP model, since it only allows one kind of zero: that coming from the Bernoulli process. However, it posits that all sites that are suitable in principle *will* be occupied and have a count greater than 0. This is not sensible biologically, since in reality a suitable site may well be unoccupied due to local chance extinction, dispersal limitation, or some other reason.

Our fifth comment is that the manner in which we have just described the ZIP model has a neat interpretation of ψ as the probability of site suitability. This makes a lot of sense when we think about this model as a species distribution model (SDM). Also, this parameterization is related to the kind of zero inflation specified in an occupancy model (see comment 2 above). However, this is *not* how a zero-inflated model is usually described in the statistics literature. There, ψ is usually defined as the probability that a zero is an excess, or structural, zero, that is, such that $w_i \sim Bernoulli(1 - \psi_i)$ and $C_i \sim Poisson(w_i \lambda_i)$, that is, where ψ is the probability that a site is unsuitable. For the rest of Section 12.3, we will adhere to the statistical parameterization of zero inflation, but we wanted to point out that to an ecologist, the other parameterization may be more intuitive.

12.3.1 **Data generation**

We generate the simplest kind of zero-inflated count data for our (Poisson) hare example. We assume different densities in arable and grassland areas and a constant zero inflation, that is, a single value of ψ for all sites, regardless of land-use or other environmental covariates.

```
set.seed(123)
psi <- 0.2                    # NOTE: here psi is probability of zero inflation
nSites <- 50
x <- gl(n = 2, k = nSites, labels = c("grassland", "arable"))
```

For each site, we flip a coin to determine whether it is suitable in principle and store the result in the latent state variable w.

```
w <- rbinom(n = 2 * nSites, size = 1, prob = 1 - psi)
```

We assume identical effects of arable and grass as before and generate expected counts *at suitable sites* as before.

```
lambda <- exp(0.69 + (0.92*(as.numeric(x)-1)))
```

We then combine (actually, multiply) the effects of both processes (suitability: Bernoulli, and abundance: Poisson) and inspect the counts we've generated. Note how all counts at unsuitable sites (with $w_i = 0$) are necessarily zero in this model.

```
C <- rpois(n = 2*nSites, lambda = w *lambda)
data.frame('habitat' = x, 'suitability' = w, 'count' = C)  # Look at data (not shown)

# Save true parameter values
truth <- c(Intercept = 0.69, arable = 0.92, psi = psi)
```

12.3.2 **Likelihood analysis with canned functions in R**

A wide range of ZIP and related models can be fitted in R using the function `zeroinfl()` in package `pscl` (Zeileis et al., 2008). We load that package, fit the simplest possible ZIP model (which is also our data-generating model), and save the estimates.

```
library(pscl)
out12.3.2 <- zeroinfl(C ~ x | 1, dist = "poisson")
summary(out12.3.2)
pscl_est <- c(coef(out12.3.2)[1:2], plogis(coef(out12.3.2)[3]))

Count model coefficients (poisson with log link):
            Estimate  Std. Error  z value  Pr(>|z|)
(Intercept)   0.7637      0.1186    6.438  1.21e-10 ***
xarable       0.7269      0.1388    5.236  1.64e-07 ***

Zero-inflation model coefficients (binomial with logit link):
            Estimate  Std. Error  z value  Pr(>|z|)
(Intercept)  -1.8043      0.3553   -5.078  3.81e-07 ***
```

Due to sampling and estimation error, the coefficients for the count model (corresponding to Equation 2 above) may not always be very close to the input values.

12.3.3 Bayesian analysis with JAGS

Next, the solution in JAGS. As always, the elementary manner of model specification using the BUGS language makes it exceedingly clear what model is fitted. Matching our data simulation, we also parameterize the model such that ψ denotes the habitat unsuitability in an SDM context, rather than habitat suitability.

```
# Bundle and summarize data
str(dataList <- list(C = C, x = as.numeric(x)-1, n = length(x)) )

List of 3
 $ C : int [1:100] 2 1 2 0 0 5 2 0 4 4 ...
 $ x : num [1:100] 0 0 0 0 0 0 0 0 0 0 ...
 $ n : int 100

# Write JAGS model file
cat(file = "model12.3.3.txt", "
model {
# Priors
psi ~ dunif(0,1)
alpha ~ dnorm(0, 0.0001)
beta ~ dnorm(0, 0.0001)

# Likelihood
for (i in 1:n) {
  w[i] ~ dbern(1-psi)                  # Habitat suitability
  C[i] ~ dpois(eff.lambda[i])
  eff.lambda[i] <- w[i] * lambda[i]
  log(lambda[i]) <- alpha + beta * x[i]   # expected abundance at suitable sites
}
}
")

# Function to generate starting values
inits <- function(){ list(alpha = rlnorm(1), beta = rlnorm(1), w = rep(1, 2 * nSites))}
```

We could easily obtain estimates of the latent state variable *w*, that is, the intrinsic suitability for brown hares of each site, by simply adding it to the list of estimated parameters, but we won't.

```
# Parameters to estimate
params <- c("alpha", "beta", "psi")

# MCMC settings
na <- 1000  ;  ni <- 12000  ;  nb <- 2000  ;  nc <- 4  ;  nt <- 10

# Call JAGS (ART <1 min), check convergence, summarize posteriors and save results
out12.3.3 <- jags(dataList, inits, params, "model12.3.3.txt", n.iter = ni,
  n.burnin = nb, n.chains = nc, n.thin = nt, n.adapt = na, parallel = TRUE)
jagsUI::traceplot(out12.3.3)                      # not shown
print(out12.3.3, 3)
jags_est <- as.numeric(out12.3.3$mean[1:3])
```

	mean	sd	2.5%	50%	97.5%	overlap0	f	Rhat	n.eff
alpha	0.758	0.119	0.517	0.760	0.985	FALSE	1	1.000	4000
beta	0.728	0.139	0.454	0.729	0.997	FALSE	1	1.000	4000
psi	0.150	0.042	0.072	0.148	0.238	FALSE	1	1.001	2438

As so very often, we find pretty similar estimates between a frequentist and a Bayesian analysis of the model for our data set.

```
# Compare likelihood with Bayesian estimates and with truth
comp <- cbind(truth = truth, pscl = pscl_est, JAGS = jags_est)
print(comp, 4)
            truth     pscl     JAGS
Intercept    0.69   0.7637   0.7575
arable       0.92   0.7269   0.7281
psi          0.20   0.1413   0.1497
```

12.3.4 Bayesian analysis with NIMBLE

NIMBLE code for the model is available on the book website.

12.3.5 Bayesian analysis with Stan

Again we use the same data set.

```
# Bundle and summarize data
str(dataList <- list(C = C, x = as.numeric(x)-1, n = length(x)) )
```

In our example, the parameter w indicates if a site is suitable or not, and thus can take on only values 1 (suitable) or 0 (not suitable). We do not observe w directly—it is not part of our data, but instead estimated by the model. It is therefore a parameter that is both discrete (takes on a defined set of values) and latent (unobserved). JAGS and NIMBLE can handle estimation of discrete latent variables directly, but Stan cannot, due to its reliance on the HMC algorithm. Thus we will have to use a different approach in the Stan code (and the DIY and TMB code as well).

To work around this limitation, we use an integrated likelihood approach similar to the one in Section 12.2.6, treating w_i as a random effect. We must calculate the likelihood for each data point i after integrating out w_i; that is, by integrating over all possible values of w_i. Fortunately, there are only two possible values of w_i, 1 and 0. Therefore, for each count C_i we just need to calculate the likelihood for each possible value of w_i and add them together. This calculation differs depending on the observed data. If we observe $C_i = 0$, then either value of w_i is possible; i.e., the 0 could be because the site wasn't suitable in principle, or because it was suitable, but the count simply was 0 by chance. On the other hand, if $C_i > 0$, we know w_i must be 1: the site must be suitable in principle if we observed a hare there! This leads to the following equation for the likelihood of λ_i and ψ_i given count C_i:

$$L\left(\lambda_i, \psi_i \,|\, C_i\right) = \begin{cases} \psi_i + \left(1 - \psi_i\right) \cdot dpois(C_i|\lambda_i), & C_i = 0 \\ \left(1 - \psi_i\right) \cdot dpois(C_i|\lambda_i), & C_i > 0 \end{cases}$$

Below, we translate this equation into Stan code, adding the log of the likelihood for each data point i to the total model log-likelihood (called `target` in Stan). Stan generally works with probability density functions (or in this case, more specifically, a probability mass function) on the log scale. The Poisson log probability mass function in Stan is `poisson_lpmf()`. We must exponentiate the result of this function to get back to the regular scale.

```
# Write Stan model
cat(file = "model12_3_5.stan", "
data{
  int n;
  array[n] int C;
  array[n] int x;
}

parameters{
  real alpha;
  real beta;
  real<lower = 0,upper = 1> psi;
}

model{
  vector[n] lambda;
  vector[n] lik;

  alpha ~ normal(0, 100);
  beta ~ normal(0, 100);

  for (i in 1:n){
    lambda[i] = exp(alpha + beta * x[i]);

    if(C[i] == 0){                          // Either w state is possible
      lik[i] = psi + (1-psi) * exp(poisson_lpmf(0 | lambda[i]));
    } else {                                // Only w = 1 is possible
      lik[i] = (1-psi) * exp(poisson_lpmf(C[i] | lambda[i]));
    }
    target += log(lik[i]);                  //target is the total log likelihood
  }
}
")

# HMC settings
ni <- 2000  ;  nb <- 1000  ;  nc <- 4  ;  nt <- 1

# Call STAN (ART 56/4 sec), assess convergence, print results table and save results
system.time(
out12.3.5 <- stan(file = "model12_3_5.stan", data = dataList,
            warmup = nb, iter = ni, chains = nc, thin = nt) )
rstan::traceplot(out12.3.5)                     # not shown
print(out12.3.5, dig = 3)                       # not shown
stan_est <- summary(out12.3.5)$summary[1:3,1]
```

12.3.6 Do-it-yourself maximum likelihood estimates

Here's the zero-inflated Poisson GLM fit using maximum likelihood. We will later see a variant of this for a binomial response, in the so-called occupancy model (Chapter 19). As with the implementation of the model in Stan, we need to integrate out the binary random effects w. Note that summing over discrete random effects corresponds to integration over continuous random effects as with the models in Chapters 10, 14, and 17 and in Section 12.2.

```
# Definition of NLL for ZIP model
NLL <- function(param, data) {
  alpha <- param[1]
  beta <- param[2]
  psi <- plogis(param[3])
  lambda <- exp(alpha + beta * data$x)

  lik <- numeric(data$n)

  for (i in 1:data$n){
    if(data$C[i] == 0){
      lik[i] <- psi + (1-psi) * dpois(0, lambda[i])
    } else {
      lik[i] <- (1-psi) * dpois(data$C[i], lambda[i])
    }
  }

  NLL <- -sum(log(lik))
  return(NLL)
}
```

```
# Minimize that NLL to find MLEs, get SEs and CIs and save estimates
inits <- c('Intercept' = 0, 'arable' = 0, 'logit.psi' = 0)
out12.3.6 <- optim(inits, NLL, data = dataList, hessian = TRUE)
get_MLE(out12.3.6, 4)
diy_est <- c(out12.3.6$par[1:2], plogis(out12.3.6$par[3]))
```

```
              MLE      ASE    LCL.95   UCL.95
Intercept   0.7638   0.1186   0.5313   0.9963
arable      0.7269   0.1388   0.4548   0.9990
logit.psi  -1.8042   0.3553  -2.5005  -1.1078
```

We note that the first two parameters are on the log and the third is on the logit scale.

12.3.7 Likelihood analysis with TMB

And finally, we fit the ZIP model using TMB as well. We can again use the same data object as before.

```
# Bundle and summarize data
str(dataList <- list(C = C, x = as.numeric(x)-1, n = length(x)) )
```

The TMB code uses integrated likelihood and looks very similar to both our Stan code and the DIY code.

```
# Write TMB model file
cat(file = "model12_3_7.cpp",
"#include <TMB.hpp>

template<class Type>
Type objective_function<Type>::operator() ()
{
  //Describe input data
  DATA_VECTOR(C);                          //response
  DATA_VECTOR(x);                          //covariate
  DATA_INTEGER(n);                         //Number of observations

  //Describe parameters
  PARAMETER(alpha);
  PARAMETER(beta);
  PARAMETER(logit_psi);

  Type psi = invlogit(logit_psi);

  Type LL = 0.0;                           //Initialize log-likelihood at 0

  for (int i=0; i<n; i++){
    Type lik;
    Type lambda = exp(alpha + beta * x(i));
    if(C(i) == 0){
      lik = psi + (1-psi) * dpois(Type(0), lambda, false);
    } else {
      lik = (1-psi) * dpois(C[i], lambda, false);
    }
    LL += log(lik);
  }

  return -LL;                              //Return negative log-likelihood
}
")

# Compile and load TMB function
compile("model12_3_7.cpp")
dyn.load(dynlib("model12_3_7"))

# Provide dimensions and starting values for parameters
params <- list(alpha = 0, beta = 0, logit_psi = 0)

# Create TMB object
out12.3.7 <- MakeADFun(data = dataList,
              parameters = params,
              DLL = "model12_3_7", silent = TRUE)

# Optimize TMB object and print results
starts <- rep(0, 3)
opt <- optim(starts, fn = out12.3.7$fn, gr = out12.3.7$gr, method = "BFGS")
(tsum <- tmb_summary(out12.3.7))                    # not shown
tmb_est <- c(tsum[1:2,1], plogis(tsum[3,1]))
```

12.3.8 Comparison of the parameter estimates

And, as you know well by now: here comes our final comparison of the point estimates produced by all our model-fitting engines, frequentist (i.e., maximum likelihood) or Bayesian ... we see point estimates that for all practical purposes are numerically identical.

```
# Compare estimates with truth
comp <- cbind(truth = truth, pscl = pscl_est, JAGS = jags_est,
  NIMBLE = nimble_est, Stan = stan_est, DIY = diy_est, TMB = tmb_est)
print(comp, 4)
```

	truth	pscl	JAGS	NIMBLE	Stan	DIY	TMB
Intercept	0.69	0.7637	0.7575	0.762	0.7615	0.7638	0.7637
arable	0.92	0.7269	0.7281	0.726	0.7240	0.7269	0.7269
psi	0.20	0.1413	0.1497	0.149	0.1479	0.1413	0.1413

12.4 Offsets

In our Poisson GLM, we assume that the expected counts are adequately described by the effects of the covariates in the model. However, frequently the "counting window" is not constant, for example, study areas aren't the same size or, in temporal samples, the duration of counting periods differs. To account for this known component of variation in the Poisson mean, we define the log of the size of the "counting window" (study area size, count duration) as an *offset* in the linear predictor. In doing so, we scale the response to a unit-size "counting window" and effectively we then model a density in space or time as a response.

Let's consider this for the hare counts using algebra. Our Poisson GLM is $C_i \sim Poisson(\lambda_i)$, that is, hare counts C_i are conditionally distributed as Poisson with expected count λ_i. When study areas differ in size, we have $C_i \sim Poisson(A_i \cdot \lambda_i)$, where A_i is the area of study area i and now λ_i is the expected count in an area of size 1. That is, now λ_i becomes a density, where the measurement scale of A_i defines the scale in which this density is expressed (e.g., it could be ha or km^2). To model variability in this density, we still adopt a log link function, that is, we may then go on to develop models for the density per unit area like $\log(\lambda_i) = \alpha + \beta x_i$, and the expected count is then simply $E(C_i) = A_i \cdot \lambda_i$. This is how we can effectively implement an offset in our generic model-fitting engines. But unfortunately this is not the way in which offsets are declared in the typical GLM jargon, including in function glm(). There, you will always have to take a log of the "counting window" size and specify that as the offset.

To understand that, note that in the presence of an offset, the linear predictor becomes $\log(A_i \cdot \lambda_i) = \log(A_i) + \log(\lambda_i)$. If we model a covariate x on the expected count, we get $\log(A_i \cdot \lambda_i) = \log(A_i) + \alpha + \beta x_i$. This is equivalent to forcing the coefficient of $\log(\text{area})$ to be equal to 1. That is, we effectively fit the model $\log(A_i \cdot \lambda_i) = \beta_0 \cdot \log(A_i) + \alpha + \beta x_i$ with $\beta_0 = 1$. The offset compensates for the additional and known variation in the response due to differing study area size.

Implicitly, use of an offset assumes that density does not depend on the size of a study area. If in reality larger areas had a greater density of hares, then $\log(\text{area})$ should not be then used as an offset in our analysis (or as a covariate with fixed coefficient 1). Instead, we should then use area as a "traditional" covariate with a coefficient that we estimate as a free parameter.

12.4.1 Data generation

We simulate a data set and once more observe how R code can serve to explain a statistical model succinctly and in a clear manner.

```
set.seed(124)
nSites <- 50
A <- runif(n = 2 * nSites, 2, 5)                    # Areas range in size from 2 to 5 km2
x <- gl(n = 2, k = nSites, labels = c("grassland", "arable"))
linear.predictor <- log(A) + 0.69 + (0.92*(as.numeric(x)-1))
lambda <- exp(linear.predictor)
lambda <- A * exp(0.69 + (0.92*(as.numeric(x)-1)))  # exactly the same !
C <- rpois(n = 2 * nSites, lambda = lambda)         # Add Poisson noise

# Save true parameter values
truth <- c(Intercept = 0.69, arable = 0.92)
```

12.4.2 Likelihood analysis with canned functions in R

We use R for an analysis with and without consideration of the differing areas. Offsets can be specified in most regression functions in R, including glm(), and must be supplied as the log-transform of the "counting window" size, or here, area.

```
glm.fit.no.offset <- glm(C ~ x, family = poisson)
glm.fit.with.offset <- glm(C ~ x, family = poisson, offset = log(A))
summary(glm.fit.no.offset)                  # not shown
summary(glm.fit.with.offset)                # not shown
```

Comparing the residual deviance of the two models clarifies that specification of an offset represents a correction for a systematic kind of OD, i.e., one that can be explained in terms of a special kind of covariate, to which the response is assumed to have a constant proportionality. Thus the residual deviance of the model ignoring the variation in study area size (i.e., without offset log (A)) is about 135, while that for the adequate model including the offset comes down to about 88.

Comparing the point estimates of the two models we see this:

```
# Compare to truth
glmNoOff <- coef(glm.fit.no.offset)
glmOff <- coef(glm.fit.with.offset)
tmp <- cbind(truth, glmNoOff = glmNoOff, glmOff = glmOff)
print(tmp, 4)

           truth   glmNoOff   glmOff
Intercept  0.69     1.9228    0.6828
arable     0.92     0.8818    0.9022
```

So, by ignoring the variable areas among sites, we are estimating the density intercept at the average area. Since in our data simulation we defined the intercept to be the density for a site with area $A = 1$, and since our simulated areas average 3.5, this estimate will greatly overestimate the true per unit-area density, actually, by a factor of 3.5. This is borne out by these estimates, since $\exp(1.9228)/\exp(0.6828)$ is indeed about 3.5.

12.4.3 Bayesian analysis with JAGS

Note how simple it is in JAGS to jump from one kind of analysis for the hare counts to another. In our own model definition code we can define the proportionality between observed counts and the area in either way mentioned above, that is, by use of A or $\log(A)$.

```
# Bundle and summarize data
str(dataList <- list(C = C,  x = as.numeric(x)-1, A = A,  n = length(x)))

List of 4
 $ C: int [1:100] 3 6 6 4 3 8 8 11 9 4 ...
 $ x: num [1:100] 0 0 0 0 0 0 0 0 0 0 ...
 $ A: num [1:100] 2.25 3.23 3.55 3.19 2.67 ...
 $ n: int 100

# Write JAGS model file
cat(file = "model12.4.3.txt", "
model {
# Priors
alpha ~ dnorm(0, 0.0001)
beta ~ dnorm(0, 0.0001)

# Likelihood
for (i in 1:n) {
  C[i] ~ dpois(lambda[i])
  log(lambda[i]) <- 1 * log(A[i]) + alpha + beta *x[i]   # Note offset
  # lambda[i] <- A * exp(alpha + beta *x[i])             # exactly the same
}
}
")

# Function to generate starting values
inits <- function(){ list(alpha = rnorm(1), beta = rnorm(1))}

# Parameters to estimate
params <- c("alpha", "beta")

# MCMC settings
na <- 1000  ;  ni <- 12000  ;  nb <- 2000  ;  nc <- 4  ;  nt <- 10

# Call JAGS (ART <1 min), check convergence, summarize posteriors and save estimates
out12.4.3 <- jags(dataList, inits, params, "model12.4.3.txt", n.iter = ni,
  n.burnin = nb, n.chains = nc, n.thin = nt, n.adapt = na, parallel = TRUE)
jagsUI::traceplot(out12.4.3)                       # not shown
print(out12.4.3, 3)
jags_est <- as.numeric(out12.4.3$mean[1:2])
```

	mean	sd	2.5%	50%	97.5%	overlap0	f	Rhat	n.eff
alpha	0.680	0.054	0.574	0.681	0.785	FALSE	1	1.000	4000
beta	0.903	0.065	0.776	0.904	1.029	FALSE	1	1.001	4000

```
# Compare likelihood with Bayesian estimates and with truth
comp <- cbind(truth, glmOff = glmOff, JAGS = jags_est)
print(comp, 4)
```

	truth	glmOff	JAGS
Intercept	0.69	0.6828	0.6802
arable	0.92	0.9022	0.9032

As so often, numerically there is practically no difference between the MLEs and the Bayesian posterior means.

12.4.4 Bayesian analysis with NIMBLE

NIMBLE code for the model is available on the book website.

12.4.5 Bayesian analysis with Stan

We can use the same data set for Stan as we bundled for JAGS and NIMBLE.

```
# Bundle and summarize data
str(dataList <- list(C = C, x = as.numeric(x)-1, A = A, n = length(x)))

# Write Stan model
cat(file = "model12_4_5.stan", "
data{
  int n;
  array[n] int C;
  array[n] int x;
  vector[n] A;
}

parameters{
  real alpha;
  real beta;
}

model{
  vector[n] lambda;

  alpha ~ normal(0, 100);
  beta ~ normal(0, 100);

  for (i in 1:n){
    lambda[i] = exp(alpha + beta * x[i] + log(A[i]));
    C[i] ~ poisson(lambda[i]);
  }
}
")

# HMC settings
ni <- 2000  ;  nb <- 1000  ;  nc <- 4  ;  nt <- 1

# Call STAN (ART 44/2 sec), assess convergence, print results table and save estimates
system.time(
out12.4.5 <- stan(file = "model12_4_5.stan", data = dataList,
            warmup = nb, iter = ni, chains = nc, thin = nt) )
rstan::traceplot(out12.4.5)                    # not shown
print(out12.4.5, dig = 3)                       # not shown
stan_est <- summary(out12.4.5)$summary[1:2,1]
```

12.4.6 **Do-it-yourself maximum likelihood estimates**

This is a particularly simple negative log-likelihood function: in the log-scale expression for the Poisson expectation (called `mu` here), we simply add `log(A)` in the sum term there or, alternatively, multiply the natural-scale `lambda` with `A` (this is the hashed out part in the function).

```
# Bundle data into list
str(dataList <- list(C = C, x = as.numeric(x) - 1, A = A, n = length(x)))
```

```
# Definition of NLL for Poisson GLM with offset
NLL <- function(param, data) {
  alpha <- param[1]
  beta <- param[2]
  mu <- exp(log(data$A) + alpha + beta * data$x)
# mu <- data$A * exp(alpha + beta * data$x)    # Exactly the same
  L <- dpois(data$C, mu)                        # Likelihood contr. for 1 observation
  LL <- log(L)                                  # Loglikelihood contr. for 1 observation
  NLL <- -sum(LL)                               # NLL for all observations
  return(NLL)
}
```

```
# Minimize that NLL to find MLEs, get SEs and CIs and save estimates
inits <- c('Intercept' = 0, 'arable' = 0)
out12.4.6 <- optim(inits, NLL, data = dataList, hessian = TRUE, method = "BFGS")
get_MLE(out12.4.6, 4)
diy_est <- out12.4.6$par
```

```
              MLE       ASE    LCL.95   UCL.95
Intercept   0.6828   0.05407   0.5769   0.7888
arable      0.9022   0.06430   0.7761   1.0282
```

12.4.7 **Likelihood analysis with TMB**

And here is the solution with TMB. We have the same data set again.

```
# Bundle data into list
str(dataList <- list(C = C, x = as.numeric(x) - 1, A = A, n = length(x)))

# Write TMB model file
cat(file = "model12_4_7.cpp",
"#include <TMB.hpp>

template<class Type>
Type objective_function<Type>::operator() ()
{
  //Describe input data
  DATA_VECTOR(C);                          //response
  DATA_VECTOR(x);                          //covariate
  DATA_VECTOR(A);                          //offset
  DATA_INTEGER(n);                         //Number of observations

  //Describe parameters
  PARAMETER(alpha);
  PARAMETER(beta);

  Type LL = 0.0;                           //Initialize log-likelihood at 0

  for (int i=0; i<n; i++){
    Type mu = exp(alpha + beta * x(i) + log(A(i)));
    LL += dpois(C[i], mu, true);           //arg true means log-lik is returned
  }

  return -LL;                              //Return negative log likelihood}
}
")

# Compile and load TMB function
compile("model12_4_7.cpp")
dyn.load(dynlib("model12_4_7"))

# Provide dimensions and starting values for parameters
params <- list(alpha = 0, beta = 0)

# Create TMB object
out12.4.7 <- MakeADFun(data = dataList,
                       parameters = params,
                       DLL = "model12_4_7", silent = TRUE)

# Optimize TMB object, print and save results
starts <- rep(0, 2)
opt <- optim(starts, fn = out12.4.7$fn, gr = out12.4.7$gr, method = "BFGS")
(tsum <- tmb_summary(out12.4.7))                # not shown
tmb_est <- tsum[1:2,1]
```

12.4.8 Comparison of the parameter estimates

And here's our grand comparison, where once again we see good agreement among all the estimates from models that accommodate the offset.

```
# Compare results with truth and previous estimates
comp <- cbind(truth, glmOff = glmOff, JAGS = jags_est,  NIMBLE = nimble_est,
   Stan = stan_est, DIY = diy_est, TMB = tmb_est)
print(comp, 4)

          truth  glmOff    JAGS  NIMBLE    Stan     DIY     TMB
Intercept  0.69  0.6828  0.6802   0.680  0.6790  0.6828  0.6828
arable     0.92  0.9022  0.9032   0.905  0.9062  0.9022  0.9022
```

12.5 Summary and outlook

Overdispersion, zero inflation, and offsets are important GLM topics that frequently occur when modeling count data, especially for Poisson responses, but also in binomial responses. Therefore, it is important to understand and be able to use these modeling figures. The specification of the associated models especially in the BUGS language is easy and clarifies the actual meaning of all three topics. This is not usually the case when fitting these models in a canned routine in R or another software. Thus, this is another example of where the simple model specification in the BUGS language enforces an understanding of the fitted model that is easily lost in other stats packages.

Although we have treated overdispersion, zero inflation, and offsets in separate sections in this chapter, they may often occur together when modeling a real data set. For instance, in a model setting such as our hare counts, we may have sites that differ in size, some sites which are completely unsuitable for hares, and additional unexplained density variation among the suitable sites. In this case, we might want to adopt a model that has an offset, zero inflation, and also accommodates overdispersion relative to the basic Poisson.

Poisson GLM with continuous and categorical explanatory variables⊛ 13

Chapter outline

13.1 Introduction

In many practical applications, Poisson regression models will have both discrete and continuous covariates. Here we examine this important variety of generalized linear model (GLM). To stress the similarity with the normal linear case, we alter the inferential setting considered in Chapter 9 only slightly. We assume that instead of measuring body mass in Asp vipers in three populations in the Pyrenees, Massif Central, and the Jura mountains, leading to a normal model, we instead assessed ectoparasite load in a dragonfly, the sombre goldenring (Fig. 13.1), leading to a Poisson model. We are particularly interested in whether the number of little red mites differs with dragonfly size (expressed as wing length), and whether this relationship differs among the three mountain ranges.

We will fit the following model to mite load C_i on dragonfly i:

1. Distribution: $C_i \sim Poisson(\lambda_i)$
2. Link function: log, i.e., $\eta_i = \log(\lambda_i) = \log(E(C_i))$
3. Linear predictor (η): $\eta_i = \beta_0 + \beta_1 \cdot x_{MC,i} + \beta_2 \cdot x_{Jura,i} + \beta_3 \cdot x_{wing,i} + \beta_4 \cdot x_{wing,i} \cdot x_{MC,i} + \beta_5 \cdot x_{wing,i} \cdot x_{Jura,i}$

⊛This book has a companion website hosting complementary materials, including all code for download. Visit this URL to access it: https://www.elsevier.com/books-and-journals/book-companion/9780443137150.

Applied Statistical Modelling for Ecologists. DOI: https://doi.org/10.1016/B978-0-443-13715-0.00013-3

FIGURE 13.1

Sombre goldenring (*Cordulegaster bidentata*), Kleinlützel, Switzerland, 1995 (Photo by Felix Labhardt).

In this model, β_0 is the expected number of mites (on the log scale) of a dragonfly with average wing length (wing length is centered, see below) in the Pyrenees, and the other parameters represent the effects of population and wing length in this effects parameterization of the model. We use subscripts both to label a covariate (e.g., as MC for Massif Central) and as an index for individual i. Note the great similarity between this model and the one we fitted to the mass of asps in Chapter 9. Apart from the link function, the other main difference is simply that for this model we do not have a dispersion term; the Poisson already comes with a built-in variability. We could model additional variability (i.e., overdispersion) in the mite counts by using a Poisson-lognormal formulation as in the previous chapter, or assume a negative binomial distribution for the counts instead, but we omit such additional complexity here.

13.2 Data generation

We need to simulate a data set.

```
set.seed(13)
nPops <- 3                              # Three populations
nSample <- 100                          # ...with 100 individuals each
n <- nPops * nSample                    # Total sample size

x <- rep(1:nPops, rep(nSample, nPops))  # Population indicator
pop <- factor(x, labels = c("Pyrenees", "Massif Central", "Jura"))

orig.length <- runif(n, 4.5, 7.0)       # Wing length (cm)
wing.length <- orig.length - mean(orig.length)  # Center values
```

We build the design matrix of an interactive combination of wing length and population:

```
Xmat <- model.matrix(~ pop * wing.length)
head(Xmat, 10)          # Look at first 10 rows of design matrix
```

Next we select the parameter values, i.e., we chose values for β_0, β_1, β_2, β_3, β_4, β_5, corresponding to the six columns of the design matrix. How did we come up with them? By trial and error: we simply chose some values and then looked at the resulting patterns (as in Fig. 13.2), repeating until we got results that we liked.

```
# Save truth for comparisons
truth <- beta.vec <- c(-2, 1, 2, 4, -2, -5)
```

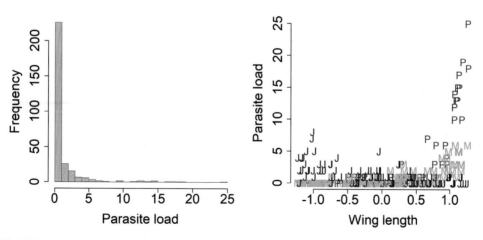

FIGURE 13.2

Left: Frequency distribution of ectoparasite load in sombre goldenrings. Right: Relationship between parasite load and wing length (deviation from mean, in cm) in three mountain ranges (P—Pyrenees, M—Massif Central, J—Jura mountains).

Here is the recipe for assembling the mite counts in three steps. Note that this is in essence also a clear description of the Poisson GLM. We like to say that data generation is usually done "from the inside out" and thus the order in which we build up the final, simulated data is the opposite of how the GLM is usually defined (as above):

1. we matrix-multiply the model matrix with our parameter vector to get the values of the linear predictor, which is the expected mite count on the log-transformed scale,
2. we exponentiate the linear predictor to get the actual value of the expected mite count on the natural scale, and
3. we add Poisson noise.

Here is the recipe converted into R code:

```
lin.pred <- Xmat[,] %*% beta.vec        # Value of linear predictor
lambda <- exp(lin.pred)                 # Poisson mean: expected count
load <- rpois(n = n, lambda = lambda)   # Add Poisson noise

# Inspect what we've created (Fig. 13.2)
par(mfrow = c(1, 2), mar = c(5,5,4,3), cex.axis = 1.5, cex.lab = 1.5)
hist(load, col = "grey", breaks = 30, xlab = "Parasite load", main = "", las = 1)
plot(wing.length, load, pch = rep(c("P", "M", "J"), each = nSample), las = 1,
  col = rep(c("Red", "Green", "Blue"), each = nSample), ylab = "Parasite load",
  xlab = "Wing length", cex = 1.2, frame = FALSE)
```

We have generated a data set where parasite load varies from 0 to 25 and increases with wing length in the South (Pyrenees, Massif Central), but decreases in the North (Jura mountains); see Fig. 13.2.

```
# Load required libraries
library(ASMbook); library(jagsUI); library(rstan); library(TMB)
```

13.3 Likelihood analysis with canned functions in R

We begin by fitting the model using the R function glm(), which results in parameter estimates similar to the simulation inputs. Obviously, with larger sample sizes the correspondence would be better still (as always, please try this yourself, e.g., by setting nSample = 1000).

```
# Fit model using least-squares and save estimates
summary(out13.3 <- glm(load ~ pop * wing.length, family = poisson))
glm_est <- coef(out13.3)
```

Coefficients:

	Estimate	Std. Error	z value	Pr(>|z|)	
(Intercept)	-1.8618	0.2761	-6.744	1.54e-11	***
popMassif Central	0.7809	0.3540	2.206	0.0274	*
popJura	1.8107	0.2994	6.047	1.48e-09	***
wing.length	3.9066	0.2576	15.167	< 2e-16	***
popMassif Central:wing.length	-1.8632	0.3542	-5.261	1.44e-07	***
popJura:wing.length	-4.9582	0.2931	-16.918	< 2e-16	***

As usual, it is important to assess model goodness-of-fit using residual analysis. We inspect results both for traditional Pearson and quantile residuals. As emphasized by Hartig (2022), the quantile residual plots are far easier to interpret than the traditional residual plots.

```
# Model assessment by checks of residuals (not shown)
plot(out13.3)                     # Check of traditional (Pearson) residuals
library(DHARMa)                   # Compute quantile residuals based on simulation
simOut <- simulateResiduals(out13.3, n = 1000, plot = TRUE)
```

13.4 Bayesian analysis with JAGS

13.4.1 Fitting the model

To fit the model with JAGS, we apply a means parameterization instead of the effects parameterization we used in the simulation and the canned function analysis. We now estimate a separate intercept (vector α) and slope (vector β) for each dragonfly population. The equation for the linear predictor becomes:

$$\log(\lambda_i) = \alpha_{pop(i)} + \beta_{pop(i)} \cdot x_{winglength,i}$$

We can simply adapt the JAGS code for the normal linear case (see Chapter 9) to the Poisson case, adding the appropriate link function to the likelihood section and removing the unneeded `sigma` parameter. The elements in each parameter vector α and β are treated as fixed effects, i.e., unknown constants that are completely unrelated to each other (see the next chapter for a variant of the model where we treat them as random effects). As with the model from Chapter 9, we will calculate the effects as derived parameters so we can compare directly to the output from `glm()` and other engines.

```
# Bundle and summarize data
str(dataList <- list(load = load, pop = as.numeric(pop), nPops = nPops,
    wing.length = wing.length, n = n) )

List of 5
 $ load       : int  [1:300] 0 0 1 0 10 0 0 0 4 0 ...
 $ pop        : num  [1:300] 1 1 1 1 1 1 1 1 1 1 ...
 $ nPops      : num  3
 $ wing.length: num  [1:300] 0.546 -0.614 -0.255 -1.001 1.176 ...
 $ n          : num  300
```

```
# Write JAGS model file
cat(file = "model13.4.txt", "
model {
# Priors
for (i in 1:nPops){              # Loop over 3 populations
  alpha[i] ~ dnorm(0, 0.0001)    # Intercepts
  beta[i] ~ dnorm(0, 0.0001)     # Slopes
}

# Likelihood
for (i in 1:n) {                 # Loop over all 300 data points
  load[i] ~ dpois(lambda[i])     # The response variable (C above)
  lambda[i] <- exp(alpha[pop[i]] + beta[pop[i]]* wing.length[i])
}                                # Note nested indexing: alpha[pop[i]] and beta[pop[i]]

# Derived quantities
# Recover effects relative to baseline level (no. 1)
a.effe2 <- alpha[2] - alpha[1]   # Intercept Massif Central vs. Pyr.
a.effe3 <- alpha[3] - alpha[1]   # Intercept Jura vs. Pyr.
b.effe2 <- beta[2] - beta[1]     # Slope Massif Central vs. Pyr.
b.effe3 <- beta[3] - beta[1]     # Slope Jura vs. Pyr.

# Custom test
test1 <- beta[3] - beta[2]       # Slope Jura vs. Massif Central
}
")
```

We need fairly long chains to get convergence for this model.

```
# Function to generate starting values
inits <- function(){list(alpha = rlnorm(nPops, 3, 1), beta = rlnorm(nPops, 2, 1))}

# Parameters to estimate
params <- c("alpha", "beta", "a.effe2", "a.effe3", "b.effe2", "b.effe3", "test1")

# MCMC settings
na <- 2000;   ni <- 25000;   nb <- 5000;   nc <- 4;   nt <- 5

# Call JAGS (ART <1 min), check convergence and summarize posteriors
out13.4 <- jags(dataList, inits, params, "model13.4.txt", n.iter = ni,
  n.burnin = nb, n.chains = nc, n.thin = nt, n.adapt = na, parallel = TRUE)
jagsUI::traceplot(out13.4)                    # not shown
print(out13.4, 3)
```

	mean	sd	2.5%	50%	97.5%	overlap0	f	Rhat	n.eff
alpha[1]	-1.881	0.277	-2.449	-1.870	-1.373	FALSE	1.000	1.002	9865
alpha[2]	-1.106	0.216	-1.553	-1.097	-0.708	FALSE	1.000	1.003	1334
alpha[3]	-0.058	0.117	-0.294	-0.054	0.166	TRUE	0.682	1.002	1181
beta[1]	3.921	0.257	3.444	3.911	4.445	FALSE	1.000	1.002	16000
beta[2]	2.063	0.235	1.621	2.058	2.546	FALSE	1.000	1.003	1424
beta[3]	-1.052	0.140	-1.333	-1.051	-0.782	FALSE	1.000	1.003	949
a.effe2	0.775	0.356	0.103	0.764	1.486	FALSE	0.990	1.002	10110
a.effe3	1.823	0.299	1.267	1.811	2.422	FALSE	1.000	1.002	16000
b.effe2	-1.858	0.351	-2.550	-1.850	-1.189	FALSE	1.000	1.002	5837
b.effe3	-4.974	0.293	-5.569	-4.965	-4.420	FALSE	1.000	1.002	2130
test1	-3.116	0.272	-3.671	-3.109	-2.595	FALSE	1.000	1.005	680

We inspect the results and compare them with truth in the data-generating random process as well as with inferences from `glm()`.

```
# Compare glm() estimes with Bayesian estimates and with truth
jags_est <- unlist(out13.4$mean)[c(1, 7, 8, 4, 9, 10)]
comp <- cbind(truth = truth, glm = glm_est, JAGS = jags_est)
print(comp, 4)
```

```
                               truth       glm      JAGS
(Intercept)                       -2   -1.8618   -1.8810
popMassif Central                  1    0.7809    0.7752
popJura                            2    1.8107    1.8231
wing.length                        4    3.9066    3.9210
popMassif Central:wing.length     -2   -1.8632   -1.8577
popJura:wing.length               -5   -4.9582   -4.9735
```

Remember that `alpha[1]` and `beta[1]` in JAGS correspond to the intercept and the `length` main effect in the analysis in R and `a.effe2`, `a.effe3`. `b.effe2`, `b.effe3` to the remaining terms of the analysis in R. As expected, we find similar estimates between the two engines, as the data set is large relative to model complexity, and because in our Bayesian analysis we use vague priors.

13.4.2 **Forming predictions**

Finally, let us summarize our main findings from the analysis in a graph. To do that, we compute the predicted relationship between expected mite count (i.e., λ_i) and wing length in each population for every posterior MCMC draw of the involved parameters and summarize the resulting derived

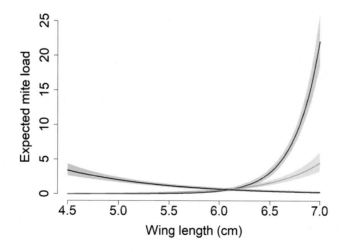

FIGURE 13.3

Plot showing the predicted relationships between expected mite load and the size of a dragonfly in three mountain ranges: Pyrenees (red), Massif Central (green), and Jura mountains (blue). Shaded areas show a 95% Bayesian credible interval for the posterior means, which are represented by the colored lines.

posteriors by taking the mean and the central 95% percentiles (Fig. 13.3). First, we generate the derived posterior distributions of predicted mite load for each of the three populations:

```
# Create a vector with 100 wing lengths
orig.wlength <- sort(orig.length)
wlength <- orig.wlength - mean(orig.length)

# Create matrices to contain prediction for each winglength and MCMC iteration
nsamp <- out13.4$mcmc.info$n.samples       # Get size of posterior sample
mite.load.Pyr <- mite.load.MC <- mite.load.Ju <- array(dim = c(nsamp, 300))

# Fill in these vectors: this is clumsy, but it works
alpha <- out13.4$sims.list$alpha       # posterior of alpha
beta <- out13.4$sims.list$beta         # posterior of beta
for (i in 1:300){
    mite.load.Pyr[,i] <- exp(alpha[,1] + beta[,1] * wlength[i])
    mite.load.MC[,i] <- exp(alpha[,2] + beta[,2] * wlength[i])
    mite.load.Ju[,i] <- exp(alpha[,3] + beta[,3] * wlength[i])
}
```

Next, we compute means and 95% credible intervals from the derived posterior for each population, and plot the results (Fig. 13.3).

```
# Compute 95% Bayesian credible intervals around predictions
LCB.Pyr    <- apply(mite.load.Pyr,   2,   quantile,   prob = 0.025)
UCB.Pyr    <- apply(mite.load.Pyr,   2,   quantile,   prob = 0.975)
LCB.MC     <- apply(mite.load.MC,    2,   quantile,   prob = 0.025)
UCB.MC     <- apply(mite.load.MC,    2,   quantile,   prob = 0.975)
LCB.Ju     <- apply(mite.load.Ju,    2,   quantile,   prob = 0.025)
UCB.Ju     <- apply(mite.load.Ju,    2,   quantile,   prob = 0.975)

# Compute posterior means
mean.rel <- cbind(exp(out13.4$mean$alpha[1] + out13.4$mean$beta[1] * wlength),
    exp(out13.4$mean$alpha[2] + out13.4$mean$beta[2] * wlength),
    exp(out13.4$mean$alpha[3] + out13.4$mean$beta[3] * wlength))
covar <- cbind(orig.wlength, orig.wlength, orig.wlength)

# Plot (Fig. 13.3)
par(mar = c(6, 6, 5, 3), cex.lab = 1.5, cex.axis = 1.5)
matplot(orig.wlength, mean.rel, col = c("red", "green", "blue"), type = "l",
    lty = 1, lwd = 2, las = 1, ylab = "Expected mite load",
    xlab = "Wing length (cm)", frame = FALSE, ylim = c(0, 25))
polygon(c(orig.wlength, rev(orig.wlength)), c(LCB.Pyr, rev(UCB.Pyr)),
    col = rgb(1,0,0,0.2), border = NA)
polygon(c(orig.wlength, rev(orig.wlength)), c(LCB.MC, rev(UCB.MC)),
    col = rgb(0,1,0,0.2), border = NA)
polygon(c(orig.wlength, rev(orig.wlength)), c(LCB.Ju, rev(UCB.Ju)),
    col = rgb(0,0,1,0.2), border = NA)
matplot(orig.wlength, mean.rel, col = c("red", "green", "blue"),
    type = "l", lty = 1, lwd = 3, add = TRUE)
```

We can clearly see in Fig. 13.3 that the lines representing the relationship between wing length and mite load for each population are not parallel, which illustrates the interaction effects between population and wing length. Comparing with Fig. 13.2, our model captures well the relationships between expected (or mean) mite counts and the two explanatory variables "population" and "wing length."

13.5 **Bayesian analysis with NIMBLE**

NIMBLE code for the model is available on the book website.

13.6 **Bayesian analysis with Stan**

The Stan model uses the same data, only replacing the dot in the names with an underline, and looks nearly identical to the JAGS and NIMBLE code. As usual we calculate the effects, i.e., the differences between pairs of slopes and intercepts, in the `generated quantities` block.

```
# Bundle and summarize data
str(dataList <- list(load = load, pop = as.numeric(pop), nPops = nPops,
  wing_length = wing.length, n = n) )

# Write Stan model
cat(file = "model13_6.stan", "
data {
  int n;
  int nPops;
  array[n] int load;
  vector[n] wing_length;
  array[n] int pop;
}

parameters {
  vector[nPops] alpha;
  vector[nPops] beta;
}

model {
  vector[n] lambda;

  for (i in 1:nPops){
    alpha[i] ~ normal(0, 100);
    beta[i] ~ normal(0, 100);
  }

  for (i in 1:n) {
    lambda[i] = exp(alpha[pop[i]] + beta[pop[i]] * wing_length[i]);
    load[i] ~ poisson(lambda[i]);
  }
}

generated quantities {
  real a_effe2 = alpha[2] - alpha[1];
  real a_effe3 = alpha[3] - alpha[1];
  real b_effe2 = beta[2] - beta[1];
  real b_effe3 = beta[3] - beta[1];
  real test1 = beta[3] - beta[2];
}
")

# HMC settings
ni <- 2000  ;  nb <- 1000  ;  nc <- 4  ;  nt <- 1
```

```
# Call STAN (ART 60/10 sec), assess convergence and print results table
system.time(
out13.6 <- stan(file = "model13_6.stan", data = dataList,
  warmup = nb, iter = ni, chains = nc, thin = nt) )
rstan::traceplot(out13.6)                         # not shown
print(out13.6, dig = 3)                           # not shown
stan_est <- summary(out13.6)$summary[c(1,7,8,4,9,10),1]  # Save estimates
```

13.7 Do-it-yourself maximum likelihood estimates

The likelihood function is very similar to the one from Chapter 11. The only difference is here we will calculate expected counts λ by matrix-multiplying the design matrix $\mathbf{X}$ and a vector of parameters β and exponentiating the result, the same way we did when simulating the data.

$$L(\beta \mid y, X) = \prod_{i=1}^{n} \frac{\lambda_i^{y_i} e^{-\lambda_i}}{y_i!}$$

$$\lambda_i = \exp(X\beta)$$

As a reminder, here is the design matrix we already created:

```
head(Xmat, 10)          # Look at first 10 rows of design matrix
```

The corresponding R function also looks similar to the do-it-yourself function from Chapter 11. The function begins by calculating `lambda` for each dragonfly by matrix-multiplying `Xmat` by the vector of parameters `beta` and exponentiating the result. This parameter vector is ordered the same way as the `beta.vec` we used for simulation, and thus our results will correspond to an effects parameterization, unlike JAGS, NIMBLE, and Stan, but like the default in `glm()`. Then we calculate the log-likelihood of each observation `y` given `lambda`, sum up, and take the negative.

```
# Define NLL for general Poisson regression
NLL <- function(beta, y, Xmat) {
  lambda <- exp(Xmat %*% beta)      # Multiply design matrix by beta
  LL <- dpois(y, lambda, log=TRUE)  # Log-likelihood of each obs
  NLL <- -sum(LL)                   # NLL for all observations in data set
  return(NLL)
}
```

```
# Minimize that NLL to find MLEs, get SEs and CIs and save estimates
# Need to use method = "BFGS" here
inits <- rep(0, 6)
names(inits) <- colnames(Xmat)
out13.7 <- optim(inits, NLL, y = load, Xmat = Xmat, hessian = TRUE, method = "BFGS")
get_MLE(out13.7, 4)
diy_est <- out13.7$par                    # Save estimates
```

	MLE	ASE	LCL.95	UCL.95
(Intercept)	-1.8619	0.2761	-2.40306	-1.321
popMassif Central	0.7802	0.3541	0.08611	1.474
popJura	1.8108	0.2995	1.22389	2.398
wing.length	3.9068	0.2576	3.40196	4.412
popMassif Central:wing.length	-1.8624	0.3542	-2.55668	-1.168
popJura:wing.length	-4.9585	0.2931	-5.53287	-4.384

13.8 **Likelihood analysis with TMB**

For the Template Model Builder (TMB) model, we start with the same data set as with Stan, and make one minor change: we must index our populations (pop) starting at 0 as per C++ conventions.

```
# Bundle and summarize data
str(tmbData <- list(load = load, pop = as.numeric(pop) - 1,
  nPops = nPops, wing_length = wing.length, n = n) )        # not shown
```

The TMB model file looks similar to the Stan code. Like JAGS, NIMBLE, and Stan, we apply a means parameterization in which we estimate separate intercepts (alpha) and slopes (beta) for each population. We also calculate derived parameters corresponding to the effects parameterization and add them to the output using ADREPORT.

```
# Write TMB model file
cat(file = "model13_8.cpp",
"#include <TMB.hpp>

template<class Type>
Type objective_function<Type>::operator() ()
{
  //Describe input data
  DATA_VECTOR(load);                      //response
  DATA_VECTOR(wing_length);               //covariate
  DATA_IVECTOR(pop);                      //population index (of integers; IVECTOR)
  DATA_INTEGER(n);                        //Number of observations

  //Describe parameters
  PARAMETER_VECTOR(alpha);
  PARAMETER_VECTOR(beta);

  Type LL = 0;                            //Initialize log-likelihood at 0

  for (int i=0; i<n; i++){
    Type lambda = exp(alpha(pop(i)) + beta(pop(i)) * wing_length(i));
    //Calculate log-likelihood of observation and add to total
    LL += dpois(load(i), lambda, true);
  }

  //Derived parameters
  Type a_effe2 = alpha(1) - alpha(0);
  Type a_effe3 = alpha(2) - alpha(0);
  Type b_effe2 = beta(1) - beta(0);
  Type b_effe3 = beta(2) - beta(0);
  Type test1 = beta(2) - beta(1);
  ADREPORT(a_effe2);
  ADREPORT(a_effe3);
  ADREPORT(b_effe2);
  ADREPORT(b_effe3);
  ADREPORT(test1);

  return -LL;                             //Return negative log likelihood
}
")
```

```
# Compile and load TMB function
compile("model13_8.cpp")
dyn.load(dynlib("model13_8"))

# Provide dimensions and starting values for parameters
params <- list(alpha = rep(0, tmbData$nPops), beta = rep(0, tmbData$nPops))

# Create TMB object
out13.8 <- MakeADFun(data = tmbData,
                     parameters = params,
                     DLL = "model13_8", silent = TRUE)

# Optimize TMB object and print results
opt <- optim(out13.8$par, fn = out13.8$fn, gr = out13.8$gr, method = "BFGS")
(tsum <- tmb_summary(out13.8))
tmb_est <- tsum[c(1, 7, 8, 4, 9, 10), 1]        # Save parameter estimates
```

13.9 Comparison of the parameter estimates

We find the typical, good agreement among the different methods for all parameter estimates.

```
# Compare results with truth and previous estimates
comp <- cbind(truth = beta.vec, glm = glm_est, JAGS = jags_est,
   NIMBLE = nimble_est, Stan = stan_est, DIY = diy_est, TMB = tmb_est)
print(comp, 3)
```

	truth	glm	JAGS	NIMBLE	Stan	DIY	TMB
(Intercept)	-2	-1.862	-1.881	-1.882	-1.885	-1.86	-1.862
popMassif Central	1	0.781	0.775	0.764	0.781	0.78	0.781
popJura	2	1.811	1.823	1.820	1.823	1.81	1.811
wing.length	4	3.907	3.921	3.922	3.925	3.91	3.907
popMassif Central:wing.length	-2	-1.863	-1.858	-1.848	-1.862	-1.86	-1.863
popJura:wing.length	-5	-4.958	-4.974	-4.978	-4.982	-4.96	-4.958

13.10 Summary

We have generalized the general linear model from the normal to the Poisson case to model the effects of a categorical and a continuous covariate for a count response. The result is a Poisson GLM with a linear model of the analysis of covariance form (Quinn & Keough, 2002). The changes involved for each engine were minor compared to the normal linear case in Chapter 9: change the distribution for the response from normal to Poisson, the link function from identity to log, and get rid of the dispersion parameter in the normal. The inclusion of further covariates is straightforward. The Poisson GLM is an important intermediate step for your understanding of the Poisson generalized linear mixed model, which we arrive at by declaring as random effects one or more sets of population-level parameters. We do this next.

Poisson generalized linear mixed model, or Poisson GLMM[⊛]

Chapter outline

14.1 Introduction

The Poisson generalized linear mixed model (GLMM) is an extension of the Poisson generalized linear model (GLM) to include at least one additional source of random variation over and above the random variation intrinsic to a Poisson distribution. Here, we adopt a Poisson GLMM to analyze a set of long-term population surveys of red-backed shrikes (Fig. 14.1).

We assume that pair counts over 30 years were available in each of 16 shrike populations and that our intent was to model population trends. First, we write down the random-coefficients model without correlation between the intercepts and slopes (see Section 10.4). We model C_i, counts of pairs of red-backed shrikes, with i running from 1 to the total sample size n. Each count is associated with one of $J = 16$ populations, indexed by j, and occurs in a particular year x taking values from 1 to 30. Each population j has a normally distributed intercept and slope value. We describe our random-effects GLM as follows:

1. Distribution for observed data: $C_i \sim \text{Poisson}(\lambda_i)$
2. Link function: log, i.e., $\eta_i = \log(\lambda_i) = \log(E(C_i))$
3. Linear predictor (η): $\eta_i = \alpha_{j(i)} + \beta_{j(i)} x_i$
4. Submodel for parameters/distribution of random effects:

$$\alpha_j \sim Normal(\mu_\alpha, \sigma_\alpha^2)$$

$$\beta_j \sim Normal(\mu_\beta, \sigma_\beta^2)$$

⊛This book has a companion website hosting complementary materials, including all code for download. Visit this URL to access it: https://www.elsevier.com/books-and-journals/book-companion/9780443137150.

Applied Statistical Modelling for Ecologists. DOI: https://doi.org/10.1016/B978-0-443-13715-0.00006-6

FIGURE 14.1

Male red-backed shrike (*Lanius collurio*), Switzerland, 2004 (Photo by Alain Saunier).

The linear model in this GLMM is identical to that in the GLM we saw in Chapter 13. In both cases, the model's geometric representation is a cluster of log-linear regressions; this is described by the first three lines of algebra above. However, we have now specified the population-level slopes and intercepts as random effects coming from normal distributions which are common to all shrike populations.

The model as written above is our preferred parameterization, and how we usually implement it in the custom model-fitting engines. However, there is another and equivalent way of writing Equations 3 and 4, which is the parameterization used in the R packages `lme4` and `glmmTMB`.

3. Linear predictor (η):
$$\eta_i = \mu_\alpha + \alpha_{j(i)} + \mu_\beta + \beta_{j(i)} x_i$$

4. Submodel for parameters/distribution of random effects:

$$\alpha_j \sim Normal(0, \sigma_\alpha^2)$$

$$\beta_j \sim Normal(0, \sigma_\beta^2)$$

In this new, yet equivalent version of the model, we moved the random-effect means μ_α and μ_β out of the random-effects distributions (4) and they are now explicitly shown in the linear predictor (3). As a result, the random effects distributions now have mean 0. The population-level random effects α_j and β_j then represent deviations from the overall mean intercept μ_α and slope μ_β, instead of population-level intercepts and slopes.

Note also that we don't add a year-specific residual to the linear predictor. This could be done to account for random year effects or overdispersion in the Poisson data distribution as in Chapter 12. In addition, this might also be a first step towards modeling serial autocorrelation, e.g., by imposing an autoregressive structure on successive random year effects (Schaub & Kéry, 2022). However, we omit such complexity here, nor will we model a correlation between the random intercepts and slopes as we did in Chapter 10.

In conducting this analysis, we implicitly make one of two assumptions (see Chapter 11.2). Either we assume that we find all shrike pairs in every year and study area or we assume that at least the proportion of pairs overlooked does not vary among years or study areas in a systematic way, i.e., that there is no time trend in detection probability. Otherwise, the interpretation of these data would be difficult. For an alternative protocol for data collection and associated analysis method that doesn't require these potentially restrictive assumptions see the binomial *N*-mixture model in the bonus Chapter 19B on our website.

14.2 **Data generation**

We generate data under the random-coefficients model, without correlation between the intercepts and slopes. When executing this code, notice once more how it reflects exactly the algebraic statements above with which we defined the Poisson GLMM.

```
set.seed(14)
nPops <- 16
nYears <- 30
n <- nPops * nYears        # n = 480
pop <- gl(n = nPops, k = nYears)
```

We standardize the year covariate to a range from 0 to 1.

```
orig.year <- rep(1:nYears, nPops)
year <- (orig.year-1)/29   # Squeeze between 0 and 1
```

We build a design matrix without the intercept (i.e., yielding a means parameterization) and look at the top 91 rows—make sure you understand what the design matrix means.

```
Xmat <- model.matrix(~ pop * year - 1 - year)
print(Xmat[1:91,], 2)      # Print top 91 rows of 480
```

Next, we draw the intercept and slope parameter values from their respective two normal distributions. We need to pick values for the hyperparameters first.

```
# Choose values for hyperparams and draw Normal random numbers
mu.alpha <- 3                    # Mean of intercepts
sigma.alpha <- 1                 # SD of intercepts
mu.beta <- -2                    # Mean of slopes
sigma.beta <- 0.6                # SD of slopes
alpha <- rnorm(n = nPops, mean = mu.alpha, sd = sigma.alpha)
beta <- rnorm(n = nPops, mean = mu.beta, sd = sigma.beta)
all.effects <- c(alpha, beta)    # All together

# Save true parameter values
truth <- c(mu.alpha = mu.alpha, mu.beta = mu.beta,
           sigma.alpha = sigma.alpha, sigma.beta = sigma.beta)
```

As in the previous chapter, we assemble the counts C_i by first computing the linear predictor using matrix multiplication, then exponentiating it, and finally adding Poisson noise. Then, we look at the data; we have $n = 480$ site-year combinations.

```
lin.pred <- Xmat[,] %*% all.effects       # Value of lin.predictor (eta)
C <- rpois(n = n, lambda = exp(lin.pred)) # Exponentiate and add Poisson noise
hist(C, col = "grey")                      # Inspect what we've created (not shown)

# Load required libraries
library(ASMbook); library(jagsUI); library(rstan); library(TMB); library(lattice);
  library(glmmTMB); library(DHARMa); library(abind); library(pracma)
```

We use a lattice graph to plot the shrike counts against time for each population (Fig. 14.2).

```
xyplot(C ~ orig.year | pop, ylab = "Red-backed shrike counts", xlab = "Year",
  pch = 16, cex = 1.2, col = rgb(0, 0, 0, 0.4))
```

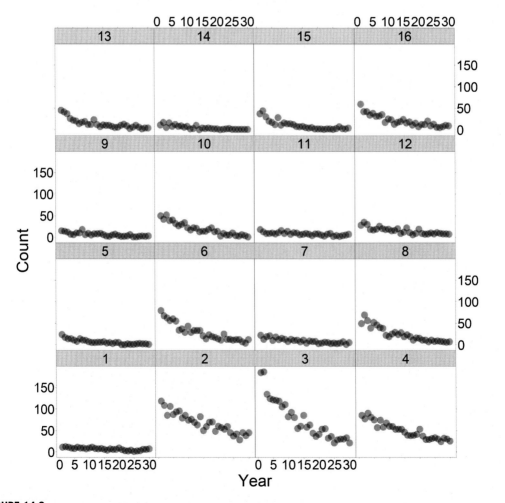

FIGURE 14.2

Trellis or lattice plot of pair counts in 16 populations of red-backed shrikes over 30 years (total sample size is 480).

We could now analyze the shrike counts under the assumption that all shrike populations have the same trend, but at different levels, corresponding to a random-intercepts model, as we did for the Gaussian case in Section 10.3. However, we only show here the analysis under a more general model with both random intercepts and random slopes. From that, you will easily be able to see how to fit the simpler model with just random intercepts if this is what you need.

14.3 Likelihood analysis with canned functions in R

As in Chapter 10, we will use the `glmmTMB` package to fit the model, but the code would be almost the same using `lme4::glmer`. As a review, the random effects structure is specified in the formula. To specify random intercepts by population, the notation is `(1 | pop)`. For both random (year) slopes and intercepts by population, it's `(year | pop)`, which is equivalent to `(1 + year | pop)`. By default the model assumes that random slopes and intercepts are correlated. Here, we don't want a correlation, so we must specify the model as `(year || pop)`.

```
gtmb.data <- data.frame(C = C, year = year, pop = pop)      # bundle data
out14.3 <- glmmTMB(C ~ year + (year || pop), data = gtmb.data, family = poisson)
summary(out14.3)                                            # Inspect results
sds <- attr(VarCorr(out14.3)$cond$pop, 'stddev')            # Save results
gtmb_est <- c(fixef(out14.3)$cond, sds)

Random effects:

Conditional model:
 Groups Name          Variance  Std.Dev.  Corr
 pop    (Intercept)   0.6301    0.7938
        Year          0.3861    0.6214    0.00
Number of obs: 480,  groups:    pop, 16

Conditional model:
              Estimate  Std. Error  z value  Pr(>|z|)
(Intercept)    3.5842     0.1995     17.96    <2e-16 ***
Year          -2.0081     0.1630    -12.32    <2e-16 ***
```

The summary output from `glmmTMB` requires some explanation (and we note that we truncated it somewhat). The first section immediately under the heading `Random effects` can be a bit confusing and ambiguous, as what is shown are not the actual random effects (in our case α_j and β_j). Instead, this section of the summary gives estimates of the variance hyperparameters for the random effects: σ^2_α and σ^2_β. There are no estimates of the mean hyperparameters μ_α and μ_β given here. As we described in Section 14.1, `glmmTMB` adopts a parameterization for mixed models in which the distributions of random effects always have mean 0; i.e., the mean is "pulled out" of the random effects distribution and put explicitly into the linear predictor. Therefore, μ_α and μ_β are equivalent to the estimates given at the end of the summary for `(Intercept)` and `year`, respectively.

In order to obtain estimates of the actual random effects α_j and β_j, we can run the function `ranef` on the model object:

```
ranef(out14.3)   # ran-ef. estimates from zero-mean distribution (not shown)
```

You should be able to tell just by looking that these random effect estimates come from distributions with mean 0. If we want to obtain "complete" estimates of population-level intercepts and slopes α_j and β_j, equivalent to our original parameterization in Section 14.1, we simply need to add together the estimates of the means μ_α and μ_β with the population-level deviations α_j and β_j. In R we can do this automatically with the function `coef()`.

```
coef(out14.3)   # 'full' ran-ef estimates (not shown)
```

As a next step in our analysis workflow, we assess goodness-of-fit (GOF) by first plotting the estimates of both sets of random effects, which the model assumes are draws from two normal distributions. We note, though, that this assessment has very little power with our small sample sizes of 16 populations only. Second, we conduct an assessment of quantile residuals and tests, using R package DHARMa (Hartig, 2022). Some tests show up significant, which is not surprising because when well calibrated, such tests are expected to sound false alarms in about 5% of the cases on average.

```
# Assess goodness-of-fit of the model (not shown)
par(mfrow = c(1, 2))    # Distribution of estimates of alpha and beta
hist(coef(out14.3)$cond$pop[,1], main = 'alpha', breaks = 12)
hist(coef(out14.3)$cond$pop[,2], main = 'beta', breaks = 12)

simOut <- simulateResiduals(out14.3, n = 1000, plot = TRUE)        # from DHARMa
```

14.4 Bayesian analysis with JAGS

In our Poisson GLMM fit in JAGS, we use a means parameterization. The model is very similar to the linear mixed model from Section 10.4.2. We just need to add the appropriate exponential inverse link function to the linear predictor (or log link function to the response). We also add code to compute the predictive distribution of the data as a preparation of residual assessments of the data using DHARMa. We condition on the random effects here; see the package vignette called "DHARMa for Bayesians" (Hartig, 2022).

```
# Bundle and summarize data
str(dataList <- list(C = C, pop = as.numeric(pop), year = year, nPops = nPops, n = n) )

List of 5
 $ C     : int [1:480] 11  12  11  9  8  11  9  9  7  9 ...
 $ pop   : num [1:480] 1  1  1  1  1  1  1  1  1  1 ...
 $ year  : num [1:480] 0  0.0345  0.069  0.1034  0.1379 ...
 $ nPops : num 16
 $ n     : num 480
```

```
# Write JAGS model file
cat(file = "model14.4.txt", "
model {
# Priors
for (i in 1:nPops){
  # Models for the sets of random effects alpha and beta
  alpha[i] ~ dnorm(mu.alpha, tau.alpha)          # Intercepts
  beta[i] ~ dnorm(mu.beta, tau.beta)             # Slopes
}
mu.alpha ~ dnorm(0, 0.0001)                      # Hyperparam. for random intercepts
tau.alpha <- pow(sigma.alpha, -2)
sigma.alpha ~ dt(0, 0.1, 1)T(0,)                 # half-Cauchy prior

mu.beta ~ dnorm(0, 0.0001)                       # Hyperparam. for random slopes
tau.beta <- pow(sigma.beta, -2)
sigma.beta ~ dt(0, 0.1, 1)T(0,)                  # half-Cauchy prior

# 'Likelihood'
for (i in 1:n){
  C[i] ~ dpois(lambda[i])
  lambda[i] <- exp(alpha[pop[i]] + beta[pop[i]] * year[i])
  # log(lambda[i]) <- alpha[pop[i]] + beta[pop[i]]* year[i]    # same
}

# Posterior predictive simulations (used in DHARMa GoF assessment)
for (i in 1:n) {
  C.new[i] ~ dpois(lambda[i])
}
}
")


# Function to generate starting values
inits <- function(){
  list(mu.alpha = rnorm(1), mu.beta = rnorm(1),
       sigma.alpha = runif(1), sigma.beta = runif(1))
}


# Parameters to estimate
params <- c("mu.alpha", "mu.beta", "sigma.alpha", "sigma.beta", "alpha", "beta", "C.new")

# MCMC settings
na <- 1000 ; ni <- 12000 ; nb <- 2000 ; nc <- 4 ; nt <- 10

# Call JAGS (ART <1 min), check convergence and summarize posteriors
out14.4 <- jags(dataList, inits, params, "model14.4.txt", n.iter = ni,
  n.burnin = nb, n.chains = nc, n.thin = nt, n.adapt = na, parallel = TRUE)
jagsUI::traceplot(out14.4)          # not shown
print(out14.4, 3)                   # first four lines shown only
```

	mean	sd	2.5%	50%	97.5%	overlap0	f	Rhat	n.eff
mu.alpha	3.590	0.229	3.142	3.588	4.055	FALSE	1	1.001	1821
mu.beta	-2.004	0.189	-2.389	-1.999	-1.631	FALSE	1	1.000	4000
sigma.alpha	0.894	0.182	0.618	0.864	1.340	FALSE	1	1.001	2095
sigma.beta	0.706	0.159	0.468	0.684	1.087	FALSE	1	1.000	4000

This GLMM converges easily. Before we compare the truth with the frequentist and the Bayesian solutions, we sketch out the code for a GOF assessment with histograms of the estimates of the random effects and assessment of quantile residuals using DHARMa.

```
# Distribution of random effects estimates (alpha, beta)
par(mfrow = c(1,2)) # Very low power with small samples...
hist(out14.4$mean$alpha, main = 'alpha', breaks = 12)
hist(out14.4$mean$beta, main = 'beta', breaks = 12)

# Do quantile residual assessments (not shown)
C.new <- out14.4$sims.list$C.new
sim <- createDHARMa(simulatedResponse = t(C.new), observedResponse = C,
   fittedPredictedResponse = apply(C.new, 2, median), integerResponse = T)
plot(sim)

# Compare likelihood with Bayesian estimates and with truth
jags_est <- unlist(out14.4$mean[1:4])
comp <- cbind(truth = truth, gtmb = gtmb_est, JAGS = jags_est)
print(comp, 4)
```

	truth	gtmb	JAGS
mu.alpha	3.0	3.5842	3.5898
mu.beta	-2.0	-2.0081	-2.0044
sigma.alpha	1.0	0.7938	0.8936
sigma.beta	0.6	0.6214	0.7061

We find very similar point estimates for the mean hyperparameters, but slight differences in the estimates of the dispersion parameters. They are likely due to the small sample size at the level of the random-effects factor pop. As a consequence, even vague priors may exert some effects on the posteriors. You can test this idea by repeating the simulation with 64 populations instead, for instance, and will then find almost identical point estimates for these dispersion parameters as well.

14.5 Bayesian analysis with NIMBLE

NIMBLE code for the model is available on the book website.

14.6 Bayesian analysis with Stan

The model in Stan looks very similar to BUGS. Recall from Chapter 11 that Stan's Poisson distribution is called poisson, rather than dpois as it is in BUGS.

```
# Bundle and summarize data
str(dataList <- list(C = C, pop = as.numeric(pop), year = year, nPops = nPops, n = n) )

# Write Stan model
cat(file = "model14_6.stan", "
data{
  int n;
  int nPops;
  array[n] int C;
  vector[n] year;
  array[n] int pop;
}

parameters{
  real mu_alpha;
  real mu_beta;
  real<lower=0> sigma_alpha;
  real<lower=0> sigma_beta;
  vector[nPops] alpha;
  vector[nPops] beta;
}

model{
  vector[n] lambda;

  mu_alpha ~ normal(0, 100);
  sigma_alpha ~ cauchy(0, sqrt(10));
  mu_beta ~ normal(0, 100);
  sigma_beta ~ cauchy(0, sqrt(10));
  for (i in 1:nPops){
    alpha[i] ~ normal(mu_alpha, sigma_alpha);
    beta[i] ~ normal(mu_beta, sigma_beta);
  }

  for (i in 1:n){
    lambda[i] = exp(alpha[pop[i]] + beta[pop[i]] * year[i]);
    C[i] ~ poisson(lambda[i]);
  }
}
")

# HMC settings
ni <- 2000  ; nb <- 1000  ; nc <- 4  ; nt <- 1

# Call STAN (ART 67/12 sec), assess convergence, show results and save estimates
system.time(
out14.6 <- stan(file = "model14_6.stan", data = dataList,
                warmup = nb, iter = ni, chains = nc, thin = nt) )
rstan::traceplot(out14.6)                            # not shown
print(out14.6, dig = 3)                              # not shown
stan_est <- summary(out14.6)$summary[1:4,1]          # Save estimates
```

14.7 Do-it-yourself maximum likelihood estimates

Remember that for the likelihood of a mixed model, we start with the joint density of the observed and unobserved random variables and then integrate out the latter. This leaves us with a joint density of observed random variables and hyperparameters; this is the integrated, or marginal, likelihood that we maximize (see Section 10.3.5). As we have seen in Chapter 10, with both random intercepts and random slopes, the math for the likelihood function gets a little more involved. To simplify things, we will first subset our complete vector of counts C into $J = 16$ subvectors C_j each containing just the counts for population j and start by writing out the likelihood math for just population j. Each population has $K = 30$ yearly counts, and we will represent a single data point k from population j with notation $C_{j,k}$ where $k = 1...K$.

We start with a basic Poisson likelihood for counts C_j with population-specific intercept α_j and slope β_j (adopting our preferred parameterization from Section 14.1):

$$\prod_{k=1}^{K} \text{dpois}(C_{j,k}|\lambda_{j,k})$$

$$\lambda_{j,k} = \exp(\alpha_j + \beta_j \cdot year_{j,k})$$

We multiply that value with the likelihoods for the population-level intercept α_j and slope β_j, which are draws from two normal distributions with hyperparameters μ_α and σ_α, and μ_β and σ_β, respectively.

$$\prod_{k=1}^{K} \left[\text{dpois}\left(C_{j,k}|\lambda_{j,k}\right)\right] \cdot \text{dnorm}(\alpha_j|\mu_\alpha, \sigma_\alpha) \cdot \text{dnorm}(\beta_j|\mu_\beta, \sigma_\beta)$$

We'll call the equation above $f(\alpha_j, \beta_j, \mu_\beta, \sigma_\beta, \mu_\alpha, \sigma_\alpha, C_j, year_j)$. Here's f as an R function:

```
f <- function(alpha_j, beta_j, mu.alpha, sig.alpha,
          mu.beta, sig.beta, C_j, year_j){
  lambda <- exp(alpha_j + beta_j * year_j)
  prod(dpois(C_j, lambda)) * dnorm(alpha_j, mu.alpha, sig.alpha) *
    dnorm(beta_j, mu.beta, sig.beta)
}
```

Now we need to integrate over all possible values of α_j and β_j, for which we will need to evaluate a double integral:

$$\iint_{-\infty}^{\infty} f(\alpha_j, \beta_j, \mu_\beta, \sigma_\beta, \mu_\alpha, \sigma_\alpha, C_j, year_j) d\alpha_i \, d\beta_i = g(\mu_\alpha, \sigma_\alpha, \mu_\beta, \sigma_\beta, C_j, year_j)$$

The resulting new function g depends only on the data and hyperparameters (i.e., we have now integrated out α_j and β_j); it represents the likelihood for the data coming from a single population. Now, to write g as an R function, in which we do double integration we'll use the `integral2` function from package `pracma`. We'll set the minimum and maximum values for each integration at $-10, 10$.

```
g <- function(mu.alpha, sig.alpha, mu.beta, sig.beta, C_j, year_j){
  integral2(fun = f, xmin = -10, xmax = 10, ymin = -10, ymax = 10,
        vectorized = FALSE,
        mu.alpha = mu.alpha, sig.alpha = sig.alpha, mu.beta = mu.beta,
        sig.beta = sig.beta, C_j = C_j, year_j = year_j)$Q
}
```

To get the value of the likelihood for the entire data set, we need to take the product of the like-lihoods for the data in all $J = 16$ populations:

$$L\left(\mu_\alpha, \mu_\beta, \sigma_\alpha, \sigma_\beta | C, year\right) = \prod_{j=1}^{J} g(\mu_\alpha, \sigma_\alpha, \mu_\beta, \sigma_\beta, C_j, year_j)$$

Here's the complete likelihood function in R using our functions *f* and *g*, where as usual rather than multiplying the likelihoods for each population, we'll add the log-likelihoods, or in this case subtract, since we're always aiming for the negative log-likelihood.

```
# Define NLL for Poisson GLMM (using custom functions f and g)
NLL <- function(pars, data) {
  mu.alpha <- pars[1]
  mu.beta <- pars[2]
  sig.alpha <- exp(pars[3])
  sig.beta <- exp(pars[4])

  nll <- 0                    # Initialize nll at 0 prior to summation in loop below

  for (j in 1:data$nPops){
    # Subset data to just pop j
    C_j <- data$C[data$pop == j]
    year_j <- data$year[data$pop == j]
    lik <- g(mu.alpha, sig.alpha, mu.beta, sig.beta, C_j, year_j)
    nll <- nll - log(lik)
  }
  return(nll)
}
```

```
# Minimize that NLL to find MLEs, get SEs and CIs and save estimates (ART 122 sec)
inits <- c(mu.alpha = 0, mu.beta = 0, log.sigma.alpha = 0, log.sigma.beta = 0)
system.time(
  out14.7 <- optim(inits, NLL, data = dataList, hessian = TRUE,
    method = 'BFGS', control = list(trace = 1, REPORT = 1)) )
get_MLE(out14.7, 4)
diy_est <- c(out14.7$par[1:2], exp(out14.7$par[3:4]))        # Save estimates
```

	MLE	ASE	LCL.95	UCL.95
mu.alpha	3.5760	0.1964	3.1909	3.96102
mu.beta	-2.0021	0.1661	-2.3277	-1.67661
log.sigma.alpha	-0.2468	0.1786	-0.5968	0.10325
log.sigma.beta	-0.4489	0.1945	-0.8301	-0.06765

14.8 Likelihood analysis with TMB

The Poisson GLMM in TMB looks very similar to the implementation of the model in Stan. It's a lot easier to specify a random intercepts and slopes model in TMB compared to doing it ourselves! As usual, we need to adjust the population index to start at 0 to make C++ happy.

```
# Bundle and summarize data
tmbData <- dataList
tmbData$pop <- tmbData$pop - 1   #indices start at 0 in TMB
str(tmbData)
```

```
# Write TMB model file
cat(file = "model14_8.cpp",
"#include <TMB.hpp>

template<class Type>
Type objective_function<Type>::operator() ()
{
  //Describe input data
  DATA_VECTOR(C);                          //response
  DATA_VECTOR(year);                       //covariate
  DATA_IVECTOR(pop);                       //population index (of integers; IVECTOR)
  DATA_INTEGER(n);                         //Number of observations
  DATA_INTEGER(nPops);                     //Number of groups

  //Describe parameters
  PARAMETER(mu_alpha);
  PARAMETER(mu_beta);
  PARAMETER(log_sigma_alpha);
  PARAMETER(log_sigma_beta);
  PARAMETER_VECTOR(alpha);
  PARAMETER_VECTOR(beta);

  Type sigma_alpha = exp(log_sigma_alpha);
  Type sigma_beta = exp(log_sigma_beta);

  Type LL = 0.0;                           //Initialize log-likelihood at 0

  //Random effects
  for (int i = 0; i<nPops; i++){
    LL += dnorm(alpha(i), mu_alpha, sigma_alpha, true);
    LL += dnorm(beta(i), mu_beta, sigma_beta, true);
  }

  for (int i = 0; i<n; i++){
    Type lambda = exp(alpha(pop(i)) + beta(pop(i)) * year(i));
    LL += dpois(C(i), lambda, true);
  }

  return -LL;
}
")
```

We need to provide dimensions for all parameters and tell TMB which parameters are random effects (i.e., the random intercepts and slopes `alpha` and `beta`).

```
# Compile and load TMB function
compile("model14_8.cpp")
dyn.load(dynlib("model14_8"))

# Provide dimensions and starting values for parameters
params <- list(mu_alpha = 0, mu_beta = 0, log_sigma_alpha = 0,
    log_sigma_beta = 0, alpha = rep(0, tmbData$nPops),
    beta = rep(0, tmbData$nPops))
```

```
# Create TMB object
out14.8 <- MakeADFun(data = tmbData,
          parameters = params,
          random = c("alpha", "beta"),
          DLL = "model14_8", silent = TRUE)

# Optimize TMB object and print and save results
opt <- optim(out14.8$par, fn = out14.8$fn, gr = out14.8$gr, method = "BFGS")
(tsum <- tmb_summary(out14.8))                     # not shown
tmb_est <- c(tsum[1:2,1], exp(tsum[3:4,1]))        # save estimates
```

14.9 Comparison of the parameter estimates

We make our usual grand comparison and find fairly similar estimates among the engines.

```
# Compare estimates with truth
comp <- cbind(truth = truth, gtmb = gtmb_est, JAGS = jags_est,
    NIMBLE = nimble_est, Stan = stan_est, DIY = diy_est, TMB = tmb_est)
print(comp, 4)
```

	truth	gtmb	JAGS	NIMBLE	Stan	DIY	TMB
mu.alpha	3.0	3.5842	3.5898	3.582	3.5847	3.5760	3.5842
mu.beta	-2.0	-2.0081	-2.0044	-2.008	-2.0088	-2.0021	-2.0081
sigma.alpha	1.0	0.7938	0.8936	0.894	0.8932	0.7813	0.7938
sigma.beta	0.6	0.6214	0.7061	0.706	0.7077	0.6383	0.6214

As discussed in Section 14.4 already, the estimates of the mean hyperparameters are identical for all practical purposes, but those for the two dispersion hyperparameters are not quite "quasi-identical." This is due to the fact that variance (and covariance) parameters are always harder to estimate than parameters that represent a mean, and that we have a small sample size at the population level. With these, even our vague half-Cauchy priors for the standard deviations of the normal distributions adopted for the random effects appear to exert some influence on the posterior distributions obtained by the Bayesian engines. As a consequence, we find small discrepancies in these estimates between the likelihood and the Bayesian solutions. With a substantially larger sample size in terms of the number of populations in the study, we would obtain virtually identical estimates also for these parameters, as you can easily ascertain by repeating the computations in this chapter with a larger number of populations.

14.10 Summary

For a Poisson generalized linear mixed model, or Poisson GLMM, we have seen how the concept of random effects, which we first met in Chapters 8 and 10 for the normal linear case, can be carried over fairly smoothly to the non-normal case. From an educational point of view, we believe that there is no better way to explain a GLMM to a non-statistician than to look at the algebra or the BUGS code for the model, or at the R code for simulating data under such a model. The only difference between a GLM as in Chapter 13 and the mixed-effects version of that model in this chapter is that here we have added a distributional assumption to each of the sets of intercepts and slopes, making both sets of parameters random effects. In a mixed model, the distribution assumed for these random effects is always a normal, and the mean and the standard deviation (or the variance) of this normal then become the basic parameters estimated in the model. Since this model is also a hierarchical model, and the normal distributions adopted for the random effects are "one level up" compared to the observed data, their parameters are also known as "hyperparameters." We will see another instance of a random-effects GLM, for a binomial response, in Chapter 17. In Chapters 19 and 19B, we will cover examples of more general hierarchical models which no longer qualify formally as mixed models. This is because their random effects are not given a normal distribution and the random effects do not appear linearly in the linear predictor for the statistical model for the observed data.

Comparing two groups in a logistic regression model[*]

15

Chapter outline

15.1 Introduction

The next three chapters deal with another common kind of count data, *binomial* data, from which we want to estimate a proportion p as a parameter. The crucial difference between binomial and Poisson random variables is the presence of a ceiling in the former: binomial counts cannot be greater than some upper limit. Therefore, we model bounded counts, where the bound is provided by trial size N. It makes sense to express a count relative to that bound and this yields a proportion. In contrast, Poisson counts lack an upper limit, at least in principle. A very common type of model associated with binomial data is called logistic regression, specifically, when we use the *logit* link function (more on the logit function later).

Modeling a binomial random variable in essence means modeling a series of coin flips. We count the number of heads, r, among the total number of coin flips (N) and from this want to estimate the general tendency of the coin to show heads. That is, we want to estimate Pr(heads). The binomial distribution describes the number of times r a coin shows heads among a number of N flips, where the coin has Pr(heads) $= p$. We can also write $r \sim \text{Binomial}(N, p)$. A special case of the binomial distribution with $N = 1$, corresponding to a single flip of the coin, is called the Bernoulli distribution. It has just a single parameter, p. Data coming from coin flip-like processes are ubiquitous in nature and include survival or the presence/absence of an organism.

As our inferential setting in this chapter, we consider a plant inventory on calcareous grasslands in the Jura mountains. Assume a total of 50 sites were visited by experienced botanists who recorded whether

---

[*]This book has a companion website hosting complementary materials, including all code for download. Visit this URL to access it: https://www.elsevier.com/books-and-journals/book-companion/9780443137150.

Applied Statistical Modelling for Ecologists. DOI: https://doi.org/10.1016/B978-0-443-13715-0.00019-4
363

they saw a species or not. The Cross-leaved gentian (Fig. 15.1) was found at 15 sites and the Chiltern gentian (see Chapter 19) at 27 sites. We wonder whether this is enough evidence, given the variation inherent in a binomial random variable, to claim that the cross-leaved gentian has a more restricted distribution in the Jura mountains than does the Chiltern gentian. We will conceptualize "distribution" by occupancy, or occurrence, probability θ, which is the probability a randomly chosen site in the Jura mountains is occupied by a species, that is, has at least one individual growing on it.

This type of data is often called "presence-absence data." However, it is more accurate to call it "detection-nondetection data," since the number of sites at which a species is *detected* depends on two entirely different things: first, the number of sites where a species is actually present, and second, the ease with which a species is detected at an occupied site (Royle & Dorazio, 2008; Kéry, 2011). Without a special kind of repeated-measures data (see Chapter 19) we have no way of distinguishing between the two component processes that generate such "presence-absence" data. All we can do is hope, pray, or claim that either both gentian species are found at every occupied site or else that the probability each species is overlooked is identical, such that a comparison of the two error-prone counts will be fair. See also Section 11.2 for similar considerations about the measurement error in Poisson counts.

For now, we ignore these real-life complications of species distribution modeling, and simply assume that every individual present is indeed detected, that is, that every occupied site is observed as such. Given such data, our question can be framed statistically by what can be called a binomial version of a *t*-test, or as a logistic regression that contrasts two groups. This is similar to the

FIGURE 15.1

Cross-leaved gentian (*Gentiana cruciata*), Aragonese Pyrenees, 2006 (Photo by Marc Kéry).

Poisson model for two groups in Chapter 11. For gentian species i, let C_i be the number of sites it was detected. A simple model for C_i is this:

1. (Statistical) Distribution: $C_i \sim \text{Binomial}(N, \theta_i)$
2. Link function: logit, that is, $\eta_i = \text{logit}(\theta_i) = \log(\frac{\theta_i}{1-\theta_i})$
3. Linear predictor (η): $\eta_i = \alpha + \beta x_i$

If x is an indicator for the Chiltern gentian, then α can be interpreted as a logit-scale parameter for the probability of occurrence (θ) of the cross-leaved gentian in the Jura mountains and β is the difference, on the logit scale, between the probability of occurrence of the Chiltern gentian and that of the cross-leaved gentian.

A binomial random variable with trial size N and success probability θ can be represented as the sum of N i.i.d. Bernoulli random variables with the same θ (Chapter 3). Thus, the counts of 15 and 27 described above can be seen as the aggregations of 50 equivalent binary detection/nondetection data y. Then, for binary datum y_i (now with site index $i = 1 \ldots 100$) and the accompanying species indicator variable x (also of length 100), the exact same model can be written as follows (but note that now i runs from 1 to 100 and is an index for sites, while above it runs from 1 to 2 only and indexes the two species):

1. (Statistical) Distribution: $y_i \sim \text{Bernoulli}(\theta_i)$
2. Link function: logit, that is, $\eta_i = \text{logit}(\theta_i) = \log(\frac{\theta_i}{1-\theta_i})$
3. Linear predictor (η): $\eta_i = \alpha + \beta x_i$

We will illustrate this equivalence by first simulating the data as binary detection/nondetections and then keeping two variants, one aggregated (which we will analyze through the rest of the chapter), and the other one disaggregated, which we will only analyze in Section 15.3 to demonstrate the equivalence of the two variants of the Bernoulli/binomial models.

15.2 **Data generation**

We simulate the data from a Bernoulli process with a parameter θ that we define such that the sample data will match those in our example above.

```
# Simulate Bernoulli variant of the data set first
set.seed(15)
N <- 50                                    # Total number of sites (binomial total)
theta.cr <- 12/50                          # Success probability Cross-leaved
theta.ch <- 38/50                          # Success probability Chiltern gentian

# Simulate 50 'coin flips' for each species
y.cr <- rbinom(N, 1, prob = theta.cr); y.cr    # Det/nondet data for cr
y.ch <- rbinom(N, 1, prob = theta.ch); y.ch    # ditto for ch
y <- c(y.cr, y.ch)                             # Merge the two binary vectors
species.long <- factor(rep(c(0,1), each = N), labels = c("Cross-leaved", "Chiltern"))
data.frame('species' = species.long, 'det.nondet' = y)    # not shown

# Aggregate the binary data to become two binomial counts
C <- c(sum(y.cr), sum(y.ch))                    # Tally up detections
species <- factor(c(0,1), labels = c("Cross-leaved", "Chiltern"))

# Save true parameter values
truth <- c(Intercept = log(theta.cr/(1-theta.cr)),
          chiltern = (log(theta.ch/(1-theta.ch)) - log(theta.cr/(1-theta.cr))))

# Required packages
library(ASMbook); library(jagsUI); library(rstan); library(TMB)
```

When you execute this code then you will see that the two variants of the simulated data, the binary detection/nondetection data y and the binomial counts C, contain exactly the same amount of information about the observed occupancy probability of the two gentians when all that we want to model is the species effects and are not interested in possible site effects. We will show this in the next section, but for the remainder of the chapter work with the binomial variant of the data set.

15.3 Likelihood analysis with canned functions in R

We begin by fitting the model with function glm() in R for both the binary (y) and the aggregated (C) variant of the simulated data to demonstrate the equivalence of the associated Bernoulli and binomial models. For a binomial response with trial size > 1, there are a number of ways to specify the response variable for a logistic regression in glm(). In this case, we specify the response as a vector of length 2, where the first value is the number of "successes" or "heads" (sites where a plant species was detected, C) and the second value is the number of sites where it was not detected (N-C). We must also set the option family = binomial. See ?binomial for other ways of specifying the response of a logistic regression with glm().

```
# Fit the Bernoulli model to the binary variant of the data set
out15.3X <- glm(y ~ species.long, family = binomial)
summary(out15.3X)
predict(out15.3X, type = "response")[c(1, 51)]   # Predict for two species

# Fit the Binomial model to the aggregated counts and save estimates
out15.3 <- glm(cbind(C, N - C) ~ species, family = binomial)
summary(out15.3)
predict(out15.3, type = "response")
glm_est <- out15.3$coef

# Bernoulli variant of the model
Coefficients:
                      Estimate  Std. Error   z value   Pr(>|z|)
(Intercept)            -0.8473      0.3086    -2.746    0.00604 **
species.longChiltern    1.0076      0.4192     2.404    0.01624 *

   1     51
0.30   0.54

# Binomial variant of the model
Coefficients:
                  Estimate  Std. Error   z value   Pr(>|z|)
(Intercept)        -0.8473      0.3086    -2.746    0.00604 **
speciesChiltern     1.0076      0.4192     2.404    0.01624 *

   1     2
0.30   0.54
```

Thus, we see that we get the exact same numerical estimates of parameters and of predicted proportion of occupied sites for the two species when analyzing the data either for the Bernoulli or the binomial variant. The estimates furthermore suggest a significant difference in the extent of the distribution of the cross-leaved and the Chiltern gentian.

Normally at this place in our workflow we would use residuals checks to assess the goodness-of-fit of our model. However, for this data set and model, we cannot do this, since no lack of fit can be tested for when the data are either strictly binary (as for the Bernoulli variant of our data set; McCullagh & Nelder, 1989, p. 120–122) or when we only have two observations, as for the binomial variant of the data.

15.4 Bayesian analysis with JAGS

The two-group binomial generalized linear model (GLM) in JAGS is very similar to the two-group Poisson GLM from Chapter 11. Here we use the `dbin` distribution instead of `dpois`. Note that the order of the parameters for `dbin` in the BUGS language is opposite to what it is in R—the binomial success probability comes first, followed by the trial size N. In the derived parameters, we calculate the proportion of sites occupied for each species, as well as the difference between the species on the probability scale.

```
# Bundle and summarize data
str(dataList <- list(C = C, species = c(0,1), N = 50, n = 2) )

List of 4
 $ C      : int [1:2] 15 27
 $ species: num [1:2] 0  1
 $ N      : num 50
 $ n      : num 2

# Write JAGS model file
cat(file = "model15.4.txt", "
model {
# Priors
alpha ~ dnorm(0, 0.0001)
beta ~ dnorm(0, 0.0001)

# Likelihood
for (i in 1:n) {
  C[i] ~ dbin(theta[i], N)                       # Note theta before N in JAGS
  logit(theta[i]) <- alpha + beta * species[i]   # species Chiltern = 1
# theta[i] <- ilogit(alpha + beta * species[i])  # same (see also STAN)
}

# Derived quantities
Occ.cross <- exp(alpha)/(1 + exp(alpha))
Occ.chiltern <- exp(alpha + beta)/(1 + exp(alpha + beta))
Occ.Diff <- Occ.chiltern - Occ.cross             # Test quantity
}
")

# Function to generate starting values
inits <- function(){ list(alpha = rlnorm(1), beta = rlnorm(1))}

# Parameters to estimate
params <- c("alpha", "beta", "Occ.cross", "Occ.chiltern", "Occ.Diff")

# MCMC settings
na <- 1000  ;  ni <- 6000  ;  nb <- 1000  ;  nc <- 4  ;  nt <- 5
```

```
# Call JAGS (ART <1 min), check convergence, summarize posteriors and save estimates
out15.4 <- jags(dataList, inits, params, "model15.4.txt",
   n.iter = ni, n.burnin = nb, n.chains = nc, n.thin = nt, n.adapt = na, parallel = TRUE)
jagsUI::traceplot(out15.4)                              # not shown
print(out15.4, 3)
jags_est <- unlist(out15.4$mean[1:2])
```

	mean	sd	2.5%	50%	97.5%	overlap0	f	Rhat	n.eff
alpha	-0.869	0.313	-1.516	-0.857	-0.281	FALSE	0.998	1.000	4000
beta	1.035	0.427	0.224	1.026	1.885	FALSE	0.993	1.001	2586
Occ.cross	0.300	0.064	0.180	0.298	0.430	FALSE	1.000	1.000	4000
Occ.chiltern	0.541	0.070	0.404	0.541	0.676	FALSE	1.000	1.001	1853
Occ.Diff	0.241	0.095	0.054	0.242	0.422	FALSE	0.993	1.001	2442

Next, we plot the posterior distribution of the (biological) distributions of the two species, and their difference, as perceived from the detection-nondetection data (Fig. 15.2).

```
# Draw Fig. 15.2
par(mfrow = c(1, 2), mar = c(5, 5, 5, 3), cex.lab = 1.5, cex.axis = 1.5, cex.main = 1.5)
hist(out15.4$sims.list$Occ.cross, main = "", col = "red", xlab = "Observed occupancy",
   xlim = c(0,1), breaks = 20, freq = FALSE, ylim = c(0, 8))
hist(out15.4$sims.list$Occ.chiltern, col = "blue", breaks = 20, freq = FALSE, add = TRUE)
legend('topright', legend = c('Cross-leaved', 'Chiltern'),
       fill=c("red", "blue"), cex = 1.5)

hist(out15.4$sims.list$Occ.Diff, col = "grey", las = 1,
   xlab = "Difference in occupancy", main = "", breaks = 30,
   freq = FALSE, xlim = c(-0.2, 0.6), ylim = c(0, 5))
abline(v = 0, col = 'grey', lty = 2, lwd = 3)
```

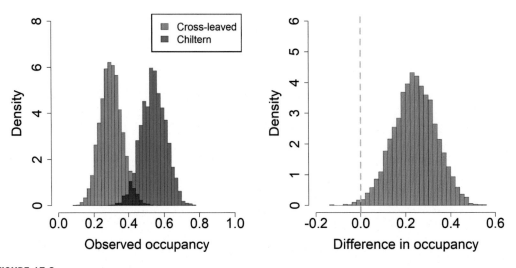

FIGURE 15.2

Posterior distributions of (left) the apparent or observed occupancy probability of cross-leaved and the Chiltern gentian, and (right) of the difference of the two species' apparent occupancy probability; vertical dashed line indicates zero and thus equality of apparent occupancy probability between the two gentian species.

The majority of the posterior distribution of the difference is above zero (Fig. 15.2, right), and the 95% credible interval is (0.054, 0.422). Therefore, we can say that the Chiltern gentian can be found at more sites than the cross-leaved gentian in the region in which we conducted our study. This is *not* necessarily equivalent to saying that the Chiltern gentian is truly more widespread in the Jura mountains than is the cross-leaved gentian, only that it is *detected* at more sites. Whether it also means the former depends on whether the assumption of perfect or at least of constant detection probability holds (Kéry & Schmidt, 2008; Kéry, 2011). Therefore, in this section we prefer to write 'apparent occupancy probabilty' rather than 'occupancy probability'. For a design and analysis method that can tease apart true occupancy and detection probability, see occupancy models in Chapter 19 and a host of literature including MacKenzie et al. (2002, 2017), Tyre et al. (2003), Royle & Dorazio (2008), and Kéry & Royle (2016, 2021).

```
# Compare likelihood with Bayesian estimates and with truth
tmp <- cbind(truth = truth, glm = glm_est, JAGS = jags_est)
print(tmp, 4)

            truth       glm      JAGS
Intercept  -1.153   -0.8473   -0.8688
chiltern    2.305    1.0076    1.0351
```

As so often, we find no substantial numerical difference between the MLEs and the Bayesian posterior means. However, we find substantial estimation error for the analysis of this small data set, that is, the point estimates are pretty far away from the true values used in data simulation. This is due to our small sample size, as you can easily convince yourself by repeating the data simulation and analysis with, say, 10 times greater sample size.

15.5 Bayesian analysis with NIMBLE

NIMBLE code for the model is available on the book website.

15.6 Bayesian analysis with Stan

The Stan model differs only slightly from the formulations in JAGS and NIMBLE. Most importantly, as with the Poisson model in Chapter 11, we must apply the inverse link function (inv_logit) to the right-hand side of the calculation of p, instead of having the link function on the left-hand side as with the BUGS code (although there we can also use the ilogit function as we have seen above). Also, the binomial distribution is called binomial in Stan instead of dbin, and the order of its parameters is opposite BUGS.

```
# Bundle and summarize data
str(dataList <- list(C = C, species = c(0,1), N = 50, n = 2) )
```

```
# Write Stan model
cat(file = "model15_6.stan", "

data{
  array[2] int C;
  array[2] int species;
  int N;
  int n;
}

parameters{
  real alpha;
  real beta;
}

model{
  vector[2] theta;

  alpha ~ normal(0, 100);
  beta ~ normal(0, 100);

  for (i in 1:n){
    theta[i] = inv_logit(alpha + beta * species[i]);
    C[i] ~ binomial(N, theta[i]);
  }
}

generated quantities{
  real Occ_cross = inv_logit(alpha);
  real Occ_chiltern = inv_logit(alpha + beta);
  real Occ_Diff = Occ_chiltern - Occ_cross;
}
")

# HMC settings
ni <- 2000   ;   nb <- 1000   ;   nc <- 4   ;   nt <- 1

# Call STAN (ART 42/1 sec), assess convergence, print and save results
system.time(
out15.6 <- stan(file = "model15_6.stan", data = dataList,
            warmup = nb, iter = ni, chains = nc, thin = nt) )
rstan::traceplot(out15.6)                    # not shown
print(out15.6, dig = 3)                       # not shown
stan_est <- summary(out15.6)$summary[1:2,1]   # save estimates
```

15.7 Do-it-yourself maximum likelihood estimates

We start with the equation of the binomial probability mass for data point y_i:

$$f(y_i|N, \theta) = \binom{N}{y_i} \theta^{y_i}(1-\theta)^{N-y_i}$$

We then convert this to a likelihood function with θ modeled as a function of species (*ilogit* is the inverse logit function):

$$L(\alpha, \beta | y, N) = \prod_{i=1}^{2} \binom{N}{y_i} \theta_i^{y_i} (1-\theta_i)^{N-y_i}$$

$$\theta_i = \text{ilogit}(\alpha + \beta \cdot \text{species}_i)$$

The equivalent negative log-likelihood function in R looks similar. We use the `dbinom` function to calculate the probability mass. The `plogis` function in R can be used to calculate the inverse logit.

```
# Definition of NLL for logistic regression with 1 factor
NLL <- function(param, data) {
  alpha <- param[1]
  beta <- param[2]
  LL <- theta <- numeric(2)
  for (i in 1:2){
    theta[i] <- plogis(alpha + beta * data$species[i])
    LL[i] <- dbinom(data$C[i], data$N, theta[i], log = TRUE)
  }
  NLL <- -sum(LL)                          # NLL for all observations
  return(NLL)
}
```

```
# Minimize that NLL to find MLEs, get SEs and CIs and save estimates
inits <- c('Intercept' = 0, 'beta.Chiltern' = 0)
out15.7 <- optim(inits, NLL, data=dataList, hessian=TRUE)
get_MLE(out15.7, 4)
diy_est <- out15.7$par                 # Save estimates
```

```
                   MLE      ASE    LCL.95    UCL.95
Intercept      -0.8473   0.3086    1.4521    0.2424
beta.Chiltern   1.0076   0.4192    0.1859    1.8292
```

15.8 Likelihood analysis with TMB

As so often, the TMB code looks very similar to our DIY likelihood function, although TMB uses a slightly different name for the inverse link function (`invlogit`). As usual we calculate our derived parameters at the end and add them to the output with `ADREPORT`.

```
# Bundle and summarize data
str(dataList <- list(C = C, species = c(0, 1), N = 50, n = 2))
```

```
# Write TMB model file
cat(file = "model15_8.cpp",
"#include <TMB.hpp>

template<class Type>
Type objective_function<Type>::operator() ()
{
  //Describe input data
  DATA_VECTOR(C);
  DATA_VECTOR(species);
  DATA_INTEGER(N);
  DATA_INTEGER(n);

  //Describe parameters
  PARAMETER(alpha);
  PARAMETER(beta);

  Type LL = 0.0;                                    //Initialize log-likelihood at 0

  for (int i=0; i<n; i++){
    Type theta = invlogit(alpha + beta * species(i));
    LL += dbinom(C(i), Type(N), theta, true);
  }

  Type Occ_cross = invlogit(alpha);
  Type Occ_chiltern = invlogit(alpha + beta);
  Type Occ_Diff = Occ_chiltern - Occ_cross;
  ADREPORT(Occ_cross);
  ADREPORT(Occ_chiltern);
  ADREPORT(Occ_Diff);

  return -LL;
}
")

# Compile and load TMB function
compile("model15_8.cpp")
dyn.load(dynlib("model15_8"))

# Provide dimensions and starting values for parameters
params <- list(alpha = 0, beta = 0)

# Create TMB object
out15.8 <- MakeADFun(data = dataList,
                     parameters = params,
                     DLL = "model15_8", silent = TRUE)

# Optimize TMB object and print results
starts <- rep(0, 2)
opt <- optim(starts, fn = out15.8$fn, gr = out15.8$gr, method = "BFGS")
(tsum <- tmb_summary(out15.8))            # not shown
tmb_est <- tsum[1:2,1]                    # save estimates
```

15.9 Comparison of the parameter estimates

Once again, our grand comparison reveals numerically virtually identical estimates among our six fitting engines. Neither of the two parameter estimates are right on target, which is unsurprising given the fairly small sample sizes. As always, remember that this discrepancy does not mean that our estimation methods (i.e., the estimators) are biased: to gauge bias, we would have to run a simulation with many replicated cycles of data generation and analysis, and then look at whether the sampling distribution of the estimates is centered on the true value (and this is another great exercise for you).

```
# Compare results with truth and previous estimates
comp <- cbind(truth = truth, glm = glm_est, JAGS = jags_est,
  NIMBLE = nimble_est, Stan = stan_est, DIY = diy_est, TMB = tmb_est)
print(comp, 4)

            truth     glm     JAGS  NIMBLE    Stan      DIY      TMB
Intercept  -1.153  -0.8473  -0.8688  -0.876  -0.861  -0.8473  -0.8473
Chiltern    2.305   1.0076   1.0351   1.046   1.027   1.0076   1.0076
```

15.10 Summary

We have introduced the binomial GLM for bounded counts and demonstrated its equivalence with a Bernoulli GLM for the corresponding disaggregated, binary data. As for all GLMs, a binomial two-group comparison is a fairly trivial generalization of the corresponding normal response model. The example in this chapter was a prototypical species distribution model, or SDM, where we compared the distributional extent of two plant species. When computing the difference in the apparent occupancy probability between the two, we have seen another example of the ease with which, in MCMC-based statistical inference, we can compute derived parameters exactly, i.e. without any approximations, and with full error propagation. Estimates of such functions of parameters can also be obtained with likelihood-based inference, as we have seen in the analysis with TMB, but in general this is much harder when the functions of estimated parameters become more complex.

Binomial GLM with continuous and categorical explanatory variables[*]

16

Chapter outline

16.1 Introduction

In the previous chapter, we fit a binomial generalized linear model (GLM) with a single categorical covariate that had two levels. Now, we will expand on this by fitting a binomial GLM that includes both a categorical and a continuous covariate in the linear predictor, i.e., what could be called a binomial analysis of covariance (ANCOVA). Once again, to stress the structural similarity with the analogous normal linear model in Chapter 9, we modify the Asp viper example from there just slightly. Instead of modeling a continuous measurement such as body mass in Chapter 9, we will model a count governed by an underlying probability; specifically, we model the proportion of black individuals in populations of the adder (*Vipera berus*). This species has an all-black and a zigzag morph, where females tend to be brown and males gray (Fig. 16.1).

We hypothesize that the black color confers a thermal advantage, and therefore the proportion of black individuals in a population should be greater in cooler or wetter habitats. We also hypothesize that the relationship between habitat and the proportion of individuals that are black morph may differ by region. We will simulate data that reflect a study designed to answer this question. First, we select three study regions: the Jura mountains, the Black Forest, and the Alps. Within each region, we will sample 10 populations, or sites. At each site i, we will capture N_i adders and count the number of black individuals C_i. For each site, we will also simulate a value for an index combining low temperature, wetness, and northerliness of the site. As always, remember that an observed count is the result of a true number (here, of the true number of black and zigzag adders

---

[*]This book has a companion website hosting complementary materials, including all code for download. Visit this URL to access it: https://www.elsevier.com/books-and-journals/book-companion/9780443137150.

Applied Statistical Modelling for Ecologists. DOI: https://doi.org/10.1016/B978-0-443-13715-0.00022-4

FIGURE 16.1

Adder (*Vipera berus*) of the zigzag morph, Switzerland, 2017 (Photo by Andreas Meyer).

in a population) and a detection probability. We ignore detection probability in the following analyses. Therefore we are making the implicit assumption that detectability does not differ by morph, nor among populations, or at least, that any differences in detection error among populations are not related to wetness.

We will model the number of black adders C_i among N_i captured animals at site i as a function of region and wetness index. Here is the description of the model, where p_i denotes the proportion of black adders at site i.

1. Distribution: $C_i \sim Binomial(p_i, N_i)$
2. Link function: logit, i.e., $\eta_i = \text{logit}(p_i) = \log\left(\frac{p_i}{1-p_i}\right)$
3. Linear predictor (η): $\eta_i = \beta_0 + \beta_1 \cdot x_{BlackF,i} + \beta_2 \cdot x_{Alps,i} + \beta_3 \cdot x_{wet,i} + \beta_4 \cdot x_{wet,i} \cdot x_{BlackF,i} + \beta_5 \cdot x_{wet,i} \cdot x_{Alps,i}$

Here, the covariates x represent region membership and wetness indicator of a site, and as usual we use subscripts both to label a covariate and to index an individual i. In this model, the number of animals captured, N_i, is not a parameter of the binomial distribution that we need to estimate, but rather it is known, since we observe it. The link function is the logit, as is customary for a binomial distribution, leading to a logistic regression. Other links are possible for binomial data such as probit or complementary log-log (McCullagh & Nelder, 1989; Dobson & Barnett, 2018). The latter has an intriguing interpretation that relates the detection/nondetecton data to an underlying, latent abundance (see Chapter 20). The linear predictor is composed of an intercept β_0 for the expected proportion, on the logit scale and at a site with average wetness, of black adders in the Jura, and of parameters β_1 and β_2 for the difference in the intercept between Black Forest and the Jura and the Alps and the Jura, respectively. The parameters β_3, β_4, and β_5 specify the slope of the logit-linear relationship between the proportion of black adders and the wetness index of a site in the Jura, and the difference from β_3 of these slopes in the Black Forest and the Alps, respectively.

16.2 **Data generation**

```
set.seed(16)
nRegion <- 3
nSite <- 10
n <- nRegion * nSite
x <- rep(1:nRegion, rep(nSite, nRegion))
region <- factor(x, labels = c("Jura", "Black Forest", "Alps"))
```

We construct a continuous wetness index: 0 denotes wet sites lacking sun and 1 is the converse. For ease of presentation, we sort this covariate within each region; this has no effect on the analysis.

```
wetness.Jura <- sort(runif(nSite, 0, 1))
wetness.BlackF <- sort(runif(nSite, 0, 1))
wetness.Alps <- sort(runif(nSite, 0, 1))
wetness <- c(wetness.Jura, wetness.BlackF, wetness.Alps)
```

We also need the number of adders examined at each site (N_i), i.e., the binomial totals, also called sample or trial size of the binomial distribution. We assume that the total number of snakes examined in each population is a random variable drawn from a discrete uniform distribution. This is just a nice bit of realism and not essential to our demonstration of this model.

```
N <- round(runif(n, 10, 50) )   # Get discrete uniform values for N
```

We build the design matrix of an interactive combination of region and wetness.

```
Xmat <- model.matrix(~ region * wetness)
print(Xmat, dig = 2)   # not shown, but make sure you understand this!
```

Select the parameter values, i.e., choose values for β_0, β_1, β_2, β_3, β_4, and β_5.

```
truth <- beta.vec <- c(-4, 1, 2, 6, 2, -5)
```

We assemble the number of black adders captured at each site i in the usual three steps. As always, remember that a crucial goal of simulating these data sets is for the R code to serve as another explanation of what the model is:

1. we add up all components of the linear model to get the value of the linear predictor η,
2. we apply the inverse logit transformation to get the expected proportion (p_i) of black adders at each site i (Fig. 16.2; left) and finally,
3. we add binomial noise, i.e., use p_i and N_i to draw binomial random numbers representing the count of black adders in each sample of N_i snakes (Fig. 16.2; right).

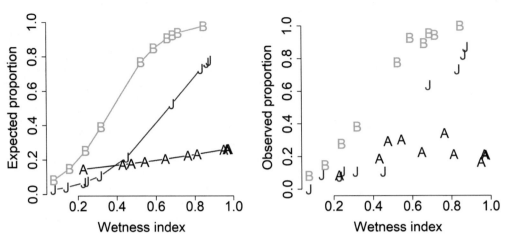

FIGURE 16.2

Relationship between the wetness index and the expected (left) and the observed or realized proportion p (right) of black adders in three regions in Western Europe (Jura mountains: red J, Black Forest: green B, Alps: blackish-blue A). The discrepancy between the two graphs is due to binomial sampling variability.

The value of the linear predictor is again obtained by matrix multiplication of the design matrix (Xmat) and the parameter vector (beta.vec).

```
lin.pred <- Xmat[,] %*% beta.vec          # Value of lin.predictor
exp.p <- exp(lin.pred) / (1 + exp(lin.pred))   # Expected proportion, note this is
                                          # same as plogis(lin.pred) in R
C <- rbinom(n = n, size = N, prob = exp.p)     # Add binomial noise
hist(C)                                   # Inspect simulated binomial counts

# Draw Fig. 16.2
par(mfrow = c(1,2), mar = c(5,5,3,1))
matplot(cbind(wetness[1:10], wetness[11:20], wetness[21:30]), cbind(exp.p[1:10],
    exp.p[11:20], exp.p[21:30]), ylab = "Expected proportion black", xlab = "Wetness index",
    col = c("red","green","blue"), pch = c("J","B","A"), lty = "solid", type = "b", las = 1,
    cex = 1.2, main = "Expected proportion", lwd = 2, frame = FALSE)
matplot(cbind(wetness[1:10], wetness[11:20], wetness[21:30]), cbind(C[1:10]/N[1:10],
    C[11:20]/N[11:20], C[21:30]/N[21:30]), ylab = "Observed proportion black",
    xlab = "Wetness index", col = c("red","green","blue"), pch = c("J","B","A"), las = 1,
    cex = 1.2, main = "Realized proportion", frame = FALSE)
```

In our simulated dataset, the proportion of black adders increases with wetness most steeply in the Black Forest, less so in the Jura, and hardly at all in the Alps (Fig. 16.2).

```
# Load required libraries
library(ASMbook); library(jagsUI); library(rstan); library(TMB)
```

16.3 Likelihood analysis with canned functions in R

As usual, the analysis in R is concise. Function `glm()` uses iteratively reweighted least-squares to produce what for this model are the maximum likelihood estimates (MLEs).

```
summary(out16.3 <- glm(cbind(C, N-C) ~ region * wetness, family = binomial))
glm_est <- coef(out16.3)                # Save estimates
```

```
Coefficients:
                         Estimate   Std. Error   z value   Pr(>|z|)
(Intercept)               -3.8682       0.3746   -10.327    <2e-16 ***
regionBlack Forest         0.7568       0.5861     1.291     0.1966
regionAlps                 2.3112       0.5233     4.416   1.00e-05 ***
wetness                    6.1042       0.5738    10.638    <2e-16 ***
regionBlack Forest:wetness 2.5472       1.1026     2.310     0.0209 *
regionAlps:wetness        -5.6947       0.7588    -7.505   6.15e-14 ***
```

As usual, we add some code for doing residual checking of the distributional assumptions of the model using both Pearson residuals and package `DHARMa` (Hartig, 2022). We do not show these results, and indeed for our simulated, "perfect" data, these are not so interesting. But you can view this code block as a placeholder for an important activity that should always follow the fitting of a model to a "real" data set (i.e., not a simulated one): checking whether your model captures the salient features of the data and is any good for learning something from your data about the underlying processes.

```
# Diagnostic checks of residuals/model GOF (not shown)
plot(out16.3)                           # Traditional Pearson residual check
library(DHARMa)                         # Compute quantile residuals based on simulation
simOut <- simulateResiduals(out16.3, n = 1000, plot = TRUE)
plotResiduals(simOut, form = wetness)   # same as right plot ...
testDispersion(out16.3)                 # Test for over- or underdispersion
```

16.4 Bayesian analysis with JAGS

The basic structure of the model in JAGS combines elements from the models we fit in Chapters 9 and 15. We fit the model using a means parameterization—i.e., we define a separate intercept and (wetness index) slope for each region using nested indexing. We calculate derived effects-parameterization parameters, so we can compare results to other engines. We also add a few lines to compute Pearson residuals. For a binomial response, these can be computed as $(C_i - N_i p_i)/\sqrt{N_i p_i (1 - p_i)}$, where C_i is the binomial count for unit i and N_i and p_i are the trial size and success probability, respectively, of the associated binomial distributions. The Pearson residual has the form of a raw residual divided by the standard deviation of unit i. For a goodness-of-fit (GOF) test, we do a posterior predictive check based on the squared Pearson residuals and we also show how output from the Bayesian analysis can be used in package `DHARMa` for an alternative GOF test.

```
# Bundle and summarize data
str(dataList <- list(C = C, N = N, nRegion = nRegion,
  region = as.numeric(region),wetness = wetness, n = n) )

List of 6
 $ C       : int [1:30] 0 4 3 5 4 3 14 33 37 20 ...
 $ N       : num [1:30] 12 45 39 46 37 28 22 45 45 23 ...
 $ nRegion : num 3
 $ region  : num [1:30] 1 1 1 1 1 1 1 1 1 1 ...
 $ wetness : num [1:30] 0.0743 0.1394 0.2294 0.2441 0.3112 ...
 $ n       : num 30

# Write JAGS model file
cat(file = "model16.4.txt", "
model {
# Priors
for (i in 1:nRegion){
  alpha[i] ~ dnorm(0, 0.0001)                         # Intercepts
  beta[i] ~ dnorm(0, 0.0001)                          # Slopes
}

# Likelihood
for (i in 1:n) {
  C[i] ~ dbin(p[i], N[i])
  logit(p[i]) <- alpha[region[i]] + beta[region[i]]*wetness[i]  # Jura is baseline
  # p[i] <- ilogit(alpha[region[i]] + beta[region[i]]*wetness[i]) # same !

# Fit assessments: Pearson residuals and posterior predictive check
  Presi[i] <- (C[i]-N[i]*p[i])/sqrt(N[i]*p[i]*(1-p[i]))  # Pearson resi
  C.new[i] ~ dbin(p[i], N[i])                            # Create replicate data set
  Presi.new[i] <- (C.new[i]-N[i]*p[i])/sqrt(N[i]*p[i]*(1-p[i]))
}

# Add up squared residual as our discrepancy measures
fit <- sum(Presi[]^2)
fit.new <- sum(Presi.new[]^2)

# Derived quantities
# Recover the effects relative to baseline level (no. 1)
a.effe2 <- alpha[2] - alpha[1]                 # Intercept Black Forest vs. Jura
a.effe3 <- alpha[3] - alpha[1]                 # Intercept Alps vs. Jura
b.effe2 <- beta[2] - beta[1]                   # Slope Black Forest vs. Jura
b.effe3 <- beta[3] - beta[1]                   # Slope Alps vs. Jura

# Custom comparison
test1 <- beta[3] - beta[2]                     # Difference slope Alps-Black Forest
}
")

# Function to generate starting values
inits <- function(){ list(alpha = rnorm(nRegion, 3, 1),  beta = rnorm(nRegion, 2, 1))}

# Parameters to estimate (add "C.new" for DHARMa GoF test)
params <- c("alpha", "beta", "a.effe2", "a.effe3", "b.effe2", "b.effe3",
  "test1", "Presi", "fit", "fit.new", "C.new")
```

Preliminary runs of the model with shorter chains suggest considerable autocorrelation within the chains. To accumulate a more "information-dense" sample from the joint posterior, we run longer chains and set the thinning rate to keep 1 of every 40 draws.

```
# MCMC settings
na <- 5000  ;  ni <- 50000  ;  nb <- 10000  ;  nc <- 4  ;  nt <- 40

# Call JAGS (ART <1 min), check convergence, summarize posteriors and save results
out16.4 <- jags(dataList, inits, params, "model16.4.txt", n.iter = ni,
  n.burnin = nb, n.chains = nc, n.thin = nt, n.adapt = na, parallel = TRUE)
jagsUI::traceplot(out16.4)                      # not shown
print(out16.4, 3)                               # shown partially
jags_est <- unlist(out16.4$mean)[c(1, 7, 8, 4, 9, 10)]
```

	mean	sd	2.5%	50%	97.5%	overlap0	f	Rhat	n.eff
alpha[1]	-3.931	0.384	-4.706	-3.924	-3.213	FALSE	1.000	1.000	4000
alpha[2]	-3.153	0.452	-4.099	-3.139	-2.292	FALSE	1.000	1.000	2806
alpha[3]	-1.567	0.376	-2.335	-1.560	-0.859	FALSE	1.000	1.001	4000
beta[1]	6.191	0.585	5.090	6.193	7.351	FALSE	1.000	1.000	4000
beta[2]	8.751	0.943	6.997	8.725	10.661	FALSE	1.000	1.001	1990
beta[3]	0.407	0.504	-0.552	0.402	1.410	TRUE	0.791	1.000	4000
a.effe2	0.777	0.601	-0.449	0.789	1.933	TRUE	0.899	1.001	3213
a.effe3	2.364	0.529	1.316	2.366	3.396	FALSE	1.000	1.000	4000
b.effe2	2.559	1.119	0.463	2.532	4.813	FALSE	0.991	1.001	1884
b.effe3	-5.784	0.762	-7.279	-5.794	-4.310	FALSE	1.000	1.000	4000
test1	-8.343	1.074	-10.496	-8.306	-6.346	FALSE	1.000	1.001	2402

To follow good statistical practice, we will next assess model fit. After all, only when a model adequately reproduces the main features in the data set should we believe what it says about the parameters. We first plot the Pearson residuals and, naturally, find no obvious remaining pattern related to order or wetness of a site (Fig. 16.3 left and middle). We also conduct a posterior predictive check for overall GOF of the model (Fig. 16.3 right), using as our discrepancy measure the sum of the squared Pearson residuals. Of course, this looks good, too, and so does the Bayesian p-value.

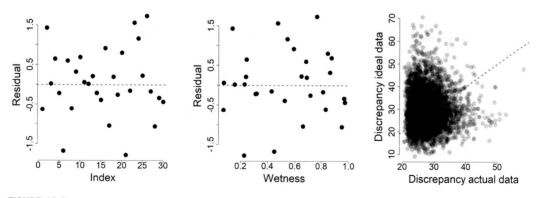

FIGURE 16.3

(left) Pearson residuals plotted against the order of each observation and (middle) against the value of the wetness indicator. (right) Posterior predictive check for the black adder analysis. The Bayesian p-value (here, 0.58) is the proportion of points above the line.

```
mean(out16.4$sims.list$fit.new > out16.4$sims.list$fit)

[1] 0.57675

# Draw Fig. 16.3
par(mfrow = c(1, 3), cex = 1.5)
plot(out16.4$mean$Presi, ylab = "Residual", las = 1, pch = 16, frame = FALSE)
abline(h = 0)
plot(wetness, out16.4$mean$Presi, ylab = "Residual", las = 1, pch = 16, frame = FALSE)
abline(h = 0)
plot(out16.4$sims.list$fit, out16.4$sims.list$fit.new, main = "",
   xlab = "Discrepancy actual data", ylab = "Discrepancy ideal data",
   col = rgb(0,0,0,0.3), pch = 16, cex = 1, frame = FALSE)
abline(0,1, lwd = 2, col = "black")
```

For illustration, we complement this assessment by a check based on quantile residuals using the DHARMa package applied to Markov Chain Monte Carlo output (Hartig, 2022). On execution of the next code block, you will find that results concur with our previous assessment that no lack of fit can be detected.

```
# Do quantile residual assessments (not shown)
C.new <- out16.4$sims.list$C.new
sim <- createDHARMa(simulatedResponse = t(C.new), observedResponse = C,
   fittedPredictedResponse = apply(C.new, 2, median), integerResponse = T)
plot(sim)
```

Hence, we feel justified to compare the results of the Bayesian analysis with the truth from the data-generating random process as well as with the frequentist inference from R's function `glm()`.

```
# Compare likelihood with Bayesian estimates and with truth
comp <- cbind(truth = beta.vec, glm = glm_est, JAGS = jags_est)
print(comp, 4)
```

	truth	glm	JAGS
(Intercept)	-4	-3.8682	-3.9306
regionBlack Forest	1	0.7568	0.7773
regionAlps	2	2.3112	2.3640
wetness	6	6.1042	6.1912
regionBlack Forest:wetness	2	2.5472	2.5595
regionAlps:wetness	-5	-5.6947	-5.7839

We get similar estimates, but both sets of estimates are somewhat off the truth, as we would expect for a small sample size.

16.5 Bayesian analysis with NIMBLE

NIMBLE code for the model is available on the book website.

16.6 Bayesian analysis with Stan

The Stan code is very similar to JAGS. Two things to recall from Chapter 15 are as follows: we need to use the inverse link function on the right-hand side of the linear predictor definition (instead of the link function on the left-hand side as with BUGS), and the parameter order for the binomial distribution in Stan is opposite to what it is in BUGS.

```r
# Bundle and summarize data
str(dataList <- list(C = C, N = N, nRegion = nRegion,
  region = as.numeric(region), wetness = wetness, n = n) )

# Write Stan model
cat(file = "model16_6.stan", "
data{
  int n;                              // Number of samples
  int nRegion;                        // Number of regions
  array[n] int C;                     // Counts for each sample
  array[n] int N;                     // Sizes of each sample
  array[n] int region;                // Region indices
  vector[n] wetness;                  // Covariate
}

parameters{
  real alpha[nRegion];                // Intercepts for each region
  real beta[nRegion];                 // Slopes for each region
}

transformed parameters{
  vector[n] p;                        // Estimated probabilities
  for (i in 1:n){
    p[i] = inv_logit(alpha[region[i]] + beta[region[i]] * wetness[i]);
  }
}

model{
  for (i in 1:nRegion){
    alpha[i] ~ normal(0, 100);
    beta[i] ~ normal(0, 100);
  }
  for (i in 1:n){
    C[i] ~ binomial(N[i], p[i]);
  }
}

generated quantities{
  real a_effe2 = alpha[2] - alpha[1];  // Intercept Black Forest vs. Jura
  real a_effe3 = alpha[3] - alpha[1];  // Intercept Alps vs. Jura
  real b_effe2 = beta[2] - beta[1];    // Slope Black Forest vs. Jura
  real b_effe3 = beta[3] - beta[1];    // Slope Alps vs. Jura
  real test1 = beta[3] - beta[2];      // Difference slope Alps-Black Forest
  array[n] int C_new;                  // New simulated dataset
  vector[n] Presi;                     // Pearson residuals for real dataset
  vector[n] Presi_new;                 // Pearson residuals for simulated dataset
  vector[n] Presi2;                    // Squared Pearson resi. for real dataset
  vector[n] Presi2_new;                // Squared Pearson resi. for simulated dataset
  real fit;                            // Sum of Pearson residuals
  real fit_new;

  for (i in 1:n){
    Presi[i] = (C[i]-N[i]*p[i])/sqrt(N[i]*p[i]*(1-p[i]));
    Presi2[i] = pow(Presi[i], 2);
    C_new[i] = binomial_rng(N[i], p[i]);
    Presi_new[i] = (C_new[i]-N[i]*p[i])/sqrt(N[i]*p[i]*(1-p[i]));
    Presi2_new[i] = pow(Presi_new[i], 2);
  }
  fit = sum(Presi2);                   // Add up squared discrepancies
  fit_new = sum(Presi2_new);
}
")
```

```
# HMC settings
ni <- 2000  ;  nb <- 1000  ;  nc <- 4  ;  nt <- 1

# Call STAN (ART 53/4 sec), assess convergence and print and save result
system.time(
out16.6 <- stan(file = "model16_6.stan", data = dataList,
              warmup = nb, iter = ni, chains = nc, thin = nt) )
rstan::traceplot(out16.6)                           # not shown
print(out16.6, dig = 3)                             # not shown
stan_est <- summary(out16.6)$summary[c(1,37,38,4,39,40),1]  # save estimates
```

16.7 Do-it-yourself maximum likelihood estimates

We will calculate the linear predictor in the likelihood by matrix-multiplying the model matrix we obtained earlier (**X** in the equation below, Xmat in our simulation code) and our vector of parameters (β below, beta.vec in the simulation code). Note also that unlike in Chapter 15, N is now indexed by i because our sample size of captured adders varies by site. Also remember that the R function plogis is just a different name for the inverse logit link function.

$$L(\beta|y,N,X) = \prod_{i=1}^{n} \binom{N_i}{y_i} p_i^{y_i}(1-p_i)^{N_i-y_i}$$

$$p = \mathrm{ilogit}(X\beta)$$

Here is the corresponding R code, which vectorizes the calculation of the log-likelihood.

```
# Define NLL for general logistic regression with Binomial response
NLL <- function(beta, y, N, Xmat) {
  p <- plogis(Xmat %*% beta)
  LL <- dbinom(y, N, p, log = TRUE)      # log-likelihood contr. for each obs
  NLL <- -sum(LL)                         # NLL for all observations in data set
  return(NLL)
}
```

```
# Minimize that NLL to find MLEs, get SEs and CIs and save estimates
inits <- rep(0, 6)
names(inits) <- names(coef(out16.3))
out16.7 <- optim(inits, NLL, y = C, N = N, Xmat = Xmat,
           hessian = TRUE, method = "BFGS")
get_MLE(out16.7, 4)
diy_est <- out16.7$par              # Save estimates
```

	MLE	ASE	LCL.95	UCL.95
(Intercept)	-3.8687	0.3746	-4.6030	-3.134
regionBlack Forest	0.7575	0.5861	-0.3914	1.906
regionAlps	2.3118	0.5234	1.2860	3.338
wetness	6.1048	0.5739	4.9801	7.230
regionBlack Forest:wetness	2.5466	1.1026	0.3854	4.708
regionAlps:wetness	-5.6953	0.7588	-7.1826	-4.208

16.8 **Likelihood analysis with TMB**

To fit the model in Template Model Builder (TMB) we use a means parameterization similar to the version in Stan, and also the TMB model in Chapter 9. As usual with TMB, we need to make sure our population indices start at 0 by subtracting 1 from all of them.

```
# Bundle and summarize data
tmbData <- dataList
tmbData$region <- tmbData$region - 1    #Indices start at 0 in TMB/C++
str(tmbData)

# Write TMB model file
cat(file = "model16_8.cpp",
"#include <TMB.hpp>

template<class Type>
Type objective_function<Type>::operator() ()
{
  // Describe input data
  DATA_VECTOR(C);
  DATA_VECTOR(N);
  DATA_IVECTOR(region);
  DATA_VECTOR(wetness);
  DATA_INTEGER(n);

  // Describe parameters
  PARAMETER_VECTOR(alpha);
  PARAMETER_VECTOR(beta);

  Type LL = 0.0;                           // Initialize log-likelihood at 0

  vector<Type> p(n);
  vector<Type> Presi(n);

  for (int i=0; i<n; i++){
    p(i) = invlogit(alpha(region(i)) + beta(region(i)) * wetness(i));
    LL += dbinom(C(i), N(i), p(i), true);
    Presi(i) = (C(i)-N(i)*p(i))/sqrt(N(i)*p(i)*(1-p(i)));
  }

  Type a_effe2 = alpha(1) - alpha(0);      // Intercept Black Forest vs. Jura
  Type a_effe3 = alpha(2) - alpha(0);      // Intercept Alps vs. Jura
  Type b_effe2 = beta(1) - beta(0);        // Slope Black Forest vs. Jura
  Type b_effe3 = beta(2) - beta(0);        // Slope Alps vs. Jura
  Type test1 = beta(2) - beta(1);          // Difference slope Alps-Black Forest
  ADREPORT(a_effe2);
  ADREPORT(a_effe3);
  ADREPORT(b_effe2);
  ADREPORT(b_effe3);
  ADREPORT(test1);
  ADREPORT(Presi);

  return - LL;
}
")
```

```
# Compile and load TMB function
compile("model16_8.cpp")
dyn.load(dynlib("model16_8"))

# Provide dimensions and starting values for parameters
params <- list(alpha = rep(0, tmbData$nRegion), beta = rep(0, tmbData$nRegion))

# Create TMB object
out16.8 <- MakeADFun(data = tmbData,
                     parameters = params,
                     DLL = "model16_8", silent = TRUE)

# Optimize TMB object and print and save results
opt <- optim(out16.8$par, fn = out16.8$fn, gr = out16.8$gr, method = "BFGS")
(tsum <- tmb_summary(out16.8))                    # not shown
tmb_est <- tsum[c(1, 7, 8, 4, 9, 10),1]           # save estimates
```

16.9 Comparison of the estimates

Here comes our grand comparison table. We get the usual similarity among the point estimates for all parameters, which is actually remarkable in view of the small sample size.

```
# Compare estimates with truth
comp <- cbind(truth = truth, glm = glm_est, JAGS =jags_est,
   NIMBLE = nimble_est, Stan = stan_est, DIY = diy_est, TMB = tmb_est)
print(comp, 3)
```

	truth	glm	JAGS	NIMBLE	Stan	DIY	TMB
(Intercept)	-4	-3.868	-3.931	-3.915	-3.92	-3.869	-3.868
regionBlack Forest	1	0.757	0.777	0.737	0.76	0.757	0.757
regionals	2	2.311	2.364	2.338	2.34	2.312	2.311
wetness	6	6.104	6.191	6.175	6.18	6.105	6.104
regionBlack Forest:wetness	2	2.547	2.559	2.637	2.60	2.547	2.548
regionAlps:wetness	-5	-5.695	-5.784	-5.749	-5.76	-5.695	-5.694

16.10 Summary

Moving from the normal and the Poisson to a binomial GLM involves only minor changes in the code of our engines. Similarly, the concepts of residuals and posterior predictive distributions carry over to this class of models. We have seen examples for both, and we have again illustrated the use for GOF assessment with quantile residuals in the useful DHARMa package for both a frequentist analysis using glm() and with our Bayesian analysis using JAGS. For "real" data sets, the same type of GOF assessment can and should be done for analyses with any of the other engines in our comparison, too.

Binomial generalized linear mixed model[*]

17

Chapter outline

17.1 Introduction

As with a Poisson generalized linear model (GLM) and generalized linear mixed model (GLMM), we can also add into a binomial GLM random variation beyond what is stipulated by the binomial distribution assumed for the observed data. We illustrate this for a slight modification of the red-backed shrike example from Chapter 14. Instead of counting the number of shrike pairs, which naturally leads to the adoption of a Poisson model, we now imagine that we study the reproductive success (success or failure) in its much rarer cousin, the glorious Woodchat shrike (Fig. 17.1). We examine the relationship between precipitation during the breeding season and breeding success; wet springs are likely to depress the proportion of successful nests. We assemble data from 16 populations studied over 10 years.

First, we write down the random-coefficients model (without intercept-slope correlation) for a binomial response. We model C_i, the number successful pairs among N_i studied pairs with i running from 1 to the total sample size n. Each count is associated with one of $J = 16$ populations, indexed by $j_{(i)}$, and occurs in a particular year x_i taking values from 1 to 10.

1. Distribution for observed data: $C_i \sim Binomial(p_i, N_i)$
2. Link function: logit, that is, $\eta_i = \text{logit}(p_i) = \log(\frac{p_i}{1-p_i})$
3. Linear predictor (η): $\eta_i = \alpha_{j(i)} + \beta_{j(i)} \cdot x_i$
4. Submodel for parameters/distribution of random effects:

$$\alpha_j \sim Normal(\mu_\alpha, \sigma_\alpha^2)$$

$$\beta_j \sim Normal(\mu_\beta, \sigma_\beta^2)$$

---

[*]This book has a companion website hosting complementary materials, including all code for download. Visit this URL to access it: https://www.elsevier.com/books-and-journals/book-companion/9780443137150.

Applied Statistical Modelling for Ecologists. DOI: https://doi.org/10.1016/B978-0-443-13715-0.00015-7

FIGURE 17.1

Woodchat shrike (*Lanius senator*), Catalonia, 2008 (Photo by Jordi Rojals).

Except for a different distribution and link function, the additional kind of data provided by the binomial totals N_i, and a different interpretation of the covariate, this model looks exactly like the Poisson GLMM in Chapter 14! The linear predictor, $\eta_i = \alpha_{j(i)} + \beta_{j(i)} \cdot x_i$, specifies a population-specific, logit-linear relationship between breeding success and precipitation. Furthermore, populations are assumed to be related in the sense that both intercepts (α_j) and slopes (β_j) come from two normal distributions with hyperparameters that we estimate. Note that it is these additional distributional assumptions in line 4 which take us from a simpler binomial GLM with a cluster of regression lines as in Chapter 16 to the binomial GLMM here.

17.2 Data generation

We generate data under the random-coefficients model, that is, with both α_j and β_j assumed to be independent sets of random effects. That is, we assume no correlation between random intercepts and slopes.

```
set.seed(17)
nPops <- 16
nYears <- 10
n <- nPops * nYears              # n = 160
pop <- gl(n = nPops, k = nYears)
```

We create a uniform covariate as an index to spring precipitation: -1 denotes little rain and 1 a lot of rain. This implicit centering of the continuous covariate leads to the desirable interpretation of the intercept as the expected value of the response at the average of the covariate.

```
precip <- runif(n, -1, 1)
```

The binomial total, N_i, is the number of nesting attempts surveyed in year i. For added realism, we assume some variability in our sample sizes by chosing a random integer between 20 and 50.

```
N <- round(runif(n, 20, 50))
```

We build the design matrix as before and again look at the top 91 rows.

```
Xmat <- model.matrix(~ pop * precip - 1 - precip)
print(Xmat[1:91,], dig = 2)                     # Print top 91 rows (not shown)
```

Next, we draw the random-effects parameters from their respective normal distributions, but first need to pick the values of the associated hyperparameters.

```
mu.alpha <- 0                                   # Select hyperparams
mu.beta <- -2
sigma.alpha <- 1
sigma.beta <- 1
alpha <- rnorm(n = nPops, mean = mu.alpha, sd = sigma.alpha)
beta <- rnorm(n = nPops, mean = mu.beta, sd = sigma.beta)
all.pars <- c(alpha, beta)                      # All parameters together

# Save vector of true parameter values
truth <- c(mu.alpha = mu.alpha, mu.beta = mu.beta,
          sigma.alpha = sigma.alpha, sigma.beta = sigma.beta)
```

We assemble the counts C_i by first computing the value of the linear predictor, then applying the inverse logit transformation and finally adding binomial noise (using N_i). We write out the inverse logit but could also have used `plogis()`.

```
lin.pred <- Xmat %*% all.pars                   # Value of lin.predictor
exp.p <- exp(lin.pred)/(1 + exp(lin.pred))      # Expected proportion
```

For each population, we plot the expected (Fig. 17.2) and the observed, or realized, breeding success (Fig. 17.3) of woodchat shrikes against standardized spring precipitation using a Trellis plot. Note that the difference between the two is due to binomial random variation.

```
library(lattice)
xyplot(exp.p ~ precip | pop, ylab = "Expected woodchat shrike breeding success ",
    xlab = "Spring precipitation index", main = "Expected breeding success", pch = 16,
    cex = 1.2, col = rgb(0, 0, 0, 0.4))

C <- rbinom(n = n, size = N, prob = exp.p)      # Add binomial variation
xyplot(C/N ~ precip | pop, ylab = "Realized woodchat shrike breeding success ",
    xlab = "Spring precipitation index", main = "Realized breeding success",
    pch = 16, cex = 1.2, col = rgb(0, 0, 0, 0.4))
```

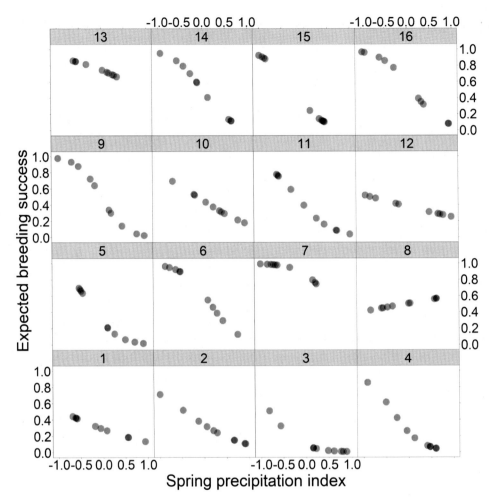

FIGURE 17.2

Trellis plot of the relationship between spring precipitation (a centered index between −1 and 1) and expected breeding success of woodchat shrikes in 16 populations over 10 years.

In the analyses that follow in the rest of the chapter, we could assume that all shrike populations have the same relationship between breeding success and standardized spring precipitation, but with different intercepts, corresponding to the random-intercepts model discussed in Section 10.3. However, unlike in Chapter 10, we directly adopt the random-coefficients model instead without correlation between intercepts and slopes. Hence, we assume that every shrike population has a specific response to precipitation, but that intercepts and slopes from each population come from common distributions.

```
# Required libraries
library(ASMbook); library(glmmTMB); library(DHARMa); library(jagsUI); library(rstan); library(TMB)
```

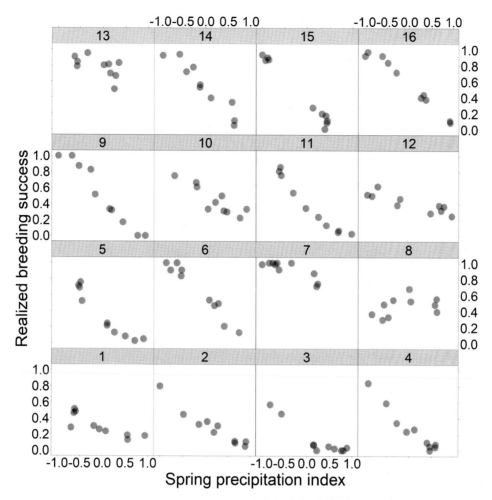

FIGURE 17.3

Trellis plot of the relationship between spring precipitation (a centered index between −1 and 1) and the realized breeding success of Woodchat shrikes in 16 populations over 10 years. The difference between this and the previous plot is due to binomial sampling variation.

17.3 Likelihood analysis with canned functions in R

We use the function glmmTMB() from the package of the same name; we note again that lmer() in lme4 has virtually the same syntax and should yield identical estimates. We specify a model with random intercepts and slopes, but without correlation between intercepts and slopes by using the || operator (see Section 14.3 for more discussion of the formula structure).

```
# Fit model, present and save estimates
gtmb.data <- data.frame(C = C, N = N, precip = precip, pop = pop)    # bundle data
out17.3 <- glmmTMB(cbind(C, N-C) ~ precip + (precip || pop), data = gtmb.data, family = binomial)
summary(out17.3)                                                     # Inspect results
sds <- attr(VarCorr(out17.3)$cond$pop, 'stddev')                     # Save results
gtmb_est <- c(fixef(out17.3)$cond, sds)

Random effects:

Conditional model:
 Groups Name        Variance  Std.Dev.  Corr
 pop    (Intercept) 0.818     0.9044
        precip      1.466     1.2107    0.00
Number of obs: 160, groups: pop, 16

Conditional model:
            Estimate  Std. Error  z value  Pr(>|z|)
(Intercept)  -0.2354     0.2293    -1.026    0.305
precip       -2.3535     0.3137    -7.503   6.26e-14 ***
```

As a reminder that goodness-of-fit (GOF) of a model should also be assessed for mixed or hierarchical models (Conn et al., 2018), we present the next code block. It produces histograms of the estimated random effects for a visual test of their distributional assumption of a normal, and an assessment of model fit using quantile residuals with package DHARMa (Hartig, 2022). We note that the former have very limited power to detect deviations from a normal in small samples such as ours, but it is good practice to inspect these plots anyway.

```
# Model goodness of fit
par(mfrow = c(1, 2))    # Distribution of estimates of alpha and beta
hist(ranef(out17.3)$cond$pop[,1], main = 'alpha', breaks = 12)
hist(ranef(out17.3)$cond$pop[,2], main = 'beta', breaks = 12)

simOut <- simulateResiduals(out17.3, n = 1000, plot = TRUE)
```

17.4 Bayesian analysis with JAGS

The binomial GLMM in JAGS is nearly identical to the Poisson GLMM in Section 14.4. We only need to make changes in the "Likelihood" section, updating the linear predictor, link function, and distribution of the data to correspond to our binomial data. We also provide for residual checks on the fitted model using DHARMa by computing posterior predictive distributions.

```
# Bundle and summarize data
str(dataList <- list(C = C, N = N, pop = as.numeric(pop), precip = precip, nPops = nPops, n = n) )

List of 6
 $ C      : int [1:160] 6  6  11  9  14  9  16  20  3  17 ...
 $ N      : num [1:160] 21  33  42  48  46  38  33  43  23  33 ...
 $ pop    : num [1:160] 1  1  1  1  1  1  1  1  1  1 ...
 $ precip : num [1:160] -0.6899  0.9368  -0.0635  0.5536  -0.1842 ...
 $ nPops  : num 16
 $ n      : num 160
```

```
# Write JAGS model file
cat(file = "model17.4.txt", "
model {
# Priors
# Models for the random effects alpha and beta
for (i in 1:nPops){
  alpha[i] ~ dnorm(mu.alpha, tau.alpha)      # Intercepts
  beta[i] ~ dnorm(mu.beta, tau.beta)         # Slopes
}
# (Hyper-)Priors for hyperparameters
mu.alpha ~ dnorm(0, 0.0001)                  # Hyperparameter for random intercepts
tau.alpha <- pow(sigma.alpha, -2)
sigma.alpha ~ dt(0, 0.1, 1)I(0,)

mu.beta ~ dnorm(0, 0.0001)                   # Hyperparameter for random slopes
tau.beta <- pow(sigma.beta, -2)
sigma.beta ~ dt(0, 0.1, 1)I(0, )

# Likelihood
for (i in 1:n) {
  C[i] ~ dbin(p[i], N[i])
  logit(p[i]) <- alpha[pop[i]] + beta[pop[i]]* precip[i]
}

# Posterior predictive simulations (used in DHARMa GoF assessment)
for (i in 1:n) {
  C.new[i] ~ dbin(p[i], N[i])
}
}
")

# Function to generate starting values
inits <- function(){ list(mu.alpha = rnorm(1, 0, 1), mu.beta = rnorm(1, 0, 1))}

# Parameters to estimate
params <- c("mu.alpha", "mu.beta", "sigma.alpha", "sigma.beta",
            "alpha", "beta", "C.new")

# MCMC settings
na <- 1000  ;  ni <- 12000  ;  nb <- 2000  ;  nc <- 4  ;  nt <- 10

# Call JAGS (ART <1 min), check convergence and summarize posteriors
out17.4 <- jags(dataList, inits, params, "model17.4.txt", n.iter = ni, n.burnin = nb,
  n.chains = nc, n.thin = nt, n.adapt = na, parallel = TRUE)
jagsUI::traceplot(out17.4)             # not shown
print(out17.4, 3)                      # shown partly, half-page down
```

The model converges quickly. Before we inspect any estimates, we sketch out the code for a GOF assessment with histograms of the estimates of the random effects and assessment of quantile residuals using DHARMa.

```
# Distribution of random effects estimates (alpha, beta)
par(mfrow = c(1,2))     # Low power with small samples...
hist(out17.4$mean$alpha, main = 'alpha', breaks = 12)
hist(out17.4$mean$beta, main = 'beta', breaks = 12)
```

```
# Do quantile residual assessments (not shown)
C.new <- out17.4$sims.list$C.new
sim <- createDHARMa(simulatedResponse = t(C.new), observedResponse = C,
   fittedPredictedResponse = apply(C.new, 2, median), integerResponse = T)
plot(sim)
```

Interestingly, for this data set the Bayesian version of the quantile residual assessment yields three significant test results out of six. Since data generation and analysis models are exactly identical, we conclude in this case that DHARMa sounds a false alarm. We proceed to inspect the estimates and compare the frequentist and the Bayesian point estimates with the truth.

	mean	sd	2.5%	50%	97.5%	overlap0	f	Rhat	n.eff
mu.alpha	-0.236	0.264	-0.740	-0.240	0.285	TRUE	0.834	1.000	3503
mu.beta	-2.357	0.350	-3.067	-2.359	-1.667	FALSE	1.000	1.000	4000
sigma.alpha	1.017	0.217	0.697	0.990	1.533	FALSE	1.000	1.000	4000
sigma.beta	1.353	0.279	0.917	1.313	2.004	FALSE	1.000	1.000	3306

```
# Compare likelihood with Bayesian estimates and with truth
jags_est <- out17.4$summary[1:4,1]
comp <- cbind(truth = truth, gtmb = gtmb_est, JAGS = jags_est)
print(comp, 4)
```

	truth	gtmb	JAGS
mu.alpha	0	-0.2354	-0.2364
mu.beta	-2	-2.3535	-2.3568
sigma.alpha	1	0.9044	1.0174
sigma.beta	1	1.2107	1.3526

This seems to work well, too. As is typical in harder-to-estimate models such as those with random effects, estimates of the mean parameters are more similar across the likelihood/Bayes comparison than are the dispersion parameter estimates. This indicates a slight effect of our vague priors in the latter, which would vanish with moderate to large sample sizes (such as 64 populations—as always, you can try that out).

17.5 Bayesian analysis with NIMBLE

NIMBLE code for the model is available on the book website.

17.6 Bayesian analysis with Stan

The Stan code is again nearly identical to the code in Section 14.6. We just need to change the linear predictor, link function, and distribution of the response.

```
# Bundle and summarize data
str(dataList <- list(C = C, N = N, pop = as.numeric(pop), precip = precip, nPops = nPops, n = n) )

# Write Stan model
cat(file = "model17_6.stan", "

data{
  int n;                              //Number of samples
  int nPops;                          //Number of populations
  array[n] int N;                     //Number of trials in each sample
  array[n] int C;                     //Successes in each sample
  vector[n] precip;                   //covariate
  array[n] int pop;                   //Population index
}

parameters{
  real mu_alpha;
  real mu_beta;
  real<lower = 0> sigma_alpha;
  real<lower = 0> sigma_beta;
  vector[nPops] alpha;
  vector[nPops] beta;
}

model{
  vector[n] p;                        //Estimated success probability

  mu_alpha ~ normal(0, 100);
  mu_beta ~ normal(0, 100);
  sigma_alpha ~ cauchy(0, sqrt(10));
  sigma_beta ~ cauchy(0, sqrt(10));

  for (i in 1:nPops){
    alpha[i] ~ normal(mu_alpha, sigma_alpha);
    beta[i] ~ normal(mu_beta, sigma_beta);
  }

  for(i in 1:n){
    p[i] = inv_logit(alpha[pop[i]] + beta[pop[i]] * precip[i]);
    C[i] ~ binomial(N[i], p[i]);
  }
}
")

# Parameters to estimate
params <- c("mu_alpha", "mu_beta", "sigma_alpha", "sigma_beta", "alpha", "beta")

# HMC settings
ni <- 2000   ;   nb <- 1000   ;   nc <- 4   ;   nt <- 1

# Call STAN (ART 50/5 sec), assess convergence, print and save results
system.time(
out17.6 <- stan(file = "model17_6.stan", data = dataList, pars = params,
                warmup = nb, iter = ni, chains = nc, thin = nt) )
rstan::traceplot(out17.6)                          # not shown
print(out17.6, dig = 3)                            # not shown
stan_est <- summary(out17.6)$summary[1:4,1]        # Save estimates
```

17.7 **Do-it-yourself maximum likelihood estimates**

As in Chapters 10 and 14, the DIY likelihood for mixed models in R requires more work than that for the simple GLMs, since we need to first define the integrated likelihood and then use that in the numerical function minimization done by `optim()`. In our model here, we have both random intercepts and slopes, so we need to evaluate a double integral in the likelihood code. Luckily, both the math and the code are very similar to Section 14.7. As in the other sections of this chapter, we just need to adjust the linear predictor, link function, and response distribution. As in Chapter 14, we'll separate the math and code into a series of separate functions for clarity. First, the function $f(\alpha_j, \beta_j, \mu_\alpha, \sigma_\alpha, \mu_\beta, \sigma_\beta, C_j, N_j, \textbf{precip}_j)$, which is the likelihood for the K_j counts of a given population j ($C_{j,k}$) at particular values of the random intercepts and slopes α_j and β_j.

$$f(\ldots) = \prod_{k=1}^{K_j} [\text{dbinom}(C_{j,k}|N_{j,k}, p_{j,k})] \cdot \text{dnorm}(\alpha_j|\mu_\alpha, \sigma_\alpha) \cdot \text{dnorm}(\beta_j|\mu_\beta, \sigma_\beta)$$

$$p_{j,k} = \text{ilogit}(\alpha_j + \beta_j \cdot precip_{j,k})$$

Again, notice the only things that are changed in function f relative to the function for the Poisson GLMM in Section 14.7 is the distribution of the counts and the link function. Here's the R version of the function for our binomial GLMM:

```
f <- function(alpha_j, beta_j, mu.alpha, sig.alpha,
        mu.beta, sig.beta, C_j, N_j, precip_j){
  p <- plogis(alpha_j + beta_j * precip_j)
  prod(dbinom(C_j, N_j, p)) * dnorm(alpha_j, mu.alpha, sig.alpha) *
    dnorm(beta_j, mu.beta, sig.beta)
}
```

We need to integrate out α and β, generating a function g that does not depend on α or β, but only on the hyperparameters and the data. Function g represents the contribution to the likelihood of population j.

$$\iint_{-\infty}^{\infty} f(\ldots) d\alpha_j \ d\beta_j = g(\mu_\alpha, \sigma_\alpha, \mu_\beta, \sigma_\beta, C_j, N_j, \textbf{precip}_j)$$

In R we'll once again rely on the `pracma` library to do the necessary double integration. The `integral2` function occasionally throws an error, so we added some code to ignore these errors, using function `tryCatch()`.

```
library(pracma)
g <- function(mu.alpha, sig.alpha, mu.beta, sig.beta, C_j, N_j, precip_j){
  tryCatch({
  integral2(fun = f, xmin = -10, xmax = 10, ymin = -10, ymax = 10,
      vectorized = FALSE,
      mu.alpha = mu.alpha, sig.alpha = sig.alpha, mu.beta = mu.beta,
      sig.beta = sig.beta, C_j = C_j, N_j = N_j, precip_j = precip_j)$Q
  }, error = function(e) return(Inf))
}
```

Now we can write the complete likelihood by multiplying together the likelihoods for all 16 populations.

$$L\left(\mu_\alpha, \mu_\beta, \sigma_\alpha, \sigma_\beta | C, N, \mathbf{precip}\right) = \prod_{j=1}^{16} g(\mu_\alpha, \sigma_\alpha, \mu_\beta, \sigma_\beta, C_j, N_j, \mathbf{precip}_j)$$

We implement this algebra in R next. As usual, since we need to return the negative log-likelihood, we initialize `nll` at 0 and then we'll decrementally subtract the log-likelihood of each population instead of multiplying the likelihoods.

```
# Define NLL for Binomial GLMM
NLL <- function(pars, data) {
  mu.alpha <- pars[1]
  mu.beta <- pars[2]
  sig.alpha <- exp(pars[3])
  sig.beta <- exp(pars[4])

  nll <- 0                                      # Initialize at zero

  for (j in 1:data$nPops){
    #Subset data to just pop j
    C_j <- data$C[data$pop == j]
    precip_j <- data$precip[data$pop == j]
    N_j <- data$N[data$pop == j]
    lik <- g(mu.alpha, sig.alpha, mu.beta, sig.beta, C_j, N_j, precip_j)
    nll <- nll - log(lik)
  }
  return(nll)
}
```

```
# Minimize that NLL to find MLEs, get SEs and CIs and save estimates (ART 100 sec)
inits <- c(mu.alpha = 0, mu.beta = 0, log.sigma.alpha = 0, log.sigma.beta = 0)
system.time(
  out17.7 <- optim(inits, NLL, data = dataList, hessian = TRUE,
    method = 'BFGS', control = list(trace = 1, REPORT = 1)) )
get_MLE(out17.7, 4)
diy_est <- c(out17.7$par[1:2], exp(out17.7$par[3:4]))    # Save estimates
```

	MLE	ASE	LCL.95	UCL.95
mu.alpha	-0.23373	0.2296	-0.6837	0.2163
mu.beta	-2.35381	0.3154	-2.9720	-1.7357
log.sigma.alpha	-0.09902	0.1847	-0.4610	0.2630
log.sigma.beta	0.19399	0.1882	-0.1748	0.5628

That's our first binomial GLMM fitted "by hand"—how cool is that?

17.8 Likelihood analysis with TMB

We again need to renumber the population index to start at 0 to make C++ happy. The model code for the binomial GLMM is nearly identical to that in Chapter 14—we just need to change the linear predictor, link function, and response distribution.

```r
# Bundle and summarize data
tmbData <- dataList
tmbData$pop <- tmbData$pop - 1   #indices start at 0 in TMB
str(tmbData)

# Write TMB model file
cat(file = "model17_8.cpp",
"#include <TMB.hpp>

template<class Type>
Type objective_function<Type>::operator() ()
{
  //Describe input data
  DATA_VECTOR(C);
  DATA_VECTOR(N);
  DATA_IVECTOR(pop);
  DATA_VECTOR(precip);
  DATA_INTEGER(nPops);
  DATA_INTEGER(n);

  //Describe parameters
  PARAMETER(mu_alpha);
  PARAMETER(mu_beta);
  PARAMETER(log_sigma_alpha);
  PARAMETER(log_sigma_beta);
  PARAMETER_VECTOR(alpha);
  PARAMETER_VECTOR(beta);

  Type sigma_alpha = exp(log_sigma_alpha);
  Type sigma_beta = exp(log_sigma_beta);

  Type LL = 0.0;                              //Initialize log-likelihood at 0

  for (int j=0; j<nPops; j++){
    LL += dnorm(alpha(j), mu_alpha, sigma_alpha, true);
    LL += dnorm(beta(j), mu_beta, sigma_beta, true);
  }

  vector<Type> p(n);
  for (int i = 0; i<n; i++){
    p(i) = invlogit(alpha(pop(i)) + beta(pop(i)) * precip(i));
    LL += dbinom(C(i), N(i), p(i), true);
  }

  return -LL;
}
")

# Compile and load TMB function
compile("model17_8.cpp")
dyn.load(dynlib("model17_8"))

# Provide dimensions and starting values for parameters
params <- list(mu_alpha = 0, mu_beta = 0, log_sigma_alpha = 0,
               log_sigma_beta = 0, alpha = rep(0, tmbData$nPops),
               beta = rep(0, tmbData$nPops))
```

```
# Create TMB object
out17.8 <- MakeADFun(data = tmbData, parameters = params,
           random = c("alpha", "beta"),
           DLL = "model17_8", silent = TRUE)

# Optimize TMB object and print and save results
opt <- optim(out17.8$par, fn = out17.8$fn, gr = out17.8$gr, method = "BFGS")
(tsum <- tmb_summary(out17.8))
tmb_est <- c(tsum[1:2,1], exp(tsum[3:4,1])) # Save estimates
```

17.9 Comparison of the parameter estimates

We're eager to compare the estimates in this model, which is one of the hardest to fit in the book. . . .

```
# Compare results with truth and previous estimates
comp <- cbind(truth = truth, gtmb = gtmb_est, JAGS = jags_est,
    NIMBLE = nimble_est, Stan = stan_est, DIY = diy_est, TMB = tmb_est)
print(comp, 4)
```

	truth	gtmb	JAGS	NIMBLE	Stan	DIY	TMB
mu.alpha	0	-0.2354	-0.2364	-0.236	-0.2363	-0.2337	-0.2354
mu.beta	-2	-2.3535	-2.3568	-2.364	-2.3530	-2.3538	-2.3536
sigma.alpha	1	0.9044	1.0174	1.016	1.0207	0.9057	0.9044
sigma.beta	1	1.2107	1.3526	1.351	1.3627	1.2141	1.2107

. . . and we find the usual pattern, with almost identical estimates for the mean hyperparameters, and very similar, but not quite identical estimates for the dispersion hyperparameters. The discrepancies in the latter are mainly such that there is a slight difference between the solutions given by maximum likelihood and the Bayesian posterior means. Since our data set is small at the population scale, it is likely that the vague priors do exert a slight effect on the Bayesian estimates here. We could test this hypothesis by analyzing a much larger copy of the same data set and ideally in a simulation, that is, for many replicate data sets.

17.10 Summary

We have introduced the binomial GLMM, which can be seen as a binomial GLM, or logistic regression, where we add (in our case) two sets of normally distributed random effects in the linear predictor. Alternatively, we can view this model as a slight variant on the normal mixed model from Chapter 10, where the data distribution is changed from normal to binomial and with the corresponding changes in the link function and the additional specification of the binomial totals N as data. As with the Poisson case, the introduction of random effects into a binomial GLM in the BUGS language (i.e., in JAGS and NIMBLE) is particularly straightforward and transparent. Fitting the resulting binomial GLMM is very helpful for your general understanding of the class of mixed models or random effects models in general.

Model building, model checking, and model selection[⊛]

18

Chapter outline

18.1 Introduction

In this chapter, we deal with several important and challenging topics, including model checking (also called model criticism or goodness-of-fit testing) and model selection (also called model choice or model validation; with the latter focusing on the assessment of predictive accuracy of a model). Model checking and selection differ in one sense, but are confusingly similar in another. Briefly, model checking asks whether a particular model is any good as a formal learning vehicle

---

[⊛]This book has a companion website hosting complementary materials, including all code for download. Visit this URL to access it: https://www.elsevier.com/books-and-journals/book-companion/9780443137150.

Applied Statistical Modelling for Ecologists. DOI: https://doi.org/10.1016/B978-0-443-13715-0.00025-X
© 2024 Elsevier Inc. All rights reserved, including those for text and data mining, AI training, and similar technologies.

about your data set. In contrast, model selection typically presupposes several models and asks which one is best. But even the best model in a set may still be rather bad; hence, model selection is not a substitute for model checking. Model checking and selection are tightly linked via the fit of the model to the data. The comparison of observed data with data expected under the model is the main basis for model checking, while most methods of model selection are based on a trade-off between model fit and model complexity. We note in passing that all of this is a completely separate issue from the convergence of an algorithm (maximum likelihood or Bayesian) used to fit a model. Convergence must be ensured before we can do model checking or selection.

Most key concepts for model checking and selection apply similarly in frequentist and Bayesian analyses. In particular, predictive simulations are a crucial method in both: here, we use our model with the parameter estimates obtained by fitting it to our actual data to generate "ideal" or "perfect" data sets, where obviously all model assumptions are met. Then, we compare these replicate data sets with our actual data set, either directly (e.g., in terms of each data point) or using some discrepancy measure or test statistic, which serves as an omnibus goodness-of-fit test of the model to the data. Such a number (e.g., a Chi-squared statistic) quantifies the distance between the observed data and their expected values, such as the mean of a normal, binomial, or Poisson response. In addition to numerical summaries of model fit to the data, visual checks can be extremely effective for judging model fit and are often used (Gelman et al., 1996; Gelman & Hill, 2007; Buja et al., 2009; Gelman & Shalizi, 2012; Conn et al., 2018).

Interestingly, such "replicate data simulations" are fundamentally a frequentist rather than a Bayesian concept. Nevertheless, many Bayesian writers strongly emphasize them as an invaluable tool for model checking, for example, Box (1980), Rubin (1984), Gelman et al. (1996), Gelman & Shalizi (2012), and Gelman et al. (2020). Indeed, Little (2006) argues that *"inferences under a particular model should be Bayesian, but model assessment can and should involve frequentist ideas."* In this chapter we use frequentist and Bayesian methods fairly interchangeably, but we highlight the differences as needed.

Model checking and selection are intimately linked with the question of *"How do I build a statistical model?"*, and this in turn cannot be separated from the question of *"Why do I build a model?"*. Surprisingly, the goal of a model is a crucial, yet commonly ignored topic. We think it is a key decision that you ought to make explicit at the start of every modeling project. Therefore, we also say something about these two topics, which to some degree determine both whether a model is good enough (i.e., whether the model fits the data) and which model to choose when we have several of them (i.e., model selection).

The plan of this chapter is as follows: in Section 18.2, we review different goals for a statistical model. In Section 18.3, we introduce a family of simulated data sets that will serve as an illustration throughout the chapter. In Section 18.4, we offer some thoughts about how to build a statistical model. Then, in Sections 18.5 and 18.6, we describe model checking and model selection, while in Sections 18.7 and 18.8, we briefly cover model averaging and regularization. The latter can be an exciting alternative to traditional model selection or averaging, when you have (too) many covariates and when prediction is the main goal of a model.

18.2 Why do we build a statistical model?

At least implicitly, every modeling project has a goal. Since there is literally an infinity of possible models for any given data set, knowing why we build a model is extremely helpful for telling us *how* to build it, and later also how to *check* it and *select* among multiple candidates. Here are five possible goals of a model, knowing that any given model may have more than one:

- Enforcing clarity of thought.
- Data summary and description.
- Data exploration and search for patterns.
- Prediction.
- Explanation, inference about mechanisms, identification of causes underlying patterns.

The first, crucial, and frequently ignored goal of a model is simply to *enforce more clarity in your thinking*. For instance, when building a model you have to make a decision about the quantities that you want to consider in a system about which you want to learn, and which ones you want to ignore. Then, you must think about how they are related to each other and by what functional forms. This is similar to a flowchart of cause and effect in a system, but is a much more precise and powerful manner of putting rigor into your thought process.

Moreover, when you translate this description into computer code, and choose values for each quantity (e.g., for covariates and model coefficients), you can already learn a lot in a *what-if* manner, for example, whether a factor of interest may indeed cause a response of some observed magnitude. This type of modeling is related to classical, data-free mathematical modeling, but it can also be done with statistical models. A variant of such a more conceptual statistical model that is even richer and more powerful for enforcing understanding is to write computer code to simulate data under a model (e.g., Chapter 4 in Kéry & Royle (2016) and Chapter 1 in this book). Data simulation has tremendous benefits for understanding in science and management, and its importance for learning as well as for teaching can hardly be overstated. Indeed, this is one reason for why in this book we use almost exclusively simulated data sets and show all the code with which to generate them.

A second goal of any statistical model is to serve as a summary or a *parsimonious description of a data set* (Burnham & Anderson, 2002). For instance, a simple linear regression may replace with just two constants, the intercept and the slope, any number of individual data points in a scatterplot. This is a huge reduction in complexity and may be very useful by itself.

A third goal is *pattern searching*, for example, for associations among variables, including between and among explanatory variables and a response. Pattern searching is a fundamental component of any science, since historically, a science typically progresses from the discovery and description of patterns to an attempt at identifying the mechanisms that cause them.

A fourth modeling goal is *prediction*, that is, estimation of system output at new or "unseen" places and times. For instance, we may want to extrapolate a modeled process in space or forecast it in time.

Finally, a fifth modeling goal is *causal explanation*, that is, the identification of mechanisms, or inference about the causes that underlie a pattern. The search for causes is arguably the most celebrated goal in science. Thus, the importance of causal explanation as a modeling goal can hardly be overstated.

There may be some confusion between prediction and explanation of a mechanism. At a "large" scale, the predictive performance of a scientific theory, which typically is about mechanisms, is usually taken as a main referee for its validity, especially with respect to that of another theory. Hence, for a theory, prediction and explanation are almost the same, since a good prediction is taken to mean a good causal explanation. However, at the "smaller" scale of a statistical model, there are much deeper differences between prediction and explanation (Shmueli, 2010).

To explain the difference between a model for prediction and one for explanation, Tredennick et al. (2021) consider a fitted regression model $\hat{\mathbf{y}} = \mathbf{X}\hat{\boldsymbol{\beta}}$, where $\hat{\mathbf{y}}$ is the predicted response, $\mathbf{X}$ a design matrix, and $\hat{\boldsymbol{\beta}}$ is a vector of estimated parameters. The goal of prediction is to produce good estimates of $\hat{\mathbf{y}}$, regardless of what the input variables in $\mathbf{X}$ may be. Hence, $\mathbf{X}$ may be related to the response via a genuine causal mechanism or by pure statistical association. In contrast, the main interest of scientific explanation in a causal-explanatory statistical model is to have the "right" input variables in $\mathbf{X}$ and then infer their effects $\hat{\boldsymbol{\beta}}$ as accurately as possible. In addition, $\mathbf{X}$ may also contain some nuisance variables that may bias, or add noise to, these direct mechanistic relationships, and for which we therefore want to adjust or correct.

The goals of pattern searching, prediction, and explanation may all overlap with the first two goals (enforcing clarity of thought and summarizing the data), but there is only limited overlap among these three (Tredennick et al., 2021). Pattern searching and prediction may overlap to some degree, as do prediction and causal explanation. In contrast, only independent studies can take you from pattern searching to causal explanation (but see Arif & MacNeil, 2022, and Section 18.4.2). In particular, you should avoid using for inference models obtained using pattern searching tools like data mining and "all subsets" model selection procedures.

A model that does a good job describing associations will often be good for prediction; this is the magic of machine learning. But such a model is not guaranteed to be good at explanation of mechanisms—that's the old adage of a correlation that's not the same as a causation. Likewise, a model with a good description of mechanisms may also be good for prediction (de Valpine, 2003; Hefley et al., 2017). However, this need not always be the case and indeed in Section 18.8 you will see an example of where the actual, true model (i.e., the one which we simulated the data with) predicts less well than a regularized and, in this sense, less true model.

Making the goals of a model explicit is crucial for our decision about how to build a model in the first place, how to judge whether the model is any good, and how to decide which model to select when we have more than one. But before we go on examining these questions, we first introduce a family of simulated data sets that will serve as an illustration throughout this chapter.

18.3 Ticklish rattlesnakes

We will use for illustration a study of how relative abundance (i.e., counts) of timber rattlesnakes in Virginia (Fig. 18.1) is related to a number of environmental variables. These include a measure of ground cover by large rocks, which provide shelter, and the frequency of oak trees, which produce acorns. The latter are a staple food for chipmunks and other rodents, both of which are favorite prey for rattlesnakes. A final environmental variable is an index of chipmunk abundance, since the snakes like to eat them. However, there is a quadratic relationship between the abundance of rattlers and chipmunks: at high chipmunk abundance, rattler abundance will decrease. The reason for this counterintuitive pattern is that, unbeknownst to current rattlesnake science, rattlers are in fact extremely ticklish! When occurring at high density, chipmunks will often inadvertently enter the rocky crevices where rattlers rest. The rattlers can't eat them all, but will suffer terrible tickling by the chipmunks moving around in the crevice. To avoid that, rattlers then tend to leave an area.

We will work with data simulation function `simDat18()` that lets us choose sample size and the coefficients of the main explanatory variables, in addition to those for an arbitrary number of additional (and unspecified) covariates affecting the relative abundance of rattlesnakes. Thus, the function lets us create any of a large "family" of data sets. Fig. 18.2 shows a typical data set generated by executing the function.

FIGURE 18.1

Timber rattlesnake (*Crotalus horridus*) in the Blue Ridge Mountains of Virginia, July 2023 (Photo: Marc Kéry).

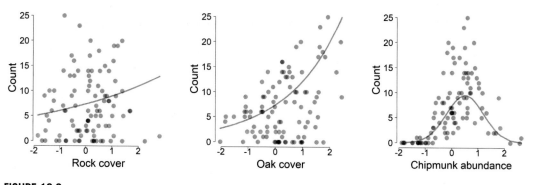

FIGURE 18.2

A typical data set produced by executing the ticklish rattlesnake simulation function `simDat18()` with default argument values. The red lines show the expected relationships given by the coefficients of the simulation model.

18.4 How do we build a statistical model?

A statistical model is an abstraction of the processes that produce some data of interest, and it must serve one or more of the modeling goals just discussed. It is a simplified mathematical representation of reality and should contain the right quantities (e.g., response and covariates), link them in an adequate way (e.g., with the right link functions and the right structural form), and comprise the right types of noise terms (e.g., random effects). So how do we achieve all this?

A first answer is simply that *we must build a model that meets our modeling goals*. Any more detailed answer will depend to a large part on your answer to the question of why you build your model, especially with respect to the three goals of pattern searching, prediction, and causal explanation (Shmueli, 2010; Tredennick et al., 2021).

A primary consideration is the rigidity and flexibility of a model. Like a human body, a statistical model has both rigid, or hard, and flexible, or soft, parts. That is, we will want to have some hard structure in it for sure, represented by specific parameters of interest and the way they are connected in the model. These are things that we know are there in the process that we want to understand or predict, and thus we pick them on purpose. Also, the hard structure may differ between different models and may represent different scientific hypotheses, which we may then test by comparing these models. On the other hand, there will typically also be additional soft structure in a model that is more of a "nuisance." We're often not particularly interested in these soft parts, but we need them to "get things right" for the rest of the model. Hence, we may not care so much about their specific form, but rather want those parts of the model to be flexible and to adapt to the specific data, without much (or even without any) input from the modeler. Covariates used for statistical adjustment, year or site random effects, or smoothing splines are examples of such "nuisance structure," which really may be just some sort of flexible stuffing in the model (but see Chapters 19 and 19B for models where the random effects are in fact of particular interest). Thus, a statistical model typically contains hard parts, which are dictated by the science, and soft parts, which we want to adapt flexibly to our data set.

The relative amount of hard and soft structure in a model will differ between causal-explanatory and predictive models. In the former, we often have much structure that we define *a priori*, while in the latter, we may not worry so much about the hard structure, provided that the soft structure does a good job at predicting a response. Sex in an evolutionary-ecological model provides a good example. In a survival analysis, we might always want to have sex as an explanatory variable, since we just *know* that the life history of the sexes differs so much for most species. Thus, we may not even ask whether including sex actually improves the predictive power of the model, but simply add it in. In contrast, in a pattern searching or predictive mode of modeling, we might only include sex if it really helps improve predictive accuracy.

To some degree, there may also be a difference in this respect between more mature sciences and newer or less theory-rich sciences. Prediction is more closely related to pattern searching and thus may be more useful at the earlier developmental stages of a science. In contrast, in a mature science with plenty of established theory, there will typically be much more known structure that we want to put into our models.

As an illustration for how the modeling goals dictate what model we build, consider a species distribution model (SDM) for a deer species. If the SDM is to be used for setting harvest regulations, then the model must be able to predict the true deer abundance as an unobserved, latent variable. That is, we must learn not just how many deer were observed, but how many are actually out there, including those that we fail to detect, and we might then build a capture-recapture, distance sampling or *N*-mixture model (Royle & Dorazio, 2008; Kéry & Royle, 2016, 2021). They all have a relatively rigid structure that relates to the sampling design and is instrumental in the ability of these models to estimate true abundance while correcting for imperfect detection. In contrast, if our aim is simply to obtain a map of the spatial variation of an index of deer abundance, then a point process model (Renner et al., 2015) or boosted regression trees (Elith et al., 2008) may be better and easier, since in them we can more flexibly deal with a multitude of inter-related covariates and produce nice maps of relative abundance.

In the next two sections, we outline a view of model building that is arguably most appropriate for causal-explanatory models, where identification of mechanisms is a major goal, though rarely the only one. These models are directly science-based and they will contain plenty of "hard" structure that is dictated by a scientific question or a management problem. In contrast, a more flexible and data-based approach is often adopted for purely predictive models, in which model building is geared towards optimized predictive success (see Section 18.6 on model selection).

18.4.1 Model expansion and iterating on a single model

For any project where causal explanation is a major goal, applied statistical modeling is best seen as an iterative process, where we go several times back and forth between model building and model checking. Under this view, model checking is an integral part of modeling (Conn et al., 2018) and every checking round may suggest additional structure to be added to the model. This iterative process has also been called *model expansion* (Draper, 1999; Gelman et al., 2014, 2020) or *iterating on a single model* (ver Hoef & Boveng, 2015). In this approach, we start from our scientific question or management problem, define the goals of our model, and get an overview about the data that we have or can get. Then, we build a simplified mathematical representation of our study system that must be just complex enough to contain all the necessary structure (i.e., parameters) to answer our scientific question or solve our management problem. At the same time, the model must also contain all the necessary nuisance structure to get unbiased estimates of the main parameters of interest. And finally, our model must include all relevant sources of variability, so that we will get valid uncertainty assessments. This is an often forgotten task of a statistical model (Cressie et al., 2009).

At first, we want to build a relatively big model where we put just enough in there, but not too much. Where to strike this balance will depend on our questions, but also on the richness of our data set. Naturally, estimating five parameters may be too much for 10 data points, but with 100,000 data points we may estimate hundreds of parameters. Finding the right level of abstraction of the system described by our model is arguably *the most crucial part of the art of modeling*. And be warned: typically, as ecologists (as opposed to, say, physicists or mathematicians) we tend to err towards wanting to put too much complexity into our model.

Once we have built a first version of our model, we must check it and see whether it is a useful basis for inference or prediction. This is where model checking comes in (Section 18.5). When we find our model deficient in a way that might compromise our modeling goals, we can improve it, usually by adding more structure, that is, additional parameters. Then, we enter the next model checking cycle, and so forth. Remember that what represents a deficiency of a model can solely be decided in view of our modeling goals, which we should therefore best articulate clearly.

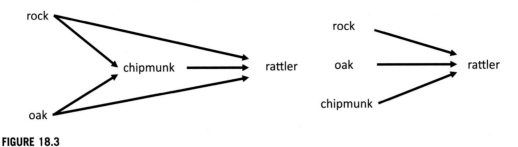

FIGURE 18.3

The true (left) and the simplified (right) causal structure in the ticklish rattlesnakes.

18.4.2 Causal, path, and structural equation models

Using the methods presented in this book, we can build wonderful models to help answer our scientific questions or solve our management problems. However, in most of current statistical modeling in ecology, we do not entertain any models with more complex, reticular causal relationships involved. But arguably, simple models of cause and effect can only be an approximation to reality. Fig. 18.3 shows on the left the true causal graph for the main explanatory variables in the ticklish rattlesnakes, while on the right is the simplified structure considered throughout this chapter. Admittedly, a mesh of causal paths is a more convincing description of how we often see nature with its reticulate cause–effect relationships. Such a model easily allows for correlated explanatory variables and thus naturally deals with collinearity (Dormann et al., 2013), enables us to distinguish direct from indirect effects (e.g., for rock or oak), and allows us to test much finer mechanistic hypotheses than the simpler, non-reticulate model.

Models that allow reticular cause-effect relationships and where some nodes in such a mesh, or graph, may be directly observed and others not, are known as causal, path or structural equation models (SEMs; Shipley, 2004; Grace, 2006; Pearl, 2009; Grace & Irvine, 2020). They are not yet widely used in ecology, but are on the increase. A proper section on this topic will be one major improvement in the second edition of this book. In the meantime, you can learn more by reading existing published articles such as Cubaynes et al. (2012), Joseph et al. (2016), Souchay et al. (2018), Davis et al. (2023), Gibson et al., (2023), and Latif et al. (2023). The `lavaan` package (Rosseel, 2012) lets you fit SEMs in R, while the related package `blavaan` (Merkle & Rosseel, 2018) is helpful as a user-friendly interface for fitting SEMs in JAGS or Stan (Merkle et al., 2021) and produces model code in either software, which you may then further modify.

18.4.3 How to build a predictive model

When prediction is the main or only goal of a statistical model, quite a different construction mode is appropriate. Here, by far the main focus is typically on the covariates: how we chose them for inclusion in a model, in what form, how we link them to each other and especially to a response, and how we estimate them. Predictive models often contain mostly soft parts, to be tuned flexibly so as to maximize predictive performance of the model. We cover variable selection, including model averaging and regularization, in Sections 18.6–18.8, all with a strong focus on the predictive accuracy of a model.

18.5 **Model checking and goodness-of-fit testing**

A typical parametric statistical model is a tool to learn about the mechanisms that generated our data set. By confronting model and data, we can estimate parameters, and inference about them lets us judge whether one data-generation mechanism is more likely than another. This may then serve to test different scientific hypotheses, or we may estimate unobserved, but potentially observable, quantities of interest, such as the size of a population or the presence or absence of a species at our study sites (Chapters 19 and 19B). In addition, we may use the model to predict as yet unseen responses, for instance at unstudied sites or sometime into the future.

All of this requires that our model is an adequate abstraction of the usually high-dimensional processes that really produced our data. For instance, this abstraction must ensure that the main covariates are included, that their functional form is specified correctly (e.g., with a quadratic effect when there is such a parabolic relationship), that the right quantities in the model are assumed as noise terms, and that we properly describe this noise. Only with an adequate model can we hope to achieve the goals of our modeling, be they hypothesis generation/pattern searching, inference, or prediction. Thus, we should always check whether our model can describe the salient features in our data set. Only when it does so should we use it further for inferring mechanisms or making predictions. This sort of model checking is called goodness-of-fit testing.

Arguably, the first and in some sense most important goodness-of-fit test is simply the question of whether the results make sense in the light of our contextual and subject-matter knowledge. We like to call this the commonsense goodness-of-fit test. Although there is a danger of circularity in it, this test can't and should not be avoided. In fact, we do it all the time when fitting a model, although perhaps mostly unconsciously. For instance, in our rattlesnake example, when your fitted model predicts rattlesnake counts in the millions at a 1-ha site, your herpetologist colleague will immediately tell you that your model is probably no good.

In this section, we deal with more formal approaches to model checking. Data simulation is a very powerful method for testing the goodness-of-fit of a model (Rubin, 1984; Gelman & Hill, 2007). We use simulation to obtain the distribution of data under our model as well as the distribution of any test, fit, or discrepancy statistic that we can compute from the data. Thus, simulation lets us gauge how data and such fit statistics look for a fitting model. It thus provides the reference distribution under the hypothesis of a fitting model. We can then compare such replicate data with our actual data set, and the fit statistics computed for them with those for the actual data set. When they don't resemble each other, we judge that our model does not fit the data set. We emphasize again the power of visual goodness-of-fit tests (Buja et al., 2009).

Throughout the book, we have encountered goodness-of-fit tests conducted with the DHARMa package (Hartig, 2022), as well as Bayesian posterior predictive checks (e.g., in Chapter 5). Here, we expand on these and present some general principles for goodness-of-fit. Central to all that is the *predictive distribution of the data*, which is the distribution of the data implied by our model, including the actual values of all parameters. We can assess the predictive data distribution using a parametric bootstrap (see Section 2.5.4): we use parameter estimates obtained from fitting our model to the actual data set, plug them back into our model along with any covariate values in the analysis, and use the model in a generative way to simulate new data sets. This is exactly how we simulate data sets throughout the book, except that now we plug into the R code parameter estimates rather than just some arbitrary coefficients that lead to data sets that look nice to us. In typical applications of the bootstrap, we will work with point estimates (e.g., the MLEs) and ignore the uncertainty associated with these estimates. Thus, this approach to the bootstrap will not represent quite the full variability-plus-uncertainty (or stochastic plus epistemiological uncertainty; Spiegelhalter, 2019) in the predictive data distribution.

The Bayesian approach to goodness-of-fit tests is different. Here, we work with the *posterior predictive distribution* of the data, in which we average over the posterior distributions of the parameters when simulating our data sets. As a result, we can obtain a better characterization of the combined variability-plus-uncertainty about future data sets under a model. Goodness-of-fit checks based on posterior predictive distributions are called posterior predictive checks (Gelman et al., 1996; Conn et al., 2018). They can be cast as a significance test, when the proportion of fit statistics computed for the simulated data sets that is greater than the fit statistics computed for the actual data set is taken as a Bayesian *p*-value to test the fit of a model to a data set (Conn et al., 2018). Regardless of our mode of inference, we often compute omnibus or global goodness-of-fit tests, which try to distill the fit of a model into a single number. Not surprisingly, this must be viewed with some caution, since a single number can hardly "say it all."

We can also compute residuals, which let us inspect the fit in different parts of the model. Thus, residuals offer a local assessment of the goodness-of-fit of a model. Beyond ensuring that a model fits an actual data set, residual checking represents a very powerful, but heavily underutilized method of learning from your data (Conn et al., 2018). It has often been argued that we can actually learn more from a non-fitting model than from a fitting one. Specifically, where and when a model breaks may provide major opportunities to learn. For this reason, we think that residual assessment is really important because this (but not an omnibus test) can highlight observations that are surprising under a given model. Trying to understand why a model fits for some data but not for others may then give us important indications for the next iteration of an improved model and just generally more insight about the processes we're interested in. Unfortunately, however, far too often we view data that don't fit just as a nuisance, and thus forego this invaluable opportunity for scientific discovery.

An important question is: "What happens when our model does not fit?" Do we then have to discard it and perhaps declare our data set as un-analyzable? This may be the best course of action in very rare cases, but most of the time will not be what we do. Rather, if a model doesn't pass these tests, we can typically improve it as described in Section 18.4.1.

The plan for Section 18.5 is as follows: in Section 18.5.1, we briefly cover traditional residual diagnostics, which are useful mostly for normal-response models. In Section 18.5.2, we cover predictive distributions of the data, inferred either by a frequentist bootstrap or by Bayesian posterior predictive simulations. They are useful, but often not very sensitive to lack of fit, since in both we use the data twice: first, we estimate the parameters such that they best represent a particular data set, and second we use these same parameters to create replicate data sets that we compare back with our actual data. This will lead to a too optimistic assessment of model fit as we will see. Therefore, in Section 18.5.3, we use cross-validation (CV) to compute predictive distributions for *new data* that the model hasn't seen. CV is a powerful tool not only for model checking, but also for model selection, as we will later see in Section 18.6. Finally, in Sections 18.5.4–18.5.6, we cover overdispersion (OD), measures of absolute model fit, and parameter estimability and robustness to violation of the assumptions of a model.

18.5.1 Traditional residual diagnostics

For normal linear models, the typical goodness-of-fit tests are based on residuals, that is, the difference between observed and expected datum, or $r_i = y_i - \mu_i$. Under i.i.d. and normality assumptions, there should be no pattern in the residuals, since they are random numbers from a zero-mean normal distribution. We can examine this using various diagnostic plots obtained as follows; see also Section 5.3.2.

```
library(ASMbook); library(jagsUI)   # Load packages
dat <- simDat9()                    # Simulate data set like in Chapter 9
fm1 <- lm(dat$mass ~ dat$pop)       # Fit model with population factor only
fm2 <- lm(dat$mass ~ dat$lengthC)   # Fit model with length only
plot(fm1)                           # Produce plots with residual diagnostics
plot(fm2)
```

These checks work well for linear models, provided the sample size is not too small. But for non-normal responses such as those with a Poisson or binomial distribution they are much less useful, in part because the residual variance is not constant, but a function of the mean. Residuals can be standardized for this as in the Pearson residuals (Chapter 11), but this may not always help that much, as we now illustrate with the ticklish rattlesnakes. We fit and test the goodness-of-fit of the data-generating model and will still see some patterns that may easily look alarming or at least hard to interpret.

```
set.seed(18)
str(dat <- simDat18(nSites = 200, beta1.vec = c(2.5, 0.2, 0.5, 1, -1),
   ncov2 = 1, beta2.vec = rnorm(1, 0, 0)))
fm <- glm(dat$C ~ dat$rock + dat$oak + dat$chip1 + dat$chip2 + dat$Xrest, family = poisson)
plot(fm)
```

The variance-mean relationship and the discrete nature of residuals in Poisson and binomial models often make meaningful interpretation of any residual patterns nearly impossible for non-normal GL(M)Ms (Hartig, 2022). Another type of residual known as a quantile residual is a powerful and more interpretable alternative. We will cover them in the next section, which is the first of two where we show how to use predictive simulations of the data for model checking.

18.5.2 Bootstrapped and posterior predictive distributions

Traditional residual tests are based on an implicit comparison between the observed data (in the form of their residuals) and hypothetical data under a model. These hypothetical data sets don't actually appear, but we know from theory how their residuals would look like. Thus, we can look at the residuals from the actual data and compute various formal statistical tests based on this implicit comparison.

We have seen in Chapter 2 how we may replace statistical theory with computation and directly see how data look under a fitting model. That is, we can use our parametric model in a generative way and use the exact parametric assumptions along with the covariate values and the parameter estimates obtained in our analysis to simulate new data sets. These serve to characterize the distribution of data under a fitting model and therefore as a reference for comparisons with our actual data set. This technique is a form of a parametric bootstrap when based on the MLEs from a fitted model (see Chapter 2.5.4), while its Bayesian version is called a posterior predictive check. The latter is based on the posterior predictive distribution of future or, more generally, "unseen" data produced by the processes embodied by our model (Gelman et al., 1996, 2014; Vehtari et al., 2017).

Simulation-based goodness-of-fit testing is extremely versatile and powerful. For instance, we can conduct targeted tests that may assess how well our model fits extreme observations by comparing the observed range of the data, or perhaps the maximum and the mean, with the value of the same statistics in the simulated data. Similarly, we can use simulated data both for omnibus tests, as when computing a Bayesian p-value (Section 5.4.2), or to examine residuals for individual data points to see where exactly our model breaks, as a powerful method for scientific discovery.

In addition, we can use the more general definition of the quantile residual or the empirical cumulative distribution function (CDF) of each datum, as a measure for how unusual each observed datum is, in the light of what we expect based on the simulation replicates (Section 5.4.2). A quantile residual is simply the proportion of simulated values that are less than or equal to the observed value of a datum, with a correction for ties in discrete responses (that's the "randomized" in the title of Dunn & Smyth, 1996). Using simulated data, quantile residuals can be computed for a much wider class of models than traditional residuals, also for mixed and hierarchical models, and they are easier to interpret for these models than traditional residuals.

Throughout the book, we have used functions in package DHARMa (Hartig, 2022) for model checking based on randomized quantile residuals, for models fit with either maximum likelihood or Bayesian inference. To better understand what the package does under the hood, we now conduct similar types of residual analyses "by hand." This will enhance your understanding of these methods and may enable you to obtain quantile residuals for model classes not supported by the package. We use the ticklish rattlesnake data to illustrate how an important lack of fit in the model can be diagnosed with quantile residuals computed from either bootstrapped replicate data or from the posterior predictive simulations. We will simulate a default data set with linear and quadratic effects of chipmunk abundance, but then fit a mis-specified model with only a linear chipmunk effect on the relative abundance of timber rattlesnakes.

```
# Create data set with major covariates plus one noise covariate (ncov2)
set.seed(18)
str(dat <- simDat18(nSites = 200, beta1.vec = c(2.5, 0.2, 0.5, 1, -1),
  ncov2 = 1, beta2.vec = rnorm(1, 0, 0.2)))

# Fit the mis-specified model without chip2
summary(fm <- glm(C ~ rock + oak + chip1, family = poisson, data = as.data.frame(dat)))

Call:
glm(formula = C ~ rock + oak + chip1, family = poisson, data = as.data.frame(dat))

Coefficients:
            Estimate   Std. Error   z value   Pr(>|z|)
(Intercept)  2.02727    0.02761     73.418    <2e-16 ***
rock         0.21674    0.02387      9.081    <2e-16 ***
oak          0.56584    0.02437     23.220    <2e-16 ***
chip1        0.25874    0.02211     11.701    <2e-16 ***
```

The first three estimates resemble their known true values, but now we estimate a strictly monotonic rise of rattlesnake counts with increasing chipmunk abundance. We know that this does not match the patterns in the data (see Fig. 18.2). As a first measure of model fit we now compute the residual variation of the data around their expectation, by summing the squared Pearson residuals. Below, we will compare this with what we get in replicates of the data simulated under model fm.

```
# Compute expected values and residual variation of data around them
mu.lambda <- predict(fm, type = 'response')          # Expected values
resi2.obs <- ((dat$C - mu.lambda) / sqrt(mu.lambda))^2  # Residual variation
(fit.obs <- sum(resi2.obs))                           # Sum over data set
sum(residuals(fm, "pearson")^2)                       # Same
```

Next, we use parametric bootstrapping to create a large number of replicate data sets under the fitted model, which we save in an object. In addition, for each simulated data set we compute the residual variation around the expectation of the data to assess the "natural variation" in the residual noise under the assumption that the fitted model describes the actual data set well.

```
# Simulate data conditional on observed covariate values, model and MLEs
simrep <- 100000                                        # Number of bootstrapped
                                                        # (replicate) data sets

# Create R objects ... to hold replicate (= bootstrapped) data sets
yrep <- array(NA, dim = c(dat$nSites, simrep))

# ...to hold squared Pearson residuals from replicated data
# ...and expectation from the model fit to the actual data
resi2.rep <- numeric(200)

# ...to hold sum of squared Pearson residuals
fit.rep <- numeric(simrep)

# Launch parametric bootstrap: produce simulated data sets
for(k in 1:simrep) {
  if(k %% 1000 == 0) cat(paste('\n*** iter', k))        # Counter
  Crep <- rpois(n = dat$nSites, lambda = mu.lambda)     # Draw Poisson RVs
  yrep[, k] <- Crep                                     # ...and save them

  # Compute squared Pearson residuals and residual variation
  resi2.rep <- ((Crep - mu.lambda) / sqrt(mu.lambda))^2
  fit.rep[k] <- sum(resi2.rep)
}
```

When we're done, we make plots of the omnibus test statistic: the residual variation in the form of the sum of the squared Pearson residuals across the data set (Fig. 18.4 left). We see clearly that the residual

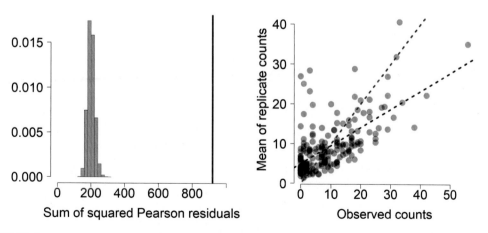

FIGURE 18.4

(left) Goodness-of-fit test based on a comparison of the observed residual variation in the actual data set (blackish blue line) with the bootstrapped distribution of that statistic under a fitting model (gray). (right) Comparison of the mean of the predictive distribution for each of the $n = 200$ data points (on the y-axis) with the observed values (on the x-axis), with regression of y on x shown in blackish blue. Compare this with the Bayesian version of such plots in Fig. 18.7.

variation under the fitted model is far greater for the observed data than what we would expect for data sets produced by this model. That the fit statistic is so *surprising* for the actual data set should make us skeptical about the fit of this model for the data set. We can quantify the magnitude of this surprise by a significance test as *Prob(fit.rep > fit.obs)* and get a value of 0. We can also plot the average of the predictive distribution of every datum against its observed value (Fig. 18.4 right). Under a fitting model we would expect this scatterplot to sit symmetrically around the red 1:1 line. However, the blue regression line indicates the general tendency of the relationship between *y* and *x* and shows a systematic pattern where for small observed data we would expect greater values under the model, while for larger observed data the opposite is true. Overall, this casts doubt as to whether the fitted model is useful for learning about the processes that generated our data set.

```
par(mfrow = c(1,2), mar = c(6,5,4,2), cex.axis = 1.5, cex.lab = 1.5, cex.main = 1.5) # Fig. 18.4
hist(fit.rep, xlim = c(0, 1000), col = 'grey', xlab = 'Sum of squared Pearson residuals',
main = 'Fit statistic for replicate data (grey)\n and for observed data (blue line)')
abline(v = fit.obs, lwd = 3, col = 'blue')

# Plot mean simulated count vs. observed counts for entire data set
plot(dat$C, apply(yrep, 1, mean), pch = 16, col = rgb(0,0,0,0.3), frame = FALSE,
   xlab = 'Observed counts', ylab = 'Mean of replicate counts',
   main = 'Comparison of expected vs. observed')
legend('bottomright', lwd = 3, lty = 3, col = c('red', 'blue'),
   legend = c("1:1 line", "Regression of y on x"), bty = "n")
abline(0, 1, col = 'red', lwd = 3, lty = 3)
abline(lm(apply(yrep, 1, mean) ~ dat$C), col = 'blue', lwd = 3, lty = 3)
```

Throughout, we emphasize the importance of local assessments of the fit of a model to a data set, both as a way of seeing where a model can be improved and, in a more general manner, to see where our understanding of a process may be limited and where therefore we ought to put more study. Predictive data simulations can easily be used for such local fit assessments of a model. Fig. 18.5 shows such a comparison for the first 100 data points with their bootstrapped predictive

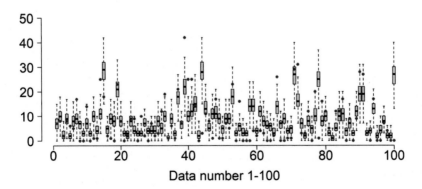

FIGURE 18.5

Example of a local goodness-of-fit assessment via comparison of observed data (red) with their bootstrapped predictive distributions (represented by the boxplots) for the first 100 data in the simulated rattlesnake data set.

distributions, represented as a boxplot. We see again the large discrepancies between the observed data (in red) and what we would expect the data to look like (the boxes) if our model was a good representation of the processes that generated them.

```
boxplot(t(yrep[1:100,]), outline = FALSE, xlab = 'Data number 1-100',
  frame = FALSE, main = 'Bootstrapped predictive distributions (boxes)
  and observed data (red)')                              # Fig. 18.5
points(1:100, dat$C[1:100], pch = 16, cex = 0.8, col = 'red')
```

We next conduct the comparison between observed data and their predictive distributions via randomized quantile residuals (Dunn & Smyth, 1996; Warton et al., 2017; Hartig, 2022). This type of residual expresses a datum as a percentile in its CDF. Technically, it is based on the *probability integral transform* (PIT, Blitzstein & Hwang, 2019): if random variable X follows a continuous distribution f with corresponding CDF F, then applying F to X will result in a new random variable U that is uniformly distributed, that is, $U \sim \text{Uniform}(0, 1)$. Applying the PIT and checking the distribution of the resulting quantile residuals for uniformity is thus a straightforward goodness-of-fit check. In addition, as with any residual, quantile residuals can also be used for local checks to see where a model breaks. Strictly, the PIT applies to continuous random variables, but for discrete ones such as our rattlesnake counts we can apply a randomization correction to make the residuals approximately continuous (Dunn & Smyth, 1996). We now compute randomized quantile residuals for our bootstrapped results for illustration (see also Conn et al., 2018, p. 534, and Algorithm 4 in its Supplemental Appendix 1). Note that this is the key technology underlying the DHARMa package (Hartig, 2022). See the package vignette for much helpful advice about how different forms of assumption violations in a model can show up as a pattern in quantile residuals (Fig. 18.6).

```
yobs <- dat$C                             # copy data
u <- numeric(dat$nSites)                  # Vector to hold quantile residual
for(i in 1:dat$nSites){                   # Loop over all data points
  u[i] <- mean((yrep[i,] < yobs[i]) + runif(1) * (yrep[i,] == yobs[i]))
}

par(mfrow = c(2,2), mar = c(6,5,4,2),     # Fig. 18.6
  cex.axis = 1.5, cex.lab = 1.5, cex.main = 1.5)
hist(u, col = 'grey', main = 'Distribution of quantile residuals (u)')
plot(dat$oak, u, xlab = 'oak', pch = 16, frame = FALSE)
plot(dat$chip1, u, xlab = 'chip1', pch = 16, frame = FALSE)
plot(dat$chip2, u, xlab = 'chip2', pch = 16, frame = FALSE)
```

For a fitting model, the distribution of u should be uniform and there should be no relationship between u and any covariate inside or outside of the model. We check this visually in Fig. 18.6 and find that the frequency distribution of the randomized quantile residuals u is clearly not uniform. In addition, there are strong patterns in u with respect to chip1, which was included in the model, and to chip2 which was not included in the model. Thus, this model clearly does not fit the data well.

We next repeat this model checking approach for the Bayesian analysis of the model: we simulate the predictive distribution of the data under the model, conduct specific graphical and

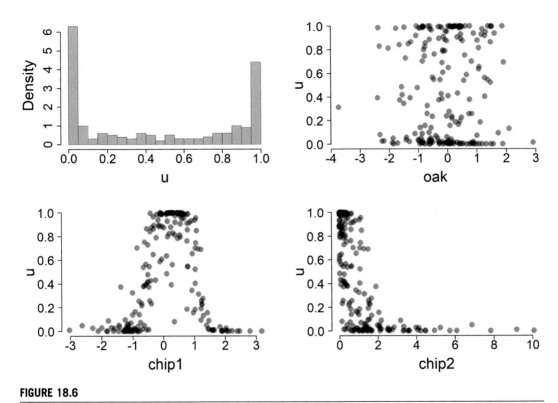

FIGURE 18.6

Goodness-of-fit of the mis-specified rattlesnake model based on bootstrapped quantile residuals. Note that covariate `chip2` is not in the fitted model.

numerical tests, and compute randomized quantile residuals. This is known as a *posterior predictive check* (Gelman et al., 1996; Conn et al., 2018). You will see that, when you ignore the fundamentally different interpretation of probability (Chapter 2), the only difference between the two approaches is that the Bayesian is more inclusive of all the uncertainties involved than our simple bootstrap. A Bayesian computes the posterior predictive distribution of the data, which includes the parametric uncertainty stemming from the fact that we don't know the parameter values, but must estimate them. In contrast, when we bootstrapped the predictive distributions above, we ignored estimation uncertainty and simply treated the MLEs as known constants. It would be possible to include estimation uncertainty also in the bootstrap, but at the price of increased conceptual and computational complexity of the procedure.

The following material is somewhat repetitive of the technique that we saw in Chapter 5 and thus we shorten somewhat. Also, we here emphasize the similarities with the bootstrap approach above. Basically we package computation of replicate data sets and of squared Pearson residuals for the actual and the replicate data all into one model fit in JAGS.

```
# Bundle and summarize data
str(dataList <- list(C = dat$C, rock = dat$rock, oak = dat$oak, chip = dat$chip1, n = length(dat$C)))

# Write JAGS model file
cat(file = "model18.1.txt", "
model {
# Priors
alpha ~ dunif(-10, 10)
for(k in 1:3){
  beta[k] ~ dunif(-5, 5)
}

# Likelihood
for (i in 1:n) {                          # Loop over all data points
  C[i] ~ dpois(lambda[i])                  # The response variable (C above)
  lambda[i] <- exp(alpha + beta[1]* rock[i] + beta[2]* oak[i] + beta[3] * chip[i])
  resi2.obs[i] <- ((C[i] - lambda[i])/sqrt(lambda[i]))^2
  # Squared Pearson residuals for the observed data
}
# Create replicate data under the same model for each data point
for (i in 1:n) {
  Crep[i] ~ dpois(lambda[i])
  resi2.rep[i] <- ((Crep[i] - lambda[i])/     # Squared Pearson residuals for the replicate data
    sqrt(lambda[i]))^2
}
fit.obs <- sum(resi2.obs)                  # Sum over all data
fit.rep <- sum(resi2.rep)                  # ditto
}
")

# Function to generate starting values
inits <- function(){list(alpha = rnorm(1), beta = rnorm(3))}

# Parameters to estimate
params <- c("alpha", "beta", "fit.obs", "fit.rep", "Crep")

# MCMC settings
na <- 1000  ;  ni <- 6000  ;  nb <- 2000  ;  nc <- 4  ;  nt <- 4

# Call JAGS (ART <1 min), check convergence and summarize posteriors
out18.1 <- jags(dataList, inits, params, "model18.1.txt", n.iter = ni, n.burnin = nb,
  n.chains = nc, n.thin = nt, n.adapt = na, parallel = TRUE)
traceplot(out18.1)                         # not shown
print(out18.1, 3)
```

Next, we can produce all the same plots that we just did with the bootstrapped replicate data. For illustration we produce the traditional plot for a posterior predictive check, where the posterior draws of some omnibus test statistic (here, the sum of squared Pearson residuals) for the replicate data are plotted against those for the same statistic when computed for the actual data (Fig. 18.7, left). We can also plot the two posterior distributions side by side (Fig. 18.7, middle). And in Fig. 18.7 (right), we plot the mean values of the replicates of each datum against their observed

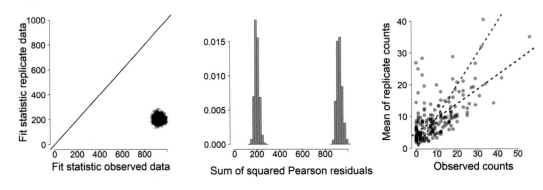

FIGURE 18.7

Bayesian goodness-of-fit assessment of the mis-specified rattlesnake model without the squared chipmunk covariate. Compare with the frequentist version of these plots in Fig. 18.4.

values. We can compute the Bayesian p-value for a numerical goodness-of-fit assessment and it turns out to be 0 or, to be more precise, smaller than 1/(number of values).

```
par(mfrow = c(1,3), mar = c(6,5,4,2), cex.axis = 1.5, cex.lab = 1.5,
  cex.main = 1.5)                                                # Fig. 18.7
# Traditional plot for posterior predictive check (replicated vs. observed)
xylim <- c(0, 1000)
plot(out18.1$sims.list$fit.obs, out18.1$sims.list$fit.rep, xlim = xylim,
  ylim = xylim,xlab = 'Fit statistic observed data',
  ylab = 'Fit statistic replicate data', frame = FALSE)
abline(0, 1)
# Compute a Bayesian p-value for goodness-of-fit
mean(out18.1$sims.list$fit.rep > out18.1$sims.list$fit.obs)       # [1] 0

# Non-traditional plot for posterior predictive check
hist(out18.1$sims.list$fit.rep, xlim = xylim, col = 'grey', xlab = 'Sum of squared Pearson residuals',
  main = 'Fit statistic for replicate data (grey)\n and observed data (blue)',
  freq = FALSE, breaks = 20)
hist(out18.1$sims.list$fit.obs, col = 'blue', freq = FALSE, add = TRUE, breaks = 20)

# Plot mean simulated count vs. observed counts for entire data set
yobs <- dat$C
plot(yobs, out18.1$mean$Crep, pch = 16, col = rgb(0,0,0,0.3), frame = FALSE,
  xlab = 'Observed counts', ylab = 'Mean of replicate counts',
  main = 'Comparison of expected vs. observed counts')
legend('bottomright', lwd = 3, lty = 3, col = c('red', 'blue'),
  legend = c("1:1 line", "Regression of y on x"), bty = "n")
abline(0, 1, col = 'red', lwd = 3, lty = 3)
abline(lm(out18.1$mean$Crep ~ dat$C), col = 'blue', lwd = 3, lty = 3)
```

We could now easily plot the posterior predictive distribution of every datum alongside its observed value, that is, produce the Bayesian version of Fig. 18.5, but we omit this to save space. Instead we show computation of the randomized quantile residuals in the Bayesian way and note that we could plot them in the same manner as in Fig. 18.6 to check the "local fit" of the model.

```
# Compute quantile residuals from Bayesian posterior predictive distributions and
# plot distribution (not shown)
yrepJ <- out18.1$sims.list$Crep          # J for JAGS
u <- numeric(dat$nSites)                 # Vector to hold quantile residual
for(i in 1:dat$nSites){
  u[i] <- mean((yrepJ[,i] < yobs[i]) + runif(1) * (yrepJ[,i] == yobs[i]))
}
hist(u, col = 'grey', main = 'Distribution of quantile residuals (u)')
```

Remember that the material in this section is the core of what package DHARMa does: using the bootstrap, for a frequentist model fit, or Bayesian simulations to evaluate the predictive distribution of the data, and from that compute randomized quantile residuals, on which a range of tests are conducted and presented. Next we show how we can import our own predictive data simulations (by bootstrap or MCMC) into DHARMa and benefit from its functionality to conduct various tests on the quantile residuals.

```
# Use of DHARMa for our own bootstrapped preditive simulations
library(DHARMa)
sim <- createDHARMa(simulatedResponse = yrep, observedResponse = yobs,
  fittedPredictedResponse = apply(yrep, 1, median))
par(mfrow = c(1,2))                       # not shown
plotQQunif(sim)
plotResiduals(sim, rank = FALSE)


# Use of DHARMa for our own Bayesian posterior preditive simulations
sim <- createDHARMa(simulatedResponse = t(yrepJ), observedResponse = yobs,
  fittedPredictedResponse = apply(yrepJ, 2, median))
par(mfrow = c(1,2))                        # not shown
plotQQunif(sim)
plotResiduals(sim, rank = FALSE)
```

Goodness-of-fit assessments with predictive data distributions as shown in this section are easy to conduct and can be very helpful to detect major problems with the fit of a model. However, they have limited sensitivity, since they use the data twice: first to estimate the parameters and compute predictive distributions and then to compare the latter with the original data. They provide a goodness-of-fit assessment that is too optimistic, since ideally "model training" and "model testing" should be conducted on separate, that is, independent data (Hastie et al., 2016). Such independence can be achieved by cross-validation, which is a tremendously powerful concept that can be widely used in modeling. We cover this next and then again in Sections 18.6 and 18.8, but there in a different context.

18.5.3 Cross-validation for goodness-of-fit assessments

Cross-validation (CV) is a family of methods that splits a data set into k non-overlapping partitions, called *folds*, and then fits a model to k-1 folds and predicts the data in the remaining fold. These latter are independent "hold-out" or "left-out" data and are thus unknown to the model. A comparison between the observed data and the data predicted by CV is very useful for model checking and model selection. By splitting the full data set into a training set and a testing set, CV solves the problem of double use of the data and the resulting overoptimism. CV lets us get closer to gauging the fit of a model not just for the particular data set at hand, but to "any" data set that the modeled process could have produced.

Two broad types of CV are k-fold CV, where k is often 5 or 10, and leave-one-out (LOO-)CV, where k is equal to the sample size n. For k-fold CV, the data are divided into k pieces. The model is then fit k times, each time holding out one of the k pieces of data, and then the fitted model is used to predict the remaining data that were held out. For LOO-CV, the model is fit $k=n$ times and each time we predict the single data point that was left out. Regular k-fold CV is computationally much cheaper than LOO-CV (though see the approximation by Vehtari et al. (2017)), but LOO-CV has advantages especially for smaller data sets. Here, we illustrate LOO-CV for model checking in a rattlesnake model, while in Section 18.6 you will see its use for model selection. See Vehtari (2023) and Yates et al. (2023) for useful overviews of CV.

As a caveat, we note that our illustration of CV deals with the simplest possible case: we consider i.i.d. data, a decent sample size, and a simple GLM without any hierarchical structure. CV becomes more challenging when there are dependencies in the data, when sample sizes are small, and for hierarchical models (Roberts et al., 2017; Valavi et al., 2019). In these cases, we will usually want to leave out not single data points, but whole groups or clusters, such as populations in Chapter 10 or sites in Chapters 19 and 19B.

The basic approach for LOO-CV shown next is almost exactly like what we did in the previous section: we use simulation to generate samples from the predictive, or posterior predictive, distributions of the data and then compare these with the observed data. Comparison may be in terms of omnibus tests or local tests using residuals. The only difference is that now we repeat these computations $k=n$ times, each time setting aside one datum, fitting the model to the remainder, and using the fitted model to predict that independent datum. We will illustrate the approach directly for a model that ostensibly does not fit and conduct LOO-CV for goodness-of-fit first using maximum likelihood and second Bayesian inference. We simulate regular ticklish rattlesnake data with both linear and squared effects of chipmunks and then fit the mis-specified model without the squared chipmunks.

```
set.seed(18)
str(dat <- simDat18(nSites = 100, beta1.vec = c(2.5, 0.2, 0.5, 1, -1),
   ncov2 = 50, beta2.vec = rnorm(50, 0, 0)))
summary(fm <- glm(C ~ rock + oak + chip1, family = poisson, data = dat))
```

The next code block conducts LOO-CV with maximum likelihood, followed by computation of CV randomized quantile residuals. This code is very similar to the parametric bootstrap in the previous section, but now each time we predict for a datum that is not included in the data set used to fit the model.

```
simrep <- 1000                          # Number of samples of predictive distribution

YrepCV <- array(NA, dim = c(dat$nSite, simrep))  # Array to hold CV-replicated data
for(i in 1:dat$nSites){                  # Loop over all data points = sites
  if(i %% 5 == 0) cat(paste('\n*** site', i))
  # Re-fit model to all data points minus 1 (i.e., minus site i)
  summary(fm_tmp <- glm(C[-i] ~ rock[-i] + oak[-i] + chip1[-i], family = poisson, data = dat))
  # Produce 'simrep' samples of the predictive distribution for the left-out datum at site i
  for(k in 1:simrep){                     # Loop over simreps of predictive distribution
    lam_i <- exp(as.numeric(cbind(1, dat$rock[i], dat$oak[i], dat$chip1[i]) %*% coef(fm_tmp)))
    YrepCV[i, k] <- rpois(n = 1, lambda = lam_i)
  }
}

# Compute quantile residuals
yobs <- dat$C
u <- numeric(dat$nSites)
for(i in 1:dat$nSites){
  u[i] <- mean((YrepCV[i,] < yobs[i]) + runif(1) * (YrepCV[i,] == yobs[i]))
}
```

We can then summarize the results using the predictive distribution of the left-out data exactly as we did in the previous section, for example, by executing the following code (the results of which we won't show).

```
par(mfrow = c(1, 2), mar = c(6,5,4,2), cex.axis = 1.5, cex.lab = 1.5, cex.main = 1.2) # Plots not shown
# Plot mean simulated datum vs. observed datum for entire data set
meanYCV <- apply(YrepCV, 1, mean)
plot(dat$C, meanYCV, pch = 16, col = rgb(0,0,0,0.3), frame = FALSE,
  main = 'Mean simulated data vs. observed data')
abline(0, 1, col = 'red', lwd = 3, lty = 3)
abline(lm(meanYCV ~ dat$C), col = 'blue', lty = 2, lwd = 3)

# Plot frequency distribution of quantile residuals
hist(u, col = 'grey', main = 'Distribution of quantile residuals')
```

Next, we move to the Bayesian version of model checking with LOO-CV. First, we set up R objects to hold results, and then fit the model 100 times, each time to a different subset of 99 sites and predicting the value of the one site that was left out. For ease of presentation, we slighty deviate from our usual workflow of a Bayesian analysis.

```
# Select number of samples from the CV-based predictive distribution
simrep <- 2000                              # Number of MCMC draws produced below
YrepCV <- array(NA, dim = c(dat$nSite, simrep)) # Array to hold CV-replicated data

# Function to generate starting values
inits <- function(){list(alpha = rnorm(1), beta = rnorm(3))}

# Parameters to estimate
params <- c("Crep")

# MCMC settings
na <- 100  ;  ni <- 1200  ;  nb <- 200  ;  nc <- 4  ;  nt <- 2

for(i in 1:dat$nSites){  # Loop over all data points = sites
  if(i %% 10 == 0) cat(paste('\n*** site', i))

  # Bundle and summarize data of length n-1 (each time with one datum used as external data !)
  dataList <- list(C = dat$C[-i], rock = dat$rock[-i],
    oak = dat$oak[-i], chip = dat$chip1[-i], n = length(dat$C)-1,
    pred.covs = cbind(dat$rock, dat$oak, dat$chip1)[i,])

  # Write JAGS model file
  cat(file = "model18.txt", "
model {
# Priors
alpha ~ dunif(-10, 10)
for(k in 1:3){
  beta[k] ~ dunif(-5, 5)
}
# Likelihood
for (i in 1:n) {                    # Loop over all data points
  C[i] ~ dpois(lambda[i])           # The response variable (C above)
  lambda[i] <- exp(alpha + beta[1]* rock[i] + beta[2]* oak[i] + beta[3] * chip[i])
}
```

```
# Predict left-out (independent) datum
lam_pred <- exp(alpha + beta[1]* pred.covs[1] + beta[2]* pred.covs[2] + beta[3] * pred.covs[3])
Crep ~ dpois(lam_pred)
}
")

# Call JAGS (ART <1 min)
out18 <- jags(dataList, inits, params, "model18.txt", n.iter = ni, n.burnin = nb,
    n.chains = nc, n.thin = nt, n.adapt = na, parallel = TRUE)

# Save samples from CV-posterior predictive distribution
    YrepCV[i,] <- out18$sims.list$Crep
}
```

We can now summarize the results from the CV-posterior predictive distributions of our data as before, and the interpretation of everything will also be as before. However, CV has the benefit of producing more powerful tests of model fit, since they are based on an independent test data set. For illustration, we use the observed data and the posterior predictive distribution of the data obtained using LOO-CV to compute randomized quantile residuals and plot their frequency distribution (Fig. 18.8). We find a strong deviation from the uniform that would be expected for a fitting model.

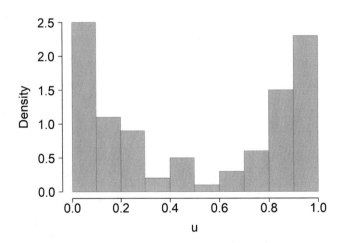

FIGURE 18.8

Bayesian goodness-of-fit assessment, using LOO-CV, for the mis-specified rattlesnake model without the squared chipmunk abundance covariate. The histogram shows the frequency distribution of randomized quantile residuals which for a fitting model should be approximately uniform.

```
# Compute quantile residuals from Bayesian LOO-CV and produce Fig. 18.8
yobs <- dat$C
u <- numeric(dat$nSites)          # Vector to hold quantile residual
for(i in 1:dat$nSites){
   u[i] <- mean((YrepCV[i,]<yobs[i]) + runif(1) * (YrepCV[i,]==yobs[i]))
}
hist(u, col = 'grey', main = 'Distribution of quantile residuals')
```

For the more specific case of hierarchical models, Conn et al. (2018) found that quantile residuals computed from predictive distributions obtained by LOO-CV were among the most sensitive methods for assessing goodness-of-fit, much more so than posterior predictive checks. CV is much more computationally expensive than these simpler methods, but if you can afford to wait, or if CV doesn't take much time as in our model, then LOO-CV is probably the best method for goodness-of-fit assessment; plus, there's the approximation by Vehtari et al. (2017) implemented in the `loo` R package. Moreover, LOO-CV is also powerful for model selection (Section 18.6). Hence, CV is definitely something worth learning.

18.5.4 Identifying and correcting for overdispersion

Overdispersion (OD) is a somewhat vague term that denotes the common situation where the residual variation in a model is greater than expected under the specified statistical distribution. This may occur in either continuous or discrete distributions. For instance, a t distribution with small degrees of freedom is overdispersed relative to a normal distribution, with greater spread and more outliers. But arguably the most typical case of OD is for models with discrete responses (see Chapter 12.2).

OD is always the result of some form of mis-specification of the model, which could be in its mean or dispersion part. Common sources of OD include some dependency in the data that is not adequately modeled, covariates that affect the response but are not in the model, or zero-inflation (see Chapter 12). Since these will be present to some degree in almost any data set, it should not come as a surprise that some amount of OD is almost the rule in Poisson or binomial models.

Consequences of unmodeled OD depend on where the OD comes from and whether you have a flat or a hierarchical model. For a flat model, if OD results from a severe mis-specification in the mean then it may render a model at least partly useless. An example of this would be the rattle-snake model without the squared chipmunk term in it. In contrast, some degree of mis-specification of correlations in the data and left-out covariates often only add noise in the data, without seriously biasing inferences about the mean. Failure to accommodate them will only rarely bias inferences about the mean of the model, but typically the computed SEs and CIs will be too small, and likewise for the Bayesian posterior SDs and CRIs. In contrast, the effects of OD may be quite different in hierarchical models, where mis-specified dispersion may also lead to bias in the mean (see e.g., Barker et al., 2018; Duarte et al., 2018, and Section 6.7 in Kéry & Royle, 2016).

Clearly, the ideal treatment for OD would be to correct the entire model specification, but for this we may simply lack the required knowledge or the data. For better or worse, a typical approach in applied statistics when faced with OD is to sweep under the rug the possibility of more serious problems in the mean structure of the model and instead simply assume that the mean of the model is right, but the dispersion is not large enough. The simplest ("cheap") correction for OD in a GLM is then the quasi-likelihood approach to a Poisson or binomial regression, where the ratio of the observed and the expected residual variation is taken as an OD correction factor c, with which the computed variance is multiplied for construction of SEs and CIs, that is, SEs are multiplied by $\sqrt{\hat{c}}$; see Chapter 12.2.2 and Section 18.7. A more explicit, model-based solution is to move from a Poisson to a negative binomial or from a binomial to a beta-binomial distribution, which both have one additional parameter for the extra variation. An alternative to the negative binomial is the Poisson-lognormal distribution that we saw in Chapter 12.2.

It is possible that OD in your analyses may more often be of the second, more benign kind than the first, that is, causing extra noise rather than a structural break-down of the model. However, simple and omnibus tests for OD, such as that for the residual variation in Sections 18.5.2 and 18.5.3, remain silent about the causes underlying the extra dispersion. A rule of thumb in capture-recapture says that the cheap OD correction should only be applied when $\hat{c}$ is no greater than about 3. This rule may often lead to reasonable results, but it is clearly *ad hoc* and lacks a theoretical basis.

We end this section with a simulation that demonstrates that the magnitude of OD does not tell us anything about whether it is acceptable or not to make the cheap OD correction. We first fit the mis-specified regular Poisson model without squared chipmunks to our regular ticklish rattlesnake counts. Second, we look at the alternative case where the model is correctly specified in terms of the main covariates, but a number of left-out covariates add somewhat harmless extra variation in the data. We first repeat the mis-specified model from Section 18.4.2.

```
# Create data set with major covariates plus one noise covariate (ncov2)
set.seed(18)
str(dat <- simDat18(nSites = 200, beta1.vec = c(2.5, 0.2, 0.5, 1, -1), ncov2 = 1,
    beta2.vec = rnorm(1, 0, 0.2)))

# Fit the mis-specified model without chip2
summary(fm <- glm(C ~ rock + oak + chip1, family = poisson, data = as.data.frame(dat)))

Coefficients:
              Estimate   Std. Error   z value   Pr(>|z|)
(Intercept)    2.02727      0.02761    73.418    <2e-16 ***
rock           0.21674      0.02387     9.081    <2e-16 ***
oak            0.56584      0.02437    23.220    <2e-16 ***
chip1          0.25874      0.02211    11.701    <2e-16 ***
---
Signif. codes: 0 '***' 0.001 '**' 0.01 '*' 0.05 '.' 0.1 ' ' 1

(Dispersion parameter for poisson family taken to be 1)

    Null deviance: 1924.3 on 199 degrees of freedom
Residual deviance: 1099.1 on 196 degrees of freedom
AIC: 1741.3
```

From theory, a well-fitting Poisson GLM has a residual deviance of the same magnitude as the residual d.f. (McCullagh & Nelder, 1989) and indeed the ratio of the two is one estimator of $\hat{c}$. In our analysis, the deviance is about 1099 on 196 d.f., yielding $\hat{c} = 5.6$. Here, it would probably not result in acceptable inferences if we simply presented the point estimates for the three main covariates in the model and stretched their SEs by $\sqrt{\hat{c}}$. At least in terms of the inferences about the chipmunk effect, this OD represents a structural failure of the model and not just extra noise.

Next, consider the case where we have the correct model for the main covariates of interest, but where an additional 45 covariates add unaccounted noise in the response.

```
# Create data set with major covariates plus 45 noise covariates
set.seed(18)
str(dat <- simDat18(nSites = 200, beta1.vec = c(2.5, 0.2, 0.5, 1, -1), ncov2 = 45,
   beta2.vec = rnorm(45, 0, 0.1)))

# Fit the correctly specified model (in terms of the main covariates )
data <- data.frame(C = dat$C, rock = dat$rock, oak = dat$oak, chip1 = dat$chip1, chip2 = dat$chip2)
summary(fm <- glm(C ~ rock + oak + chip1 + chip2, family = poisson, data = data))

Coefficients:
             Estimate   Std. Error   z value   Pr(>|z|)
(Intercept)   2.71584     0.03002      90.46    <2e-16 ***
dat$rock      0.26514     0.02090      12.69    <2e-16 ***
dat$oak       0.65479     0.02265      28.91    <2e-16 ***
dat$chip1     1.04022     0.04624      22.50    <2e-16 ***
dat$chip2    -1.09453     0.04482     -24.42    <2e-16 ***
---
Signif. codes: 0 '***' 0.001 '**' 0.01 '*' 0.05 '.' 0.1 ' ' 1

(Dispersion parameter for poisson family taken to be 1)

    Null deviance: 4060.5 on 199 degrees of freedom
Residual deviance: 1218.3 on 195 degrees of freedom
AIC: 1864.1
```

This model has point estimates for the main coefficients of interest that are fairly close to the truth, despite an OD of magnitude $\hat{c} = 6.2$ (i.e., 1218/195). Thus, in this particular case stretching SEs and CI limits might well be acceptable, even though $\hat{c}$ is far greater than what the above rule of thumb says: we get the mean of the model about right, but should adequately account for the extra uncertainty due to the OD.

Our take-home message is this: we can band-aid a model with OD by stretching the SEs/CIs, moving to a distribution with more variability than a Poisson or binomial or, similarly, adding additional random effects. These common approaches often lead to acceptable inferences, but they do assume that the source of the OD simply adds noise, rather than corrupting the part of the model that is typically of main interest, that is, the mean. In addition, mis-specified dispersion in hierarchical models can have much more insidious effects (Duarte et al., 2018), and simply adding random effects to soak up the extra dispersion may lead to models that fit, but yield terrible estimates (e.g., the "good fit/bad prediction" dilemma; see Section 6.9 in Kéry & Royle, 2016). Thus, OD corrections have to be applied with caution. If possible, it is always better to diagnose the source of OD (e.g., correlations in the data, additional covariates required in a model) and include this additional structure into the model.

18.5.5 Measures of the absolute fit of a model

So far we have dealt with the question of whether our model is any good at all, so that we can use it, for example, for explanation or prediction. But can we also say *how good* it is? Is there a useful measure of its absolute fit to the data? Well, yes and no. For normal-response models, there is the coefficient of determination, R^2 ("R-squared"), a concept that has also been adapted to non-normal responses. This measure of absolute model fit is fairly popular, but we admit that we are not big fans of it. Here, we offer some comments on the topic of absolute fit.

For a normal linear model fit with least squares, R^2 is given by the ratio of the sum of squares explained by the model and the total sum of squares of a model that contains only an intercept (Quinn & Keough, 2002). Its value ranges from 0 to 1 (or 0% to 100%); thus, it has the neat interpretation as a proportion (or percentage) of the variance in the data that is "explained" by the model. On the surface, this seems like a reasonable measure and interpretation, but there are at least two problems. First, R^2 is only a measure of association strength, and it does not tell us anything about any causal explanations. Second, it would be tempting to believe that a high R^2 must indicate a good experiment or a more relevant study than a low R^2. But look at its construction:

$$R^2 = \frac{\text{explained}}{\text{total}} = \frac{\text{explained}}{\text{explained} + \text{rest}}$$

Thus, we can essentially obtain any value of R^2 by simply tuning the magnitude of the residual variation, or "rest," in the study. This noise background may be affected by a myriad of things that have absolutely nothing to do with how good a model or study is. True, sometimes a small background variation ("rest") may indicate a good study or a well-controlled experiment. But not always. Indeed, sometimes a large background variation and thus a smaller R^2 may indicate a better study, such as when a greater scope in space or time will naturally increase the amount of background variation (i.e., the "rest"). In general, it will rarely make sense to compare values of R^2 across studies.

Absolute measures of model fit such as R^2 may sometimes provide useful information, but we think it is important to keep several caveats in mind when we interpret them. The concept of R^2 has variously been extended to non-normal responses such as a Poisson, to models fit using Bayesian inference (Gelman & Pardoe, 2006; Gelman et al., 2020) and to GLMMs (Nakagawa & Schielzeth, 2013). Measures of R^2 for non-normal and mixed models can be computed using functions in the R package MuMIn (Barton, 2023). Finally, note that variance ratios like R^2 computed for measured animal or plant traits are the basis for the study of heritability in quantitative genetics. Our criticism here of what we sometimes perceive as "naive" use of R^2 does not extend to this field.

18.5.6 Parameter identifiability and robustness to assumption violations

In this final section on model checking, we briefly cover two really big topics. The first is parameter identifiability and the latter the robustness of a model to assumption violations.

Identifiability is the opposite of nonidentifiability: when a parameter cannot be estimated from a data set, and we distinguish intrinsic from extrinsic nonidentifiability. In Chapter 3.4.5, we encountered an example of extrinsic nonidentifiability, where given the right data our model would have permitted estimation of all its parameters. However, because not all cells in our cross-classification were observed, we could not estimate all interaction parameters. In contrast, with intrinsic nonidentifiability, it is impossible in principle to estimate a certain parameter with a given *type* of data, regardless of their amount. Below, we will show an example of this.

Parameter identifiability in the types of models in this book is not normally such a big issue, since it is relatively well understood (e.g., in analysis of variance models for cross-classified factors; Quinn & Keough, 2002) and lack of identifiability may be more easy to diagnose. However, in more complex models it may be much harder to diagnose, especially when using Bayesian inference (Knape, 2008; Auger-Méthé et al., 2016; Stoudt et al., 2023). Hence, it is important that when in doubt, you have a strategy for evaluating parameter identifiability.

There are formal methods for this (Cole, 2021), but these can be challenging for ecologists. A less rigorous and formal, but far easier and extremely flexible, approach to the study of parameter identifiability is again … simulation! Here we give an illustration. The basic strategy is to simulate many data sets with varying values of the parameter in question, fitting your model to each, and then seeing whether the estimates obtained resemble the known true values used to simulate the data. We can plot the estimates vs. true values, and for an identifiable parameter we expect a scatterplot concentrated around the 1:1 line. (As an accessible alternative to simulation for this task, see also data cloning; Lele et al., 2007.)

In our example, we consider a variant of a binomial *N*-mixture model (Chapter 19B). In this model, repeated counts at a spatial sample of sites are required to enable one to separately estimate a parameter for abundance and another for detection probability. Let's imagine that we didn't know this and had asked whether we can estimate both abundance and detection from single counts from a set of sites, that is, from data like our rattlesnake counts. We next repeat the following a large number of times:

- We randomly pick a value of expected abundance (`lambda`) and detection probability (`p`) on some sensible range and save these as our truth.
- Then, we simulate unreplicated count data for each of 100 sites using the traditional Poisson/binomial form of the *N*-mixture model (see Chapter 19B for more information about this model).
- Finally, we fit to each data set the data-generating form of the *N*-mixture model, that is, with only an intercept for abundance and detection.

```
library(unmarked)

# Choose simreps and create arrays to hold true values and estimates
simrep <- 1000
esti <- true.vals <- array(NA, dim = c(simrep, 2), dimnames = list(NULL, c("lambda", "p")))

# Launch simulation to study parameter identifiability
system.time(                                      # Set timer
for(k in 1:simrep){                               # Loop over simreps
  cat(paste("*** iter", k, "\n"))                 # Counter
  lam <- runif(1, 0, 10)                          # Pick a value for lambda within some range
  p <- runif(1, 0, 1)                             # Same for p
  true.vals[k,] <- c(lam, p)                      # Save true values for both
  N <- rpois(n = 100, lambda = lam)               # Simulate latent abundances (N)
  C <- matrix(rbinom(n = 100, size = N, prob = p), ncol = 1) # Simulate observed counts (C)
  umf <- unmarkedFramePCount(y = C)               # Bundle data for unmarked
  fm <- pcount(~1 ~1, data = umf, se = FALSE)     # Fit model
  esti[k, ] <- c(exp(coef(fm)[1]), plogis(coef(fm)[2])) # Save estimates
}
)
```

This takes only about 5 minutes, and we get our answer from looking at the following plots (which, however, we don't actually show).

```
# Test for identifiability: compare estimates with truth
par(mfrow = c(1, 2))                      # Plots not shown
plot(true.vals[,1], esti[,1], xlab = 'True', ylab = 'Estimated',
  main = 'Abundance (lambda)', frame = FALSE)
abline(0, 1, col = 'red', lwd = 2)        # Red signifies truth
plot(true.vals[,2], esti[,2], xlab = 'True', ylab = 'Estimated',
  main = 'Detection (p)', frame = FALSE)
abline(0, 1, col = 'red', lwd = 2)
```

The visual test for identifiability of the two parameters in the N-mixture model fit to such "single-visit" count data consists in checking whether the scatterplots cluster near the 1:1 line, red in our plots. That the scatter is all over the place reveals that we cannot estimate these parameters with unreplicated data. This shows how the analysis of simulated data can give you insight whether you can estimate the parameters in your model at all. You could of course also customize such simulations, for instance, to the specific sample sizes or expected values of parameters in your study.

The second big topic in this section is the assessment of the robustness of a model to its assumptions or, put in another way, its sensititivity to assumption violations. This is important to know, since model assumptions almost never hold exactly. This is not necessarily a problem, since a model is not reality but merely a tool to help us learn something about reality. But there is a problem when these assumption violations have such big effects on the particular inferences we desire from our model that they jeopardize the whole use of the model for the desired goal. Thus, we may want to know how robust our model is at least to violations of its key assumptions, or perhaps those which we are most worried about might be violated in a big way. We can also say we want to know how sensitive our inferences, predictions, etc. are to the assumptions of our model, which we know right from the start are never exactly true.

Two broad approaches are robustness assessment and sensitivity analysis. In the first, we repeat an analysis while varying one or more model assumptions and seeing how much the inferences of main interest change. The second one again uses simulation and generates data under different conditions, representing different types of assumption violations. Then, we fit our model to each and see how these known assumption violations affect the inferences.

There are plenty of assumptions in any statistical modeling project, and they start right with the question about whether a data set at hand is really relevant for the intended task of a model (Gelman et al., 2014). But more specifically, a likelihood is nothing but a set of assumptions, for example, about the distributions of random variables, link functions, and linear predictors (Stoudt et al., 2023). We could and sometimes should assess the robustness of our models to these (DiRenzo et al., 2023). In addition, when we use Bayesian inference to fit our model, we combine the likelihood with priors and these are yet another set of modeling assumptions. Moreover, the validity of our assumptions about the priors cannot really be tested directly (but see Gelman & Shalizi, 2012; Gelman et al., 2020). Thus, a sensitivity analysis of our inferences to different prior choices in a Bayesian analysis is particularly useful. In our models for the ticklish rattlesnakes, this might include the following:

- Fit the model with uniform or "flat normal" priors for the regression coefficients and compare the estimates.
- Fit the model with different data distributions, for example, a normal and a t, to check for sensitivity to outliers and extreme values.

Of course, in this particular case, this would not really make any sense, since we know exactly what the right modeling assumptions are. But in real-world analyses, we never know the true model. Both such sensitivity analyses and also simulations for the study of the effects of known assumption violations on the inferences from a model are an important part of the methods that an applied statistical modeler in ecology ought to master.

18.6 **Model selection**

Model selection is a big topic in modeling. There is a sense in which we could view it as synonymous with model building. Then, *the most important criterion for selecting, or building, a model would be that your model must be useful for your scientific question or management problem.* To us, this is the most fundamental and most important part of model selection when understood in this broad way: we must choose the model that is best for our objectives. This will include decisions about the type of model, for example, what class of SDM to work with in the deer example in Section 18.4. Then, within a given class of model, we might make different choices about hierarchical structure to accommodate temporal and spatial dependencies. And finally, we typically have a number of covariates and there may be a host of possible combinations in which we could add them in our model, for example, as different subsets, as polynomials, or as splines.

Now, for better or worse, model selection is frequently understood as meaning only variable selection (Hooten & Hobbs, 2015; Tredennick et al., 2021). Hence, from here on we adhere to this definition and denote by it the question of which covariates to put into a model and possibly in what form. The topics of model averaging and regularization also belong under this heading (Hooten & Hobbs, 2015), but we cover them in Sections 18.7 and 18.8.

18.6.1 **When to do model selection ... and when not**

Much suffering has been caused in ecology by model selection. When faced with the question of what to do with a set of covariates in an analysis, people may try them in many different combinations, often without too much scientific thought (e.g., fitting models also that don't make sense biologically). Even if there are just a few covariates, the large number of possible combinations may yield many dozens or more models. The result is then a lot of agonizing over what to do with all of them, how to interpret the results, and how to present them in a paper. This is the Black of Hole of Statistics quoted by Tredennick et al. (2021).

One seemingly straightforward method of variable selection is stepwise selection of covariates, for example, forward selection, backward elimination, or a combination of the two, based on significance tests. However, such stepwise selection has serious shortcomings (Whittingham et al., 2006), including the "watered down" significance test levels owing to heavy multiple testing on which the decisions for variable inclusion/exclusion rest in this data-dredging enterprise. So, for instance, when our software yields $p = 0.04$ for some coefficient in the final, "dredged" model, then the real value of p may in fact be 0.38... or 0.95, if you could properly account for all the degrees of freedom spent during the dredging. Thus, stepwise variable selection should be avoided or at least minimized (e.g., in the context of model expansion; Section 18.4).

To avoid the black hole of variable selection, it is helpful to recognize additional structure in the model selection decision problem. First of all, the goal of a modeling project is an important guide for how to do model selection. In a causal-explanatory mode, we should best fit only one model or at most a small handful of models (Section 18.4). In contrast, in a prediction mode, we will often consider a much larger number of models, while an even larger number of models may be built in the exploratory mode of modeling (Tredennick et al., 2021).

Second, covariates in a model may have different roles (and sometimes more than one). They may:

- Embody a scientific hypothesis ("causal covariates").
- Be used for adjustment as a statistical way of standardization ("adjustment" or "nuisance covariates").
- Help predict observed or latent responses ("prediction covariates").

A causal-explanatory model may contain all three types, but typically not in the same model parts. It will naturally have causal covariates in its hard parts, since these are almost a defining feature of a causal-explanatory model. In addition, it may have adjustment covariates in its hard parts and prediction covariates in its soft parts. In contrast, a pure predictive model will not usually contain any causal covariates, but for the most part only prediction covariates. In addition, we emphasize that not any old covariate that you can get from the internet or that you have measured deserves to be incorporated into your model. Rather, at least in a causal-explanatory model one ought to only consider covariates that have some meaningful interpretation (especially those that represent your scientific hypotheses), or which may indeed have some association with a modelled response. All of this advice somewhat narrows down the problem of model selection. In addition, model averaging (Section 18.7) and regularization (Section 18.8) are both somewhat automated ways of dealing with even large numbers of models (i.e., covariate combinations). They may save us from the black hole as well. Building on Tredennick et al. (2021), we offer the following advice.

In the causal-explanatory mode of modeling:

- Build either just a single large model that contains all effects of interest or else just a very small handful of typically nested models that represent different hypotheses about mechanisms. Tredennick et al. (2021) write *"If you have more than a handful of candidate models, it is likely that you are actually still in the exploration phase, or you may be more interested in prediction than inference."*
- Test your hypotheses using Null hypothesis significance tests or by the analogous approach of checking whether 0 is in- or outside a 95% confidence or credible interval (see Chapter 2). Be sure you really understand the limitations of significance testing (Wasserstein & Lazar, 2016). Try to avoid model selection in the main parts of interest of the model (i.e., in its hard parts) as much as possible, since any changes of the original model you had in mind with the associated multiple testing will water down the probability associated with your tests in the final model, be they frequentist or Bayesian. This will reduce your inferential power and may lead to the erroneous "detection" of spurious patterns.
- You may also use information criteria (IC) such as the Akaike information criterion (AIC) to discriminate among a small handful of models at this stage, though that chooses models that are optimal for prediction and not necessarily models that are also good for inference (see Section 18.6).
- Some degree of model expansion will typically be desirable (Section 18.4.1), although even this may weaken the confirmatory strength of the final model. In addition, a causal-explanatory model may also contain soft parts, for which some model selection (as for a predictive model; see next) may be less problematic, since we're not interested in testing the associated parameters, but want to optimize them in some way.

In the predictive mode of modeling, and for nuisance covariate structure in the soft parts of any model:

- Optimize prediction performance using a measure of prediction accuracy (see below).
- Consider automatic shrinkage or smoothing methods (Section 18.8).
- Remember this: *"Predictive modeling often requires extensive model selection ... The parameter uncertainty resulting from extensive model selection is not accounted for in the p values from the final models"* (Tredennick et al., 2021). Thus, tests and confidence or credible intervals from the final models should not be taken at face value, but at best only in a descriptive way.

To reiterate: the process of fitting, comparing, and combining many model formulations is an appropriate strategy when you are working in the predictive modeling mode, or when you are specifying the *soft parts* of models in the causal explanation modeling mode. In the rest of this section, we cover the assessment of the predictive performance of a model.

18.6.2 Measuring the predictive performance of a model

To measure the predictive performance of a model, we compare some data $\mathbf{y}$ with a prediction $\hat{\mathbf{y}}$ made by a model, that is, the fitted values under the model. There are two issues involved: about the kind of data used for this validation and about the metric of comparison. As an aside, note that by "model validation" we always mean the assessment of the prediction performance of a model and *not* a general validation of the model in the sense of its goodness-of-fit for instance. But there really are many ways in which a model may be invalidated, beyond just inferior prediction performance.

First, we might want to check how well our model explains the actual data set from which we have estimated the parameters. R^2 is such a measure of prediction accuracy, or model fit (Section 18.5.5). However, gauging prediction accuracy on the same data from which we estimated the parameters will lead to an overly optimistic assessment (Hastie et al., 2016). Instead, for an honest assessment of how well a model can in principle predict data we want to test predictive performance on an independent data set (Gelman et al., 2014b; Hooten & Hobbs, 2015; Vehtari et al., 2017). Thus, in the context of model validation we distinguish within-sample and out-of-sample data, also called training and testing data. We fit the model to the former and validate it on the latter. If we do both on the same data set instead, then we must correct for the optimism stemming from this double use of the data.

Second, a measure for the similarity between a probabilistic prediction and a validation datum is called a scoring rule (Hooten & Hobbs, 2015). A widely used scoring rule for quantitative responses is the mean squared prediction error (MSPE), $\frac{1}{n}\sum_{i=1}^{n}\left(y_{i,oos}-\hat{y}_i\right)^2$. That is, the average of the squared differences between out-of-sample validation data ($y_{i,oos}$) and their prediction under the model ($\hat{y}_i$) over n data points. A model with lower MSPE is a better-predicting model.

A more general and widely used measure of predictive accuracy is the logarithm of the predictive density of data $\mathbf{y}_{oos}$ under the model, that is, $\log(p(\mathbf{y}_{oos}|\hat{\boldsymbol{\theta}}))$. This is also called the *log score*. This expression is algebraically identical to that of the log-density of the data or the log-likelihood of the model for the data to which we fit it. Here, however, we plug in an estimate of the parameters and normally want to evaluate it for out-of-sample data $\mathbf{y}_{oos}$. Models with a higher log-predictive density predict better.

We can work with the log-predictive density as a measure of predictive performance with either likelihood or Bayesian inference. In the former, $\hat{\theta}$ will typically be the MLEs, while in Bayesian inference we may either work with the posterior mean of the density (as for the deviance information criterion [DIC]) or with the full posterior predictive distribution, which averages over the posterior distribution of the parameters (for the Watanabe-Akaike information criterion [WAIC]); see Section 18.6.4.

Remember that the most honest assessment of the predictive performance of a model is obtained for new data which were not used for parameter estimation. Thus, ideally we could average over "all possible" future data sets. Of course, this is not possible, but there are several options to obtain an approximation. First, we can estimate the out-of-sample predictive performance from another ("testing") data set that is completely distinct from the training data set, that is, that is not in the end put back into the fitted model. This is called an external validation approach. It is best for validation, but can be wasteful for parameter estimation, since you will then not use all the available data for estimation. Hence, in practice you will only do this when you have really large data sets.

The second and third options both make do just with a single data set. In the second option, you use CV (see Section 18.5.3), that is, you partition your data set into k nonoverlapping groups ("folds"), use k-1 of them as a training set and the k-th fold as the testing set and repeat this for all k folds. With $k=5$ or $k=10$, we speak of k-fold CV, while when $k=n$, we have LOO-CV. k-fold CV is computationally much cheaper, but will somewhat underestimate validation performance of a model that is fit to all n data points.

In the third option for estimating out-of-sample prediction performance of a model, we only fit the model once, but use an information criterion (IC) such as AIC, DIC, or WAIC (Akaike, 1973; Spiegelhalter et al., 2002; Watanabe, 2010). This attempts to correct for the optimism in the assessment of predictive accuracy that arises from evaluating it on the same data set from which we estimate the parameters. We do this by adjusting the estimate of the log-predictive density obtained from the same data set with a bias-correction term (see end of Section 18.6 for some comments on IC and CV in hierarchical models).

We now illustrate model selection by external validation, CV, and IC for the ticklish rattlesnake data. We start by simulating two data sets from the same process, that is, using the same parameter values. We don't show the whole code, but refer to the book website for a complete script.

```
set.seed(18)
beta2.vec <- rnorm(10, 0, 0.1)
trainDat <- simDat18(nSites = 50, beta1.vec = c(2, 0.2, 0.5, 1, -1),
   ncov2 = 10, beta2.vec = beta2.vec, show.plot = TRUE)    # Training data
testDat <- simDat18(nSites = 50, beta1.vec = c(2, 0.2, 0.5, 1, -1),
   ncov2 = 10, beta2.vec = beta2.vec, show.plot = TRUE)    # Testing data
```

18.6.3 Fully external model validation using "out-of-sample" data

Fully external validation is the gold standard for judging the predictive performance of a model (Hooten & Hobbs, 2015). However, for the small to moderate data sets common in ecology, it may be too wasteful, since not using all the available data for parameter estimation may be too much of a cost. In contrast, for many modern data collection methods such as autonomous recording units (ARUs) and camera traps, data sets may be almost too large anyway, and then external validation may become a realistic option. We illustrate this with our rattlesnake data by training (i.e., fitting) five models of increasing complexity to `trainDat`. Then, we compute the log-predictive density for `testDat`, and for comparison also to the same `trainDat`. In this and the next section, we work with maximum likelihood because it's faster, but we could have chosen Bayesian inference just as well.

```
# Bundle training data
dat <- data.frame(C = trainDat$C, rock = trainDat$rock,  oak = trainDat$oak,
  chip1 = trainDat$chip1, chip2 = trainDat$chip2)

# Fit 5 models of increasing complexity to training data set
fm1 <- glm(C ~ 1, family = poisson, data = dat)
fm2 <- glm(C ~ rock, family = poisson, data = dat)
fm3 <- glm(C ~ rock + oak, family = poisson, data = dat)
fm4 <- glm(C ~ rock + oak + chip1, family = poisson, data = dat)
fm5 <- glm(C ~ rock + oak + chip1 + chip2, family = poisson, data = dat)

# Compute log predictive density for the training data
# Compute the predictions for each data point
predlam1 <- predict(fm1, type = 'response')
predlam2 <- predict(fm2, type = 'response')
predlam3 <- predict(fm3, type = 'response')
predlam4 <- predict(fm4, type = 'response')
predlam5 <- predict(fm5, type = 'response')

# 'Validation' with the same training set, i.e. "in-sample" or "is"
lpd1is <- sum(dpois(dat$C, predlam1, log = TRUE))
lpd2is <- sum(dpois(dat$C, predlam2, log = TRUE))
lpd3is <- sum(dpois(dat$C, predlam3, log = TRUE))
lpd4is <- sum(dpois(dat$C, predlam4, log = TRUE))
lpd5is <- sum(dpois(dat$C, predlam5, log = TRUE))
```

Then, we do the same for the validation, or testing, data set: for each data point we compute the expected value based on the parameters estimated from the training data. Then, we use that to compute the log-predictive density of the whole testing data under each model.

```
# Compute log predictive density for the testing (= validation) data
# Bundle testing data
datoos <- data.frame(C = testDat$C, rock = testDat$rock,
 oak = testDat$oak, chip1 = testDat$chip1, chip2 = testDat$chip2)

# Compute the predictions for each data point in the test data set
predlam1oos <- predict(fm1, type = 'response', newdata = datoos)
predlam2oos <- predict(fm2, type = 'response', newdata = datoos)
predlam3oos <- predict(fm3, type = 'response', newdata = datoos)
predlam4oos <- predict(fm4, type = 'response', newdata = datoos)
predlam5oos <- predict(fm5, type = 'response', newdata = datoos)

# True model validation with this independent test data set
lpd1oos  <- sum(dpois(datoos$C, predlam1oos, log = TRUE))
lpd2oos  <- sum(dpois(datoos$C, predlam2oos, log = TRUE))
lpd3oos  <- sum(dpois(datoos$C, predlam3oos, log = TRUE))
lpd4oos  <- sum(dpois(datoos$C, predlam4oos, log = TRUE))
lpd5oos  <- sum(dpois(datoos$C, predlam5oos, log = TRUE))
```

Finally, we compare the two sets of log-predictive densities for our five models, once computed for the same (training) data set ("is," for "in-sample") from which we estimated the parameters and then for the independent validation (or test) data set ("oos," or "out-of-sample"), remember code is on the website. Also note that the direct comparison between these scores for training and testing data sets requires that they have the same size.

```
           lpd_is   lpd_oos
Null       -201.8   -251.5
 +rock     -182.7   -262.7
 +oak      -179.5   -234.0
 +chip1    -162.9   -210.1
 +chip2    -113.3   -181.9
```

Not surprisingly, we see that the most complex model is the best-predicting one by both scores. But we also see that when we test the predictive accuracy of our model "in-sample" (on the left), the log-predictive density is about 50 units too high compared to the more honest assessment that we get when assessing it "out-of-sample" (on the right). This difference is also known as optimism or overfitting (Hastie et al., 2016). It is a key concept for the optimization of the predictive accuracy of a model.

18.6.4 Approximating external validation for "in-sample" data with cross-validation

In most analyses, we will not have enough data to conduct a fully external validation of our models. Then, we may use CV, which we introduced in Section 18.5.3 for model checking and there showed both a frequentist and a Bayesian version. Here now, we use CV for model validation and illustrate frequentist LOO-CV as well as five-fold CV (see the book website for code).

Remember that LOO-CV works like this: we cycle through all n data points, drop one of them from the data, fit the model to the remaining n-1 data points, predict the expected value of the left-out data point, and from that compute its log-predictive density of the data point under the model. The combined log score for prediction is then the sum of this over all n data points. For fivefold CV, we partition the data set of 50 points into 5 nonoverlapping subsets. Then, we set aside one of the folds, fit the model to the remaining k-1 folds, and predict the observed responses in the left-out fold and compute the log-predictive density, which we sum over all 10 data points in the fold. We repeat this five times, and the combined log score is again given by the sum of the scores in each fold. In the end, we compare the log scores for all four assessments up to now.

```
           lpd_is    lpd_oos   lpd_loocv   lpd_fivefold
Null       -201.83   -251.49    -208.00      -206.54
 +rock     -182.68   -262.67    -190.25      -192.61
 +oak      -179.51   -234.05    -191.96      -195.78
 +chip1    -162.87   -210.13    -179.59      -181.06
 +chip2    -113.33   -181.88    -128.81      -130.23
```

We see that both LOO-CV and five-fold CV achieve some correction of the optimism of the in-sample estimate of the log score, though not quite reaching the score of the external validation. Thus, CV achieves some correction for using the training set for model validation, but it remains tied to that data set and is only correct in expectation (Gelman et al., 2014b, p. 1001). We note also that there is a lot of stochasticity, or noise, in these computations. For a more formal comparison of these methods, we would have to run a simulation with replicate data sets.

18.6.5 Approximating external validation for "in-sample" data with information criteria: AIC, DIC, WAIC, and LOO-IC

CV is often the method of choice for assessment of the predictive accuracy of a model (Hooten & Hobbs, 2015). However, for big data sets and complex models, it may be prohibitively expensive computationally, since even in the cheaper k-fold CV, a model must be fit k times. A different approach is to fit a model only once and evaluate the log-predictive density for the actual, training, data (i.e., "in-sample"), but then apply an adjustment for the resulting overoptimism. This adjustment is a penalty for the double use of the data in parameter estimation and validation. This approach is an attempt at assessing the out-of-sample predictive accuracy of a model using only a single data set, that is, using solely "in-sample" information. For historical reasons, it is known as "information criterion", of which we cover four here: AIC, DIC, WAIC, and LOO-IC. This section draws heavily on the insightful papers by Gelman et al. (2014b) and Vehtari et al. (2017).

All these information criteria are approximations to different forms of LOO-CV (Stone, 1977; Shibata, 1989; Watanabe, 2010) and have a common form as:

$$IC = model\ fit + penalty.$$

Here, model fit is a measure for the fit of the model to the training data, while the penalty is a bias correction for the overoptimism that results from not using a proper testing data set. All ICs provide an estimate of the expected log-predictive density (*elpd*) for new data, averaged over all possible new data sets. But they differ both in the measure of model fit and in the penalty. Note that in the actual ICs, the *elpd* is typically multiplied by -2 to bring it to the same scale as the deviance, that is, twice the negative log-likelihood. As a result, a higher value of *elpd* indicates a better-predicting model, but the reverse is true for the ICs.

The AIC (Akaike, 1973) starts with the deviance as a measure of model fit, that is, twice the negative log-likelihood of the data evaluated at the MLEs. Through the work of Burnham & Anderson (e.g., their 2002 book), AIC has become very widely known and used in ecology. Different versions of AIC exist, for example, AIC$_c$ has a correction for small sample size and QAIC corrects for OD. R packages `AICcmodavg` (Mazerolle, 2023) and `MuMIn` (Barton, 2023) give plenty of options to compute various forms of AIC and also to combine the inferences from multiple models by model averaging (Section 18.7).

For our rattlesnake models fit with function `glm()` the regular AIC (i.e., not AICc or QAIC) is output by default when printing the fitted model object, but this is no fun. Instead, let's compute the AIC by hand, using DIY-MLEs obtained by defining a function for the negative log-likelihood of each model and then using function `optim()` to minimize it for the observed data. We will see that we can read off the two ingredients of the regular AIC from the `optim` output. The full code for all five models is again available on the website. For illustration, we here compute the AIC for model 2, which has `rock` as the only explanatory variable.

```
Xmat <- cbind(1, dat$rock, dat$oak, dat$chip1, dat$chip2)

# Define NLL for Poisson regression with covariates (models 2–5)
NLL <- function(beta, C, Xmat) {
  lambda <- exp(Xmat %*% beta)
  LL <- dpois(C, lambda, log = TRUE)
  -sum(LL)
}
```

```
fit2 <- optim(c(1, 0), NLL, C = dat$C, Xmat = Xmat[,1:2], method = "BFGS")
fit2                                        # print fit
2 * fit2$value + 2 * length(fit2$par)       # Compute AIC

$par
[1] 1.562731 0.385681

$value
[1] 182.6784

$counts
function gradient
      32       10

$convergence
[1] 0

$message
NULL


> 2 * fit2$value + 2 * length(fit2$par)    # Compute AIC
[1] 369.3568
```

We have presented function `optim()` and its output in detail in Chapters 2 and 4, so here we simply remind you that `$value` now contains the NLL function value at its minimum, that is, at the MLEs, or $\log p(y|\hat{\theta}_{MLE})$. In the AIC, the estimate of the expected out-of-sample predictive accuracy (*elpd*) is given by $\log p(y|\hat{\theta}_{MLE}) - k$, where k is the number of parameters. The latter is a measure of model complexity, but it may also be viewed as the penalty for not evaluating prediction accuracy with proper out-of-sample test data. The actual value of AIC is given by $-2 \cdot \log p(y|\hat{\theta}_{MLE}) + 2k$, where the first term is the deviance of the model (Gelman et al., 2014b, p. 1001). Remember you can use functions in `AICcmodavg` (Mazerolle, 2023) or `MuMIn` (Barton, 2023) if you need any of a number of variants of the regular AIC. For a simple example of AIC-based model averaging, see Section 18.7.

We won't show the AICs for all five models here, but instead move on to the next information criterion: the DIC (Spiegelhalter et al., 2002). Gelman et al. (2014b, p. 1002) describe it as a *"somewhat Bayesian version of AIC that takes [the equation for AIC] and makes two changes, replacing the maximum likelihood estimate $\hat{\theta}$ with the posterior mean $\hat{\theta}_{Bayes} = E(\theta|y)$ and replacing k with a data-based bias correction."* Thus, as a measure for predictive accuracy (*elpd*) of a model, the DIC uses $\log p(y|\hat{\theta}_{Bayes}) - p_D$, and similar as for the AIC, the actual value of the DIC is given by $-2 \cdot \log p(y|\hat{\theta}_{Bayes}) + 2p_D$, that is, as twice the deviance evaluated at the posterior mean of the parameters, plus twice the effective number of parameters.

The first term is straightforward, but multiple variants have been suggested for p_D (see e.g., Gelman et al., 2014b, p. 1002). The one currently output by `jagsUI` is based on what was used in the earlier R2WinBUGS package (Sturtz et al., 2005). With this method, p_D is estimated as half the variance of the deviance samples within a single MCMC chain and is then also sometimes called p_V. If you have multiple MCMC chains, you can calculate this value for each chain and average them to get a single estimate for p_D. A probably better way of computing p_D is implemented in package `rjags` (Plummer, 2023a) via function `dic.samples()`. It outputs DIC scores with a penalty computed according to Plummer (2002, 2008).

For illustration, we compute the DIC for model 2 with `rocks`. We fit the model with JAGS and will draw samples from the posterior predictive distribution of the data. This is trivial: we simply add in the model another node called `Crep` which is given a Poisson distribution with parameters that are estimated from the training data set. Below we will need these samples for use in package `loo` (Vehtari et al., 2024) to compute the loo information criterion, or LOO-IC. In addition, although the deviance is by default always monitored as a node in the model, we will explicitly define it as node `myDeviance`, just to clarify what the deviance is for this model. We note there is a lot of MC error in the calculation of the DIC. Thus, we run fairly long chains.

```
# Bundle and summarize data for each analysis
str(dataList2 <- list(n = length(dat$C), C = dat$C, X = Xmat[,1:2]))

# Write JAGS model file for model 2
cat(file="model2.txt", "
model {
# Priors
for(k in 1:2){
  beta[k] ~ dnorm(0, 0.0001)
}
# Likelihood, and computation of log-density and replicate data Crep
for (i in 1:n) {
  lambda[i] <- exp(inprod(X[i,], beta))
  C[i] ~ dpois(lambda[i])
  logdens[i] <- logdensity.pois(C[i], lambda[i])
  Crep[i] ~ dpois(lambda[i])
}
# Define our own deviance for fun
myDeviance <- -2 * sum(logdens)
}
")

# Don't need to actually give starting vals for simple models
inits <- NULL

# Parameters to estimate
params <- c("beta", "logdens", "Crep", "myDeviance")

# Long MCMC settings (since DIC has lot of MC error)
na <- 1000  ;  ni <- 32000  ;  nb <- 2000  ;  nc <- 4  ;  nt <- 30

# Call JAGS and summarize posteriors
out2 <- jags(dataList2, inits, params, "model2.txt", n.iter = ni,
  n.burnin = nb, n.chains = nc, n.thin = nt, n.adapt = na, parallel = T)
print(out2, 2)                          # not shown
```

In our run of the model, the value of the DIC is given by `jagsUI` as 370.96 and that of p_D as 3.1. Here, the latter is an overestimate of the number of parameters, which we know is 2. Let's repeat this calculation by hand.

```
# Calculate pD manually using Gelman's approximation
dev2 <- out2$samples[,"deviance"]   # Extract -2 * log p (y | theta_hat)

# Calculate the variance of deviance samples in each chain, divide by 2
pd2 <- sapply(dev2, var)/2

# And average them over all chains
mean(pd2)

# Calculate the mean of deviance samples in each chain and add pD
dic2 <- sapply(dev2, mean) + pd2

# and average them
(dic2_jagsUI <- mean(dic2))
```

Package `rjags` (Plummer, 2023a) offers two variants of the DIC (Plummer, 2002, 2008). They don't work for a model fit in parallel, so we must re-run the model.

```
# This does not work with parallel output so have to re-run model
out2.np <- jags(dataList2, inits, params, "model2.txt", n.iter = ni,
    n.burnin = nb, n.chains = nc, n.thin = nt, n.adapt = na, parallel = F)
dic2_rjags <- rjags::dic.samples(out2.np$model, n.iter = 1000, type = "pD")
(dic2_rjags)

Mean deviance: 368
penalty 2.071
Penalized deviance: 370.1
```

We compare the DIC scores and the values of p_D between `jagsUI` and `rjags` (see the ASM book website for full code). We see fairly comparable values of the DIC, but more variability in the estimated number of effective parameters (p_D), which in this case at least is much better in the `rjags` output, in the sense of being much closer to the actual number of parameters estimated in each model.

	DIC_jagsUI	pD_jagsUI	DIC_rjags	pD_rjags
1	405.5821	0.9250468	405.7537	1.044954
2	370.9609	3.0577671	369.7980	1.907052
3	366.5821	4.1006406	365.6748	3.046952
4	337.9955	6.7054118	335.0307	4.292559
5	236.7365	4.7220331	236.3742	4.844784

For many models, DIC is a suitable predictive criterion for model selection, though not for models with discrete random effects, such as capture-recapture, occupancy or N-mixture models (Millar, 2009; Hooten & Hobbs, 2015). In addition, the DIC is not fully Bayesian, because it uses point estimates of the parameters rather than propagating the full estimation uncertainty into the predictive density.

The WAIC (Watanabe, 2010) improves on the DIC in both respects. As a measure of model fit, it summarizes the full log posterior predictive distribution of the data instead of the log-predictive distribution conditional on a point estimate of the parameters, as do AIC and DIC. As for the DIC, different variants have been suggested for the bias-correction term p_W, of which the most commonly used one is simply the sum (over all data) of the variance (over the posterior draws) of the log-predictive density of the data (Gelman et al., 2014b, p. 1003). For the measure of predictive accuracy of a model, WAIC uses the log pointwise predictive density of the data, which is the log-predictive density of the data averaged over the posterior distribution of the parameters (see Gelman et al., 2014b, p. 1000), and from this subtracts the bias-correction term p_W, that is, $\log\prod_{i=1}^{n} p_{post}(y_i) - p_W$. The actual value of WAIC is then given by $-2 \times \log\prod_{i=1}^{n} p_{post}(y_i) + 2 \times p_W$. We now illustrate computation of WAIC for model 2; the full code for all models is again found on the website. We also note that NIMBLE can output the WAIC if requested (this is also shown in the full code on the website).

```
# Extract posterior draws of the log-density of C at each site
ld.samples2 <- out2$sims.list$logdens

# Compute log-pointwise-predictive-density over data set
lppd2 <- sum(log(colMeans(exp(ld.samples2))))

## Compute penalty
pW2 <- sum(apply(ld.samples2, 2, var))

# Return WAIC
(waic2 <- -2*(lppd2-pW2) )   # not shown, but it yields approximately 375
```

Finally, Vehtari et al. (2017) have developed an approximation to Bayesian LOO-CV that can simply be computed from predictive simulations of a single run of a model. The associated measure of predictive model accuracy is called LOO-IC and is implemented in package loo (Vehtari et al., 2024). We briefly illustrate it, using the simulation draws of the log-density of each data point produced by JAGS above.

```
# Get LOO-IC from log-density draws produced by JAGS
library(loo)
(loo_ic2 <- loo(out2$sims.list$logdens))

Computed from 4000 by 50 log-likelihood matrix

            Estimate      SE
elpd_loo      -187.5    23.9
p_loo            6.8     1.9
looic          375.0    47.7
------
Monte Carlo SE of elpd_loo is 0.1.
```

Finally, we produce a table for the Grand Comparison of all the methods used in this section for assessing the predictive accuracy of the five models in our example. For these simple models, we don't see great differences in the ranking between the methods, though we caution that this may be different for many other model types or for different ratios of sample size to model complexity.

	Null	+rock	+oak	+chip1	+chip2
lpd_is	-201.83	-182.68	-179.51	-162.87	-113.33
lpd_oos	-251.49	-262.67	-234.05	-210.13	-181.88
lpd_loocv	-208.00	-190.25	-191.96	-179.59	-128.81
lpd_fivefold	-206.54	-192.61	-195.78	-181.06	-130.23
AIC	405.67	369.36	365.02	333.74	236.67
DIC	405.80	370.93	366.58	338.00	236.74
WAIC	410.89	374.98	373.85	348.56	244.13
LOO-IC	411.12	374.98	374.14	350.45	245.53

Note also the different scale of the direct assessments of predictive density in the first four rows of the table and of the IC in the last four, which are all on the scale of the deviance. To make all metrics directly comparable, we can simply multiply the first four rows by -2 and then get everything on the same scale.

	Null	+rock	+oak	+chip1	+chip2
lpd_is	404	365	359	326	227
lpd_oos	503	525	468	420	364
lpd_loocv	416	381	384	359	258
lpd_fivefold	413	385	392	362	260
AIC	406	369	365	334	237
DIC	406	371	367	338	237
WAIC	411	375	374	349	244
LOO-IC	411	375	374	350	246

External validation, CV, or IC can all be used to choose a predictive model or to optimize a selection of covariates in the soft parts of a model (Section 18.4). Note though that we have illustrated them for a simple GLM. Recent research suggests that for mixed and hierarchical models, model selection should best be based on the marginal versions of IC, where the random effects have been summed out or integrated over (Millar, 2018; Merkle et al., 2019; Ariyo et al., 2020, 2022). Briefly, in a mixed model as in Chapter 10, with multiple snakes studied in each of a number of populations, all the methods shown in this section will optimize model selection for the prediction task to predict a new snake in an existing population. However, good prediction of new snakes in *new populations* may be more what one is interested in. Similarly, when using CV for assessing predictive accuracy for this task, one then must leave out entire populations, while the predictive density of the data used in the ICs must be computed with the population effects integrated out. This is outside the scope of this book, but we note that NIMBLE has functions to compute WAIC at different levels in a hierarchical model (Hug & Paciorek, 2021).

18.7 **Model averaging**

In the previous section, we have seen methods to rank models (or rather, combinations of covariates in these models) in terms of their predictive accuracy on new data. The various scores presented may all be used to pick one model over another, either for inference or, especially, for prediction (Gerber et al., 2015). Picking just one model will be most satisfactory when it really stands out from the rest. But often two or more models will be nearly tied and then we may want to base inferences or predictions on several or even all of the models. This is called model averaging (Hoeting et al., 1999; Burnham & Anderson, 2002; Link & Barker, 2006; Dormann et al., 2018). In model averaging we compute a weighted average of either the coefficients or of the predictions over a set of models. Better-predicting models get higher weights, although sometimes equal-weight averaging is also done (e.g., Monneret et al., 2018). Bolker (2023) makes a strong point that model averaging is for prediction, but may be far less optimal when the goal of a model is causal explanation.

Coefficients may in principle be averaged, but this can be risky, since they may mean very different things in different models (Cade, 2015). Thus, averaging coefficients is perhaps best avoided. In contrast, averaging predictions can be very useful and can lead to improvements and greater robustness (Dormann et al., 2018). Two specialized R packages for model averaging are `AICcmodavg` (Mazerolle, 2023) and `MuMIn` (Barton, 2023). Below, we give a very short illustration of the former. Multimodel inference can also be done with Bayesian inference, but is arguably harder. For an entry point into the Bayesian literature for ecologists, see Link & Barker (2006), Barker & Link (2013, 2015), Hobbs & Hooten (2015) and Hooten & Hobbs (2015).

We will simulate another ticklish rattlesnakes data set, but then assume that we had no clue they would be tickled by the chipmunks and thus didn't even consider a model with quadratic chipmunks. Instead, we will fit all eight combinations of `rock`, `oak`, and `chip1`, including the Null model. The full code is again on the book website; here we show excerpts of code and output only.

```
set.seed(18)
dat <- simDat18(nSites = 100, beta1.vec = c(0.5, 0.3, 0.2, 0.3, -0.4),
    ncov2 = 50, beta2.vec = rnorm(50, 0, 0.1), show.plot = TRUE)

# Fit 8 models: all combos of rock, oak and chip1
data <- data.frame(C = dat$C, rock = dat$rock, oak = dat$oak, chip1 = dat$chip1, chip2 = dat$chip2)

fm1 <- glm(C ~ 1, family = poisson, data = data)
fm2 <- glm(C ~ rock, family = poisson, data = data)
fm3 <- glm(C ~ oak, family = poisson, data = data)
fm4 <- glm(C ~ chip1, family = poisson, data = data)
fm5 <- glm(C ~ rock + oak, family = poisson, data = data)
fm6 <- glm(C ~ rock + chip1, family = poisson, data = data)
fm7 <- glm(C ~ oak + chip1, family = poisson, data = data)
fm8 <- glm(C ~ rock + oak + chip1, family = poisson, data = data)

summary(fm8)    # Look at most complex model (not shown)
```

We note some OD in the most complex model (fm8): the residual deviance is 222 on 96 residual d.f.s, yielding an estimate of the OD factor $\hat{c}$ of 2.3. This is not surprising, since we failed to include in the model the effects of chip2 and those of 50 nuisance covariates. Below, we will accommodate the resulting OD by making the SEs and the CIs of the model-averaged predictions longer (under the caveats mentioned for this approach in Section 18.5.4).

We next rank all the models by their AIC_c, noting that AICcmodavg computes by default the small-sample version rather than the regular AIC.

```
library(AICcmodavg)
cand.set <- list(fm1, fm2, fm3, fm4, fm5, fm6, fm7, fm8)
modnames <- c('Null', 'rock', 'oak', 'chip1',
              'rock+oak', 'rock+chip1', 'oak+chip1', 'rock+oak+chip1')
aictab(cand.set = cand.set, modnames = modnames)

Model selection based on AICc:
```

	K	AICc	Delta_AICc	AICcWt	Cum.Wt	LL
rock+chip1	3	366.42	0.00	0.64	0.64	-180.09
rock+oak+chip1	4	367.81	1.39	0.32	0.95	-179.69
rock	2	373.28	6.86	0.02	0.97	-184.58
rock+oak	3	373.32	6.90	0.02	1.00	-183.54
chip1	2	377.25	10.83	0.00	1.00	-186.56
oak+chip1	3	378.14	11.72	0.00	1.00	-185.95
oak	2	382.61	16.18	0.00	1.00	-189.24
Null	1	383.18	16.75	0.00	1.00	-190.57

We see that nonzero AIC_c weight (up to two decimal places) is distributed over four models and that the AIC_c-best model is not the one that is closest to the true model (which contains chip2 and is not in the model set). Next, for illustration we produce predictions of a profile of expected rattlesnake counts against covariate rock. For this, we have to create first a data frame with new data at which we want to predict using the model set. We choose 100 values of rock spaced evenly across the observed range of that covariate (−3, 3) and values of 0 for the two other covariates. Since all covariates are standardized, this corresponds to predictions made for varying values of rock at the average observed values of `oak` and `chip1`. The R function in `AICcmodavg` for averaging predictions has an option to "stretch" SEs and CIs by an estimated OD factor `c.hat`, and we use our above estimate for this.

```
newdata <- data.frame(rock = seq(-3, 3, length.out = 100), oak = 0, chip1 = 0)
# Predictions where we accommodate overdispersion
c.hat <- fm8$devi/fm8$df.resi                       # Our estimate of overdispersion
ma_pred <- modavgPred(cand.set = cand.set, modnames = modnames, newdata = newdata, c.hat = c.hat)
ma_pred                                             # print out predictions (not shown)
plot(newdata$rock, ma_pred$mod.avg.pred, xlab = 'Rock',
  ylab = 'Expected count', type = 'l', lwd = 3,
  ylim = c(0, 9), frame = FALSE)                    # not shown
polygon(c(newdata$rock, rev(newdata$rock)),
  c(ma_pred$lower, rev(ma_pred$upper)),
  border = NA, col = rgb(0,0,0,0.1))                # not shown
```

We show code to produce (but don't show) a plot of these model-averaged predictions of the effect of `rock` on the expected rattlesnake count, and we could do likewise for the other two covariates. You can set the value of the argument `c.hat` to 1 to inspect the effect of our "cheap" OD correction on the uncertainty of these predictions.

18.8 Regularization: penalization, shrinkage, ridge, and lasso

Model averaging based on AIC weights is a well-known and widely used method to deal with the problem when you have "too many covariates" and don't know what to do with them (we note though that the best first step is always to drop from consideration those covariates that are irrelevant for your modeling goals). Model averaging yields a sort of "weighted, brute-force inference" based on all the models in your set. Regularization, also called penalization, may be viewed as another solution to the same problem, that is, the black hole of statistics. Here, we fit a single, big model that contains all covariates of interest, but we pull their coefficients some way towards zero and each other, that is, we shrink them. Besides dealing somewhat automatically with lots of covariates, penalization/regularization/shrinkage tends to lead to more stable estimates, especially when using least squares or maximum likelihood (Moreno & Lele, 2010; Clipp et al., 2021). In addition, it confers increased robustness against the effects of collinearity among covariates, which can pose big challenges in models with lots of covariates (Dormann et al., 2013). In addition, some forms of regularization achieve what is called subset selection, that is, the identification of a smaller subset of covariates that predicts about as well as the full set, while setting coefficients of all the others to zero. In this section, we give a brief introduction to regularization methods, which we believe are greatly underutilized by ecologists. Good entry points into the literature for ecologists include especially Reineking & Schröder (2006), and then Hooten & Hobbs (2015), Gerber et al. (2015), and Tredennick et al. (2017).

Traditionally, penalization/regularization means to put a constraint on an optimization problem, such as when minimizing a negative log-likelihood function (Hooten & Hobbs, 2015). In a regression model with J covariates with coefficients β_j for $j = 1, ..., J$, this means that we minimize the sum of the negative log-likelihood plus a penalty term, which can be interpreted as a measure of model complexity (Reineking & Schröder, 2006). A general expression of a penalty is $\lambda \sum_{j=1}^{J} |\beta_j|^q$. Interestingly, when we set $q = 0$ we get a count of the parameters, and if in addition we set $\lambda = 2$, then we arrive exactly at the penalty for the AIC. This exemplifies the tight relationship between model selection and regularization (Hooten & Hobbs, 2015). For values other than 0, the penalty depends both on the number of coefficients and their values. Reineking & Schröder (2006) write *"higher values of q make large parameter estimates more 'costly' and therefore favour small absolute parameter estimates."* Two common penalties are called *ridge*, when we set $q = 2$ (Hoerl & Kennard, 1970), and *lasso* for $q = 1$ (Tibshirani, 1996). Ridge assumes many nonzero coefficients, most of which are small, while the lasso assumes few nonzero coefficients, which may be large, and assumes the others are exactly 0. Lasso is an acronym for "least absolute shrinkage and selection operator"; thus, it does shrinkage, but also "switches off" some covariates entirely by setting their coefficient to zero. Hence, it is also a method of subset selection. In terms of improving prediction accuracy, ridge and lasso often appear to perform similarly (Tredennick et al., 2017), but the lasso may have interpretational advantages, because setting some coefficients to exactly zero simplifies the model. Ridge and lasso and a hybrid approach called elastic net may be applied to a wide range of regression models with methods in package `glmnet` (Friedman et al., 2010; Tay et al. 2023).

In Bayesian inference, the prior is a natural regulator (Hooten & Hobbs, 2015): a wider prior will allow more extreme coefficients, while a narrow prior will lead to their shrinkage towards zero. Fascinatingly, both prototypes of classical regularization can be specified dead simply in a Bayesian analysis with a number of regression coefficients β_j: to obtain a ridge penalty, we simply specify a zero-mean normal prior with variance $1/\lambda$ (Kyung et al., 2010). For a Bayesian lasso, we specify a double-exponential (or Laplace) prior for the β_j's and then use as a point estimator the posterior mode instead of the mean or the median—because with the lasso, some coefficients must be able to become exactly zero, and only the mode achieves that for a Bayesian point estimate. See Appendix 1 in Gerber et al. (2015) for a picture of how the normal and the double-exponential priors compare.

By now you may wonder how we choose the all-important value of the regulator λ? After all, if we set it to zero, then we don't have any regularization at all, while if we choose for it a very large value, all coefficient values will be estimated at zero, as we will see below. The answer is once again cross-validation: we try out a large number of values for the regulator λ, fit the model to the training split, assess prediction accuracy on the testing split, and eventually choose the value of λ that leads to the highest prediction accuracy. A Bayesian CV is an empirical Bayes method (Carlin & Louis, 2009), that is, we use in-sample information to choose a parameter in a prior (here, λ; Hooten & Hobbs, 2015).

We now illustrate regularization with a ridge example implemented in Bayes, for another simulated rattlesnake data set. You find the full code on the website. We simulate count data from 100 sites, with effects of all main covariates (`rock`, `oak`, `chip1`, `chip2`) and in addition with effects of 60 additional, unspecified covariates. That is, our model is fairly heavily parameterized, with more than one estimated parameter per two data points. Note that covariates for which such a shrinkage prior is adopted ought to be on the same scale, that is, we would typically normalize them all. However, in the rattlesnake data, all covariates are already created as zero-mean normal variables, so we don't need this here.

We fit a penalized version of the data-generating model, that is, we estimated all five main parameters in the model (i.e., the intercept and the four coefficients of `rock`, `oak`, `chip1`, and `chip2`), plus the coefficients of the 60 "noise" covariates. As is customary, we did not penalize the intercept, but for all 64 slope parameters we specified a Normal(0, variance $= 1/\lambda$) prior, which in

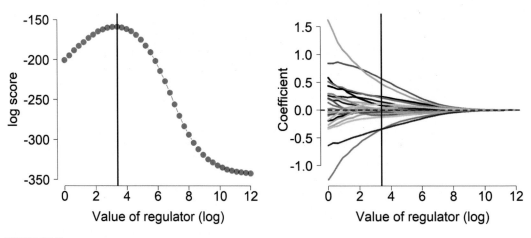

FIGURE 18.9

Ridge regularization illustrated with a big Poisson GLM fit to simulated rattlesnake data. Left: The trace of the cross-validated log score as the value of the regulator is varied between exp(0) and exp(12). Right: Trace of the parameter estimates as a function of the value of the regulator, with high values inducing more penalization and greater shrinkage than low values of λ. The drab blue vertical line indicates the optimal value of the regulator for prediction.

JAGS became a Normal(0, precision $= \lambda$) prior. We tried out 40 regularly spaced values of the regulator λ between $exp(0) = 1$ and $exp(12) = 162755$. For each, we looped over all 100 data points and used LOO-CV to assess the prediction accuracy of the full model for each value of λ.

Fig. 18.9 (left) shows the trace of the log score as the value of the regulator is varied over a wide range from essentially no penalization on the left to essentially fixed coefficients (at 0) on the right. We clearly see that the unconstrained, data-generating model (i.e., with $\lambda = 0$) would have a much inferior predictive accuracy than a penalized model. The value of the ridge regulator that maximized the CV performance score was $exp(3.4) \approx 30$. Fig. 18.9 (right) shows the trace of the 64 coefficient estimates as we move from an almost unconstrained model on the left to a super-strongly constrained model on the right, where all estimates become equal to 0. A set of intermediate coefficients is best when we want to use this model for prediction.

18.9 Summary and outlook

In this chapter, we have covered several really big topics for applied statistical modeling: model evaluation in terms of its fit to the data, model validation in the sense of variable selection, and how to combine the inferences from multiple models, for example, by averaging them or by regularization. We have emphasized how decisions in all these topics cannot be made without considering the fundamental question of the goal of a model. In addition, all of this has relevance also for the question of how to build a model.

As did Yoccoz et al. (2001) for the design of biodiversity surveys, we repeatedly stressed the importance of model goals, especially the two goals of causal explanation and prediction. There is often some overlap and most modeling goals may in fact lie along some gradient between these two extremes. Nevertheless, this dichotomy is important for the question of how we build models, how we check them, and also for how we select among them, if we have more than one. We have also described the "hard" and the "soft" parts of a model, where the hard structure is typically dictated by our science, while the soft structure is more flexible and adapts to the actual data at hand. Causal-explanatory models will have relatively more hard parts, while purely predictive models may have little hard parts.

We admit that we have written this chapter (and indeed the whole book) from the perspective of a fan of the principled approach of model building advocated so forcefully by Royle & Dorazio (2008) (and also Berliner, 1996, 2003; Wikle, 2003, and others in that same tradition), where scientific questions or management problems all have a strong say about how we build a model. Nevertheless, we think that also in this type of model, there will often be soft parts and for these flexible parts of a model, optimization of prediction accuracy makes a lot of sense.

In this chapter, we have happily stepped back and forth between classical inference using maximum likelihood (or iteratively reweighted least-squares, IRLS, with `glm()`) and Bayesian inference. We did this because the topics in this chapter are equally important for both inference schools and at the same time we show also how so many fundamental concepts in applied statistical modeling transcend the choice of inference paradigm. Also, we find it quite fascinating how predictive simulations (see below), which are so hugely important for most of what we covered in this chapter, are fundamentally a non-Bayesian idea. And yet, many influential applied Bayesian scientists (e.g., Box 1980; Rubin, 1984; Little, 2006; Gelman et al., 2020) emphasize them. We don't think it would be wise to forego all of this just to stick to the purity of the Bayesian paradigm, which says that all inference must be based on the single data set analyzed.

Simulation is a huge topic throughout this chapter: we use it to generate a "family" of data sets, and we use it to infer predictive distributions of the data, which are a key in both model checking and model selection. Also, simulation is arguably the main general and practical method for an ecologist for assessing parameter identifiability. Finally, simulation is also a key for assessing the robustness of a model to assumption violations. This is exactly why in this book we work with plenty of simulated data sets and always show the R code to do so. We believe that every applied statistical modeler must become a proficient data simulator.

We have seen that the predictive distribution of the data under a model is a key concept for both model checking and model selection. Algebraically, it is tightly related to the same old joint density function of the data and the likelihood function used in statistical inference (see Chapter 2). This is one more reason for why it is so important that you as an ecological modeler become conversant in likelihood functions, at least (or to start with) for simple models such as the GLM considered in the rattlesnake examples in this chapter.

We admit that by choosing a simple nonhierarchical GLM for all the illustrations, we have spared us much headache. The reason is that most topics in this chapter, especially model checking and selection, become more involved in mixed or more generally in hierarchical models. This is because, in hierarchical models, there are different ways in which we can simulate replicate data for predictive simulations for model checking: we can either simulate them conditional on the values of the random effects estimated in the analysis of our actual data set, or we can also simulate new values for these random effects and then simulate new values for the observed response based on them. The difference is referred to as "conditional" versus "marginal" simulation of the new data. Marshall & Spiegelhalter (2003) and Conn et al. (2018) have some good explanation of this for model checking, as does the DHARMa vignette (Hartig, 2022). The upshot is that it seems to be best to simulate also new values for the random effects, that is, to choose a marginal mode of simulation, rather than the simpler conditional one.

Similarly, for model selection in mixed and hierarchical models, the focus of interest is arguably more often on new replicates of the groupings that define the random effects, rather than on new data from within existing groups. Groups would be the populations in mixed models as in Chapter 10 or the sites in occupancy or N-mixture models in Chapters 19 and 19B. In these cases, the predictive density must be expressed for these higher-level units and the associated random effects must be summed or integrated out (Millar, 2018; Merkle et al., 2019); see Chapters 10 and 19 and the comments at the end of Section 18.6 for more information.

Occupancy models⊛

Chapter outline

19.1 Introduction

We have now seen a wide range of random-effects (also called hierarchical) models, including normal, Poisson, and binomial generalized linear mixed models, or GLMMs. In these models some sets of parameters are represented as realizations from a random process. With the exception of the zero-inflated model in Chapter 12, we have always used a normal or a multivariate normal distribution as our description of this additional source of randomness in the response. This is part of the definition of a GLMM: that there are one or more additional sets of Gaussian random effects that appear in an additive manner in the linear predictor.

However, nothing constrains us to use only the normal distribution for random effects. Sometimes, other distributions will be more appropriate for some parameters in a model with multiple sources of variation. In this chapter and the subsequent Chapter 19B (which you can get from our book website), we showcase two models with *discrete random effects*: specifically, we will assume the random effects have a Bernoulli or a Poisson distribution. Importantly, these random effects have a precise biological meaning in these models: they correspond to the true, but imperfectly observed, state of occurrence of a species at a site (this chapter) or of abundance (Chapter 19B). This is in sharp contrast to the random effects in traditional GLMMs, which seldom

---

⊛This book has a companion website hosting complementary materials, including all code for download. Visit this URL to access it: https://www.elsevier.com/books-and-journals/book-companion/9780443137150.

Applied Statistical Modelling for Ecologists. DOI: https://doi.org/10.1016/B978-0-443-13715-0.00010-8
447

have any tangible meaning, and simply serve to accommodate correlations in the data (Royle & Dorazio, 2008).

The modeling of animal and plant distributions, via use of species distribution models (SDMs), is a huge and very active area of ecological research and applications. In the past, one of the most frequently applied SDM was the binomial generalized linear model (GLM), or logistic regression, of the type we have seen in Chapters 15–17 (although nowadays the classes of models used for SDMs have become many and varied; Elith et al., 2006, 2008). Here, the probability that an organism is found at a sample unit is modeled from what often, and misleadingly, are called "presence/absence" data. These are binary indicators y for whether a species was found ($y = 1$) or not ($y = 0$) in a spatial sample unit, and the effect of covariates is modeled via the logit link function. There are many variants of this approach, but the basic principle is usually the same: so-called "presence/absence" data are directly modeled as coming from a Bernoulli distribution and the Bernoulli success parameter is interpreted as the probability of occurrence, or occupancy probability.

However, a fundamental and widely overlooked issue in species distribution modeling is that detectability (p) of most species is imperfect: typically, a species will not always be detected where it occurs (Kéry & Schmidt, 2008; Kéry, 2011; Kellner & Swihart, 2014; Kéry & Royle, 2016, 2021; MacKenzie et al., 2017). In other words, detection probability is typically less than 1, i.e., $p < 1$. This basic fact is very well known to all field naturalists and reasonably widely understood for animals, and it also applies to populations of sessile organisms such as plants or lichens (Kéry, 2004; Kéry et al., 2006; Chen et al., 2013; Perret et al., 2023; von Hirschheydt et al., 2024). However, it seems to have been overlooked by many professional ecologists dealing with species distribution data (Araújo & Guisan, 2006; Elith et al., 2006). As a consequence, arguably only a minority of studies generated by current species distribution modelers actually model the true occurrence of a species. The remaining studies are usually modeling the product of the probabilities of species occurrence and of detection (Kéry, 2011). (There is also the converse type of measurement error, leading to false positives, but we ignore these here. See Royle & Link (2006), Miller et al. (2011) and Chapter 7 in Kéry & Royle (2021) for an introduction to false positives in the context of occupancy models.)

The widespread confusion about what is actually being modeled in most species distribution models has three main consequences when p is not corrected for (MacKenzie et al., 2002; Tyre et al., 2003; Gu & Swihart, 2004; Royle & Dorazio, 2008; Kéry & Schmidt, 2008; Kellner & Swihart, 2014; Kéry & Royle, 2016, 2021; MacKenzie et al., 2017):

1. The extent of a species distribution will be underestimated whenever $p < 1$,
2. estimates of covariate relationships will be biased towards zero whenever $p < 1$,
3. and factors that affect the difficulty with which a species is detected may end up in predictive models of species occurrence or, conversely, mask genuine relationships between species occurrence and the environment.

The first is easy to understand, but the second has not been widely recognized, although Tyre et al. (2003), Gu & Swihart (2004), and MacKenzie *et al.* (2006, pp. 34–35) already described this intriguing effect. Interestingly, and in contrast to a detection-naïve analysis of count data in the presence of imperfect detection (cf. simple Poisson regression examples in Chapters 12–15), in distribution modeling, even a constant $p < 1$ will produce an attenuating bias in the slope estimate of the relationship between occurrence probability and a covariate: that is, negative slopes will be overestimated and positive slopes underestimated. Presumably, this includes also a time covariate, that is, situations where changes in distribution over time are modeled. As an example for the third effect, if a species has a higher probability to be detected near roads, perhaps because near roads more people are likely to stumble upon it, then obviously roads or habitat types associated with roads will show up as important for that species unless detection probability is accounted for. As an extreme example, when a species distribution map is constructed from road-kill records only, then no matter how much roads might actually be avoided by that species in reality, the resulting distribution map will emphasize the great positive effect of roads on the distribution of the species! On the other hand, if a species really likes dense bush habitat, but is far harder to observe there, then a detection-ignorant SDM may fail to identify dense bush as a preferred habitat.

In contrast, a class of models with the rather peculiar name "site-occupancy models", or occupancy models for short (MacKenzie et al., 2002, 2003; Tyre et al., 2003; MacKenzie et al., 2006, 2017; Kéry & Royle, 2016, 2021; Guillera-Arroita 2017) is able to estimate the true distribution of animals or plants free of any distorting effects of the difficulty with which they are found. This chapter deals with this important class of models.

As a motivating example, we consider an inventory of the beautiful Chiltern gentian (*Gentianella germanica*; Fig. 19.1) conducted in 150 patches of calcareous grassland in the wild and wonderful Jura mountains. Our aim is to estimate the number or the proportion of occupied sites and to identify environmental factors related to the presence or the absence of the gentian at a site. Interestingly, *G. germanica* is typical of nutrient-poor soils, which are often rather dry. However, within the class of nutrient-poor grasslands, this gentian preferentially occurs on the wetter sites. These sites often have a higher and denser vegetation cover, so the gentian which is rather small in size (5–40 cm height) may more frequently be overlooked at these better sites (we ignore here the fact that better sites may also hold larger populations, which would then again make them less easy to overlook). This effect could mask to some degree the gentian's preference for wetter sites. Apart from occupancy models, none of the currently widespread SDM methods such as GLM, generalized additive models (GAM), boosted regression trees, random forests, or point processes as implemented in program Maxent (Elith et al., 2006, 2008; Valavi et al., 2022) are able to tease apart the effects of a covariate that influences both the occurrence of a species and the ease with which it is detected.

We will use occupancy models (MacKenzie et al., 2002; Tyre et al., 2003) to separately estimate the gentian's probability of occurrence (called occupancy or species distribution) and its probability to be detected at occupied sites (detection probability), along with covariate effects on either occurrence or detection. The price to be paid for this improved inference is a repeated-measures design. Some sites need to be visited at least twice, preferably more frequently, and visits need to be conducted within such a short time period that we can assume population closure, that is, that the presence or absence state of a site does not change during the course of these visits. It is from the pattern of detection or nondetection at sites visited multiple times that we obtain the bulk of the information about detection probability, separately from occupancy probability. See MacKenzie & Royle (2005), MacKenzie et al. (2006, 2017), Bailey et al. (2007), Guillera-Arroita et al. (2010), Reich (2020), and von Hirschheydt et al. (2023) for design considerations relevant to this model.

A balanced design, that is, an equal number of visits to all sites, is not required for occupancy models; it simply makes things easier to simulate and present. Therefore, in our inventory of *G.*

FIGURE 19.1

The wonderful Chiltern gentian (*Gentianella germanica*), Slovenia, 2007 (Photo by Milan Vogrin).

germanica, we assume that each site is visited three times by an independent botanist, who notes whether at least one plant of *G. germanica* is detected or not during each visit. The result of these surveys may be summarized in a so-called detection history, that is, a binary string, such as 010 for a site where a gentian is detected during the second, but not during the first or the third survey. Generally, for a species surveyed T times at each of M sites, survey results are summarized in an M-by-T matrix containing a 1 when the species is detected at site i during survey t and a 0 when it is not.

The detection/nondetection observation $y_{i,t}$ at site i during survey t is naturally described by a hierarchical model that contains one submodel for the only partially observed true state (occupancy, the result of the biological process), and another submodel for the actual observations. The actual observations result from both the particular realization of the biological process and the observation process, which makes the resulting model hierarchical.

$$z_i \sim Bernoulli(\psi)$$ Biological process yields true state
$$y_{i,t} \sim Bernoulli(z_ip)$$ Observation process yields observations

Hence, true occupancy z_i of *G. germanica* at site i is a Bernoulli random variable governed by the parameter ψ, which is occupancy probability: this is exactly the parameter that most distribution modelers wish they were modeling. The actual gentian observation $y_{i,t}$, detection or not at site i during survey t (or "presence/absence" datum $y_{i,t}$), is another Bernoulli random variable with a success probability that can be expressed as the product of the actual occupancy of *G. germanica* at that site, z_i, and detection probability p. Hence, at a site where the gentian doesn't occur, $z = 0$, and y must be 0. Conversely, at an occupied site we have $z = 1$, and *G. germanica* is detected with probability p. That is, in the occupancy model, detection probability is expressed *conditional on occupancy*, and the two parameters ψ and p are separately estimable if data from replicate visits (also called "occasions") are available.

One way to look at this model is in terms of the GLM framework that features so prominently in this book. We could then describe it as a hierarchical, coupled logistic regression, where one logistic regression describes true occupancy and the other describes detection, given that the species does occur. Indeed, a majority of hierarchical models in ecology can be viewed as a combination of multiple GLMs that are linked. Another description would be as a model that is similar in a sense to a binomial GLMM, but with a binary distribution for the random effects—occupied or not occupied—instead of the normal distributions that we assume for the random effects in the true binomial GLMM in Chapter 18 as well as all other mixed models in previous chapters. Finally, the occupancy model is also zero-inflated binomial model (Tyre et al., 2003) and thus related to the zero-inflated Poisson model that we saw in Section 12.2.

The above equations describe the simplest kind of an occupancy model; they can readily be extended to many more complex cases, and we list a number of them in the final section of this chapter. However, one crucial extension that we are going to cover here is the ability to model covariate effects in either the occupancy (ψ) or the detection (p) parts of the model in a simple GLM fashion. That is, in the algebra from above we can make occupancy probability site-dependent by adding an index i to it and then can add to the model a statement like this:

$$\text{logit}(\psi_i) = \alpha + \beta x_i,$$

where x_i is the value of some occupancy-relevant covariate measured at site i and α and β are parameters. Similarly, we can make detection probability dependent on site, occasion or on both by indexing it by i and t. Then we can model $p_{i,t}$ with a logit link function using either a site covariate (x_i) or a survey, also called an observational, covariate ($x_{i,t}$). Of course we can easily model the effects of multiple explanatory variables, of polynomial terms, or even of splines to accommodate wiggly relationships with covariates (see Section 10.14 in Kéry & Royle (2016)), as well as various formulations of spatial autocorrelation (Gelfand et al., 2005; Johnson et al., 2013; Knaus et al., 2018; Doser et al., 2022).

Occupancy models can be fit with a variety of canned routines, usually with maximum likelihood. This includes MARK (White & Burnham, 1999) which can be run from R via `Rmark` (Laake, 2013) and PRESENCE (Hines, 2006), for which an R interface also exists. Occupancy models with spatial autocorrelation can be fit using the R packages `stocc` (Johnson et al., 2013), `multiocc` (Hepler & Erhardt, 2023), and `spOccupancy` (Doser et al., 2022), all using Bayesian inference.

Here, we use the specialized R package `unmarked` (Fiske & Chandler, 2011; Kellner et al., 2023; see also Kéry & Royle, 2016, 2021), which provides a large number of canned functions in R for fitting many different types of occupancy models and related models for abundance of the *N*-mixture type (see Chapter 19B). In addition, we will use the R package `ubms` (Kellner et al., 2022a), which stands for "*unmarked Bayesian models with Stan.*" This is an R user interface for Stan and allows Bayesian inference for many of the models that can be fit using `unmarked`. This is a very useful program and hence, in this chapter and in Chapter 19B, we also have a section where we present Bayesian solutions using canned functions in R from this package.

19.2 **Data generation**

We now simulate the data for our inventory of *G. germanica*. We assume that the 150 sites visited form a sample of a larger number of calcareous grassland sites in the Jura. The study objective is to use these data to learn about all calcareous grassland sites in the Jura and to see how site humidity affects the distribution of *G. germanica*.

```
set.seed(19)
nSites <- 150                    # 150 sites visited
```

We create an arbitrary continuous index for soil humidity, where –1 means dry and 1 wet.

```
humidity <- runif(n = nSites, min = -1, max = 1)
```

Next, we create the positive true relationship between occupancy probability of *G. germanica* and soil humidity (Fig. 19.2 left). We do this by a logit-linear regression as customary for binomial responses. We choose the intercept and the slope for this relationship so that on average about 50% of all sites end up being occupied by the gentian.

```
alpha.occ <- 0                   # Logit-linear intercept for humidity on occupancy
beta.occ <- 2                    # Logit-linear slope for humidity on occupancy
occ.prob <- plogis(alpha.occ + beta.occ * humidity)
```

```
# Fig. 19.2 left
par(mfrow = c(1, 2), mar = c(5,5,4,2), cex.lab = 1.5, cex.axis = 1.5)
plot(humidity, occ.prob, ylim = c(0,1), xlab = "Humidity index", ylab = "Occupancy probability",
  main = "State process", las = 1, pch = 16, cex = 2, col = rgb(0,0,0,0.3), frame = FALSE)

z <- rbinom(n = nSites, size = 1, prob = occ.prob)
z                        # Look at the true occupancy state of each site
(true_Nz <- sum(z))      # Number of occupied sites among the visited sites

> z                      # Look at the true occupancy state of each site
  [1]  0 0 1 0 1 0 0 1 1 1 0 0 1 1 0 1 0 0 1 0 1 0 0 1 1 1 1 0 1 1
 [31]  0 0 1 1 1 1 1 0 1 1 0 0 1 0 1 1 1 0 0 0 1 1 0 0 1 0 1 0 0
 [61]  1 1 1 1 1 0 1 0 1 1 0 1 1 1 0 0 0 0 0 0 1 0 1 1 1 0 0 0 0
 [91]  1 1 1 1 0 1 1 1 0 0 1 1 1 0 1 1 0 0 1 1 1 0 0 1 1 0 0 1 0 0
[121]  0 1 0 1 0 1 0 0 1 1 0 0 0 0 0 0 0 0 1 0 1 1 1 1 1 1 1 0 0 0

> (true_Nz <- sum(z)) # Number of occupied sites among the visited sites
[1] 77
```

This is the *true state* of the gentian system we are studying, that is, the realization of the sto-chastic biological process we're interested in. This state is only imperfectly observable in nature, even for plant populations (Kéry et al., 2006; Chen et al., 2013; Perret et al., 2023). However, it is what we wish we could observe, and what we would like to relate to habitat variables such as humidity in our distribution model for *G. germanica*.

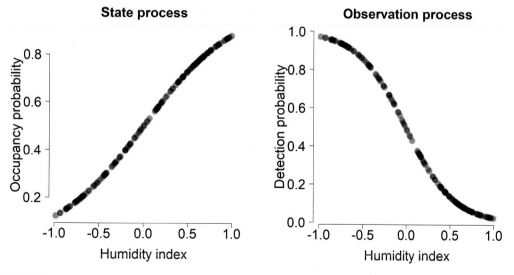

FIGURE 19.2

(Left) The unobserved true relationship between humidity and occupancy probability of the Chiltern gentian at 150 simulated sites. (Right) Relationship between site humidity and detection probability in the Chiltern gentian in our simulated data set.

Unfortunately, virtually always, we can only observe a "degraded image" of that true state of nature, where the degradation is due to the fact that *G. germanica* may be overlooked, especially at the wetter sites with higher vegetation. Next, we simulate this effect to obtain our actual "presence/absence" observations or, more accurately, detection/nondetection data. After simulating the biological process (which resulted in 150 realizations of true occurrence z_i), we now model the observation process that resides between the true biological process ("truth") and our observations (Fig. 19.2 right).

```
alpha.p <- 0                      # Logit-linear intercept for humidity on detection
beta.p <- -3                      # Logit-linear slope for humidity on detection
lp <- alpha.p + beta.p * humidity # The linear predictor for detection
p <- plogis(lp)                   # Get p on the probability scale

# Save true parameter values into a vector
truth <- c(alpha.occ = alpha.occ, beta.occ = beta.occ,
           alpha.p = alpha.p, beta.p = beta.p)

# Fig. 19.2 right
plot(humidity, p, ylim = c(0,1), xlab = "Humidity index", ylab = "Detection probability",
    main = "Observation process", las = 1, pch = 16, cex = 2, col = rgb(0,0,0,0.3), frame = FALSE)
```

Assuming no false-positive errors, i.e., that no other species is erroneously identified as *G. germanica*, the Chiltern gentian can only be detected at sites where it occurs. Hence, the "effective detection probability" can be expressed as the product of true occupancy (z_i) and detection probability (p):

```
eff.p <- z * p   # Zeroes out detection at sites where gentian is absent
```

Importantly, this effective detection probability, or apparent occupancy probability, is precisely the quantity modeled by conventional species distribution models (e.g., Elith et al., 2006)! Its expectation is the product of occupancy probability and detection probability, that is, $\psi \cdot p$ (Kéry, 2011).

We store the results of each survey, 1 (gentian detected) or 0 (no gentian detected), in an *M*-by-*T* matrix and fill it by simulating coin flips (i.e., drawing Bernoulli trials) with the effective detection probabilities just computed. We note that this is the first time in the book that we use a two-dimensional array for our response data. Consequently, you will later see a nested for loop in the BUGS code to analyze these data.

```
nVisits <- 3
y <- array(dim = c(nSites, nVisits))

# Simulate results of first through last surveys
for(t in 1:nVisits){
   y[,t] <- rbinom(n = nSites, size = 1, prob = eff.p)
}
```

Hence, y now contains the results of our simulated surveys to find *G. germanica* at the 150 studied sites. The following table compares the true states of presence and absence and the three (error-prone) measurements of presence/absence at each site. It is crucial for your understanding of biological survey of species distributions.

```
y                                        # Look at the data
# Apparent number of occupied sites among the 150 visited sites
(obs_Nz <- sum(apply(y, 1, sum) > 0))         # 39

cbind('True state' = z, 'Obs Visit 1' = y[,1],   # Look true state and at three
   'Obs Visit 2' = y[,2], 'Obs Visit 3' = y[,3]) # observed states at each site
```

	True state	Obs Visit 1	Obs Visit 2	Obs Visit 3
[1,]	0	0	0	0
[2,]	0	0	0	0
[3,]	1	0	0	0
[4,]	0	0	0	0
[5,]	1	0	0	0
[6,]	0	0	0	0
[7,]	0	0	0	0
[8,]	1	0	0	0
[9,]	1	0	0	0
[10,]	1	0	0	0
.....				
[140,]	1	0	0	0
[141,]	0	0	0	1
[142,]	1	0	0	0
[143,]	1	0	0	0
[144,]	1	0	0	0
[145,]	1	0	0	0
[146,]	1	0	0	0
[147,]	1	1	0	0
[148,]	0	0	0	0
[149,]	0	0	0	0
[150,]	0	0	0	0

On average (if we simulate this stochastic system many times), our parameter values yield about 59% detected gentian populations. Also note that we may sometimes fail to detect the species on all three visits at a site where it occurs, but we never have a detection at a site where in fact it does not occur, since in this chapter we don't account for the possibility of false positives (Royle & Link, 2006; Miller et al., 2011).

Let's see what a detection-naïve analysis of these observations would tell us about the relationship between humidity and the occupancy of *G. germanica*. We call this analysis naïve because it omits an important system component, the observation process. The simplest way to analyze this relationship is by a logistic regression of an indicator for "ever detected" (here called obsZ) on humidity. Note the use of sort() and order() to bring the data into proper order for plotting (Fig. 19.3).

```
obsZ <- as.numeric(apply(y, 1, sum) > 0)   # 'Observed presence/absence'
naive.analysis <- glm(obsZ ~ humidity, family = binomial)
summary(naive.analysis)
lpred.naive <- predict(naive.analysis, type = 'link', se = TRUE)
pred.naive <- plogis(lpred.naive$fit)
LCL.naive <- plogis(lpred.naive$fit-2*lpred.naive$se)
UCL.naive <- plogis(lpred.naive$fit+2*lpred.naive$se)

# Fig. 19.3
par(mfrow = c(1, 1), mar = c(5,5,4,2), cex.lab = 1.5, cex.axis = 1.5)
plot(humidity, pred.naive, ylim = c(0, 0.6), xlab = "Humidity index",
  ylab = "Apparent occupancy prob.", main = "Confounding of state and observation processes",
  las = 1, pch = 16, cex = 2, col = rgb(0,0,0,0.4), frame = FALSE)
polygon(c(sort(humidity), rev(humidity[order(humidity)])), c(LCL.naive[order(humidity)],
  rev(UCL.naive[order(humidity)])), col = rgb(0,0,0, 0.2), border = NA)
```

We see that in a detection-naïve analysis the gentian's preference for wetter sites is completely masked; indeed, we even predict a slight (though not statistically significant) avoidance of the

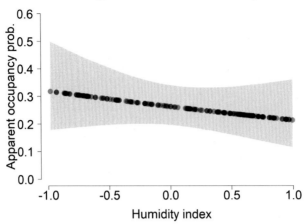

FIGURE 19.3

A detection-naïve analysis of the apparent relationship between occupancy of the Chiltern gentian and humidity. This conventional species distribution model ignores detection probability and therefore suffers from the effects of the confounding of state and observation processes. Compare with the truth in the left panel of Fig. 19.2.

wetter sites (Fig. 19.3; though in different realizations of the data-generating process this may vary). Let's see whether an occupancy model can do better.

```
# Required packages
library(ASMbook); library(jagsUI); library(rstan); library(TMB)
```

19.3 Likelihood analysis with canned functions in the R package unmarked

The R package unmarked (Fiske & Chandler, 2011; Kellner et al., 2023) has a large number of functions for fitting models of distribution and abundance. We use its function occu() to fit our static occupancy model.

```
# Load unmarked, format data into unmarked data frame and summarize
library(unmarked)
summary( umf <- unmarkedFrameOccu(y = y,  siteCovs = data.frame(humidity = humidity)) )

unmarkedFrame Object

150 sites
Maximum number of observations per site: 3
Mean number of observations per site: 3
Sites with at least one detection: 39

Tabulation of y observations:
  0   1
376  74
```

```
Site-level covariates:
    humidity
Min.   :      -0.98214
1st Qu.:      -0.43475
Median :       0.14756
Mean   :       0.07093
3rd Qu.:       0.58044
Max.   :       0.99589

# Fit model and extract estimates
# Detection covariates follow first tilde, then occupancy covariates
summary(out19.3 <- occu(~humidity ~humidity, data = umf))
unm_est <- coef(out19.3)                   # Save estimates

Call:
occu(formula = ~humidity ~ humidity, data = umf)

Occupancy (logit-scale):
              Estimate      SE       z    P(>|z|)
(Intercept)     -0.226   0.363  -0.621      0.534
humidity         1.489   0.757   1.969      0.049

Detection (logit-scale):
              Estimate      SE       z    P(>|z|)
(Intercept)      0.175   0.273   0.641   5.22e-01
humidity        -3.548   0.541  -6.560   5.39e-11

AIC: 292.1851
Number of sites: 150
optim convergence code: 0
optim iterations: 32
Bootstrap iterations: 0

# Get estimates of latent occurrence from unmarked output with ranef()
unm_Nz <- round(sum(ranef(out19.3)@post[,2,1]), 2)
tmp <- c(truth = true_Nz, observed = obs_Nz, unmarked = unm_Nz)
print(tmp, 2)                              # not shown
```

We make predictions, store them in vectors, and sort these vectors in a suitable manner for presentation. We could have avoided this sorting business if we had sorted the humidity vector in ascending order during data simulation. However, this is only possible in some cases and not in general; hence, what we do here is more general and hence safer (Fig. 19.4).

```
# Make predictions
state.pred <- predict(out19.3, type = 'state')
det.pred <- predict(out19.3, type = 'det')
p.pred <- matrix(det.pred[,1], nrow = nSites, byrow = TRUE)    # reformat
p.LCL <- matrix(det.pred[,3], nrow = nSites, byrow = TRUE)     # reformat
p.UCL <- matrix(det.pred[,4], nrow = nSites, byrow = TRUE)     # reformat

# Predict humidity relation for state and observation processes
# (Fig. 19.4)
ooo <- order(humidity)                  # Get the order of the humidity values
par(mfrow = c(1, 2), mar = c(5,5,4,2), cex.lab = 1.5, cex.axis = 1.5)
```

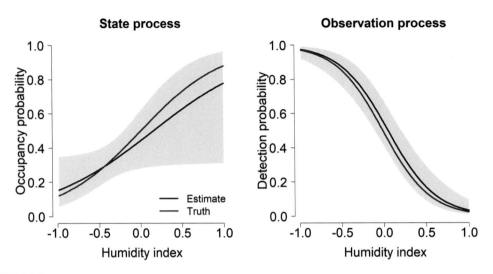

FIGURE 19.4

Comparison of the true relationships with humidity (red) and those estimated from the occupancy model fit with maximum likelihood in R package `unmarked` (blackish-blue; shaded region is 95% CI).

```
# Occupancy
plot(humidity[ooo], state.pred[ooo,1], xlab = 'Humidity index', ylab = 'Occupancy probability',
    frame = FALSE, col = 'blue', lwd = 3, main = 'State process', type = 'l', ylim = c(0, 1))
lines(humidity[ooo], occ.prob[ooo], lwd = 3, col = 'red')
polygon(c(humidity[ooo], rev(humidity[ooo])), c(state.pred[ooo,3], rev(state.pred[ooo,4])),
    col = rgb(0,0,1, 0.1), border = NA)
legend('bottomright', legend = c('Estimate', 'Truth'), lwd = 2, col = c('blue', 'red'),
    bty = 'n', cex = 1.2)

# Detection
plot(humidity[ooo], p.pred[ooo,1], xlab = 'Humidity index', ylab = 'Detection probability',
    frame = FALSE, col = 'blue', lwd = 3, main = 'Observation process', type = 'l', ylim = c(0, 1))
lines(humidity[ooo], p[ooo], lwd = 3, col = 'red')
polygon(c(humidity[ooo], rev(humidity[ooo])), c(p.LCL[ooo,1], rev(p.UCL[ooo,1])),
    col = rgb(0,0,1, 0.1), border = NA)
```

It is already clear that in our case we do much better with an occupancy model than with a detection-naïve SDM: we get the occupancy-humidity relationships right!

19.4 Bayesian analysis with JAGS

Next, we use JAGS to fit the occupancy model. As usual, notice the similarities between the BUGS code and the code we used to simulate the dataset. Also note the nested loop to define the likelihood in the BUGS language, since we have formatted our detection/nondetection response data as a matrix. At the end of the model code we've added a derived quantity, `occ.fs`, which represents the total number of occupied sites (i.e., the total number of sites where $z = 1$).

```
# Bundle and summarize data
str(dataList <- list(y = y, humidity = humidity, nSites = nSites, nVisits = nVisits) )

List of 4
 $ y        : int [1:150, 1:3]  0  0  1  0  1  0  0  0  0  0 ...
 $ humidity: num [1:150]  -0.7657  -0.0319  0.3024  -0.8632  -0.2694 ...
 $ nSites  : num 150
 $ nVisits : num 3

# Write JAGS model file
cat(file = "model19.4.txt", "
model {
# Priors
occ.int ~ dunif(0, 1)                       # Occupancy intercept on prob. scale
alpha.occ <- logit(occ.int)
beta.occ ~ dnorm(0, 0.0001)
p.int ~ dunif(0, 1)                         # Detection intercept on prob. scale
alpha.p <- logit(p.int)
beta.p ~ dnorm(0, 0.0001)

# Likelihood
for (i in 1:nSites) {                       # start initial loop over sites
  # True state model for the partially observed true state
  z[i] ~ dbern(psi[i])                      # True occupancy z at site i
  logit(psi[i]) <- alpha.occ + beta.occ * humidity[i]

  for (t in 1:nVisits) {                    # start a second loop over visits
    # Observation model for the actual observations
    y[i,t] ~ dbern(eff.p[i,t])              # Detection-nondetection at i and t
    eff.p[i,t] <- z[i] * p[i,t]
    logit(p[i,t]) <- alpha.p + beta.p * humidity[i]
  }
}

# Derived quantities
# Finite-sample inference on number occ. sites in our sample of 150
occ.fs <- sum(z[])
}
")
```

It is essential to choose initial values that do not induce a contradiction between the model and the data. Thus, we initialize the latent variables, or random effects, z (i.e., the latent presence/absence indicators) at the observed value in the data. An alternative might be to initialize them all at 1 (meaning an occupied site). Without that, JAGS will crash (you can try that out).

```
# Function to generate starting values
zst <- apply(y, 1, max)          # Get *observed* presence/absence status
inits <- function(){list(z = zst, occ.int = runif(1), beta.occ = runif(1, -3, 3),
  p.int = runif(1), beta.p = runif(1, -3, 3))}

# Parameters to estimate
params <- c("alpha.occ","beta.occ", "alpha.p", "beta.p", "occ.fs", "occ.int", "p.int")

# MCMC settings
na <- 5000  ;  ni <- 50000  ;  nb <- 10000  ;  nc <- 4  ;  nt <- 10
```

```
# Call JAGS (ART 60 sec), check convergence, summarize posteriors and save results
out19.4 <- jags(dataList, inits, params, "model19.4.txt", n.iter = ni, n.burnin = nb,
  n.chains = nc, n.thin = nt, n.adapt = na, parallel = TRUE)
jagsUI::traceplot(out19.4)                    # not shown
print(out19.4, 3)
jags_est <- out19.4$summary[1:4,1]
```

	mean	sd	2.5%	50%	97.5%	overlap0	f	Rhat	n.eff
alpha.occ	-0.140	0.376	-0.800	-0.167	0.676	TRUE	0.678	1	6185
beta.occ	1.694	0.791	0.303	1.641	3.402	FALSE	0.995	1	4846
alpha.p	0.154	0.270	-0.369	0.150	0.691	TRUE	0.714	1	16000
beta.p	-3.603	0.541	-4.658	-3.597	-2.560	FALSE	1.000	1	16000
occ.fs	73.326	10.983	52.000	74.000	94.000	FALSE	1.000	1	6416
occ.int	0.466	0.090	0.310	0.458	0.663	FALSE	1.000	1	6208
p.int	0.538	0.066	0.409	0.538	0.666	FALSE	1.000	1	16000

We compare the known true values from the data-generating process with what the occupancy analysis has recovered and those with the MLEs from unmarked.

```
# Compare likelihood with Bayesian estimates and with truth
comp <- cbind(truth = truth, unmarked = unm_est, JAGS = jags_est)
print(comp, 4)
```

	truth	unmarked	JAGS
alpha.occ	0	-0.2257	-0.1405
beta.occ	2	1.4894	1.6940
alpha.p	0	0.1750	0.1539
beta.p	-3	-3.5482	-3.6030

Note that what may look like a semi-substantial discrepancy on the logit link scale often amounts to much less of a discrepancy on the probability scale. Let's back-transform the occupancy and detection intercept estimates from unmarked and JAGS for comparison.

```
# Back-transform the intercept estimates
plogis(unm_est[c(1,3)])       # Estimates from unmarked (MLEs)
plogis(jags_est[c(1,3)])      # Estimates from JAGS (posterior means)

> plogis(unm_est[c(1,3)])     # Estimates from unmarked (MLEs)
psi(Int) p(Int)
0.4438091 0.5436481

> plogis(jags_est[c(1,3)])    # Estimates from JAGS (posterior means)
alpha.occ alpha.p
0.4649423 0.5383989
```

One more important estimated quantity for us to compare is the number of occupied sites. As a benchmark, we note that the species was actually observed at only 39 sites.

```
# Get estimate of the number of occupied sites
jags_Nz <- round(out19.4$mean$occ.fs,2)
comp <- c(truth = true_Nz, observed = obs_Nz, unmarked = unm_Nz, JAGS = jags_Nz)
print(comp, 2)
```

truth	observed	unmarked	JAGS
77	39	71	73

Thus, the occupancy SDM succeeds in recovering the true relationships between humidity and occupancy, and humidity and detection probability, which we built into the data. Particularly impressive is its ability to estimate the true number of occupied sites: gentians were only detected at 39 of the 77 sites where they actually occurred, and the occupancy model estimates this number at 73.3, with a 95% CRI of 52–94.

Finally, we will graphically compare the results of the detection-naïve and site-occupancy analyses with truth by plotting the true and estimated relationships between occupancy probability of the Chiltern gentian and site humidity (Fig. 19.5). For this, we first have to produce the predictions from the occupancy model. We could have done this inside of the model in JAGS, but here do it outside. This illustrates that we often can compute such derived quantities either directly inside of a JAGS run or outside in R, using MCMC draws produced by JAGS. We use again vector `ooo` which gives the order of the values in the `humidity` vector.

```
# Get predictions
str(post.draws <- out19.4$sims.list)           # Grab posterior draws and look
nsamp <- length(post.draws[[1]])               # Check number of posterior draws
pred.occ <- array(NA, dim = c(length(humidity), nsamp))
for(i in 1:length(humidity)){                  # Posterior predictive distribution
  pred.occ[i,] <- plogis(post.draws$alpha.occ + post.draws$beta.occ * humidity[i])
}
pm <- apply(pred.occ, 1, mean)                 # Posterior mean
CRI <- apply(pred.occ, 1,
   function(x) quantile(x, c(0.025, 0.975)))   # Central 95% percentiles
```

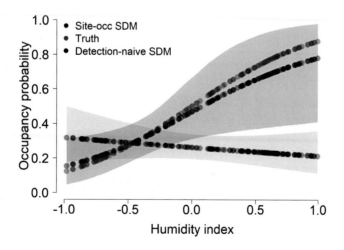

FIGURE 19.5

Comparison of true and estimated relationship between occupancy probability and humidity in the Chiltern gentian (*G. germanica*) under an occupancy model (blackish-blue) and under the naïve approach that ignores detection probability (black). Truth is shown in red, and shaded polygons show 95% prediction intervals.

```
# Fig. 19.5
par(mar = c(5,5,4,2), cex.lab = 1.5, cex.axis = 1.5)
plot(humidity, pred.naive, ylim = c(0, 1), xlab = "Humidity index",
  ylab = "Occupancy probability", main = "", las = 1, pch = 16, cex = 1.6,
  col = rgb(0, 0, 0, 0.4), frame = FALSE)                    # Detection-naive analysis
polygon(c(humidity[ooo], rev(humidity[ooo])), c(LCL.naive[ooo], rev(UCL.naive[ooo])),
  col = rgb(0, 0, 0, 0.2), border = NA)
points(humidity, occ.prob, pch = 16, cex = 1.6, col = rgb(1, 0, 0, 0.4)) # Truth
points(humidity, pm, pch = 16, cex = 1.6, col = rgb(0, 0, 1, 0.4)) # Esti
polygon(c(humidity[ooo], rev(humidity[ooo])), c(CRI[1,ooo], rev(CRI[2,ooo])),
  col = rgb(0, 0, 1, 0.2), border = NA)
legend('topleft', legend = c('Site-occ SDM', 'Truth', 'Detection-naive SDM'),
  pch = 16, col = c('blue', 'red', 'black'), bty = 'n', cex = 1.4)
```

It is evident from Fig. 19.5 that not accounting for detection in species distribution models may sometimes lead one astray quite spectacularly. However, some might argue that we have simulated a pathological case, and that one would rarely find such situation in nature. This may be true. However, we don't know whether this is so until we have conducted the right analysis. And, we know that even a constant detection probability < 1 not only biases the apparent occupancy probability in conventional models, but also biases low the strength of covariate relationships. Given suitable data the site-occupancy distribution model can correct for that. This should make this hierarchical model a serious candidate for species distribution modeling when replicate observations are available from at least some sites.

19.5 Bayesian analysis with NIMBLE

NIMBLE code for the model is available on the book website.

19.6 Bayesian analysis with Stan

With JAGS and NIMBLE, we worked with the hierarchical definition of the model likelihood, which includes the unobserved latent occupancy states z (the random effects). These latent occupancy states are discrete, that is, they can only take on values 0 and 1. As we noted in Section 12.3.5, Stan cannot sample discrete parameters. Therefore, for Stan we must define the integrated likelihood to fit this model, in which the z are integrated out. Integrated likelihood for models with discrete-valued parameters is also how we fit these models when using unmarked, DIY-MLE or TMB.

Our approach will be very similar to the one in Section 12.3.5. Again, we only have two possible states for z: 0 or 1. For each site i we will calculate the likelihoods corresponding to each state and add them together. For a site i where the species occurs, $z_i = 1$, the contribution to the likelihood is

$$\psi_i \cdot \text{dbinom}(nd_i | \text{T}, p_i)$$

The first part ψ_i corresponds to the state process and is the probability site i is occupied. The second part represents the detection process and is the likelihood we detect the species nd_i times out of T visits with a "coin flip" probability of p_i. Here nd_i is equal to the sum of the elements of the detection-nondetection vector y_i, also known as the detection frequency. This is equivalent to the total probability of getting nd_i "heads" from T individual Bernoulli trials each with probability

p_i. Using the binomial distribution instead of a series of Bernoulli trials simplifies the calculation. We are only able to make this simplification because, in this example, the probability of detection p_i at each site is constant across visits (i.e., we do not have any observation-level covariates).

For a site where the gentian is absent, $z_i = 0$, the contribution to the likelihood is:

$$(1-\psi_i) \cdot \text{dbinom}(0|T,p_i).$$

However, we can simplify this further: because the site is unoccupied, we cannot possibly detect the species and thus under our assumption of no false positives there cannot be any detections out of T visits (i.e., we *must* have $nd_i = 0$). This equation therefore simplifies to just $(1-\psi_i) \cdot 1$, or $(1-\psi_i)$.

For each site, the calculation of its contribution to the total likelihood of the dataset differs based on the observed data, specifically whether or not we never detected the species at site i (call this no_detect_i). If we never observe the species at site i ($no\_detect_i = 1$), then z_i could be either 0 or 1 and we must add the likelihoods for each possible state together to get the total likelihood for site i. On the other hand, if we observe the species at site i in at least one sampling occasion ($no\_detect_i = 0$), we *know* the site must be occupied ($z_i = 1$, again invoking the assumption that there are no false positives) and the likelihood corresponding to $z_i = 0$ must therefore be 0. Then the likelihood for site i becomes just the likelihood for $z_i = 1$. Here's the above expressed mathematically:

$$L(\psi_i, p_i | nd_i, T, no\_detect_i) = \begin{cases} \psi_i \cdot \text{dbinom}(nd_i|T,p_i), & no\_detect_i = 0 \\ \psi_i \cdot \text{dbinom}(nd_i|T,p_i) + (1-\psi_i), & no\_detect_i = 1 \end{cases}$$

This can be expressed in a single equation:

$$L(\psi_i, p_i | nd_i, J, no\_detect_i) = \psi_i \cdot \text{dbinom}(nd_i|T,p_i) + no\_detect_i \cdot (1-\psi_i)$$

Before writing the Stan code, we need to make some modifications to the input data. Specifically, we need to calculate the vector `nd`, which contains the total number of detections at each site, and `no_detect`, the vector indicating if each site had no detections.

```
# Bundle and summarize data
str(dataList <- list(y = y, humidity = humidity, nSites = nSites, nVisits = nVisits) )

# Add new variable indicating if there were 0 detections at a site
dataList$no_detect <- 1 - apply(dataList$y, 1, max)

# Add new variable for total # of detections at each site
dataList$nd <- apply(dataList$y, 1, sum)
str(dataList)

List of 6
 $ y         : int [1:150, 1:3]  0  0  1  0  1  0  0  0  0  0 ...
 $ humidity  : num [1:150]  -0.7657  -0.0319  0.3024  -0.8632  -0.2694 ...
 $ nSites    : num  150
 $ nVisits   : num  3
 $ no_detect : num [1:150]  1  1  0  1  0  1  1  0  1  1 ...
 $ nd        : int [1:150]  0  0  2  0  3  0  0  2  0  0 ...
```

In the `model` section below, we convert our final likelihood equation into Stan code. Note that Stan's binomial distribution function calculates the value on the log scale, so we need to exponentiate it. We then add the log of the total likelihood for each site to `target` as we did in Section 12.3.5. In the `generated quantities` section, we calculate the expected number of occupied sites. This is much more difficult in Stan than with BUGS; see https://github.com/stan-dev/example-models/issues/99 for a discussion of this code, written by our late colleague Mike Meredith.

```
# Write Stan model
cat(file = "model19_6.stan", "

data{
  int nSites;
  int nVisits;
  array[nSites, nVisits] int y;
  vector[nSites] humidity;
  array[nSites] int no_detect;
  array[nSites] int nd;
}

parameters{
  real occ_int;
  real beta_occ;
  real p_int;
  real beta_p;
}

transformed parameters{
  vector[nSites] psi;
  vector[nSites] p;

  real alpha_occ = logit(occ_int);
  real alpha_p = logit(p_int);

  for (i in 1:nSites){
    psi[i] = inv_logit(alpha_occ + beta_occ * humidity[i]);
    p[i] = inv_logit(alpha_p + beta_p * humidity[i]);
  }
}

model{
  vector[nSites] lik;                //Likelihood for each site

  //priors
  occ_int ~ uniform(0, 1);
  beta_occ ~ normal(0, 100);
  p_int ~ uniform(0, 1);
  beta_p ~ normal(0, 100);

  for (i in 1:nSites){
    //Calculate occupancy likelihood
    lik[i] = exp(binomial_lpmf(nd[i] | nVisits, p[i])) * psi[i] +
             no_detect[i] * (1-psi[i]);
    target += log(lik[i]);           //Add to total likelihood
  }
}

generated quantities{
  real qT;
  vector[nSites] psi_con;            //P(z=1 | Number of detects = 0)
  array[nSites] int zpost;
  real occ_fs;                       //Number of occupied sites

  for (i in 1:nSites){
    if(no_detect[i] == 0){
      zpost[i] = 1;                  //If species was detected at least once, z=1
    } else {
      qT = pow(1-p[i], nVisits);     //P(never detected | z = 1)
      psi_con[i] = psi[i] * qT/(psi[i] * qT + (1-psi[i]));
      zpost[i] = bernoulli_rng(psi_con[i]);
    }
  }
  occ_fs = sum(zpost);
}
")
```

```
# Inits
inits <- function(){ list(occ_int = runif(1,0,1), beta_occ = runif(1, -5, 5),
   p_int = runif(1,0,1), beta_p = runif(1, -5, 5))}

# Parameters to estimate
params <- c("alpha_occ", "beta_occ", "alpha_p", "beta_p", "occ_fs")

# HMC settings
ni <- 3000   ;   nb <- 1000   ;   nc <- 4   ;   nt <- 1

# Call STAN (ART 75/26 sec), assess convergence and print results table
system.time(
out19.6 <- stan(file = "model19_6.stan", data = dataList, pars = params,
             warmup = nb, iter = ni, chains = nc, thin = nt,
             init = inits, control = list(adapt_delta = 0.99)))
rstan::traceplot(out19.6)                       # not shown
print(out19.6, dig = 3)                         # not shown
stan_est <- summary(out19.6)$summary[1:4,1] # Save estimates
stan_Nz <- round(summary(out19.6)$summary[5,1],2)
```

19.7 Bayesian analysis with canned functions in the R package ubms

The ubms package (Kellner et al., 2022a) can fit a variety of hierarchical models (Royle & Dorazio, 2008), including occupancy models. The interface is conveniently designed to be almost identical to unmarked, but the models are fit with Bayesian inference using Stan instead of maximum likelihood. Leave-one-out cross-validation scores (Vehtari et al., 2017) are available for model selection for each supported model (see Chapter 18). We can use the same input dataset as we did with our earlier unmarked analysis.

```
library(ubms)   # will also load unmarked as a dependency

# Format data and summarize data (exactly the same as in Section 19.3)
summary( umf <- unmarkedFrameOccu(y = y, siteCovs = data.frame(humidity = humidity)) )
```

Single-season, or static, occupancy models are fit in ubms using the function stan_occu(). The arguments are almost identical to our previous analysis in unmarked. We will manually set some priors in the arguments so they more closely match those in our previous engines.

```
# Fit model in parallel
options(mc.cores = 4)          # number of parallel cores to use
system.time(                   # ART 28 sec
  out19.7 <- stan_occu(~ humidity ~ humidity, umf, chains = 4,
                  prior_coef_state = normal(0, 10),
                  prior_coef_det = normal(0, 10))
)
out19.7
ubms_est <- coef(out19.7)   # Save estimates

Call:
stan_occu(formula = ~humidity ~ humidity, data = umf, prior_coef_state = normal(0,10),
  prior_coef_det = normal(0, 10), chains = 4)
```

```
Occupancy (logit-scale):
              Estimate     SD    2.5%    97.5%   n_eff   Rhat
(Intercept)     -0.148   0.363  -0.791   0.642    1797     1
humidity         1.683   0.767   0.336   3.354    1837     1


Detection (logit-scale):
              Estimate     SD    2.5%    97.5%   n_eff   Rhat
(Intercept)      0.153   0.259  -0.347   0.672    2158     1
humidity        -3.610   0.530  -4.700  -2.576    2400     1

LOOIC: 292.464
```

To compute the sum of the realized values of the random effects z, that is, to obtain an estimate of the actual number of sites occupied in our sample of 150, ubms uses similar calculations under the hood as what we did when using Stan (see Section 19.6).

```
# Get estimate of the number of occupied sites and save
postz_ubms <- posterior_predict(out19.7, "z")   # Get posterior of latent z
post_Nz_ubms <- apply(postz_ubms, 1, sum)        # Calculate sumN for each draw
ubms_Nz <- mean(post_Nz_ubms)
```

19.8 Do-it-yourself maximum likelihood estimates

Let's now see what goes on under the hood of unmarked, which uses maximum likelihood for fitting this model. In fact, we have already done the hard math work for this in Section 19.6—we just need to translate the integrated likelihood code from the model in Stan into an R likelihood function. We'll use the same list of data as in Section 19.6.

```
# Same data as section 19.6
str(dataList)   # not shown
```

Here's the likelihood function, which looks nearly identical to the Stan code in its model section.

```
# Definition of NLL for a simple static site-occupancy model
# with one single site covariate humidity and fit to detection frequency data
NLL <- function(param, data) {
  alpha.lpsi <- param[1]              # Occupancy intercept (logit scale)
  beta.lpsi <- param[2]               # Occupancy slope on humidity
  alpha.lp <- param[3]                # Detection intercept (logit scale)
  beta.lp <- param[4]                 # Detection slope on humidity
  # Linear predictor for occupancy
  psi <- plogis(alpha.lpsi + beta.lpsi * data$humidity)
  # Linear predictor for detection
  p <- plogis(alpha.lp + beta.lp * data$humidity)
  L <- numeric(data$nSites)
  for (i in 1:data$nSites){
    # Likelihood contribution for 1 observation
    L[i] <- psi[i] * dbinom(data$nd[i], data$nVisits, p[i]) +
          data$no_detect[i] * (1-psi[i])
  }
  LL <- log(L)                        # Log-likelihood contr. for each observation
  NLL <- -sum(LL)                     # NLL for all observations
  return(NLL)
}
```

```
# Minimize that NLL to find MLEs and also get SEs
inits <- rep(0, 4)
names(inits) <- names(truth)
out19.8 <- optim(inits, NLL, data = dataList, hessian = TRUE, method = "BFGS")
get_MLE(out19.8, 5)
diy_est <- out19.8$par                          # Save estimates
```

```
                MLE       ASE       LCL.95      UCL.95
alpha.occ   -0.22572   0.36331   -0.9377996    0.48637
beta.occ     1.48939   0.75656    0.0065337    2.97224
alpha.p      0.17504   0.27308   -0.3601959    0.71027
beta.p      -3.54823   0.54092   -4.6084275   -2.48803
```

Excitingly, though perhaps not *entirely* surprisingly, these look exactly like the estimates provided by unmarked.

19.9 Likelihood analysis with TMB

Once again we can use the same data as we did with Stan.

```
# Same data as section 19.6
str(dataList)    # not shown
```

The TMB model looks nearly identical to our DIY likelihood in the previous section.

```
# Write TMB model file
cat(file = "model19_9.cpp",
"#include <TMB.hpp>

template<class Type>
Type objective_function<Type>::operator() ()
{
  //Describe input data
  DATA_VECTOR(nd);
  DATA_VECTOR(no_detect);
  DATA_VECTOR(humidity);
  DATA_INTEGER(nSites);
  DATA_INTEGER(nVisits);

  //Describe parameters
  PARAMETER(alpha_occ);
  PARAMETER(beta_occ);
  PARAMETER(alpha_p);
  PARAMETER(beta_p);

  Type LL = 0.0;                                //Initialize log-likelihood at 0

  vector<Type> p(nSites);
  vector<Type> psi(nSites);
  vector<Type> L(nSites);

  for (int i=0; i<nSites; i++){
    psi(i) = invlogit(alpha_occ + beta_occ * humidity(i));
    p(i) = invlogit(alpha_p + beta_p * humidity(i));

    L(i) = dbinom(nd(i), Type(nVisits), p(i), false) * psi(i) +
           no_detect(i) * (1 - psi[i]);
    LL += log(L(i));
  }

  return -LL;
}
")
```

```
# Compile and load TMB function
compile("model19_9.cpp")
dyn.load(dynlib("model19_9"))

# Provide dimensions and starting values for parameters
params <- list(alpha_occ = 0, beta_occ = 0, alpha_p = 0, beta_p = 0)

# Create TMB object
out19.9 <- MakeADFun(data = dataList, parameters = params,
                     DLL = "model19_9", silent = TRUE)

# Optimize TMB object and print results
starts <- rep(0, 4)
opt <- optim(starts, fn = out19.9$fn, gr = out19.9$gr, method = "BFGS")
(tsum <- tmb_summary(out19.9))
tmb_est <- tsum[1:4,1]                    # Save estimates
```

To estimate the number of occupied sites in the actual sample of 150 sites in our study, we can use a bootstrap. This code is very similar to that in the `generated quantities` section of the Stan model in Section 19.6, just translated to R.

```
# Estimate number of occupied sites with a bootstrap

# Generate samples of parameter vector
nsims <- 10000
library(MASS)# for mvrnorm()
sdr <- sdreport(out19.9)
Beta <- sdr$par.fixed
Sigma <- sdr$cov.fixed
param_samples <- mvrnorm(nsims, Beta, Sigma)

# Calculate estimated sum of sites occupied for each parameter vector sample
# conditional on observed data
occ_fs_sims <- rep(NA, nsims)
for (i in 1:nsims){
  beta <- param_samples[i,]
  psi <- plogis(beta[1] + beta[2] * dataList$humidity)
  p <- plogis(beta[3] + beta[4] * dataList$humidity)
  qT <- (1-p)^dataList$nVisits

  zpost <- rep(NA, dataList$nSites)
  for (j in 1:dataList$nSites){
    if(dataList$no_detect[j] == 0){ #If detected at least once at site j
      zpost[j] <- 1
    } else{ # If never detected at site j
      psi_con <- psi[j] * qT[j]/(psi[j] * qT[j] + (1-psi[j]))
      zpost[j] <- rbinom(1, 1, psi_con)
    }
  }
  occ_fs_sims[i] <- sum(zpost)
}
```

```
# Plot distribution of bootstrap samples and compare truth and estimate (not shown)
hist(occ_fs_sims, main = "Distribution of occ_fs", xlab = "occ_fs")
abline(v = mean(occ_fs_sims), col = 'blue', lwd = 4)
abline(v = true_Nz, col = 'red', lwd = 4)
legend("topleft", legend = c("Estimate", "Truth"), col = c("blue","red"), lwd = 3,
bty = 'n')
tmb_Nz <- mean(occ_fs_sims)
```

19.10 **Comparison of the parameter estimates**

We conduct our usual grand comparison, here with a set of estimates from one further engine (from the Stan wrapper package ubms; Kellner et al., 2022a).

```
# Compare results with truth and previous estimates
comp <- cbind(truth = truth, unmarked = unm_est, JAGS = jags_est,
    NIMBLE = nimble_est, Stan = stan_est, ubms = ubms_est, DIY = diy_est, TMB = tmb_est)
print(comp, 3)
```

	truth	unmarked	JAGS	NIMBLE	Stan	ubms	DIY	TMB
alpha.occ	0	-0.226	-0.140	-0.122	-0.13	-0.148	-0.226	-0.226
beta.occ	2	1.489	1.694	1.736	1.72	1.683	1.489	1.489
alpha.p	0	0.175	0.154	0.150	0.15	0.153	0.175	0.175
beta.p	-3	-3.548	-3.603	-3.619	-3.60	-3.610	-3.548	-3.548

```
# Compare estimates of the number of occupied sites
comp <- c(truth = true_Nz, observed = obs_Nz, unmarked = unm_Nz, JAGS = jags_Nz,
    NIMBLE = nimble_Nz, Stan = stan_Nz, ubms = ubms_Nz, TMB = tmb_Nz)
print(comp, 2)
```

truth	observed	unmarked	JAGS	NIMBLE	Stan	ubms	TMB
77	39	71	73	74	74	73	71

Overall, we see the good agreement between the point estimates across engines that we have come to expect by now. Remember that what may appear to be a larger difference on the logit scale typically reduces to fairly small differences on the probability scale.

In addition, we highlight the power of occupancy species distribution models to tease apart patterns in what is usually our real interest (the true species occupancy probability) from what is usually just a nuisance parameter (detection probability). In our simulated dataset there was a slight (though not significant) negative relationship between the *observed* gentian occupancy (i.e., the detection/nondetection data) and the humidity at a site. This pattern was entirely spurious and due to a failure to account for imperfect detection. In contrast, the occupancy model was able to detect that this was an artifact due to a negative relationship between detection and humidity. In addition, the occupancy model did a surprisingly good job at estimating the actual number of sites in our sample of 150 that were actually occupied by the Chiltern gentian.

19.11 **Summary and outlook**

The occupancy model that we covered in this chapter is an extended logistic regression that can estimate true occupancy probability (ψ) and the factors affecting it, while correcting for imperfect detection. The extension is represented by the model component for detection probability (p): conventional logistic regression is a special case when $p = 1$. Occupancy models are currently the only framework for inference about species distributions that model true rather than apparent distributions; the latter is the product of occupancy and detection probability (i.e., $\psi \cdot p$; Kéry, 2011). Our example shows that not accounting for detection probability may lead to erroneous inferences about the distribution of a species under a conventional, detection-naïve SDM, especially when both the state (true presence/absence) and the observation processes (represented by p) are affected by the same covariates. In contrast, the occupancy model applied to the replicated detection/nondetection data was far better able to estimate the true system state, i.e., site occupancy along with the covariate relationship and the true number of occupied sites.

In practice, this very positive conclusion about the model needs to be moderated somewhat. Performance of the model will not be quite as good when there is less information (e.g., fewer sites, a smaller proportion of occupied sites, lower detection probability, or fewer replicate visits) and also in the presence of important unmeasured covariates (Welsh et al., 2013; Guillera-Arroita et al., 2014). Remember that our comparison presents the best-case scenario where the data-generating and the data-analysis models match exactly. In reality, this will never be the case (all models are approximations!) and therefore results will often not be as excellent as the ones we present here. Hence, for suitable data, occupancy models are great, but as model complexity increases, simulations to check the quality of the inference may be useful.

In statistical terms, occupancy models can also be described as a special kind of random-effects model, where unlike the mixed models described in this book so far, the random effects are discrete, don't appear in the linear predictor of a GLM in an additive manner, and are not given a normal distribution. Rather, they are binary and are therefore given a Bernoulli distribution. This serves as an important lesson: traditional mixed models as we may fit them with functions in R packages such as `lme4` (Bates et al., 2014), `glmmTMB` (Brooks et al., 2017), or `MCMCglmm` (Hadfield, 2010) are just a small subset of the far more general class of hierarchical or random-effects models (Berliner, 1996; Royle & Dorazio, 2008; Cressie et al., 2009). There may be many instances in your work as an applied ecological modeler where you will need to use such general hierarchical models. Indeed, all ecological models for distribution, abundance, and demographic rates (such as survival) that explicitly model false-positive or false-negative measurement error processes will require the modeling of discrete random effects. Thus, they will require you to go beyond the much simpler mixed-model type of hierarchical models. Examples of such general hierarchical models (or models that can be represented as general hierarchical models) are featured extensively in textbooks including Buckland et al. (2001, 2004, 2015); Borchers et al. (2002); Williams et al. (2002); Royle & Dorazio (2008); King et al. (2009); Kéry & Schaub (2012); McCrea & Morgan (2014); Kéry & Royle (2016, 2021); MacKenzie et al. (2017); Seber & Schofield (2019, 2024); Schaub & Kéry (2022).

In this chapter, we have revisited the topic of an integrated, or marginal, likelihood, where the discrete random effects (here, the z's) are eliminated by summation (and for real-valued random effects we would use integration; see Chapter 2 in Kéry & Royle (2016) as well as Chapters 10, 14 and 17 in this book). With maximum likelihood inference or Bayesian inference using Hamiltonian Monte Carlo (HMC; e.g., in Stan, `ubms` and also optionally in NIMBLE), the likelihood will have to be specified in this manner. Compared with the much more intuitive hierarchical specification of the likelihood that we worked with for JAGS and NIMBLE, the integrated likelihood is a bit of a pain in the backside for many non-statisticians. However, it may well be worth your effort of learning this, since you may also fit these models with the integrated likelihood in JAGS and NIMBLE. That may greatly speed up run times of a model run times, since the discrete random effects no longer need to be estimated by the MCMC algorithms (Joseph, 2020b; Yackulic et al., 2020; Turek et al., 2021).

An important conceptual difference between mixed models and what here we call a general hierarchical model is that in virtually all mixed models, the random effects are merely statistical constructs that serve to accommodate correlations in the data; they don't have any physical meaning. In contrast, in an occupancy model, the binary random effects have the very important interpretation as an indicator of presence and absence of a species at a site. Similarly, in the related N-mixture model (Royle, 2004), which we introduce in the bonus Chapter 19B (available on the book website), you will see another non-negative discrete random variable that will be modeled as a random effect. In the context of that model, these random effects will be given a Poisson (or another discrete) distribution and they will have the eminently tangible meaning of the abundance of the modeled species at each site. Owing to the explicit meaning of the random effects in these types of models, Royle & Dorazio (2008) called these more general hierarchical models "explicit" hierarchical models, to distinguish them from "implicit" hierarchical models, where the random effects have no real-world interpretation and in a way are simply a modeling trick or indeed are nuisance parameters that accommodate correlations in the data.

Occupancy models were independently discovered just over 20 years ago by two research groups (MacKenzie et al., 2002; Tyre et al., 2003). Since then, they have experienced vigorous development with many extensions and generalizations to the simplest model described in this chapter. Some of the more major of these extensions include:

- open populations, range dynamics (MacKenzie et al., 2003; Royle & Kéry, 2007),
- false-positives in addition to false-negatives (Royle & Link, 2006; Miller et al., 2011; Chambert et al., 2015),
- simultaneous distribution modeling of many species (MacKenzie et al., 2004; Rota et al., 2016; Kellner et al., 2022b) and communities (Dorazio & Royle, 2005; Gelfand et al., 2005),
- Modeling of species interactions (MacKenzie et al., 2004; Rota et al., 2016; Doser et al., 2022; Kellner, Fowler et al., 2022),
- multiple states of occupancy (e.g., "occupied by a pair without breeding" versus "occupied by a breeding pair"; Royle & Link (2005), Nichols et al. (2007), MacKenzie et al. (2009)),
- model-based links between presence/absence an underlying abundance process (Royle & Nichols, 2003),
- multiple spatial or temporal scales in occupancy (Mordecai et al., 2011; Nichols et al., 2008),
- occupancy models with spatial autocorrelation (Johnson et al., 2013; Doser et al., 2022; Hepler et al., 2022; Mohankumar & Hefley, 2022),
- model-based link between presence/absence and an underlying point process (Dorazio, 2014),
- various integrated models that combine detection/nondetection data with other data types that contain information about variation in distribution or abundance (Dorazio, 2014; Koshkina et al., 2017; Pacifici et al., 2017); see also Chapter 20 that follows.

Most of these newer developments are described in the recent syntheses by MacKenzie et al. (2017) and Kéry & Royle (2016, 2021), and many of them can be fit with canned functions in MARK (White & Burnham, 1999), PRESENCE (Hines, 2006), `unmarked` (Kellner et al., 2023), `ubms` (Kellner et al., 2022a), `spOccupancy` (Doser et al., 2022), and `spAbundance` (Doser et al., 2024). And of course you can fit all of these models in our general model-fitting engines featured in this book if you learn how to use them.

Integrated models⊛

20

Chapter outline

20.1 Introduction

Throughout this book, we have stressed the role of the generalized linear model (GLM) as a fundamental building block in parametric statistical modeling. Sometimes a modeling problem may be simple enough so that a GLM may be what you need. However, often we want to accommodate additional latent structure in the data and for this invoke latent variables or random effects, either in the simpler mixed models of Chapters 7, 10, 14, and 17 or in the form of a more general hierarchical model as for instance in Chapters 19 and 19B. Either type of hierarchical model can be viewed as the combination of two or more GLMs that are linked in sequence. That is, the latent response of one random process appears in the model for another response which may be latent or observed, and these subprocesses are organized according to conditional probability (Royle & Dorazio, 2008; Cressie et al., 2009; Hobbs & Hooten, 2015; Hooten & Hefley, 2019).

In this final modeling chapter in the book, we show another way in which we may combine multiple GLMs to build a more complex model. In this class of model, the component GLMs are not necessarily arrayed in sequence, as in a hierarchical model, but rather branch off from some shared underlying component. These are models that combine the information contained in multiple and disparate data sets. We call

---

⊛This book has a companion website hosting complementary materials, including all code for download. Visit this URL to access it: https://www.elsevier.com/books-and-journals/book-companion/9780443137150.

Applied Statistical Modelling for Ecologists. DOI: https://doi.org/10.1016/B978-0-443-13715-0.00001-7

them integrated models (IMs), because they integrate (or combine, fuse, assimilate) the information in multiple data sets. In contrast to a repeated-measures design as with an occupancy- or N-mixture model, in an IM there is always some degree of disparity in the different modeled data. That is, we have multiple data sets that each contain information about a shared process with one or more shared parameters. However, at first it may not be evident how to formally combine them in a single model. Our task then in developing an IM is (1) to identify the shared process and shared parameters and (2) write a model, or develop a likelihood, that links these shared parameters with the different types of observations in each data set.

Most of the time in an IM, one or more parameters in the latent state process are shared, and then we define different observation models for each data set. "Observation model" here must be understood in a broad sense. For instance, the underlying state may be population abundance, and the different data sets may be produced by a distance-sampling, removal-sampling, or capture-recapture protocol (Royle & Dorazio, 2008), at different spatial or temporal observation scales (requiring change-of-support modeling, Pacifici et al., 2019), or by various truncation, censoring, or aggregation processes.

IMs have become a mega-trend in ecological statistics during the past 30 years, e.g., in demographic estimation, population modeling, species distribution modeling, or movement modeling; for reviews, see Schaub & Abadi (2011), Maunder & Punt (2013), Zipkin & Saunders (2018), Miller et al. (2019), chapter 10 in Kéry & Royle (2021), and Schaub & Kéry (2022). The principles of IMs in these different subfields are very similar, but this is arguably not as widely understood as it should be, and IMs in each field are often viewed as something separate. Probably the most widely popularized kind of IM in ecology has been *integrated population models* (IPMs; Besbeas et al., 2002; Schaub & Kéry, 2022; Schaub et al., 2024). However, IPMs are really just a special case of IMs in general, which all share the same basic building principles.

Fig. 20.1 shows the typical case of an integrated species distribution model (SDM) (IM$_1$) and of an IPM (IM$_2$). In the former, we typically have a single shared parameter (θ), e.g., an abundance intensity in a point process model (Dorazio, 2014; Koshkina et al., 2017), expected abundance (Zipkin et al., 2017), or occupancy probability (Landau et al., 2022). The typical setting of an IPM is where we have a time series of population counts (data1), which contains information about all components of population dynamics including initial density (λ), survival (ϕ), and recruitment (γ), and then add another data set, for example, a capture–recapture data set, which contains only information about the survival (ϕ) part of population dynamics (Besbeas et al., 2002).

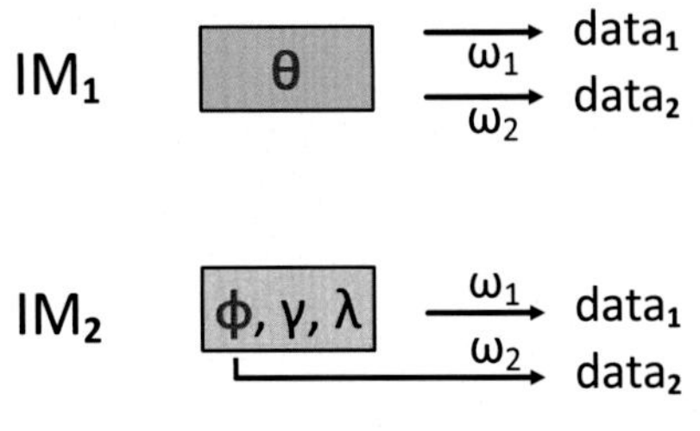

FIGURE 20.1

Schema of two kinds of integrated models where two disparate data sets inform about the same (or part of the same) underlying process (gray box) with shared parameter(s) shown in red (In IM$_1$, both data sets are informative about the same parameter(s) in the shared process, while in IM$_2$ data set 1 contains information about all three parameters, but data set 2 only about ϕ, which is the shared parameter. The arrows each denote a data-set-specific observation process, or likelihood, with parameters that are usually not shared (modified from Fig. 10.1 in Kéry & Royle, 2021).

In the context of SDMs, Pacifici et al. (2017) have discussed three different ways in which the information in different data sets may be combined and which they call the "covariate," "correlation," and "shared" approaches. The last one is what we here call an IM, and it consists in defining a joint likelihood for all parameters that govern the probability densities of all data sets. Under the simplifying assumption of statistical independence among the data sets, this joint likelihood is simply the product of the likelihoods of each individual data set, exactly analogous to how the likelihood of all the data points in a single data set is, under independence, just the product of the densities of each data point, as we have seen throughout the book. Maximizing the joint likelihood then identifies parameter values that are most likely for all data sets simultaneously.

Why should we build IMs? There are three good reasons for this: (1) It just makes tremendous sense to use all the information that is available in some estimation task. (2) More data sets can be viewed as akin to a larger sample size. Thus, it will not come as a surprise that parameter estimates in an IM are usually more precise than those obtained from each component data set individually. (3) Depending on the model and the kinds of data that are combined, parameters may become estimable in the IM which may not be identifiable from each data set alone (Schaub & Kéry, 2022).

But should we always build an IM when we have several data sets that we feel contain information about something shared? Our answer here would be, as so often, that it depends. Most of the times we would probably try to exploit the information of all data sets in an IM, but not always. For instance, in the context of an IPM, if you're most interested just in the population trends, then a simpler model applied only to the population count data may be more what you want. In contrast, if you're most interested in obtaining estimates of survival rates, then a Cormack-Jolly-Seber model (chapter 7 in Kéry & Schaub, 2012) applied to the capture–recapture data alone may be most appropriate for you. Building an IM incurs a cost: more parameters must be included in the model (typically in the observation models for each data set) and if the information about the target parameter is weak in an additional data set, then the inclusion of that data set in an IM may not be warranted.

Our goal in the rest of this chapter will be to examine the key features of an IM and to understand the principles of how we can build such a model. For this, we will develop an IM which is a very simple example of an SDM. We assume we have three data sets that contain information about spatial patterns in the distribution or abundance of common swifts (Fig. 20.2) in some region. Data set 1 comprises regular counts, but the other two data sets suffer from some loss of information relative to regular counts. In data set 2, we assume that the lazy bird-watchers had "forgotten" to record the nondetections and only recorded counts equal to 1 or greater. Such data are called zero truncated (ZT), since no zeroes occur in the data set. In data set 3, we only distinguish between a count of 0 (a nondetection) and a count of 1 or greater (i.e., a detection). In other words, these are what many people call "presence/absence data" and, statistically speaking, can also be called counts that are censored at 1. This chapter builds partly on Section 2.8 in Schaub & Kéry (2022).

Normally, our preference would be for an SDM that contains an explicit representation of the false-negative errors that always occur when collecting such data in the field. That is, for a type of occupancy or N-mixture model as shown in Chapters 19 and 19B. However, to introduce the ideas of IMs in a setting as simple as possible, we here ignore this complication of reality and assume that either detection probability is perfect or else that it does not covary with our comparison of interest, which will be an elevation gradient in abundance, and that we are happy with inferences about apparent or relative abundance.

FIGURE 20.2

A flock of common swifts (*Apus apus*) above the village of Glovelier, Swiss Jura mountains (Photo by Alain Georgy).

We will also make another assumption: that the three data sets we work with are statistically independent. Independence in IMs is a somewhat elusive concept. It has often been defined as meaning that no (or only few) individuals, or sites, are shared among the data sets to be combined in an IM (Besbeas et al., 2009; Abadi et al., 2010; Schaub & Kéry, 2022). Such "lack of sampling overlap" may often be a suitable indicator for statistical independence of the data sets, but may not be enough. Statistically speaking, independence is defined such that the joint density of some data sets is given by the product of the (joint) densities of the individual data sets (Blitzstein & Hwang, 2019). Independence in the context of an IM, the consequences of its violation, and the proper accommodation of nonindependence are areas of ongoing research (Besbeas et al., 2009; Abadi et al., 2010; Plard et al., 2021; Weegman et al., 2021; Schaub & Kéry, 2022; Barraquand et al., 2024). Here, we will assume that the three sets contain distinct sets of sites and thus will index sites with i, j, and k below. However, if there was some or even complete overlap among sites, then this could also be accommodated in an IM.

As a starting point, we might think that natural probability models for data sets 1 and 2 would be a Poisson GLM and a Bernoulli GLM with logit link (i.e., a logistic regression) for data set 3. Thus, in data set 1, for the count $y_i^{(1)}$ at site i, and where we indicate the data set by a superscript, we could write:

$$y_i^{(1)} \sim Poisson(\lambda_i)$$
$$\log(\lambda_i) = \alpha^{(1)} + \beta^{(1)} \cdot elev_i$$

Here, the elevation of site i affects the expected, or mean, count λ through a log-linear model with the two parameters $\alpha^{(1)}$ and $\beta^{(1)}$. Note that we like to simulate the effects of a covariate in our models in this chapter, but the data integration in this chapter would work just as well with a simpler intercept-only model.

As for data set 2, we might want to fit the same model also to the zero-truncated counts. But this would be wrong, since this would fail to account for the fact that zeros cannot be observed from the process that produced these data. Thus, the parameters estimated in such a naive application of a Poisson GLM to the zero-truncated data would not be comparable to the parameters estimated from the model for the "intact" data set 1. A proper probability model for the ZT Poisson data must recognize that zero is not in the sample space of the random experiment that produces data set 2. Thus, the proper probability mass function for the ZT data set 2 is a Poisson probability mass function (PMF) for counts >0 that we must renormalize using the sum of the probability of all possible outcomes, that is, of the nonzero counts only. Doing this ensures that the PMF sums to 1 over all possible outcomes, as required for a valid PMF. Once we do this, we can again go on to model the expected zero-truncated count with the same linear predictor as in the model for data set 1.

Let's look at this in algebra. The regular Poisson PMF is $p(y|\lambda) = \lambda^y e^{-\lambda}/y!$ (see Section 2.2.2.). The probability for a zero observation is $p(0|\lambda) = \lambda^0 e^{-\lambda}/0!$ which simplifies to $p(y=0|\lambda) = e^{-\lambda}$. Thus, the probability of a nonzero observation is $p(y>0|\lambda) = 1 - e^{-\lambda}$. Therefore, the PMF of the ZT Poisson is $p(y|\lambda) = \lambda^y e^{-\lambda}/((1-e^{-\lambda})y!)$. Below, we will implement this custom PMF for our do-it-yourself (DIY)-MLEs.

For data set 3, the binary detection/nondetection data $y_k^{(3)}$ at site k, we might write:

$$y_k^{(3)} \sim Bernoulli\left(\psi_k\right)$$

$$logit\left(\psi_k\right) = \alpha^{(3)} + \beta^{(3)} \cdot elev_k$$

Here, we model occurrence, or relative occupancy, probability on the logit scale as a linear function of site elevation, with parameters $\alpha^{(3)}$ and $\beta^{(3)}$.

Both the Poisson GLMs (with and without zero truncation) and the Bernoulli GLM offer a valid characterization of the spatial variation of the distribution of the study species: one in terms of relative abundance here denoted λ (Johnson, 2008) and the other in terms of relative occupancy probability here denoted ψ (Kéry, 2011). Note that the terms "relative" or "apparent" mean that true abundance or occurrence is confounded with imperfect detection, as described above. Intuitively, it seems obvious that patterns of occurrence also contain a signal of the underlying patterns of abundance. But would there be a possibility of combining the two models, for abundance and for occurrence, and thereby make use of all information about abundance that our data sets contain?

Key to data integration using a joint likelihood is always that you must describe the data sets using some "common currency." That is, with at least one parameter that is shared among two or more of the data sets combined. In our case, we can recognize that the detection/nondetection data are simply an information-poor variant of counts: they only distinguish between a count of 0 and one that is 1 or greater, or they are censored at 1. Thus, we can express the detection/nondetection data as a summary of an underlying abundance process. There are multiple ways in which such a link between abundance and the binary detection/nondetection data can be formulated (Royle & Dorazio, 2008). Here, we specify the linkage via that vastly underused link function for Bernoulli or binomial data: the complementary log-log or cloglog link (Section 3.3.6 in Kéry & Royle, 2016; Scharf et al., 2022). As we will see, by specifying that link function for the apparent occurrence probability ψ in the model for the binary detection/nondetection data, we describe these binary data in terms of an underlying Poisson abundance process. Thus, the parameters in the linear model for ψ on the cloglog scale will describe the log-linear covariate relationship for the expectation of an assumed Poisson random variable that underlies the binary data.

Does this sound like magic? Well, we think it does a little ... but let's look at this with algebra. When we model the binary data as a summary of an underlying abundance or count, then the Bernoulli parameter is the probability to obtain a count greater than 0 under a Poisson abundance model. From the Poisson PMF, we can express this as

$$\psi = P(y = 1) = P(N > 0) = 1 - P(N = 0) = 1 - \exp(-\lambda).$$

We could specify this relationship simply as a link function in a GLM, but for your understanding it is useful to derive the complementary log-log link by hand. Specification of a linear model for a Bernoulli parameter at the scale of a cloglog link transformation, that is, $\mathrm{cloglog}(\psi) = \log(-\log(1 - \psi))$, is equivalent to modeling linear effects in $\log(\lambda)$, which we can see by making the following re-arrangements:

$$\psi = 1 - \exp(-\lambda), \text{ and therefore}$$
$$\exp(-\lambda) = 1 - \psi, \text{ therefore}$$
$$\log(\exp(-\lambda)) = \log(1 - \psi), \text{ therefore}$$
$$-\lambda = \log(1 - \psi), \text{ therefore}$$
$$\lambda = -\log(1 - \psi), \text{ and therefore we finally get this:}$$
$$\log(\lambda) = \log(-\log(1 - \psi)).$$

What we have on the right-hand side of the last line is the cloglog transformation for occupancy probability ψ.

Thus, for our three SDM data sets we have achieved the two typical tasks in integrated modeling: (1) we have identified abundance (λ) as a shared process underlying all three, and (2) we have chosen observation models for the three such that the response data in each are formally linked to this underlying, shared process.

We will simulate three data sets as just described by first simulating three sets of Poisson random variables. The first one will be left unchanged and will be our data set 1. In the second set, we will toss out all zeros and the remainder will be our data set 2. Finally, we quantize (or censor at 1) the third set resulting in detection/nondetection data; this will become our data set 3.

Then, we will do what is always a wise approach in an IM: fit the component models first to each data set alone. We will do this using canned functions (when available) in R, with a DIY likelihood function that we maximize with `optim()`, and with JAGS for Bayesian posterior inference. This will help us understand the full model, since to understand an IM, you must also understand all of its component models. In addition, since it is so easy to make coding mistakes, it is always good to build up more complex code from smaller sets of code that you think work correctly. Finally, strong changes in parameter estimates between the simple and the IM may indicate some problem.

Regardless of whether we fit an IM by maximum likelihood or by Bayesian posterior inference, we work with the same joint likelihood, which is the product of the likelihoods of each individual data set (under the important assumption of statistical independence!). The magic especially of BUGS-language software for IMs is that these models are trivially easy to specify: within the same model statement we simply describe the likelihood for each data set and then define one or more shared parameters that have the same name and meaning in two or more of these component models. The wonderful ease of fitting even complex IMs in the BUGS language is the main reason for why Bayesian posterior inference has become the dominant mode of inference for IPMs (Schaub & Kéry, 2022). However, the concept of the joint likelihood as a product of the likelihoods of the individual data sets is buried somewhat in the Bayesian approach to fitting these models. In contrast, when doing likelihood inference, we will explicitly define the joint likelihood as such a product. Thus, it is conceptually valuable to see the IM fit in both ways. This is what we do in Section 20.4.

20.2 Data generation: simulating three abundance data sets with different observation/aggregation models

We start by simulating three count data sets with identical parameters for a log-linear regression of the expected counts on a covariate "elevation." Then, we degrade the information in data sets 2 (truncated Poisson) and 3 (detection/nondetection) by, respectively, discarding all zeroes and one-censoring the counts so they become binary detection/nondetections, or DND data. In real life, both of these "degraded" data types are logistically cheaper to obtain. Hence, we assume we have regular counts from 500 sites, zero-truncated counts from 1000 and detection/nondetection data from 2000. We sort the covariate values within each data set, which has no effect on the calculations, but makes plotting easier (Fig. 20.3).

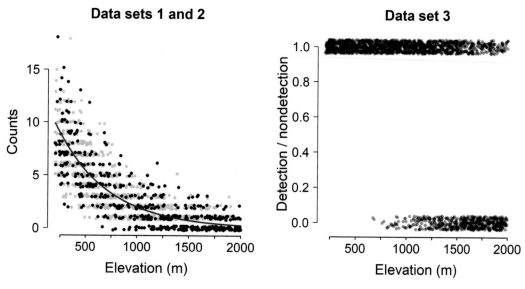

FIGURE 20.3

Simulated count (left) and detection/nondetection data (right) as a function of elevation. Original data are Poisson counts (left) with 500 sites in data set 1 (black) and 1000 sites minus 272 Poisson zeroes in the zero-truncated data set 2 (gray) and where the common expectation as a function of elevation is shown by the red curve. The detection/nondetection data (right) stem from 2000 sites. In the integrated model, we combine all three data sets in a single expression for the abundance-elevation relationship.

```
set.seed(20)

# Simulate the two data sets and plot them
# Choose sample size and parameter values for both data sets
nsites1 <- 500                                    # Sample size for count data
nsites2 <- 1000                                   # Sample size for zero-truncated counts
nsites3 <- 2000                                   # Sample size for detection/nondetection data
mean.lam <- 2                                     # Average expected abundance (lambda) per site
beta <- -2                                        # Coefficient of elevation covariate on lambda
truth <- c("log.lam" = log(mean.lam), "beta" = beta) # Save truth

# Simulate elevation covariate for all three and standardize to mean of 1000 and
# standard deviation also of 1000 m
elev1 <- sort(runif(nsites1, 200, 2000))          # Imagine 200-2000 m a.s.l.
elev2 <- sort(runif(nsites2, 200, 2000))
elev3 <- sort(runif(nsites3, 200, 2000))
selev1 <- (elev1 - 1000)/1000                     # Scaled elev1
selev2 <- (elev2 - 1000)/1000                     # Scaled elev2
selev3 <- (elev3 - 1000)/1000                     # Scaled elev3

# Create three regular count data sets with log-linear effects
C1 <- rpois(nsites1, exp(log(mean.lam) + beta * selev1))
C2 <- rpois(nsites2, exp(log(mean.lam) + beta * selev2))
C3 <- rpois(nsites3, exp(log(mean.lam) + beta * selev3))

table(C1)    # Tabulate data set 1
```

C1

0	1	2	3	4	5	6	7	8	9	10	11	12	13	14	15	18
150	97	62	38	39	29	21	22	11	11	6	3	5	2	2	1	1

```
# Create data set 2 (C2) by zero-truncating (discard all zeroes)
ztC2 <- C2                    # Make a copy
ztC2 <- ztC2[ztC2 > 0]        # Tossing out zeroes yields zero-truncated data
table(C2); table(ztC2)        # tabulate both original and ZT data set
```

C2

0	1	2	3	4	5	6	7	8	9	10	11	12	13	14	15	18
272	204	136	105	59	56	43	35	36	15	16	10	6	4	1	1	1

ztC2

1	2	3	4	5	6	7	8	9	10	11	12	13	14	15	18
204	136	105	59	56	43	35	36	15	16	10	6	4	1	1	1

```
# Turn count data set 3 (C3) into detection/nondetection data (y)
y <- C3                     # Make a copy
y[y > 1] <- 1               # Squash to binary
table(C3)   ;   table(y)    # tabulate both original counts and DND
```

C3

0	1	2	3	4	5	6	7	8	9	10	11	12	13	14	15	16	17
528	428	288	175	140	101	70	68	51	56	36	28	14	8	4	2	2	1

Y

0	1
528	1472

```
# Plot counts/DND data in all data sets (Fig. 20.3)
library(scales)
par(mfrow = c(1, 2), mar = c(5, 5, 4, 1), cex = 1.2, cex.lab = 1.5, cex.axis = 1.5, las = 1)
plot(elev2[C2 > 0], jitter(ztC2), pch = 16, xlab = 'Elevation (m)', ylab = 'Counts',
   frame = FALSE, ylim = range(c(C1, ztC2)), col = alpha('grey80', 1), main = 'Data sets 1 and 2')
points(elev1, jitter(C1), pch = 16)
lines(200:2000, exp(log(2) -2 * ((200:2000)-1000)/1000 ), col = 'red', lty = 1, lwd = 2)
axis(1, at = c(250, 750, 1250, 1750), tcl = -0.25, labels = NA)
plot(elev3, jitter(y, amount = 0.04), xlab = 'Elevation (m)', ylab = 'Detection/nondetection',
   axes = FALSE, pch = 16, col = alpha('grey60', 0.3), main = 'Data set 3')
axis(1)
axis(1, at = c(250, 750, 1250, 1750), tcl = -0.25, labels = NA)
axis(2, at = c(0, 1), labels = c(0, 1))

# Required libraries
library(ASMbook); library(jagsUI); library(rstan); library(TMB)
```

20.3 Fitting models to the three individual data sets first

In an IM, we will virtually always fit the component models to each individual data set first. We do this here, obtaining both DIY-MLEs in R and Bayesian posterior inference in JAGS. We also use canned functions in R where available.

20.3.1 Fitting a standard Poisson GLM in R and JAGS to data set 1

We start by fitting a Poisson GLM to data set 1 to see how well we can recover the two parameters of the abundance model. We do this with the iteratively reweighted least-squares method in R, which yields the MLEs for these models. We find estimates that are pretty close to the known truth.

```
# Get MLEs for individual data sets
# Data set 1: Poisson GLM with log link for counts
summary(fm1 <- glm(C1 ~ selev1, family = poisson(link = "log")))
exp(coef(fm1)[1])    # Estimate of lambda on natural scale from counts
```

```
Coefficients:
              Estimate   Std. Error   z value   Pr(>|z|)
(Intercept)    0.69493      0.03668     18.94    <2e-16 ***
selev1        -1.96130      0.06700    -29.27    <2e-16 ***

(Intercept)
   2.003559
```

Next, our DIY likelihood solution for the regular Poisson counts.

```
# Definition of negative log-likelihood (NLL) for Poisson GLM
NLL1 <- function(param, y, x) {
  alpha <- param[1]
  beta <- param[2]
  lambda <- exp(alpha + beta * x)
  LL <- dpois(y, lambda, log = TRUE)      # LL contribution for each datum
  return(-sum(LL))                        # NLL for all data
}
```

```
# Minimize NLL
inits <- c(alpha = 0, beta = 0)          # Need to provide initial values
sol1 <- optim(inits, NLL1, y = C1, x = selev1, hessian = TRUE, method = 'BFGS')
tmp1 <- get_MLE(sol1, 4)
```

```
            MLE      ASE   LCL.95    UCL.95
alpha    0.6949  0.03668    0.623    0.7668
beta    -1.9613  0.06700   -2.093   -1.8300
```

Next, the same in JAGS ...

```
# Bundle data
str(bdata <- list(C1 = C1, nsites1 = nsites1, selev1 = selev1))
```

```
List of 3
 $ C1     : int [1:500] 8 7 6 8 12 18 5 8 7 8 ...
 $ nsites1: num 500
 $ selev1 : num [1:500] -0.8 -0.797 -0.795 -0.788 -0.782 ...
```

```
# Write JAGS model file
cat(file = "model1.txt", "
model {
# Priors and linear models
alpha ~ dunif(-10, 10)                    # Abundance intercept on log scale
mean.lam <- exp(alpha)                    # Abundance intercept on natural scale
beta ~ dnorm(0, 0.0001)                   # Slope on elevation

# Likelihood for data set 1
for (i in 1:nsites1){
  C1[i] ~ dpois(lambda1[i])
  log(lambda1[i]) <- alpha + beta * selev1[i]
}
}
")
```

```
# Initial values
inits <- function(){list(alpha = runif(1), beta = rnorm(1))}
```

```
# Parameters monitored
params <- c("mean.lam", "alpha", "beta")
```

```
# MCMC settings
ni <- 6000; nb <- 2000; nc <- 4; nt <- 4; na <- 1000
```

```
# Call JAGS from R (ART <1 min), check convergence and summarize posteriors
out1 <- jags(bdata, inits, params, "model1.txt", n.iter = ni, n.burnin = nb, n.chains = nc,
  n.thin = nt, n.adapt = na, parallel = TRUE)
jagsUI::traceplot(out1)                         # Not shown
print(out1, 4)
```

	mean	sd	2.5%	50%	97.5%	overlap0	f	Rhat	n.eff
mean.lam	2.0034	0.0725	1.8620	2.0034	2.1476	FALSE	1	1.0006	3371
alpha	0.6942	0.0362	0.6216	0.6949	0.7643	FALSE	1	1.0006	3541
beta	-1.9599	0.0669	-2.0910	-1.9589	-1.8311	FALSE	1	0.9999	4000

```
# Compare truth with likelihood and Bayesian solutions
comp <- cbind(truth = truth, tmp1[,1:2], out1$summary[2:3, 1:2])
colnames(comp)[3:5] <- c("SE(MLE)", "Post.mean", "Post.sd")
print(comp, 3)
```

	truth	MLE	SE(MLE)	Post.mean	Post.sd
log.lam	0.693	0.695	0.0367	0.694	0.0362
beta	-2.000	-1.961	0.0670	-1.960	0.0669

We find practically matching estimates, as we would expect for this simple model, our use of vague priors, and the large sample size.

20.3.2 Fitting the zero-truncated Poisson GLM in R and JAGS to data set 2

Next, we want to fit the same model to the zero-truncated data set (number 2). This model could be fit using R package VGAM (Yee et al., 2008), but here we just define the zero-truncated Poisson likelihood function ourselves and maximize it. This likelihood is formed as a re-normalized Poisson likelihood without the zero class. That is, we must divide the standard Poisson PMF by 1 minus the probability of a zero count under that same PMF. The latter bit is the total probability of the zero-truncated random variable under the Poisson with same parameters.

```
# Definition of negative log-likelihood (NLL) for the ztPoisson GLM
NLL2 <- function(param, y, x) {
  alpha <- param[1]
  beta <- param[2]
  lambda <- exp(alpha + beta * x)
  L <- dpois(y, lambda)/(1-ppois(0, lambda))    # L. contribution for each datum
  return(-sum(log(L)))                          # NLL for all data
}
```

```
# Minimize NLL
inits <- c(alpha = 0, beta = 0)                 # Need to provide initial values
sol2 <- optim(inits, NLL2, y = ztC2, x = selev2[C2>0], hessian = TRUE, method = 'BFGS')
tmp2 <- get_MLE(sol2, 4)
```

	MLE	ASE	LCL.95	UCL.95
alpha	0.6461	0.03852	0.5706	0.7216
beta	-2.0351	0.07089	-2.1740	-1.8961

As an aside, we quickly compare with the solution when we ignore the truncation and erroneously fit the regular Poisson model. We see that we greatly overestimate abundance when we ignore the zero truncation by fitting a standard Poisson distribution to the zero-truncated data.

```
# Minimize 'wrong' Poisson NLL which ignores the zero truncation
sol2a <- optim(inits, NLL1, y = ztC2, x = selev2[C2>0], hessian = TRUE, method = 'BFGS')
get_MLE(sol2a, 4)
```

	MLE	ASE	LCL.95	UCL.95
alpha	0.9276	0.02755	0.8736	0.9816
beta	-1.5275	0.05402	-1.6334	-1.4216

```
# Compare intercepts of right and wrong Poisson NLL on the natural scale
exp(sol2$par[1]) ; exp(sol2a$par[1])

   alpha
1.908095
   alpha
2.528512
```

Next, we fit the zero-truncated Poisson model in JAGS, where it's particularly easy to fit this model, using the T() syntax. We are reminded that by zero truncating we lost the data from almost 300 sites in our simulated data set. We index sites by j, to emphasize that we assume a different set of sites in each of the three data sets.

```
# Bundle data
str(bdata <- list(C2 = ztC2, nsites2 = length(ztC2), selev2 = selev2[C2 > 0]))

List of 3
 $ C2     : int [1:728] 11 13 5 9 8 7 14 6 7 6 ...
 $ nsites2: int 728
 $ selev2 : num [1:728] -0.799 -0.798 -0.798 -0.797 -0.797 ...

# Write JAGS model file
cat(file = "model2.txt", "
model {
# Priors and linear models
alpha ~ dunif(-10, 10)              # Abundance intercept on log scale
mean.lam <- exp(alpha)             # Abundance intercept on natural scale
beta ~ dnorm(0, 0.0001)           # Slope on elevation

# Zero-truncated Poisson likelihood for data set 2
for (j in 1:nsites2){
  C2[j] ~ dpois(lambda1[j])T(1,)       # truncation is accommodated easily
  log(lambda1[j]) <- alpha + beta * selev2[j]
}
}
")

# Initial values
inits <- function(){list(alpha = runif(1), beta = rnorm(1))}

# Parameters monitored
params <- c("mean.lam", "alpha", "beta")

# MCMC settings
ni <- 6000; nb <- 2000; nc <- 4; nt <- 4; na <- 1000

# Call JAGS from R (ART <1 min), check convergence and summarize posteriors
out2 <- jags(bdata, inits, params, "model2.txt", n.iter = ni, n.burnin = nb,
  n.chains = nc, n.thin = nt, n.adapt = na, parallel = TRUE)
jagsUI::traceplot(out2)                  # Not shown
print(out2, 2)
```

	mean	sd	2.5%	50%	97.5%	overlap0	f	Rhat	n.eff
mean.lam	1.91	0.07	1.76	1.90	2.05	FALSE	1	1	1219
alpha	0.64	0.04	0.57	0.64	0.72	FALSE	1	1	1226
beta	-2.04	0.07	-2.18	-2.04	-1.90	FALSE	1	1	1887

We compare the likelihood and the Bayesian estimates.

```
# Compare truth with likelihood and Bayesian solutions
comp <- cbind(truth = truth, tmp2[,1:2], out2$summary[2:3, 1:2])
colnames(comp)[3:5] <- c("SE(MLE)", "Post.mean", "Post.sd")
print(comp, 3)
```

```
           truth     MLE    SE(MLE)   Post.mean   Post.sd
log.lam    0.693    0.646    0.0385      0.644     0.0391
beta      -2.000   -2.035    0.0709     -2.039     0.0719
```

That looks good!

20.3.3 Fitting the cloglog Bernoulli GLM in R and JAGS to data set 3

Finally, we fit the Bernoulli model with cloglog link to the detection/nondetection data. This means that the parameters of the model describe the relationship between an underlying abundance process and the elevation covariate.

```
# Data set 3: Bernoulli GLM with cloglog link for detection/nondetection
summary(fm3 <- glm(y ~ selev3, family = binomial(link = "cloglog")))
exp(coef(fm3)[1])    # Estimate of lambda on natural scale from binary data
```

```
Coefficients:
               Estimate   Std. Error   z value   Pr(>|z|)
(Intercept)     0.71081     0.04169      17.05    <2e-16 ***
selev3         -1.98144     0.09499     -20.86    <2e-16 ***
---

(Intercept)
  2.035632
```

And the DIY likelihood solution.

```
# Definition of NLL for Bernoulli GLM with cloglog link
NLL3 <- function(param, y, x) {
  alpha <- param[1]
  beta <- param[2]
  lambda <- exp(alpha + beta * x)
  psi <- 1-exp(-lambda)
  LL <- dbinom(y, 1, psi, log = TRUE)         # L. contribution for each datum
  return(-sum(LL))                            # NLL for all data
}
```

```
# Minimize NLL
inits <- c(alpha = 0, beta = 0)
sol3 <- optim(inits, NLL3, y = y, x = selev3, hessian = TRUE, method = 'BFGS')
tmp3 <- get_MLE(sol3, 4)
```

```
           MLE       ASE      LCL.95    UCL.95
alpha    0.7108    0.04214    0.6282    0.7934
beta    -1.9815    0.09682   -2.1712   -1.7917
```

And finally JAGS. In the model, we now index sites by k.

```
# Bundle data
str(bdata <- list(y = y, nsites3 = nsites3, selev3 = selev3))

List of 3
 $ y       : num [1:2000] 1 1 1 1 1 1 1 1 1 1 ...
 $ nsites3 : num 2000
 $ selev3  : num [1:2000] -0.798 -0.794 -0.794 -0.794 -0.792 ...

# Write JAGS model file
cat(file = "model3.txt", "
model {
# Priors and linear models
alpha ~ dunif(-10, 10)                     # Abundance intercept on log scale
mean.lam <- exp(alpha)                     # Abundance intercept on natural scale
beta ~ dnorm(0, 0.0001)                    # Slope on elevation

# Likelihood for data set 3
for (k in 1:nsites3){
  y[k] ~ dbern(psi[k])
  cloglog(psi[k]) <- alpha + beta * selev3[k]

  # Alternative implementation of same model for data set 3
  # y[k] ~ dbern(psi[k])
  # psi[k] <- 1 - exp(-lambda3[k])
  # log(lambda3[k]) <- alpha + beta * selev3[k]
}
}
")

# Initial values
inits <- function(){list(alpha = runif(1), beta = rnorm(1))}

# Parameters monitored
params <- c("mean.lam", "alpha", "beta")

# MCMC settings
ni <- 6000; nb <- 2000; nc <- 4; nt <- 4; na <- 1000

# Call JAGS from R (ART 1.2 min), check convergence and summarize posteriors
out3 <- jags(bdata, inits, params, "model3.txt", n.iter = ni, n.burnin = nb,
  n.chains = nc, n.thin = nt, n.adapt = na, parallel = TRUE)
jagsUI::traceplot(out3)                    # Not shown
print(out3, 3)
```

	mean	sd	2.5%	50%	97.5%	overlap0	f	Rhat	n.eff
mean.lam	2.040	0.084	1.885	2.038	2.211	FALSE	1	1.002	1439
alpha	0.712	0.041	0.634	0.712	0.794	FALSE	1	1.002	1413
beta	-1.986	0.095	-2.177	-1.985	-1.804	FALSE	1	1.002	1240

```
# Compare truth with likelihood and Bayesian solutions
comp <- cbind(truth = truth, tmp3[,1:2], out3$summary[2:3, 1:2])
colnames(comp)[3:5] <- c("SE(MLE)", "Post.mean", "Post.sd")
print(comp, 3)
```

	truth	MLE	SE(MLE)	Post.mean	Post.sd
log.lam	0.693	0.711	0.0421	0.712	0.0411
beta	-2.000	-1.981	0.0968	-1.986	0.0948

We find a nice numerical agreement.

20.4 Fitting the integrated model to all three data sets simultaneously

Now that we understand the component models for each data set and know how to code them up, we are ready to fit the IM. We fit the full integrated SDM which assumes a shared abundance process, but accommodates the three different types of observation processes in the three simulated data sets. There is no canned R function for us to fit this IM, so we work directly with our own negative log-likelihood function that we then maximize for the three data sets simultaneously. After that, we fit the same model with JAGS, NIMBLE, Stan, and TMB.

20.4.1 Fitting the IM with a DIY likelihood function

Under an assumption of statistical independence the joint likelihood for the three submodels is simply the product of the three single-data likelihoods:

$$L_{joint} = L_1 L_2 L_3$$

Of course, we usually work with the negative of the log of the likelihood and for this, we have a sum rather than a product. Hence, the objective function to minimize for our IM is this:

$$NLL_{joint} = -(LL_1 + LL_2 + LL_3) = -LL_1 - LL_2 - LL_3$$

Let's try this out.

```
# Definition of the joint NLL for the integrated model
NLLjoint <- function(param, y1, x1, y2, x2, y3, x3) {
  # Definition of elements in param vector (shared between data sets)
    alpha <- param[1]                                       # log-linear intercept
    beta <- param[2]                                        # log-linear slope
  # Likelihood for the Poisson GLM for data set 1 (y1, x1)
    lambda1 <- exp(alpha + beta * x1)
    L1 <- dpois(y1, lambda1)
  # Likelihood for the ztPoisson GLM for data set 2 (y2, x2)
    lambda2 <- exp(alpha + beta * x2)
    L2 <- dpois(y2, lambda2)/(1-ppois(0, lambda2))
  # Likelihood for the cloglog Bernoulli GLM for data set 3 (y3, x3)
    lambda3 <- exp(alpha + beta * x3)
    psi <- 1-exp(-lambda3)
    L3 <- dbinom(y3, 1, psi)
  # Joint log-likelihood and joint NLL: here you can see that sum!
    JointLL <- sum(log(L1)) + sum(log(L2)) + sum(log(L3))   # Joint LL
    return(-JointLL)                                        # Return joint NLL
}
```

```
# Minimize NLLjoint
inits <- c(alpha = 0, beta = 0)
solJoint <- optim(inits, NLLjoint, y1 = C1, x1 = selev1, y2 = ztC2, x2 = selev2[C2>0],
  y3 = y, x3 = selev3, hessian = TRUE, method = 'BFGS')
```

```
# Get MLE and asymptotic SE and print and save
(tmp4 <- get_MLE(solJoint, 4))
diy_est <- tmp4[,1]
```

```
                MLE         ASE        LCL.95        UCL.95
alpha     0.6860979   0.01851556    0.6498074     0.7223884
beta     -1.9703732   0.03522001   -2.0394044    -1.9013420
```

Nice... our first IM "by hand"!

20.4.2 Fitting the IM with JAGS

For the IM in JAGS, we simply stack the three GLMs inside of the same BUGS model statement and choose the same names for the parameters `alpha` and `beta` in all of them. This is what defines the joint likelihood as a product of the single-data likelihoods. Hence, even though you don't see it so clearly, the likelihood of this model is $L_{joint} = L_1L_2L_3$, where L_1, L_2, and L_3 are, respectively, the likelihoods of the Poisson GLM for the count data set 1, of the zero-truncated Poisson GLM for data set 2, and of the Bernoulli GLM for the detection/nondetection data set 3. Note again the different indices used for sites in the three data sets to emphasize that we assume they do not overlap.

```
# Bundle data
str(dataList <- list(C1 = C1, C2 = ztC2, y = y, nsites1 = nsites1, nsites2 = length(ztC2),
  nsites3 = nsites3, selev1 = selev1, selev2 = selev2[C2>0], selev3 = selev3))

List of 9
 $ C1     : int  [1:500]  8  7  6  8  12  18  5  8  7  8 ...
 $ C2     : int  [1:728]  11  13  5  9  8  7  14  6  7  6 ...
 $ y      : num  [1:2000]  1  1  1  1  1  1  1  1  1  1 ...
 $ nsites1: num  500
 $ nsites2: int  728
 $ nsites3: num  2000
 $ selev1 : num  [1:500]  -0.8  -0.797  -0.795  -0.788  -0.782 ...
 $ selev2 : num  [1:728]  -0.799  -0.798  -0.798  -0.797  -0.797 ...
 $ selev3 : num  [1:2000]  -0.798  -0.794  -0.794  -0.794  -0.792 ...

# Write JAGS model file
cat(file = "model4.txt", "
model {
# Priors and linear models: shared for models of all three data sets
alpha ~ dunif(-10, 10)              # Abundance intercept on log scale
mean.lam <- exp(alpha)              # Abundance intercept on natural scale
beta ~ dnorm(0, 0.0001)             # Slope on elevation

# Joint likelihood: Note identical alpha and beta for all data sets
# Likelihood portion for data set 1: regular counts
for (i in 1:nsites1){
  C1[i] ~ dpois(lambda1[i])
  log(lambda1[i]) <- alpha + beta * selev1[i]
}
# Likelihood portion for data set 2: zero-truncated counts
for (j in 1:nsites2){
  C2[j] ~ dpois(lambda2[j])T(1,)
  log(lambda2[j]) <- alpha + beta * selev2[j]
}
# Likelihood portion for data set 3: detection/nondetection
for (k in 1:nsites3){
  y[k] ~ dbern(psi[k])
  cloglog(psi[k]) <- alpha + beta * selev3[k]
}
}
")
```

```
# Initial values
inits <- function(){list(alpha = runif(1), beta = rnorm(1))}

# Parameters monitored
params <- c("mean.lam", "alpha", "beta")

# MCMC settings
na <- 1000; ni <- 6000; nb <- 2000; nc <- 4; nt <- 4

# Call JAGS from R (ART 170 sec), check convergence and summarize posteriors
out4 <- jags(dataList, inits, params, "model4.txt", n.iter = ni, n.burnin = nb,
  n.chains = nc, n.thin = nt, n.adapt = na, parallel = TRUE)
jagsUI::traceplot(out4)
print(out4, 2)
jags_est <- out4$summary[2:3,1]
```

	mean	sd	2.5%	50%	97.5%	overlap0	f	Rhat	n.eff
mean.lam	1.99	0.04	1.91	1.99	2.06	FALSE	1	1	4000
alpha	0.69	0.02	0.65	0.69	0.72	FALSE	1	1	4000
beta	-1.97	0.04	-2.04	-1.97	-1.90	FALSE	1	1	4000

We compare the likelihood and the Bayesian solutions with truth.

```
# Compare truth with likelihood and Bayesian solutions
comp <- cbind(truth = truth, tmp4[,1:2], out4$summary[2:3, 1:2])
colnames(comp)[3:5] <- c("SE(MLE)", "Post.mean", "Post.sd")
print(comp, 3)
```

	truth	MLE	SE(MLE)	Post.mean	Post.sd
log.lam	0.693	0.686	0.0185	0.686	0.0184
beta	-2.000	-1.970	0.0352	-1.971	0.0353

As almost always, we get numerically extremely similar estimates from maximum likelihood and Bayesian posterior inference for a model with vague priors and provided that the sample size is large with respect to the complexity of the model.

20.4.3 Fitting the IM with NIMBLE

NIMBLE code for the model is available on the book website.

20.4.4 Fitting the IM with Stan

Also for Stan, we can use the same bundled data. Here's the model code, which looks very similar to our JAGS model.

```
cat(file = "model4.stan",
"data {
  int nsites1;
  int nsites2;
  int nsites3;
  array[nsites1] int C1;
  array[nsites2] int C2;
  array[nsites3] int y;
  vector[nsites1] selev1;
  vector[nsites2] selev2;
  vector[nsites3] selev3;
}
parameters {
  real alpha;
  real beta;
}
model {
  vector[nsites1] lambda1;
  vector[nsites2] lambda2;
  vector[nsites3] psi;

  // Priors
  alpha ~ uniform(-10, 10);
  beta ~ normal(0, 100);

  // Likelihood
  for (i in 1:nsites1){
    lambda1[i] = exp(alpha + beta * selev1[i]);
    C1[i] ~ poisson(lambda1[i]);
  }
  for (j in 1:nsites2){
    lambda2[j] = exp(alpha + beta * selev2[j]);
    C2[j] ~ poisson(lambda2[j]) T[1, ];
  }
  for (k in 1:nsites3){
    psi[k] = inv_cloglog(alpha + beta * selev3[k]);
    y[k] ~ bernoulli(psi[k]);
  }
}
generated quantities {
  real mean_lam = exp(alpha);
}
" )

# Parameters monitored
params <- c("mean_lam", "alpha", "beta")

# HMC settings
ni <- 2000 ; nb <- 1000 ; nc <- 4 ; nt <- 1

# Call STAN (ART 90/45 sec), assess convergence and print results table
system.time(
out4.stan <- stan(file = "model4.stan", data = dataList,
                  warmup = nb, iter = ni, chains = nc, thin = nt) )
rstan::traceplot(out4.stan)                    # not shown
print(out4.stan, dig = 3)                      # not shown
stan_est <- summary(out4.stan)$summary[1:2,1]
```

20.4.5 **Fitting the IM with TMB**

Again we can use the same data bundle.

```
cat(file = "model4.cpp",
"#include <TMB.hpp>

template<class Type>
Type objective_function<Type>::operator() ()
{
  //Describe input data
  DATA_INTEGER(nsites1);
  DATA_INTEGER(nsites2);
  DATA_INTEGER(nsites3);
  DATA_VECTOR(C1);
  DATA_VECTOR(C2);
  DATA_VECTOR(y);
  DATA_VECTOR(selev1);
  DATA_VECTOR(selev2);
  DATA_VECTOR(selev3);

  //Describe parameters
  PARAMETER(alpha);
  PARAMETER(beta);

  Type LL = 0.0;                                  //Initialize log-likelihood at 0

  vector<Type> lambda1(nsites1);
  vector<Type> lambda2(nsites2);
  vector<Type> psi(nsites3);

  for (int i=0; i<nsites1; i++){
    lambda1(i) = exp(alpha + beta * selev1(i));
    LL += dpois(C1(i), lambda1(i), true);
  }
  for (int j = 0; j<nsites2; j++){
    lambda2(j) = exp(alpha + beta * selev2(j));
    // Truncated Poisson
    LL += log(dpois(C2(j), lambda2(j))/(1 - ppois(Type(0), lambda2(j))));
  }
  for (int k=0; k<nsites3; k++){
    psi(k) = 1 - exp(-exp(alpha + beta * selev3(k))); //inverse cloglog
    LL += dbinom(y(k), Type(1), psi(k), true);
  }

  Type mean_lam = exp(alpha);
  ADREPORT(mean_lam);

  return -LL;
}
")
```

```
# Compile and load TMB function
compile("model4.cpp")
dyn.load(dynlib("model4"))

# Provide dimensions and starting values for parameters
params <- list(alpha = 0, beta = 0)

# Create TMB object
out4.tmb <- MakeADFun(data = dataList, parameters = params,
                      DLL = "model4", silent = TRUE)

# Optimize TMB object, print and save results
opt <- optim(out4.tmb$par, fn = out4.tmb$fn, gr = out4.tmb$gr, method = "BFGS")
(tsum <- tmb_summary(out4.tmb))
tmb_est <- tsum[1:2,1]
```

20.4.6 Comparison of the parameter estimates for the IM

We compare the parameter estimates for the IMs among the engines. Remember that we have fewer engines now, since there is no canned function in R to fit our IM.

```
# Compare point estimates from the five engines
comp <- cbind(truth = truth, DIY = diy_est, JAGS = jags_est, NIMBLE = nimble_est,
  Stan = stan_est, TMB = tmb_est)
print(comp, 4)
```

	truth	DIY	JAGS	NIMBLE	Stan	TMB
log.lam	0.6931	0.6861	0.6855	0.686	0.6856	0.6861
beta	-2.0000	-1.9704	-1.9710	-1.970	-1.9715	-1.9704

We see what we see so often: numerically, these estimates are practically indistinguishable.

20.5 What do we gain by analyzing the joint likelihood in our analysis?

We compare all the MLEs for models 1–4 to see how point estimates and asymptotic standard errors change for different data types and for the single-data models and the joint-likelihood model. Since we have already seen the similarity between the likelihood and the Bayesian inferences, we're not going to bother doing this for both, but restrict this comparison to our likelihood inferences.

```
# Compare truth with MLEs only from all 4 models (stacked sideways)
print(cbind(truth = truth, "MLE(Poisson)" = tmp1[,1], "MLE(ZTPois)" = tmp2[,1],
  "MLE(cloglogBern)" = tmp3[,1], "MLE(integrated)" = tmp4[,1]), 3)

# Compare ASEs from all 4 models (stacked sideways)
print(cbind("ASE(Poisson)" = tmp1[,2], "ASE(ZTPois)" = tmp2[,2], "ASE(cloglogBern)" = tmp3[,2],
  "ASE(integrated)" = tmp4[,2]), 3)
```

```
## Point estimates
           truth   MLE(Poisson)   MLE(ZTPois)   MLE(cloglogBern)   MLE(integrated)
log.lam    0.693          0.695         0.646              0.711             0.686
beta      -2.000         -1.961        -2.035             -1.981            -1.970

## Standard errors
         ASE(Poisson)   ASE(ZTPois)   ASE(cloglogBern)   ASE(integrated)
alpha          0.0367        0.0385             0.0421            0.0185
beta           0.0670        0.0709             0.0968            0.0352
```

We don't see much difference among the point estimates. However, we see that the model for the regular counts has the greatest precision among the single-data models, followed by that for the zero-truncated counts, while that for the detection/nondetection data comes last. This makes intuitive sense, since we reduce the information by going from regular counts to tossing out the zeroes only and finally quantizing the data. In our simulation, this information loss is not compensated for by our simulated increase in sample size.

The IM has the greatest precision of all, since it uses the information from all three data sets in combination. We also make a plot to compare the point estimates and CRIs of models 1–4 (Fig. 20.4).

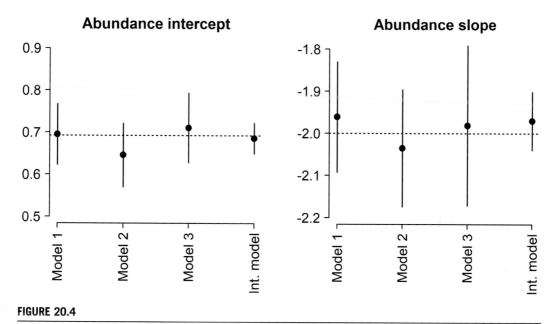

FIGURE 20.4

Point estimates (MLEs) and 95% CIs for the intercept and the slope parameter in the abundance model in all four models (dotted horizontal lines show truth). For both parameters, the precision of the estimates is greatest when we use all the information available, which is what we achieve by the integrated model.

```
# Compare MLEs and ASEs from all 4 models (Fig. 20.4)
par(mfrow = c(1, 2), mar = c(12, 6, 5, 3), cex.lab = 1.5, cex.axis = 1.5, cex.main = 1.8)
# Plot for abundance intercept (on log scale)
all.mles <- c(tmp1[1,1], tmp2[1,1], tmp3[1,1], tmp4[1,1])
all.lower.CL <- c(tmp1[1,3], tmp2[1,3], tmp3[1,3], tmp4[1,3])
all.upper.CL <- c(tmp1[1,4], tmp2[1,4], tmp3[1,4], tmp4[1,4])
plot(1:4, all.mles, pch = 16, xlab = '', ylab = '', axes = FALSE, frame = FALSE,
  main = 'Abundance intercept', cex = 2, ylim = c(0.5, 0.9))
axis(1, at = 1:4, c('Model 1', 'Model 2', 'Model 3', 'Integrated model'), las = 2)
segments(1:4, all.lower.CL, 1:4, all.upper.CL, lwd = 1.5)
axis(2, las = 1)
abline(h = log(2), lwd = 1, lty = 3)

# Plot for abundance slope (on log scale)
all.mles <- c(tmp1[2,1], tmp2[2,1], tmp3[2,1], tmp4[2,1])
all.lower.CL <- c(tmp1[2,3], tmp2[2,3], tmp3[2,3], tmp4[2,3])
all.upper.CL <- c(tmp1[2,4], tmp2[2,4], tmp3[2,4], tmp4[2,4])
plot(1:4, all.mles, pch = 16, xlab = '', ylab = '', axes = FALSE, frame = FALSE,
  main = 'Abundance slope', cex = 2, ylim = c(-2.2, -1.8))
axis(1, at = 1:4, c('Model 1', 'Model 2', 'Model 3', 'Integrated model'), las = 2)
segments(1:4, all.lower.CL, 1:4, all.upper.CL, lwd = 1.5)
axis(2, las = 1)
abline(h = -2, lwd = 1, lty = 3)
```

In Fig. 20.4, we see that the joint likelihood model is best in terms of the precision for both parameters in our model. We note that all estimators (i.e., all models) should produce unbiased estimates. Hence, it would be wrong to say that the differences between point estimates and the true values represented by the dotted lines represent bias. To gauge the unbiasedness of an estimator, we would have to repeat the whole data simulation/data analysis cycle a large number of times (e.g., 100 or 1000 times) and then look at the distribution of these estimates. Such a simulation could easily be done, especially with MLEs, since that method is much faster than MCMC.

The IM in this chapter does not illustrate another common advantage of this model class: namely that additional parameters may become estimable. For instance, in an IPM with a time series of counts that is combined with capture–recapture data we can estimate productivity, which is not an identifiable parameter in separate analyses for either data set alone (Besbeas et al., 2002; Schaub & Kéry, 2022).

20.6 **Summary and outlook**

Now you understand the principle of data integration by working with a joint likelihood: we first identify some underlying process that is shared among the different data sets and can be expressed by one or several parameters, and then describe the differences between the different data sets that typically lie in the observation process. In our case, even though at the start we had abundance data (i.e., counts) and detection/nondetection data, we assumed an underlying abundance process that is shared for all three data sets. Then, we chose a likelihood that represented the different observation processes that underlie the three different data sets: a regular Poisson PMF for the proper counts, a zero-truncated Poisson PMF for the data produced by the "lazy birders," and a Bernoulli PMF with cloglog link for the detection/nondetection data. Assuming statistical independence, the joint likelihood is the product of the likelihoods of each individual data set. Maximizing it results in parameter estimates that are most likely with respect to all data sets in the model.

Seeing that the joint likelihood under independence is a product of the single-data likelihoods is easiest when working with DIY maximum likelihood, because here the joint likelihood is very clearly apparent. However, in practice for most non-statisticians it is much easier to specify that same joint likelihood for an IM in the BUGS language or even with Stan or TMB, where we simply describe the individual-data likelihoods within the same model statement.

Opportunities for adopting such IMs are superabundant in ecology and wildlife management. Clearly, we expect to see more and more models that apply the principles of integrated modeling. The power and flexibility of the BUGS language makes it almost trivially easy to define the joint likelihood for such an IM. IMs are almost always custom models to some degree and it is here where the new-found modeling freedom that especially BUGS software gives you will be particularly exciting for you.

Conclusion⊛

21

Chapter outline

In the final chapter, we look back and ahead. Looking back, we summarize the major themes of the book: parametric statistical models and their basis in probability, maximum likelihood estimation, Bayesian posterior inference, and linear modeling. We also review the battery of general purpose model-fitting engines we apply in the book (R, JAGS, NIMBLE, Stan, and TMB), and which give you a tremendous freedom to specify exactly the model needed to answer your scientific questions or solve your management problems.

Looking ahead, we hope this book has helped you fledge as a modeler: you are no longer chained to the exact model set that "canned functions" in R enable you to fit, but instead are free to go wherever your questions and modeling tasks require you to go. This may include to boldly go where nobody has ever gone before. You have acquired this modeling freedom by understanding the fundamentals of statistical modeling and model building, and by your mastery of the powerful model-fitting engines introduced in this book.

We end this chapter by making some comments about specific model classes and topics that we think would be beneficial for you to study in the future, including study design, using machine learning techniques along with parametric modeling, do-it-yourself MCMC, structural equation modeling, and especially the overarching topic of hierarchical modeling, which offers an extremely powerful, principled, and scientific approach to statistical modeling.

21.1 Looking back

In this book, we have tried to give you the mathematical, statistical, and computational tools required to fit your own parametric statistical models to answer your research questions and to inform your management, conservation, or monitoring decisions. We have done so by providing a

---

⊛This book has a companion website hosting complementary materials, including all code for download. Visit this URL to access it: https://www.elsevier.com/books-and-journals/book-companion/9780443137150.

Applied Statistical Modelling for Ecologists. DOI: https://doi.org/10.1016/B978-0-443-13715-0.00005-4
497

gentle introduction to the concepts, methods, and software for modern applied statistical modeling. Here we summarize the most salient points.

- We have presented key mathematical concepts from probability, especially random variables and probability density functions. Most ecologists don't learn much about them in their training, and yet, these are the fundamental pieces of statistical models. Thus, we think that more "random variable thinking" or "probability density thinking" will be very beneficial for ecologists to take their statistical modeling to the next level.

- We have discussed the two common interpretations of probability. The narrower one uses probability solely to quantify variability and views probability as the limit of a relative frequency of observable quantities. Consequently, the associated school of statistics is known as frequentist. In contrast, the Bayesian school adopts a broader interpretation of probability: it includes both the use as a description of variability and as a quantification of uncertainty. The use of probability as a general measure of uncertainty is the hallmark of Bayesian statistics.

- We have emphasized the fundamental concept of the likelihood function, on which both frequentist and Bayesian inference are based. We have seen that for frequentist inference, we work with the likelihood function alone and numerically maximize its value over the parameter space to yield maximum likelihood estimates (MLEs). We have also covered in detail several methods for obtaining uncertainty assessments around the MLEs. The likelihood function and the inference based directly on it is a key concept in parametric modeling and hence, we covered it extensively in Chapter 2 and then especially in all the "Do-it-yourself MLE" sections throughout the book. We believe that seeing maximum likelihood estimation "in action" in these sections, especially for the simpler nonhierarchical models, is a great way for you to obtain an understanding of this foundational method of statistical inference.

- Bayesian inference is based on the posterior distribution, which from Bayes theorem is proportional to the product of the likelihood function and the priors of the model parameters. The prior distributions contain an assessment, in terms of a probability distribution, of how likely different values of a parameter are a priori. They enable us to readily incorporate into the estimation external information ("informative priors") if we wish to do so. However, throughout the book we have done what is mostly done in applied statistical modeling: we have specified so-called vague priors, which assign about the same probability for a large range of possible values of a parameter (see Gelman et al., 2020, for a different perspective where substantially more thought is put into the choice of priors). In this case, virtually all of the information to estimate a parameter comes from the data alone. For reasonable sample sizes, the resulting Bayesian estimates then numerically resemble very closely those obtained using the method of maximum likelihood, as we have seen throughout the book. This supports our belief that modern ecological modelers should understand well both methods of inference. Indeed, this is a premise of this book.

- We have focused on parametric statistical models, and primarily on three main concepts: linear models to describe how covariates are associated with a response, the generalized linear model (GLM), and random effects. It is important to obtain a good understanding of all three, since they can be combined to obtain an astonishing diversity of statistical models, including all those in this book, that is, linear models, GLMs, linear mixed models, generalized linear mixed models, general hierarchical models, and integrated models. These models comprise the majority of statistical models currently in wide use in ecology.

- Finally, computation- and software-wise, our book adopts a Rosetta stone approach, where each model is not only fit using a single model-fitting engine, but instead with six of them: three giving frequentist and three more giving Bayesian inference. We do this because we think that it is often useful to master more than a single fitting method or software, and that seeing code for all engines side by side will be helpful if you know one engine already and want to learn a new one. As a result, the ASM book has many layers and you could use it in many different ways, including as *"Gentle introduction for ecologists to the method of maximum likelihood," "Gentle introduction for ecologists to Bayesian inference," "... to program JAGS," "... to program NIMBLE," "... to program Stan."* and *"... to program TMB."*
- Chapter 18 covered model checking and model selection and also dealt with a number of related topics that are frequently forgotten, especially the question about the goal of a model. We discuss five such goals, and two particularly important ones are causal explanation and prediction. Depending on your objectives, a very different type of model may be "best." Parametric statistical models often emphasize causal explanation, though prediction is usually also important. For these models, a premium is on the interpretability of parameters, at least some of which ideally have a well-defined scientific meaning.

Our approach to teaching in this book necessarily includes some mathematics. For instance, we describe every model in algebra. We think that knowing how to write a statistical model in algebra will enforce a greater understanding of the model. In addition, we have seen throughout the book how the code to fit these models in several of our model-fitting engines is very similar to how we describe these models in algebra, especially for JAGS and NIMBLE. One key method to explain what a parametric model is, and which we use throughout the book, is data simulation in R. We are convinced that the R code to simulate data under a model is a highly intuitive manner to explain a statistical model, especially to nonstatisticians.

A strong focus on data simulation is one of the main features of our general approach that we advocate for ecologists in order for them to understand statistics: we call it "experimental statistics." By this we mean that even if we may not fully understand the theory underlying some statistical method, we may often get a good intuitive understanding by just "trying things out" and "seeing how things are." This will almost always include data simulation, as we have argued already in Chapter 1. We believe strongly that ecologists need a far better formal grounding in mathematics and statistics. However, we also believe that lacking that formal training does not necessarily prevent us from using advanced methods in an informed and adequate manner. "Experimental statistics" plays a key role in acquiring such an understanding.

Along the way of teaching you how to build, fit, and use parametric statistical models, we have also covered many useful topics that are often missing from the statistics curricula for ecologists. These include the bootstrap, cross-validation, the inner workings of maximum likelihood, quantile residuals, regularization, the goals of a model, and how this influences how we build, check, and select a model, model checking and model selection, and the distinction between what we call the hard and the soft parts of a model.

21.2 Looking ahead

We hope that this book contributes to a change in how you conduct statistical analyses, from a procedure-driven activity where you try to squeeze your data sets and questions into a finite set of boxes which represent the particular kinds of analyses that your software allows you to run, and towards a creative, science-based endeavor where you build your own custom models in a principled manner, guided by a scientific question or by your management objective. That is, we hope to contribute towards freeing the ecological modeler in you.

We end this book with a brief discussion of four topics which we envision as important for ecologists who are in the process of acquiring the modeling freedom that we try to teach with the ASM book. These topics are: study design, MCMC, machine learning, and hierarchical models. In earlier parts of the book, we have already touched upon all of them, but here we summarize each.

21.2.1 Study design

This book is about the fitting and understanding of models for data that you already have. Thereby we omit a field that is important for all empirical sciences: study design. This is concerned with the question of how to best collect data to learn about nature, either in a nonmanipulative context (e.g., a sample survey) or in the context of a manipulative experiment. Major concepts of experimental design, such as controls, blocking, and randomization, are employed to make the estimates of key parameters of interest as precise as possible and render their interpretation as unambiguous as possible. Much of the origins of experimental design lie in agriculture, which has been a cradle of statistical methodology for over a century, starting with the giant Fisher and of course many others. However, the same experimental principles apply everywhere, even for computer-based (i.e., simulation) experiments.

It seems to us that over the last decades, study design in ecology has moved a little to the background. We can envision two reasons for this trend. On the one hand, in the era of Big Data, data sets for analysis seem to be getting more easily available, for example, with citizen-science schemes such as http://www.ornitho.ch or http://www.ebird.org, or from any number of computerized databases. Therefore, people may think less about how to actually collect their data. On the other hand, classical experimental design imposed a strict set of rules so that the resulting data could be analyzed by the more restricted analytical possibilities of earlier times. These were often based on analysis of variance models using least-squares fitting techniques (Steel et al., 1996; Zar, 1998). Modern parametric modeling based on maximum likelihood or Bayesian inference is far more flexible. To some degree, this flexibility may have made a little obsolete the former design orthodoxy.

However, we believe that it is still crucial for ecologists to learn the principles of study design. Most importantly, this enables us to design and conduct proper experiments: a good experiment is still the gold standard for revealing cause and effect, and we arguably ought to be conducting more of them. Furthermore, even when working with existing data, an understanding of study design will allow us to recognize and account for common sources of nonindependence, such as blocking, in our models. Finally, thinking carefully about study design helps us to understand another crucial concept in modeling: statistical power. Statistical power is the ability of our analysis to detect an effect (e.g., of a covariate) when it exists. Power is driven primarily by sample size (a key component of study design), and is a concept too often ignored in ecological research. We believe that reading up on the important topics of experimental and sampling design is often a good investment for an ecologist. Good entry points to the essentials of the design of observational and experimental studies include Quinn & Keough (2002), Thompson (2002), Williams et al. (2002), and Mead et al. (2012).

Associated with classical experimental designs such as factorial, nested, and randomized block designs are certain forms of the underlying linear model. Often, these are more complex than what we have dealt with in this book. Books such as Quinn & Keough (2002) and Mead et al. (2012) present in algebra the linear models underlying a large number of experimental designs. They should allow you to implement such more complex linear models also in the model-fitting engines covered in this book. One particularly common design element is "blocking," which leads to nested random effects, which we have not covered in the book. See Schielzeth & Nakagawa (2013) for nesting in general, and Qian & Shen (2007), and Li et al. (2017) for BUGS code for nested designs.

21.2.2 Do-it-yourself Markov chain Monte Carlo

Throughout the book, we have presented both frequentist and Bayesian inference for statistical models, with their associated workhorse algorithms of numerical likelihood maximization for obtaining the MLEs, and MCMC to obtain the posterior distributions. In Chapter 2, we covered simple examples of both algorithms in some detail, but in the rest of the book, we only show DIY-MLE, but not DIY-MCMC algorithms. Instead, for Bayesian inference, we relied exclusively on the JAGS, NIMBLE, and Stan software. However, there may be advantages for you to learn to develop your own MCMC algorithms and Hooten & Hefley (2019) is the go-to source for ecologists (and see also the upcoming book by Zhao (2024), about spatial population ecology, which presents plenty of custom MCMC code in R).

21.2.3 Machine learning

Machine learning comprises a diverse array of data analysis methods that originally arose for the most part in computer science rather than in statistics, which accounts for many differences in jargon. Machine-learning models are built to maximize predictive performance. Key features are rigorous performance assessment on out-of-sample data, for instance using cross-validation, and models with few hard and mostly soft parts in the terminology of Chapter 18. Machine learning methods in wide use in ecology include generalized additive models (Wood, 2017), boosted regression trees and random forests (Elith et al., 2006; Hastie et al., 2016), and neural networks (Joseph, 2020a). These models typically have a *very* large number of parameters, which may go into the millions in the case of deep neural networks (Pichler & Hartig, 2023). Often, few if any parameters in these models have any direct meaning in terms of the science they are used for. Rather, such models are geared to accurate prediction and indeed often excel for this goal.

So, if machine learning methods with their thousands or millions of parameters are so powerful at prediction, then why should you learn the much simpler parametric statistical models? We believe that there are several important reasons for why scientists ought to be well trained in parametric statistical models. First, a key activity in science is the study of mechanisms, and parametric models excel at representing mechanisms and for testing scientific hypotheses about those mechanisms. Second, in terms of interpretability, parametric models are superb, while most machine learning methods are complete black boxes. Third, there are fundamental limits to the good prediction performance of most machine learning methods: if conditions change abruptly, a machine learning model trained on the old conditions may become useless (Pearl, 2021). In contrast, a more mechanistic, process-based model may in this case predict far better (Hefley et al. 2017). Fourth, most machine learning methods require very large data sets to estimate their parameters. It is true that recent technology such as autonomous recording units (ARUs) or camera traps may help us collect such very large data sets. However, most ecological studies will probably continue to deal mainly with relatively small data sets, for which parametric statistical models may be much better than most data-greedy machine learning methods.

In this book, we have briefly discussed model averaging in Section 18.7, which is a widely used idea in machine learning, as is regularization (Section 18.8). But otherwise, we have focused on classical parametric statistical models. Machine learning methods are very powerful for prediction purposes, but widely used methods are, currently at least, far more limited when the goal of your modeling is understanding of causal effects. This is due to the lack of hard structure in machine learning models, that is, the very large number of parameters in these models don't usually have any direct scientific or subject-matter interpretation. Indeed, most machine learning models are "prediction black boxes." But as we discussed in Chapter 18, prediction *is* an important part of science. Hence, investing some time to become conversant in common machine learning techniques is surely a good time investment for many quantitative ecologists.

Perhaps not so surprisingly, there are attempts at combining the strengths of parametric statistical models (i.e., their grounding in scientific knowledge and superior interpretability) with those of machine learning (their superior prediction power); see for instance Joseph (2020a), Mohankumar & Hefley (2022), and Wikle & Zammit-Mangion (2023). In addition, there are no doubt many attempts at adding more hard structure into machine learning models (see for instance Fink et al., 2023), but we don't know this literature well enough to say much about it. All in all, it is likely that in the future parametric models will become more flexible in their soft parts, by incorporating typical machine learning ideas such as regularization, while machine learning models will get more hard parts, to make them more useful for causal interpretation.

21.2.4 Hierarchical models

We have encountered hierarchical models throughout the book, for example, in Section 3.5 and in Chapters 7, 10, 14, 17, 19, and 19B. We like to emphasize how hierarchical models can often be viewed as two or more GLMs put in sequence, or in a hierarchy, where each GLM represents a stochastic process that adds random variation into the observed response. Typically, some of these processes produce output that is latent and which is called a latent variable or a random effect. Thus, we have seen that all the traditional mixed models (in Chapters 7, 10, 14, and 17) are examples of hierarchical models, where the random effects are given a normal distribution. A common motivation for specifying random effects is to accommodate hidden structure in the data, which will induce dependencies. These are particularly common in space and in time; that is, they show up as residual spatial or temporal autocorrelation. Thus, the modeling of time series and of spatial data (e.g., to produce species distribution maps) will typically involve hierarchical models with temporally or spatially correlated random effects (Cressie & Wikle, 2011). The parametric random-effects models in this book may be a suitable entry point for ecologists to the literature covering such models.

However, mixed models with correlated random effects often lack explicit mechanisms. Hence, their random effects rarely have any clear scientific interpretation. Instead, they are mere statistical devices for interpolation and for getting right the dispersion part of the model. However, hierarchical models have always been described as a natural framework for putting the subject-matter science into statistical modeling. They lead to science-based statistical models, where known or hypothesized mechanisms form part of the hierarchical structure that we specify (Berliner, 1996; Berliner et al., 2003). As a consequence, such "explicit" hierarchical models (Royle & Dorazio, 2008) have at least some parameters with a clear scientific interpretation. Simpler parametric models as the ones in this book are a natural stepping stone to more complex hierarchical models with such mechanistic structure.

One example for more mechanistic statistical models are structural equation models and other models for causal inference. These are naturally specified as hierarchical models (e.g., Cubaynes et al., 2012; Joseph et al., 2016; Gibson et al., 2023; Latif et al., 2023). These distinctly parametric models are very powerful for detailed investigations into hypothesized causal mechanisms underlying our data (Grace & Irvine, 2020). Thus, what you have learned in this book will again be useful for you if you want to explore those topics in more detail.

Another outstanding example of where more explicitly mechanistic hierarchical models are frequently used in ecology is in the fields of species distribution modeling, population ecology, and demography. In Chapters 19 and 19B, we have given an introduction to two specialized species distribution models for presence/absence and for abundance, respectively: one an occupancy model (MacKenzie et al., 2002; Tyre et al., 2003; MacKenzie et al. 2017) and the other an N-mixture model (Royle, 2004). Both have a detailed, mechanistic representation of the typical measurement error processes that always affect the data used for species distribution modeling, especially imperfect detection. In this respect, these models are extremely similar to the vast class of capture–recapture models which are also widely used in ecology to infer demographic rates such as survival or population growth (Williams et al., 2002; McCrea & Morgan, 2014; Seber & Schofield, 2019, 2024) or animal abundance and density (e.g., Borchers et al., 2002; Buckland et al., 2015). They can all be formulated as hierarchical models, and indeed, the following books focus on hierarchical models for population abundance, occurrence, and demography using BUGS software: Royle & Dorazio (2008), King et al. (2009), Link & Barker (2010), Kéry & Schaub (2012), Royle et al. (2014), Kéry & Royle (2016, 2021), and Schaub & Kéry (2022). All of what you have learned in the ASM book will greatly help you understand and apply these very powerful hierarchical models in ecology and related fields.

Bibliography

References

Abadi, F., Gimenez, O., Arlettaz, R., & Schaub, M. (2010). An assessment of integrated population models: Bias, accuracy, and violation of the assumption of independence. *Ecology*, 91, 7–14.

Akaike, H. (1973). Information theory and an extension of the maximum likelihood principle. In: *Proc. 2nd Int. Symp. Information Theory, Supp. to Problems of Control and Information Theory* (pp. 267–281).

Aldrich, J. (1997). R.A. Fisher and the making of maximum likelihood 1912–1922. *Statistical Science*, 12, 162–176.

Araújo, M. B., & Guisan, A. (2006). Five (or so) challenges for species distribution modelling. *Journal of Biogeography*, 33, 1677–1688.

Arif, S., & MacNeil, M. A. (2022). Utilizing causal diagrams across quasi-experimental approaches. *EcoSphere*, 13(4), e4009.

Ariyo, O., Lesaffre, E., & Verbeke, G. (2022). Bayesian model selection for longitudinal count data. *Sankhya*, 84-B, 516–547.

Ariyo, O., Quintero, A., Muñoz, J., Verbeke, G., & Lesaffre, E. (2020). Bayesian model selection in linear mixed models for longitudinal data. *Journal of Applied Statistics*, 47, 890–913.

Auger-Méthé, M., Field, C., Albertsen, C. M., Derocher, A. E., Lewis, M. A., Jonsen, I. D., & Mills Flemming, J. (2016). State-space models' dirty little secrets: Even simple linear Gaussian models can have estimation problems. *Scientific Reports*, 6, 26677.

Bailey, L. L., Hines, J. E., Nichols, J. D., & MacKenzie, D. I. (2007). Sampling design trade-offs in occupancy studies with imperfect detection: Examples and software. *Ecological Applications*, 17, 281–290.

Barker, R. J., & Link, W. A. (2013). Bayesian multimodel inference by RJMCMC: A Gibbs sampling approach. *The American Statistician*, 67(3), 150–156.

Barker, R. J., & Link, W. A. (2015). Truth, models, model sets, AIC, and multimodel inference: A Bayesian perspective. *Journal of Wildlife Management*, 79(5), 730–738.

Barker, R. J., Schofield, M. R., Link, W. A., & Sauer, J. R. (2018). On the reliability of *N*-mixture models for count data. *Biometrics*, 74, 369–377.

Barraquand, F. et al. (2024). Independence in IPMs, in prep.

Bartoń, K. (2023). MuMIn: Multi-Model Inference. R package version 1.47.5, Available from <https://CRAN.R-project.org/package=MuMIn>.

Bates, D., Mächler, M., Bolker, B.M., & Walker, S.C. (2014). Fitting linear mixed-effects models using lme4. Available from http://arxiv.org/abs/1406.5823.

Berger, J. O., Liseo, B., & Wolpert, R. K. (1999). Integrated likelihood methods for eliminating nuisance parameters. *Statistical Science*, 14, 1–22.

Berliner, L. M. (1996). Hierarchical Bayesian time series models. In K. Hanson, & R. Silver (Eds.), *Maximum entropy and Bayesian methods* (pp. 15–22). Dordrecht, The Netherlands: Kluwer Academic Publishers.

Berliner, L. M. (2003). Physical-statistical modeling in geophysics. *Journal of Geophysical Research*, 108 (D24), 8776.

Berliner, L. M., Milliff, R. F., & Wikle, C. K. (2003). Bayesian hierarchical modeling of air-sea interaction. *Journal of Geophysical Research*, 108(C4), 3104.

Besbeas, P., Borysiewicz, R. S., & Morgan, B. J. T. (2009). Completing the ecological jigsaw. In D. L. Thomson, E. G. Cooch, & M. J. Conroy (Eds.), *Modeling demographic processes in marked populations* (pp. 513–539). New York: Springer.

Besbeas, P., Freeman, S. N., Morgan, B. J. T., & Catchpole, E. A. (2002). Integrating mark-recapture-recovery and census data to estimate animal abundance and demographic parameters. *Biometrics*, 58, 540–547.

Blitzstein, J. K., & Hwang, J. (2019). *Introduction to probability (Second edition)*. CRC Press.

Bolker, B. M. (2008). *Ecological models and data in R*. Princeton, New Jersey: Princeton University Press.

Bolker, B.M. (2023). Multimodel approaches are not the best way to understand multifactorial systems. Available from https://ecoevorxiv.org/repository/view/5722/ (Preprint).

Bolker, B. M., Brooks, M. E., Clark, C. J., Geange, S. W., Poulsen, J. R., Stevens, M. H. H., & White, J. S. (2009). Generalized linear mixed models: A practical guide for ecology and evolution. *Trends in Ecology and Evolution*, 24, 127–135.

Borchers, D. L., Buckland, S. T., & Zucchini, W. (2002). *Estimating animal abundance*. London: Springer.

Box, G. E. P. (1980). Sampling and Bayes' inference in scientific modelling and robustness. *Journal of the Royal Statistical Society, Series A*, 143, 383–430.

Breimann, L. (2001). Statistical modeling: The two cultures. *Statistical Science*, 16, 199–231.

Brooks, M. E., Kristensen, K., Van Benthem, K. J., Magnusson, A., Berg, C. W., Nielsen, A., Skaug, H. J., Machler, M., & Bolker, B. M. (2017). glmmTMB balances speed and flexibility among packages for zero-inflated generalized linear mixed modeling. *The R Journal*, 9(2), 378–400.

Brooks, S. P. (2003). Bayesian computation: A statistical revolution. *Philosophical Transactions of the Royal Society A*, 361, 2681–2697.

Brooks, S. P., & Gelman, A. (1998). Alternative methods for monitoring convergence of iterative simulations. *Journal of Computational and Graphical Statistics*, 7, 434–455.

Buckland, S. T., Anderson, D. R., Burnham, K. P., Laake, J. L., Borchers, D. L., & Thomas, L. (2001). *Introduction to distance sampling*. Oxford: Oxford University Press.

Buckland, S. T., Anderson, D. R., Burnham, K. P., Laake, J. L., Borchers, D. L., & Thomas, L. (Eds.), (2004). *Advanced distance sampling*. Oxford: Oxford University Press.

Buckland, S. T., Rexstad, E. A., Marques, T. A., & Oedekoven, C. S. (2015). *Distance sampling: Methods and applications*. Cham, Switzerland: Springer.

Buja, A., Cook, D., Hofmann, H., Lawrence, M., Lee, E.-K., Swaynes, D. F., & Wickham, H. (2009). Statistical inference for exploratory data analysis and model diagnostics. *Phil. Trans. R. Soc. A*, 367, 4361–4383.

Bürkner, P.-C. (2017). brms: An R package for Bayesian multilevel models using Stan. *Journal of Statistical Software*, 80, 1–28.

Burnham, K. P., & Anderson, D. R. (2002). *Model selection and multimodel inference: A practical information theoretic approach*. New York: Springer.

Cade, B. S. (2015). Model averaging and muddled multimodel inferences. *Ecology*, 96, 2370–2382.

Carlin, B. P., & Louis, T. A. (2009). *Bayesian methods for data analysis*. Boca Raton: CRC Press/Taylor & Francis Group.

Carpenter, B., Gelman, A., Hoffman, M. D., Lee, D., Goodrich, B., Betancourt, M., Brubaker, M., Guo, J., Li, P., & Riddell, A. (2017). Stan: A probabilistic programming language. *Journal of Statistical Software*, 76(1), 1–32.

Casella, G., & Berger, R. L. (2002). *Statistical inference (second edition)*. Pacific Grove, California: Duxbury Press.

Chambert, T., Miller, D. A. W., & Nichols, J. D. (2015). Modeling false positive detections in species occurrence data under different study designs. *Ecology*, 96, 332–339.

Chen, G., Kéry, M., Plattner, M., Ma, K., & Gardner, B. (2013). Imperfect detection is the rule rather than the exception in plant distribution studies. *Journal of Ecology*, 101, 183–191.

Chiquet, J., Mariadassou, M., & Robin, S. (2018). Variational inference for probabilistic Poisson PCA. *The Annals of Applied Statistics*, 12, 2674–2698.

Clark, J. S., Carpenter, S. R., Barber, M., Collins, S., Dobson, A., Foley, J. A., Lodge, D. M., Pascual, M., Pielke, R., Jr., Pizer, W., Pringle, C., Reid, W. V., Rose, K. A., Sala, O., Schlesinger, W. H., Wall, D. H., & Wear, D. (2001). Ecological forecasts: An emerging imperative. *Science*, 293(5530), 657–660.

Clipp, H. L., Evans, A. L., Kessinger, B. E., Kellner, K., & Rota, C. T. (2021). A penalized likelihood for multispecies occupancy models improves predictions of species interactions. *Ecology*, 102, e03520.

Cochran, W. G., & Cox, G. M. (1992). *Experimental designs (second edition)*. Wiley.

Cole, D. (2021). *Parameter redundancy and identifiability*. Boca Raton: Chapman & Hall/CRC Press.

Conn, P. B., Johnson, D. S., Williams, P. J., Melin, S. R., & Hooten, M. B. (2018). A guide to Bayesian model checking for ecologists. *Ecological Monographs*, 88(4), 526–542.

Cooch, E., & White, G. 2021. *Program MARK: A gentle introduction*. Available in pdf format for free download at http://www.phidot.org/software/mark/docs/book.

Cox, D. R., & Snell, E. J. (1968). A general definition of residuals (with discussion). *Journal of the Royal Statistical Society: Series B*, 30, 248–275.

Cressie, N., Calder, C. A., Clark, J. S., Ver Hoef, J. M., & Wikle, C. K. (2009). Accounting for uncertainty in ecological analysis: The strengths and limitations of hierarchical statistical modeling. *Ecological Applications: A Publication of the Ecological Society of America*, 19, 553–570.

Cressie, N., & Wikle, C. K. (2011). *Statistics for spatio-temporal data*. Wiley.

Cubaynes, S., Doutrelant, C., Grégoire, A., Perret, P., Faivre, B., & Gimenez, O. (2012). Testing hypotheses in evolutionary ecology with imperfect detection: Capture–recapture structural equation modeling. *Ecology*, 93, 248–255.

Davis, C. L., Walls, S. C., Barichivich, W. J., Brown, M. E., & Miller, D. A. W. (2023). Disentangling direct and indirect effects of extreme events on coastal wetland communities. *Journal of Animal Ecology*, 92, 1135–1148.

Denwood, M. J. (2016). runjags: An R package providing interface utilities, parallel computing methods and additional distributions for MCMC models in JAGS. *Journal of Statistical Software*, 71(9), 1–25. Available from http://cran.r-project.org/web/packages/runjags/.

de Valpine, P. (2003). Better inferences from population-dynamics experiments using Monte Carlo state-space likelihood methods. *Ecology*, 84, 3064–3077.

de Valpine, P., Turek, D., Paciorek, C. J., Anderson-Bergman, C., Lang, D. T., & Bodik, R. (2017). Programming with models: Writing statistical algorithms for general model structures with NIMBLE. *Journal of Computational and Graphical Statistics*, 26(2), 403–413.

DiRenzo, G. V., Hanks, E., & Miller, D. A. (2023). A practical guide to understanding and validating complex models using data simulations. *Methods in Ecology and Evolution*, 14, 203–217.

Dixon, P. M. (2006). *Bootstrap resampling*. Wiley. Available from http://doi.org/10.1002/9780470057339.vab028.

Dobson, A., & Barnett, A. (2018). *An introduction to generalized linear models (Fourth Edition)*. Boca Raton: CRC/Chapmann & Hall.

Dorazio, R. M. (2014). Accounting for imperfect detection and survey bias in statistical analysis of presence-only data. *Global Ecology and Biogeography*, 23, 1472–1484.

Dorazio, R. M., & Royle, J. A. (2005). Estimating size and composition of biological communities by modeling the occurrence of species. *Journal of the American Statistical Association*, 100, 389–398.

Dorfman, R. (1938). A note on the δ-method for finding variance formulae. *The Biometric Bulletin*, 1, 129–137.

Dormann, C. F., et al. (2013). Collinearity: A review of methods to deal with it and a simulation study evaluating their performance. *Ecography*, 36, 27–46.

Dormann, C. F., et al. (2018). Model averaging in ecology: A review of Bayesian, information-theoretic and tactical approaches for predictive inference. *Ecological Monographs*, 88, 485–504.

Doser, J. W., Finley, A. O., Kéry, M., & Zipkin, E. F. (2022). spOccupancy: An R package for single species, multispecies, and integrated occupancy models. *Methods in Ecology and Evolution*, 13, 1670–1678.

Doser, J. W., Finley, A. O., Kéry, M., & Zipkin, E. F. (2024). spAbundance: An R package for single-species and multi-species spatially-explicit abundance models. *Methods in Ecology and Evolution*, 15, 1024–1033.

Draper, D. (1999). Model uncertainty yes, discrete model averaging maybe, Comment on: Hoeting JA, Madigan D, Raftery AE, Volinsky CT (1999) Bayesian model averaging: a tutorial. *Statistical Science*, 14, 405–409.

Duarte, A., Adams, M. J., & Peterson, J. T. (2018). Fitting *N*-mixture models to count data with unmodeled heterogeneity: Bias, diagnostics, and alternative approaches. *Ecological Modelling*, 374, 51–59.

Dunn, P. K., & Smyth, G. K. (1996). Randomized quantile residuals. *Journal of Computational and Graphical Statistics*, 5, 236–244.

Edwards, A. W. F. (1992). *Likelihood*. Baltimore: Johns Hopkins University Press.

Efron, B. (1979). Bootstrap methods: Another look at the jackknife. *Annals of Statistics*, 7, 1–26.

Efron, B. (1986). Why isn't everyone a Bayesian? *The American Statistician*, 40, 1–5.

Efron, B., & Tibshirani, R. J. (1993). *An introduction to the bootstrap*. Chapman & Hall/CRC Press.

Elith, J., et al. (2006). Novel methods improve prediction of species' distributions from occurrence data. *Ecography*, 29, 129–151.

Elith, J., & Leathwick, J. R. (2009). Species distribution models: Ecological explanation and prediction across space and time. *Annual Review of Ecology, Evolution, and Systematics*, 40, 677–697.

Elith, J., Leathwick, J. R., & Hastie, T. (2008). A working guide to boosted regression trees. *The Journal of Animal Ecology*, 77, 802–813.

Fink, D. A., Johnston, M., Strimas-Mackey, T., Auer, W. M., Hochachka, S., Ligocki., Oldham Jaromczyk, L., Robinson, O., Wood, C., Kelling, S., & Rodewald, A. D. (2023). A double machine learning trend model for citizen science data. *Methods in Ecology and Evolution*, 14, 2435–2448.

Fisher, R. A. (1922). On the mathematical foundations of theoretical statistics. *Philosophical Transactions of the Royal Society of London, Series A*, 222, 309–368.

Fiske, I., & Chandler, R. (2011). unmarked: An R package for fitting hierarchical models of wildlife occurrence and abundance. *Journal of Statistical Software*, 43, 1–23.

Fournier, D. A., Skaug, H. J., Ancheta, J., Ianelli, J., Magnusson, A., Maunder, M. N., & Sibert, J. (2012). AD Model Builder: Using automatic differentiation for statistical inference of highly parameterized complex nonlinear models. *Optimization Methods and Software*, 27(2), 233–249.

Friedman, J., Tibshirani, R., & Hastie, T. J. (2010). Regularization paths for generalized linear models via coordinate descent. *Journal of Statistical Software*, 33, 1–22.

Gelfand, A. E., Schmidt, A. E., Wu, S., Silander, J. A., Jr., Latimer, A., & Rebelo, A. G. (2005). Modelling species diversity through species level hierarchical modelling. *Applied Statistics*, 54, 1–20.

Gelfand, A. E., & Smith, A. F. (1990). Sampling-based approaches to calculating marginal densities. *Journal of the American Statistical Association*, 85, 398–409.

Gelman, A. (2005). Analysis of variance: Why is it more important than ever (with discussion). *Annals of Statistics*, 33, 1–53.

Gelman, A. (2006). Prior distributions for variance parameters in hierarchical models. *Bayesian Analysis*, 1, 515–534.

Gelman, A., Carlin, J. B., Stern, H. S., Dunson, D. B., Vehtari, A., & Rubin, D. B. (2014a). *Bayesian data analysis (Third Edition)*. Boca Raton: CRC/Chapman & Hall.

Gelman, A., & Hill, J. (2007). *Data analysis using regression and multilevel/hierarchical models*. Cambridge: Cambridge University Press.

Gelman, A., Hwang, J., & Vehtari, A. (2014b). Understanding predictive information criteria for Bayesian models. *Statistics and Computation*, 24, 997–1016.

Gelman, A., Meng, X.-L., & Stern, H. S. (1996). Posterior predictive assessment of model fitness via realized discrepancies (with discussion). *Statistica Sinica*, 6, 733–807.

Gelman, A., & Pardoe, I. (2006). Bayesian measures of explained variance and pooling in multilevel (hierarchical) models. *Technometrics*, 48, 241–251.

Gelman, A., & Rubin, D. B. (1992). Inference from iterative simulation using multiple sequences. *Statistical Science*, 7, 457–511.

Gelman, A., & Shalizi, C. R. (2012). Philosophy and the practice of Bayesian statistics. *British Journal of Mathematical and Statistical Psychology*, 66, 8–38.

Gelman, A., Vehtari, A., Simpson,D., Margossian, C.C., Carpenter, B., Yao, Y., Kennedy, L., Gabry, J., Bürkner, P.-C., & Modrák, M. (2020). Bayesian Workflow. arXiv. Available from https://arxiv.org/abs/2011.01808.

Geman, S., & Geman, D. (1984). Stochastic relaxion, Gibbs distributions, and the Bayesian restoration of images. *IEEE Transactions on Pattern Analysis and Machine Intelligence*, 6, 721–741.

Gerber, B. D., Kendall, W. L., Hooten, M. B., Dubovsky, J. A., & Drewien, R. C. (2015). Optimal population prediction of sandhill crane recruitment based on climate-mediated habitat limitations. *Journal of Animal Ecology*, 84, 1299–1310.

Gibson, D., Riecke, T. V., Catlin, D. H., Hunt, K. L., Weithman, C. E., Koons, D. N., Karpanty, S. M., & Fraser, J. D. (2023). Climate change and commercial fishing practices codetermine survival of a long-lived seabird. *Global Change Biology*, 29, 324–340.

Gilks, W. R., Thomas, A., & Spiegelhalter, D. J. (1994). A language and program for complex Bayesian modelling. *The Statistician: Journal of the Institute of Statisticians*, 43, 169–177.

Gimenez, O. (2025). *Bayesian analysis of capture-recapture data with hidden Markov models: Theory and case studies in R and NIMBLE*. Chapman & Hall/CRC.

Goudie, R. J. B., Turner, R. M., De Angelis, D., & Thomas, A. (2020). MultiBUGS: A parallel implementation of the BUGS modeling framework for faster Bayesian inference. *Journal of Statistical Software*, 95, 1–20.

Grace, J. B. (2006). *Structural equation modeling and natural systems*. Cambridge, United Kingdom: Cambridge University Press.

Grace, J. B., & Irvine, K. M. (2020). Scientist's guide to developing explanatory statistical models using causal analysis principles. *Ecology*, 101, e02962.

Gu, W., & Swihart, R. K. (2004). Absent or undetected ? Effects of non-detection of species occurrence on wildlife-habitat models. *Biological Conservation*, 116, 195–203.

Guillera-Arroita, G. (2017). Modelling of species distributions, range dynamics and communities under imperfect detection: Advances, challenges and opportunities. *Ecography*, 40, 281–295.

Guillera-Arroita, G., Lahoz-Monfort, J. J., MacKenzie, D. I., Wintle, B. A., & McCarthy, M. A. (2014). Ignoring imperfect detection in biological surveys is dangerous: A response to 'Fitting and Interpreting Occupancy Models'. *PLoS One*, 9(7), e99571.

Guillera-Arroita, G., Ridout, M. S., & Morgan, B. J. T. (2010). Design of occupancy studies with imperfect detection. *Methods in Ecology and Evolution*, 1, 131–139.

Hadfield, J. D. (2010). MCMC methods for multi-response generalized linear mixed models: The MCMCglmm R Package. *Journal of Statistical Software*, 33(2), 1–22.

Hartig, F. (2022). DHARMa: Residual Diagnostics for Hierarchical (Multi-Level/Mixed) Regression Models. R Package Version 0.4.6, Available from https://CRAN.R-project.org/package=DHARMa.

Hastie, T. J., Tibshirani, R. J., & Friedman, J. (2016). *The elements of statistical learning. Data mining, inference, and prediction (Second edition)*. Berlin: Springer.

Hefley, T. J., Hooten, M. B., Russell, R. E., Walsh, D. P., & Powell, J. A. (2017). When mechanism matters: Bayesian forecasting using models of ecological diffusion. *Ecology Letters*, 20, 640–650.

Hepler, S., & Erhardt, R. (2023). multiocc: An R package for spatio-temporal occupancy models for multiple species. *The R Journal*, 15/4, 1–16

Hines, J.E. (2006). *PRESENCE 3.1 Software to estimate patch occupancy and related parameters*. Available from http://www.mbr-pwrc.usgs.gov/software/presence.html.

Hobbs, N. T., & Hooten, M. B. (2015). *Bayesian models: A statistical primer for ecologists*. Princeton, NJ: Princeton University Press.

Hobert, J. P. (2000). Hierarchical models: A current computational perspective. *Journal of the American Statistical Association*, 95, 1312–1316.

Hoerl, A. E., & Kennard, R. W. (1970). Ridge regression: Biased estimation for nonorthogonal problems. *Technometrics*, 12, 55–67.

Hoeting, J. A., Madigan, D., Raftery, A. E., & Volinsky, C. T. (1999). Bayesian model averaging: A tutorial. *Statistical Science*, 14, 382–417.

Hooke, R. (1980). Getting people to use statistics properly. *The American Statistician*, 34, 39–42.

Hooten, M. B., & Hefley, T. J. (2019). *Bringing Bayesian models to life*. Boca Raton: CRC Press.

Hooten, M. B., & Hobbs, N. T. (2015). A guide to Bayesian model selection for ecologists. *Ecological Monographs*, 85, 3–28.

Hosmer, D. W., Lemeshow, S., & Sturdivant, R. X. (2013). *Applied logistic regression (Third edition)*. Wiley.

Hug, J. E. & Paciorek,C. J. (2021). A numerically stable online implementation and exploration of WAIC through variations of the predictive density, using NIMBLE. *arXiv* e-print. https://arxiv.org/abs/2106.13359.

Inchausti, P. (2023). *Statistical modeling with R: A dual frequentist and Bayesian approach for life scientists*. Oxford, UK: Oxford University Press.

Johnson, D. H. (2008). In defense of indices: The case of bird surveys. *Journal of Wildlife Management*, 72, 857–868.

Johnson, D. S., Conn, P. B., Hooten, M. B., Ray, J. C., & Pond, B. A. (2013). Spatial occupancy models for large data sets. *Ecology*, 94, 801–808.

Joseph, L. N., Elkin, C., Martin, T. G., & Possingham, H. (2009). Modeling abundance using *N*-mixture models: The importance of considering ecological mechanisms. *Ecological Applications*, 19, 631–642.

Joseph, M. B. (2020a). Neural hierarchical models of ecological populations. *Ecology Letters*, 23, 734–747.

Joseph, M. B. (2020b). A step-by-step guide to marginalizing over discrete parameters for ecologists using Stan. Available from <https://mbjoseph.github.io/posts/2020-04-28-a-step-by-step-guide-to-marginalizing-over-discrete-parameters-for-ecologists-using-stan/> Accessed 16.11.23.

Joseph, M. B., Preston, D. L., & Johnson, P. T. J. (2016). Integrating occupancy models and structural equation models to understand species occurrence. *Ecology*, 97, 765–775.

Kellner, K., & Meredith, M. (2021). Package jagsUI. A Wrapper Around rjags to streamline JAGS analyses. R Package Version 1.5.2.

Kellner, K., Parsons, A. W., Kays, R., Millspaugh, J. J., & Rota, C. T. (2022). A two-species occupancy model with a continuous-time detection process reveals spatial and temporal interactions. *Journal of Agricultural, Biological, and Environmental Statistics*, 24, 321–338.

Kellner, K. F., Fowler, N. L., Petroelje, T. R., Kautz, T. M., Beyer, D. E., Jr., & Belant, J. L. (2022). ubms: An R package for fitting hierarchical occupancy and N-mixture abundance models in a Bayesian framework. *Methods in Ecology and Evolution*, 13, 577–584.

Kellner, K. F., Smith, A. D., Royle, J. A., Kéry, M., Belant, J. L., & Chandler, R. B. (2023). The unmarked R package: Twelve years of advances in occurrence and abundance modeling in ecology. *Methods in Ecology and Evolution*, 14, 1408–1415.

Kellner, K. F., & Swihart, R. K. (2014). Accounting for imperfect detection in ecology: A quantitative review. *PLoS One*, 9, e111436.

Kéry, M. (2002). Inferring the absence of a species - a case study of snakes. *Journal of Wildlife Management*, 66, 330–338.

Kéry, M. (2004). Extinction rate estimates for plant populations in revisitation studies: Importance of detectability. *Conservation Biology*, 18, 570–574.

Kéry, M. (2010). *Introduction to WinBUGS for Ecologists. - A Bayesian approach to regression, ANOVA, mixed models and related analyses*. Burlington: Academic Press.

Kéry, M. (2011). Towards the modeling of true species distributions. *Journal of Biogeography*, 38, 617–618.

Kéry, M., & Royle, J. A. (2016). *Applied hierarchical modeling in ecology—Modeling distribution, abundance and species richness using R and BUGS (Volume 1: Prelude and Static Models)*. Elsevier/Academic Press.

Kéry, M., & Royle, J. A. (2021). *Applied hierarchical modeling in ecology—Modeling distribution, abundance and species richness using R and BUGS (Volume 2: Dynamic and Advanced Models)*. Elsevier / Academic Press.

Kéry, M., & Schaub, M. (2012). *Bayesian population analysis using WinBUGS: A hierarchical perspective*. Academic Press.

Kéry, M., & Schmidt, B. R. (2008). Imperfect detection and its consequences for monitoring for conservation. *Community Ecology*, 9, 207–216.

Kéry, M., Spillmann, J. H., Truong, C., & Holderegger, R. (2006). How biased are estimates of extinction probability in revisitation studies? *Journal of Ecology*, 94, 980–986.

King, R., Morgan, B. J. T., Gimenez, O., & Brooks, S. P. (2009). *Bayesian analysis for population ecology*. Boca Raton: Chapmann & Hall.

Knape, J. (2008). Estimability of density dependence in models of time series data. *Ecology*, 89, 2994–3000.

Knaus, P., Antoniazza, S., Wechsler, S., Guélat, J., Kéry, M., Strebel, N., & Sattler, T. (2018). *Brutvogelatlas 2013–2016. Bestandsentwicklung der Brutvögel der Schweiz und des Fürstentums Liechtensteins (Swiss Breeding Bird Atlas 2013–2016)*. Sempach: Schweizerische Vogelwarte.

Koshkina, V., Wang, Y., Gordon, A., Dorazio, R. M., White, M., & Stone, L. (2017). Integrated species distribution models: Combining presence-background data and site-occupancy data with imperfect detection. *Methods in Ecology and Evolution*, 8, 420–430.

Kristensen, K. (2023). RTMB: 'R' Bindings for 'TMB'. *R package version 1*, 4. Available from https://cran.r-project.org/package=RTMB.

Kristensen, K., Nielsen, A., Berg, C. W., Skaug, H., & Bell, B. M. (2016). TMB: Automatic differentiation and laplace approximation. *Journal of Statistical Software*, 70, 1–21.

Kruschke, J. K. (2015). *Doing Bayesian data analysis (Second Edition)*. Academic Press / Elsevier.

Kyung, M., Gill, J., Ghosh, M., & Casella, G. (2010). Penalized regression, standard errors, and Bayesian lassos. *Bayesian Analysis*, 5, 369–412.

Laake, J.L. (2013). RMark: An R interface for analysis of capture-recapture data with MARK. AFSC Processed Rep 2013-01, 25p. Alaska Fish. Sci. Cent., NOAA, Natl. Mar. Fish. Serv., 7600 Sand Point Way NE, Seattle WA 98115.

Lambert, P. C., Sutton, A. J., Burton, P. R., Abrams, K. R., & Jones, D. R. (2005). How vague is vague? A simulation study of the impact of the use of vague prior distributions in MCMC using WinBUGS. *Statistics in Medicine*, 24, 2401–2428.

Landau, V. A., Noon, B. R., Theobald, D. M., Hobbs, N. T., & Nielsen, C. K. (2022). Integrating presence-only and occupancy data to model habitat use for the northernmost population of jaguars. *Ecological Applications*, 32, e2619.

Latif, Q. S., Van Lanen, N. J., Chabot, E. J., & Pavlacky, D. C., Jr. (2023). Causal mechanisms for negative impacts of energy development inform management triggers for sagebrush birds. *Ecosphere*, 14(4), e4479.

Lee, Y., & Nelder, J. A. (2006). Double hierarchical generalized linear models. *Applied Statistics*, 55, 139–185.

Lee, Y., Nelder, J. A., & Pawitan, Y. (2017). *Generalized linear models with random effects. Unified analysis via h-likelihood (Second Edition)*. Boca Raton, FL: Chapman & Hall/CRC.

Lele, S. R., Dennis, B., & Lutscher, F. (2007). Data cloning: Easy maximum likelihood estimation for complex ecological models using Bayesian Markov chain Monte Carlo methods. *Ecology Letters*, 10, 551–563.

Lele, S. R., Moreno, M., & Bayne, E. (2012). Dealing with detection error in site occupancy surveys: What can we do with a single survey? *Journal of Plant Ecology*, 5, 22–31.

Li, X., Pei, K., Kéry, M., Niklaus, P. A., & Schmid, B. (2017). Decomposing functional trait associations in a Chinese subtropical forest. *PLoS One*, 12(4), e0175727.

Lindley, D. V. (2006). *Understanding uncertainty*. Hoboken, New Jersey: Wiley.

Link, W. A., & Barker, R. J. (2006). Model weights and the foundations of multimodel inference. *Ecology*, 87, 2626–2635.

Link, W. A., & Barker, R. J. (2010). *Bayesian inference with ecological applications*. London: Academic Press.

Link, W. A., & Eaton, M. J. (2012). On thinning of chains in MCMC. *Methods in Ecology and Evolution*, 3, 112–115.

Little, R. J. A. (2004). To model or not to model? Competing modes of inference for finite population sampling. *Journal of the American Statistical Association*, 99, 546–556.

Little, R. J. A. (2006). Calibrated Bayes: A bayes/frequentist roadmap. *The American Statistician*, 60, 213–223.

Little, R. J. A., & Rubin, D. B. (2002). *Statistical analysis with missing data (Second edition)*. New York: Wiley.

Lunn, D., Jackson, C., Best, N., Thomas, A., & Spiegelhalter, D. (2013). *The BUGS Book: A Practical Introduction to Bayesian Analysis*. Chapman & Hall/CRC.

Lunn, D. J., Spiegelhalter, D., Thomas, A., & Best, N. (2009). The BUGS project: Evaluation, critique and future directions. *Statistics in Medicine*, 28, 3049–3067.

MacKenzie, D. I., Bailey, L. L., & Nichols, J. D. (2004). Investigating species co-occurrence patterns when species are detected imperfectly. *Journal of Animal Ecology*, 73, 546–555.

MacKenzie, D. I., Nichols, J. D., Hines, J. E., Knutson, M. G., & Franklin, A. B. (2003). Estimating site occupancy, colonization, and local extinction when a species is detected imperfectly. *Ecology*, 84, 2200–2207.

MacKenzie, D. I., Nichols, J. D., Lachman, G. B., Droege, S., Royle, J. A., & Langtimm, C. A. (2002). Estimating site occupancy rates when detection probabilities are less than one. *Ecology*, 83, 2248–2255.

MacKenzie, D. I., Nichols, J. D., Royle, J. A., Pollock, K. H., Hines, J. E., & Bailey, L. L. (2006). *Occupancy estimation and modeling: inferring patterns and dynamics of species occurrence (First edition)*. San Diego: Elsevier.

MacKenzie, D. I., Nichols, J. D., Royle, J. A., Pollock, K. H., Hines, J. E., & Bailey, L. L. (2017). *Occupancy estimation and modeling: Inferring patterns and dynamics of species occurrence (Second Edition)*. San Diego: Elsevier.

MacKenzie, D. I., Nichols, J. D., Seamans, M. E., & Gutierrez, R. J. (2009). Modeling species occurrence dynamics with multiple states and imperfect detection. *Ecology*, 90, 823–835.

MacKenzie, D. I., & Royle, J. A. (2005). Designing occupancy studies: General advice and allocating survey effort. *Journal of Applied Ecology*, 42, 1105–1114.

Manly, B. F. J. (2006). *Randomization, bootstrap and Monte Carlo methods in biology (Third Edition)*. Boca Raton: CRC Press.

Marshall, E. C., & Spiegelhalter, D. J. (2003). Approximate cross-validatory predictive checks in disease mapping models. *Statistics in Medicine*, 22, 1649–1660.

Martin, T. G., Wintle, B. A., Rhodes, J. R., Kuhnert, P. M., Field, S. A., Low-Choy, S. J., Tyre, A. J., & Possingham, H. P. (2005). Zero tolerance ecology: Improving ecological inference by modelling the source of zero observations. *Ecology Letters*, 8, 1235–1246.

Maunder, M. N., & Punt, A. E. (2013). A review of integrated analysis in fisheries stock assessment. *Fisheries Research*, 142, 61–74.

Mazerolle, M.J. (2023). AICcmodavg: Model selection and multimodel inference based on (Q)AIC(c). R package version 2.3.2, Available from https://cran.r-project.org/web/packages/AICcmodavg/index.html.

McCarthy, M. A., & Masters, P. (2005). Profiting from prior information in Bayesian analyses of ecological data. *Journal of Applied Ecology*, 42, 1012–1019.

McCrea, R. S., & Morgan, B. J. T. (2014). *Analysis of capture-recapture data*. Boca Raton, FL, USA: Chapman & Hall/CRC Press.

McCullagh, P., & Nelder, J. A. (1989). *Generalized linear models*. London: Chapman & Hall.

McCulloch, C. E., & Searle, S. R. (2001). *Generalized, linear, and mixed models*. New York: Wiley.

McElreath, R. (2020). *Statistical rethinking: A Bayesian course with examples in R and STAN (Second Edition)*. Boca Raton: CRC Press.

Mead, R., Gilmour, S. G., & Mead, A. (2012). *Statistical principles for the design of experiments: Applications to real experiments*. Cambridge: Cambridge University Press.

Merkle, E., Furr, D., & Rabe-Hesketh, S. (2019). Bayesian Comparison of Latent Variable Models: Conditional Versus Marginal Likelihoods. *Psychometrika*, 84, 802–829.

Merkle, E. C., Fitzsimmons, E., Uanhoro, J., & Goodrich, B. (2021). Efficient Bayesian structural equation modeling in Stan. *Journal of Statistical Software*, 100(6), 1–22.

Merkle, E. C., & Rosseel, Y. (2018). blavaan: Bayesian structural equation models via parameter expansion. *Journal of Statistical Software*, 85(4), 1–30.

Metropolis, N., Rosenbluth, A. W., Rosenbluth, M. N., Teller, A. H., & Teller, E. (1953). Equation of state calculations by fast computing machines. *Journal of Chemical Physics*, 21, 1087–1092.

Metropolis, N., & Ulam, S. (1949). The Monte Carlo method. *Journal of the American Statistical Association*, 44, 335–341.

Miller, D. A., Nichols, J. D., McClintock, B. T., Grant, E. H. C., Bailey, L. L., & Weir, L. (2011). Improving occupancy estimation when two types of observational errors occur: Non-detection and species misidentification. *Ecology*, 92, 1422–1428.

Miller, D. A. W., Pacifici, K., Sanderlin, J. S., & Reich, B. J. (2019). The recent past and promising future for data integration methods to estimate species' distributions. *Methods in Ecology and Evolution*, 10, 22–37.

Millar, R. B. (2009). Comparison of hierarchical Bayesian models for overdispersed count data using DIC and Bayes' factors. *Biometrics*, 65, 962–969.

Millar, R. B. (2011). *Maximum likelihood estimation and inference: With examples in R, SAS, and ADMB*. Wiley.

Millar, R. B. (2018). Conditional vs marginal estimation of the predictive loss of hierarchical models using WAIC and cross-validation. *Statistics and Computing*, 28, 375–385.

Mohankumar, N. M., & Hefley, T. J. (2022). Using machine learning to model nontraditional spatial dependence in occupancy data. *Ecology*, 103, e03563.

Monnahan, C. C., & Kristensen, K. (2018). No-U-turn sampling for fast Bayesian inference in ADMB and TMB: Introducing the adnuts and tmbstan R packages. *PLoS One*, 13, e0197954.

Monnahan, C. C., Thorson, J. T., & Branch, T. A. (2017). Faster estimation of Bayesian models in ecology using Hamiltonian Monte Carlo. *Methods in Ecology and Evolution*, 8, 339–348.

Monneret, R.-J. (2006). *Le faucon pèlerin*. Paris, France: Delachaux and Niestlé.

Monneret, R.-J., Ruffinoni, R., Parish, D., Pinaud, D., & Kéry, M. (2018). The Peregrine population study in the French Jura mountains 1964–2016: Use of occupancy modeling to estimate population size and analyze site persistence and colonization rates. *Ornis Hungarica*, 26, 69–90.

Mordecai, R. S., Mattsson, B. J., Tzilkowski, C. J., & Cooper, R. J. (2011). Addressing challenges when studying mobile or episodic species: Hierarchical Bayes estimation of occupancy and use. *Journal of Applied Ecology*, 48, 56–66.

Moreno, M., & Lele, S. R. (2010). Improved estimation of site occupancy using penalized likelihood. *Ecology*, 91, 341–346.

Müller, C. (2021). Der Bienenfresser *Merops apiaster* in der Schweiz – Paradebeispiel für die Ausdehnung einer wärmeliebenden Art. *Vogelwarte*, 59, 301–312.

Mutshinda, C. M., O'Hara, R. B., & Woiwod, I. P. (2011). A multispecies perspective on ecological impacts of climatic forcing. *Journal of Animal Ecology*, 80, 101–107.

Nakagawa, S., & Schielzeth, H. (2013). A general and simple method for obtaining R^2 from generalized linear mixed-effects models. *Methods in Ecology and Evolution*, 4(2), 133–142.

Nelder, J. A., & Wedderburn, R. W. M. (1972). Generalized linear models. *Journal of the Royal Statistical Society, Series A*, 370–384.

Nichols, J. D., Bailey, L. L., O'Connell, A. F., Talancy, N. W., Grant, E. H. C., Gilbert, A. T., Annand, E. M., Husband, T. P., & Hines, J. E. (2008). Multi-scale occupancy estimation and modelling using multiple detection methods. *Journal of Applied Ecology*, 45, 1321–1329.

Nichols, J. D., Hines, J. E., MacKenzie, D. I., Seamans, M. E., & Gutierrez, R. J. (2007). Occupancy estimation and modeling with multiple states and state uncertainty. *Ecology*, 88, 1395–1400.

Northrup, J. M., & Gerber, B. D. (2018). A comment on priors for Bayesian occupancy models. *PLoS One*, 13 (2), e0192819.

Ntzoufras, I. (2009). *Bayesian modeling using WinBUGS*. Hoboken, New Jersey: Wiley.

Oldham, K. B., Myland, J., & Spanier, J. (2016). *An Atlas of functions: With equator, the Atlas function calculator*. Springer.

Pacifici, K., Reich, B. J., Miller, D. A. W., Gardner, B., Stauffer, G., Singh, S., McKerrow, A., & Collazo, J. A. (2017). Integrating multiple data sources in species distribution modeling: A framework for data fusion. *Ecology*, 98, 840–850.

Pacifici, K., Reich, B. J., Miller, D. A. W., & Pease, B. S. (2019). Resolving misaligned spatial data with integrated species distribution models. *Ecology*, 100(6), e02709.

Pawitan, Y. (2013). *In all likelihood*. Oxford, UK: Oxford University Press.

Pearl, J. (2009). *Causality: Models, reasoning, and inference (Second edition)*. New York: Cambridge University Press.

Pearl, J. (2021). Radical empiricism and machine learning research. *Journal of Causal Inference*, 9, 78–82.

Perret, J., Besnard, A., Charpentier, A., & Papuga, G. (2023). Plants stand still but hide: Imperfect and heterogeneous detection is the rule when counting plants. *Journal of Ecology*, 111, 1483–1496.

Petchey, O. L., et al. (2015). The ecological forecast horizon, and examples of its uses and determinants. *Ecology Letters*, 18, 597–611.

Pichler, M., & Hartig, F. (2023). Machine learning and deep learning—A review for ecologists. *Methods in Ecology and Evolution*, 14, 994–1016.

Pinheiro, J. C., & Bates, D. M. (1996). Unconstrained parametrizations for variance-covariance matrices. *Statistics and Computing*, 6, 289–296.

Pinheiro, J. C., & Bates, D. M. (2000). *Mixed-effects models in S and S-Plus*. New York: Springer.

Pishro-Nik, H. (2014). *Introduction to probability, statistics, and random processes*. Google Books.

Pizarro Muñoz, J. A., Kéry, M., Martins, P., & Ferraz, G. (2018). Age effects on survival of Amazon birds and the latitudinal gradient in bird survival. *The Auk*, 135, 299–313.

Plard, F., Turek, D., & Schaub, M. (2021). Consequences of violating assumptions of integrated population models on parameter estimates. *Environmental and Ecological Statistics*, 28, 667–695.

Plummer, M. (2002). Discussion of the paper by Spiegelhalter et al. *Journal of the Royal Statistical Society Series B*, 64, 620.

Plummer, M. (2003). JAGS: A program for analysis of bayesian graphical models using Gibbs sampling. In K. Hornik, F. Leisch, & A. Zeileis (Eds.), *Proceedings of the 3rd International Workshop in Distributed Statistical Computing (DSC 2003)* (pp. 1–10). Vienna, Austria: Technische Universität, March 20–22.

Plummer, M. (2008). Penalized loss functions for Bayesian model comparison. *Biostatistics (Oxford, England)*, 9, 523–539.

Plummer, M. (2017). JAGS Version 4.3.0 user manual. Available from https://sourceforge.net/projects/mcmc-jags/files/Manuals/4.x/jags_user_manual.pdf/download.

Plummer, M. (2023a). rjags: Bayesian Graphical Models using MCMC. R package. version 4–15. Available from <https://CRAN.R-project.org/package = rjags>.

Plummer, M. (2023b). Simulation-based Bayesian analysis. *Annual Review of Statistics and Its Application*, 10, 401–425.

Powell, L. A. (2007). Approximating variance of demographic parameters using the delta method: a reference for avian biologists. *Condor*, 109, 949–954.

Qian, S. S., & Shen, Z. (2007). Ecological applications of multilevel analysis of variance. *Ecology*, 88, 2489–2495.

Quinn, G. P., & Keough, M. J. (2002). *Experimental design and data analysis for biologists*. Cambridge, UK: Cambridge University Press.

R Core Team. (2023). R: A language and environment for statistical computing. Vienna, Austria: R Foundation for Statistical Computing URL. Available from https://www.R-project.org/.

Reich, H. T. (2020). Optimal sampling design and the accuracy of occupancy models. *Biometrics*, 76, 1017–1027.

Reineking, B., & Schröder, B. (2006). Constrain to perform: Regularization of habitat models. *Ecological Modelling*, 193, 675–690.

Renner, I. W., Elith, J., Baddeley, A., Fithian, W., Hastie, T., Phillips, S. J., Popovic, G., & Warton, D. I. (2015). Point process models for presence-only analysis. *Methods in Ecology and Evolution*, 6, 366–379.

Riecke, T. V., Sedinger, B. S., Williams, P. J., Leach, A. G., & Sedinger, J. S. (2019). Estimating correlations among demographic parameters in population models. *Ecology and Evolution*, 9, 13521–13531.

Roberts, D. R., Bahn, V., Ciuti, S., Boyce, M. S., Elith, J., Guillera-Arroita, G., Hauenstein, S., Lahoz-Monfort, J. J., Schröder, B., Thuiller, W., Warton, D. I., Wintle, B. A., Hartig, F., & Dormann, C. F. (2017). Cross-validation strategies for data with temporal, spatial, hierarchical, or phylogenetic structure. *Ecography*, 40(8), 913–929.

Rosseel, Y. (2012). lavaan: An R package for structural equation modeling. *Journal of Statistical Software*, 48(2), 1–36.

Rota, C. T., Ferreira, M. A. R., Kays, R. W., Forrester, T. D., Kalies, E. L., McShea, W. J., Parsons, A. W., & Millspaugh, J. J. (2016). A multi-species occupancy model for two or more interacting species. *Methods in Ecology and Evolution*, 7, 1164–1173.

Royle, J. A. (2004). *N*-mixture models for estimating population size from spatially replicated counts. *Biometrics*, 60, 108–115.

Royle, J. A., Chandler, R. B., Sollmann, R., & Gardner, B. (2014). *Spatial capture-recapture*. Academic Press.

Royle, J. A., & Dorazio, R. M. (2008). *Hierarchical modeling and inference in ecology. The analysis of data from populations, metapopulations and communities*. New York: Academic Press.

Royle, J. A., & Kéry, M. (2007). A Bayesian state-space formulation of dynamics occupancy models. *Ecology*, 88, 1813–1823.

Royle, J. A., & Link, W. A. (2005). A general class of multinomial mixture models for anuran calling survey data. *Ecology*, 86, 2505–2512.

Royle, J. A., & Link, W. A. (2006). Generalized site occupancy models allowing for false positive and false negative errors. *Ecology*, 87, 835–841.

Royle, J. A., & Nichols, J. D. (2003). Estimating abundance from repeated presence-absence data or point counts. *Ecology*, 84, 777–790.

Rubin, D. B. (1976). Inference and missing data. *Biometrika*, 63, 581–592.

Rubin, D. B. (1984). Bayesianly justifiable and relevant frequency calculations for the applied statistician. *Annals of Statistics*, 12, 1151–1172.

Sauer, J. R., & Link, W. A. (2002). Hierarchical modeling of population stability and species group attributes from survey data. *Ecology*, 86, 1743–1751.

Scharf, H. R., Lu, X., Williams, P. J., & Hooten, M. B. (2022). Constructing flexible, identifiable and interpretable statistical models for binary data. *International Statistical Review*, 90, 328–345.

Schaub, M., & Abadi, F. (2011). Integrated population models: A novel analysis framework for deeper insights into population dynamics. *Journal of Ornithology*, 152 (Suppl. 1), 227–237.

Schaub, M., & Kéry, M. (2022). *Integrated population models—A Bayesian hierarchical perspective using R and BUGS*. Elsevier/Academic Press.

Schaub, M., Maunder, M. N., Kéry, M., Thorson, J. T., Jacobson, E. K., & Punt, A. E. (2024). Lessons to be learned by comparing integrated fisheries stock assessment models (SAMs) with integrated population models (IPMs). *Fisheries Research, 272,* 106925, 1–27.

Schielzeth, H., & Nakagawa, S. (2013). Nested by design: Model fitting and interpretation in a mixed model era. *Methods in Ecology and Evolution*, 4, 14–24.

Seber, G. A. F., & Schofield, M. R. (2019). *Capture-recapture: Parameter estimation for open animal populations.* Springer.

Seber, G. A. F., & Schofield, M. R. (2024). *Estimating presence and abundance of closed populations.* Springer.

Shibata, R. (1989). Statistical aspects of model selection. In J. C. Willems (Ed.), *From data to model* (pp. 215–240). Berlin: Springer.

Shipley, B. (2004). *Cause and correlation in biology: a user's guide to path analysis, structural equations and causal inference.* Cambridge, UK: Cambridge University Press.

Shmueli, G. (2010). To explain or to predict. *Statistical Science*, 25, 289–310.

Snedecor, G. W., & Cochran, W. G. (1980). *Statistical methods (7th Edition).* Ames: Iowa State University Press.

Sólymos, P. (2010). dclone: Data cloning in R. *The R Journal*, 2, 29–37. Available from http://journal.r-project.org/.

Souchay, G., van Wijk, R. E., Schaub, M., & Bauer, S. (2018). Identifying drivers of breeding success in a long-distance migrant using structural equation modelling. *Oikos*, 127, 125–133.

Spiegelhalter, D. J. (2019). *The art of statistics. Learning from data.* London: Pelican books.

Spiegelhalter, D. J., Best, N. G., Carlin, B. P., & van der Linde, A. (2002). Bayesian measure of model complexity and fit. *Journal of the Royal Statistical Society Series B*, 64, 583–639.

Spiegelhalter, D., Thomas, A., Best, N., & Lunn, D. (2003). *WinBUGS user manual, version 1.4.* MRC Biostatistics Unit, Cambridge, UK.

Spiegelhalter, D., Thomas, A., Best, N., & Lunn, D. (2007). *OpenBUGS user manual, version 3.0. 2.* Cambridge, UK: MRC Biostatistics Unit.

Steel, R. G. D., Torrie, J. H., & Dickey, D. A. (1996). *Principles and procedures of statistics: A biometrical approach.* McGraw-Hill.

Stone, M. (1977). An asymptotic equivalence of choice of model cross-validation and Akaike's criterion. *Journal of the Royal Statistical Society Series B*, 36, 44–47.

Stoudt, S., de Valpine, P., & Fithian, W. (2023). Nonparametric identifiability in species distribution and abundance models: Why it matters and how to diagnose a lack of it using simulation. *Journal of Statistical Theory and Practice*, 17, 39. (2023). Available from https://doi.org/10.1007/s42519-023-00336-5.

Sturtz, S., Ligges, U., & Gelman, A. (2005). R2WinBUGS: A package for running WinBUGS from R. *Journal of Statistical Software*, 12, 1–16.

Tanner, M. A., & Wong, W. H. (1987). The calculation of posterior distributions by data augmentation. *Journal of the American Statistical Association*, 82, 528–540.

Tay, J. K., Narasimhan, B., & Hastie, T. (2023). Elastic net regularization paths for all Generalized Linear Models. *Journal of Statistical Software*, 106, 1–31.

Thompson, S. K. (2002). *Sampling.* New York: Wiley.

Thorson, J.T., Kristensen, K. 2024. *Spatio-Temporal Models for Ecologists.* Chapman & Hall/CRC applied environmental statistics.

Tibshirani, R. (1996). Regression shrinkage and selection via the lasso. *Journal of the Royal Statistical Society: Series B*, 58, 267–288.

Tredennick, A. T., Hooker, G., Ellner, S. P., & Adler, P. B. (2021). A practical guide to selecting models for exploration, inference, and prediction in ecology. *Ecology*, 102, e03336.

Tredennick, A. T., Hooten, M. B., & Adler, P. B. (2017). Do we need demographic data to forecast plant population dynamics? *Methods in Ecology and Evolution*, 8, 541–551.

Turek, D., de Valpine, P., & Paciorek, C. J. (2016). Efficient Markov chain Monte Carlo sampling for hierarchical hidden Markov models. *Environmental and Ecological Statistics*, 23, 549–564.

Turek, D.P., de Valpine, P., Paciorek, C.J. 2024. nimbleHMC: Hamiltonian Monte Carlo and Other Gradient-Based MCMC Sampling Algorithms for 'nimble'. R package version 0.2.1

Turek, D., Milleret, C., Ergon, T., Brøseth, H., Dupont, P., Bischof, R., & de Valpine, P. (2021). Efficient estimation of large-scale spatial capture–recapture models. *Ecosphere*, 12, e03385.

Tyre, A. J., Tenhumberg, B., Field, S. A., Niejalke, D., Parris, K., & Possingham, H. P. (2003). Improving precision and reducing bias in biological surveys: Estimating false-negative error rates. *Ecological Applications*, 13, 1790–1801.

Valavi, R., Elith, J., Lahoz-Monfort, J. J., & Guillera-Arroita, G. (2019). blockCV: An R package for generating spatially or environmentally separated folds for k-fold cross-validation of species distribution models. *Methods in Ecology and Evolution*, 10, 225–232.

Valavi, R., Guillera-Arroita, G., Lahoz-Monfort, J. J., & Elith, J. (2022). Predictive performance of presence-only species distribution models: A benchmark study with reproducible code. *Ecological Monographs*, 92 (1), e01486.

Vehtari, A. (2023). Cross-validation FAQ. Available from: <avehtari.github.io/modelselection/CV-FAQ.html> Accessed on 18.11.23.

Vehtari, A., Gabry, J., Magnusson, M., Yao, Y., Bürkner, P., Paananen, T., & Gelman, A. (2024). loo: Efficient leave-one-out cross-validation and WAIC for Bayesian models. R package version 2.6.0, Available from <https://mc-stan.org/loo/>.

Vehtari, A., Gelman, A., & Gabry, J. (2017). Practical Bayesian model evaluation using leave-one-out cross-validation and WAIC. *Statistics and Computation*, 27, 1413–1432.

ver Hoef, J. M., & Boveng, P. L. (2015). Iterating on a single model is a viable alternative to multimodel inference. *Journal of Wildlife Management*, 79, 719–729.

von Hirschheydt, G., Kéry, M., Ekman, S., Stofer, S., Dietrich, M., Keller, C., & Scheidegger, C. (2024). Occupancy models reveal limited detectability of lichens in a standardised large-scale monitoring. *Journal of Vegetation Science*, 35, e13255, 1–11.

von Hirschheydt, G., Stofer, S., & Kéry, M. (2023). "Mixed" occupancy designs: When do additional single-visit data improve the inferences from standard multi-visit models? *Basic and Applied Ecology*, 67, 61–69.

Warton, D. I., Stoklosa, J., Guillera-Arroita, G., MacKenzie, D. I., & Welsh, A. H. (2017). Graphical diagnostics for occupancy models with imperfect detection. *Methods in Ecology and Evolution*, 8, 408–419.

Wasserstein, R. L., & Lazar, N. A. (2016). The ASA Statement on *p*-Values: Context, Process, and Purpose. *The American Statistician*, 70, 129–133.

Watanabe, S. (2010). Asymptotic equivalence of Bayes cross validation and widely applicable information criterion in singular learning theory. *Journal of Machine Learning Research: JMLR*, 11, 3571–3594.

Weegman, M. D., Arnold, T. W., Clark, R. G., & Schaub, M. (2021). Partial and complete dependency among data sets has minimal consequence on estimates from integrated population models. *Ecological Applications*, 31, e2258.

Welham, S., Cullis, B., Gogel, B., Gilmour, A., & Thompson, R. (2004). Prediction in linear mixed models. *Australian & New Zealand Journal of Statistics*, 46, 325–347.

Welsh, A. H., Lindenmayer, D. B., & Donnelly, C. F. (2013). Fitting and interpreting occupancy models. *PLoS One*, 8, e52015.

White, G. C., & Burnham, K. P. (1999). Program MARK: Survival estimation from populations of marked animals. *Bird Study*, 46, 120–139.

Whittingham, M. J., Stephens, P. A., Bradbury, R. B., & Freckleton, R. P. (2006). Why do we still use stepwise modelling in ecology and behaviour? *Journal of Animal Ecology*, 75, 1182-118.

Wikle, C. K. (2003). Hierarchical Bayesian models for predicting the spread of ecological processes. *Ecology*, 84, 1382–1394.

Wikle, C. K., & Zammit-Mangion, A. (2023). Statistical deep learning for spatial and spatiotemporal data. *Annual Review of Statistics and Its Application*, 10, 247–270.

Wilkinson, G. N., & Rogers, C. E. (1973). Symbolic description of factorial models for analysis of variance. *Applied Statistics*, 22, 392–399.

Williams, B. K., Nichols, J. D., & Conroy, M. J. (2002). *Analysis and management of animal populations*. San Diego: Academic Press.

Wood, S. N. (2017). *Generalized additive models. An introduction with R (Second Edition)*. Boca Raton, USA: Chapman & Hall/CRC.

Yackulic, C. B., Dodrill, M., Dzul, M., Sanderlin, J. S., & Reid, J. A. (2020). A need for speed in Bayesian population models: A practical guide to marginalizing and recovering discrete latent states. *Ecological Applications*, 30, e02112.

Yates, L. A., Aandahl, Z., Richards, S. A., & Brook, B. W. (2023). Cross validation for model selection: A review with examples from ecology. *Ecological Monographs*, 93(1), e1557.

Yee, T. W. (2008). The VGAM package. *R News*, 8(2), 28–39.

Yoccoz, N. G., Nichols, J. D., & Boulinier, T. (2001). Monitoring biological diversity in space and time. *Trends in Ecology and Evolution*, 16, 446–453.

Youngflesh, C. (2018). MCMCvis: Tools to visualize, manipulate, and summarize MCMC output. *Journal of Open Source Software*, 3(24), 640. Available from https://doi.org/10.21105/joss.00640.

Zar, J. H. (1998). *Biostatistical analysis*. London, UK: Pearson.

Zeileis, A., Kleiber, C., & Jackman, S. (2008). Regression models for count data in R. *Journal of Statistical Software*, 27(8), 1–25. Available from https://doi.org/10.18637/jss.v027.i08.

Zellweger-Fischer, J., Kéry, M., & Pasinelli, G. (2011). Population trends of brown hares in Switzerland: The role of land-use and ecological compensation areas. *Biological Conservation*, 144, 1364–1373.

Zhao, Q. (2024). *Bayesian analysis of spatially structured population dynamics*. Springer.

Zipkin, E. F., Rossman, S., Yackulic, C. B., Wiens, J. D., Thorson, J. T., Davis, R. J., & Grant, E. H. C. (2017). Integrating count and detection–nondetection data to model population dynamics. *Ecology*, 98, 1640–1650.

Zipkin, E. F., & Saunders, S. P. (2018). Synthesizing multiple data types for biological conservation using integrated populations models. *Biological Conservation*, 217, 240–250.

Index

Printed in the United States
by Baker & Taylor Publisher Services